4th International Conference on Science and Engineering in Mathematics, Chemistry and Physics (ScieTech 2016)

Journal of Physics: Conference Series Volume 710

Bali, Indonesia
30 – 31 January 2016

Editor:

Ford Lumban Gaol

ISBN: 978-1-5108-2421-8
ISSN: 1742-6588

Printed from e-media with permission by:

Curran Associates, Inc.
57 Morehouse Lane
Red Hook, NY 12571

Some format issues inherent in the e-media version may also appear in this print version.

Printed by Curran Associates, Inc. (2016)

For permission requests, please contact the Institute of Physics
at the address below.

Institute of Physics
Dirac House, Temple Back
Bristol BS1 6BE UK

Phone: 44 1 17 929 7481
Fax: 44 1 17 920 0979

techtracking@iop.org

Additional copies of this publication are available from:

Curran Associates, Inc.
57 Morehouse Lane
Red Hook, NY 12571 USA
Phone: 845-758-0400
Fax: 845-758-2633
Email: curran@proceedings.com
Web: www.proceedings.com

Table of contents

Volume 710

4th International Conference on Science & Engineering in Mathematics, Chemistry and Physics 2016 (ScieTech 2016)
30–31 January 2016, Bali, Indonesia

Accepted papers received: 6 April 2016
Published online: 5 May 2016

Preface

Papers

Journal of Physics: Conference Series **710** (2016) 011001 doi:10.1088/1742-6596/710/1/011001

The 4th International Conference on Science & Engineering in Mathematics, Chemistry and Physics 2016 (ScieTech 2016)

Discovery Room, Discovery Kartika Plaza Hotel, Bali, Indonesia

30–31 January 2016

Preface

The 4th International Conference on Science & Engineering in Mathematics, Chemistry and Physics 2016 (ScieTech 2016) was held at the Discovery Room, Discovery Kartika Plaza Hotel, Bali, Indonesia from 30–31 January 2016.

The ScieTech 2016 conference is aimed at bringing together researchers, engineers and scientists in the field from around the world, and promoting interaction between the theoretical, experimental, and applied communities, so that a high level exchange is achieved in new and emerging areas within mathematics, chemistry and physics.

The theme of ScieTech 2016 is "The interaction of Mathematics, Chemistry and Physics from the theoretical, experimental, and applied communities to enhance Smart Society"

This year, we received 197 papers and after rigorous review, 45 papers were accepted.

The participants come from 14 countries. There are six parallel sessions and three keynote speakers.

It is an honour to present this volume of *Journal of Physics: Conference Series* (JPCS) and we deeply thank the authors for their enthusiastic and high-grade contribution.

We would like to express our sincere gratitude to all in the Technical Program Committee who have reviewed the papers and developed a very interesting Conference Program, as well as thanking the invited and plenary speakers.

Finally, we would like to thank the conference chairmen, the members of the steering committee, the organizing committee, the organizing secretariat and the financial support from the conference sponsors that allowed ScieTech 2016 to be a success.

The Editor of the Scietech 2016 Proceedings
Dr. Ford Lumban Gaol
Bina Nusantara University

Journal of Physics: Conference Series **710** (2016) 011002 doi:10.1088/1742-6596/710/1/011002

The 4th International Conference on Science & Engineering in Mathematics, Chemistry and Physics 2016 (ScieTech 2016)
Discovery Room, Discovery Kartika Plaza Hotel, Bali, Indonesia
30–31 January 2016

Keynote Speakers

1. 

Prof. Dr Binayak Samader Choudhury, Indian Institute of Engineering Science and Technology, Shibpur

Title: Discontinuous Functions in Metric Fixed Point Theory

Discontinuous functions have featured in metric fixed point theory since the theory originated in the first quarter of the 20th century. Today these functions occupy a large portion of this expanding branch of mathematics. The present talk gives an account of the recent developments of this area of research along with some of its implications. The broad subject area of the talk is functional analysis.

Short Bio:

Professor Binayak S. Choudhury is Professor at the Department of Mathematics, Indian Institute of Engineering Science and Technology, Shibpur, West Bengal, India for the last 12 years. He has a brilliant academic career. He obtained his Master of Science and Doctorate degrees from the University of Calcutta, India. He has held several administrative posts and was entrusted with various responsibilities at a governmental level. He served his Institute as Head of the Department of Mathematics (2005–2008) and as the Dean of the Faculty of Science (2013–2015). His research interests are in pure and applied mathematics and theoretical physics ranging over subject areas like topology, functional analysis, real analysis, probability theory, nonlinear dynamics and chaos, equilibrium problems, relativity and cosmology, quantum information theory, foundations of physics, mathematical biology, fuzzy systems and optimization and game theory, in each of which he has guided Ph.D. students. He has published nearly 200 research papers in reputed journals in addition to a good number of works published in conference proceedings. Until now 13 research students have obtained their Ph. D. degrees under his supervision and several other students are working with him. He is the recipient of several awards. He has delivered lectures in different parts of the world on topics both from theoretical physics and mathematics. He has keen interest in philosophy, language and literature, especially in English poetry and Sanskrit poetry (Kavya). He is also actively associated with the movement of the popularization of Sanskrit language.

2. 

Yasuo Kadono, Graduate School of Technology Management of Ritsumeikan University, Japan

Title: Management of Software Engineering Innovation in Japan

The talk will be on the achievements of the software engineering discipline as represented by IT vendors in Japan in order to deepen understanding of the mechanisms of how software engineering capabilities relate to IT vendors' business performance and business environment from the perspective of innovation and engineering management. Based on the concepts of service science and science for society, the volume suggests how to improve the sophistication of services between the demand side, i.e., IT user companies, and the supply side, i.e., IT vendors, simultaneously.

The speaker will talk about a structural model including innovational paths, such as service innovation, product innovation and process innovation, and a measurement model including the seven software engineering capabilities: deliverables, project management, quality assurance, process improvement, research and development, human resource development and customer contact. They then designed research on software engineering excellence and administered it with the Japanese Ministry of Economy, Trade and Industry and Information-Technology Promotion Agency. Through statistical analyses of the results, they found that human resource development and R&D are significant fundamental conditions to improve the quality of the deliverables and that IT firms with large outputs, derived from high levels of human resource development, quality assurance, project management and process improvement, tend to sustain high profitability.

Short Bio:

Yasuo Kadono is a professor at the Graduate School of Technology Management of Ritsumeikan University, and a visiting professor at Tokyo University of Foreign Studies, Japan. He received his PhD in business administration at Tsukuba University. He also received his master's and bachelor's degrees in applied mathematics and physics at Kyoto University. He has extensive practical experience with McKinsey & Company, Accenture, Sumitomo Metal Industries, and Management Science Institute. He has also produced important research projects in academic, business, and government circles. Consistent with his academic and business career to date, his teaching and research interests include the management of technology, competitive strategy, and information technology.

3. 

Prof. I Nyoman Suprapta Winaya, PhD., Mechanical Department, University of Udayana, Bali, Indonesia

Title: Fluidized bed technology using biomass and wastes fuels

The increased need to reduce carbon dioxide emissions to prevent global warming has led to an interest in biomass and solid waste as fuel sources. As a potential renewable energy resource, biomass and solid waste materials are receiving more attention worldwide. One feature of biomass fuels is their high volatile matter content, however, due to the high heat transfer from the bed material to the fuel, volatile matter evolution occurs very rapidly when fuel is fed into the bed. This indicates that the fuels are easier to ignite and to burn. Although the great ability of fluidized beds to conduct the combustion/gasification process has been confirmed by many researchers as well as several successful operating industrial plants, more research is still required to see extensive and flourishing progress, especially in how to minimize the effect of volatile matter. Some recent studies have been presented in the talk.

Short Bio:

I Nyoman Suprapta Winaya is a professor at Udayana University, Bali, Indonesia. He currently serves as Head of the Mechanical Engineering Department. Winaya received his bachelor's degree from Udayana University, his master's from Dalhousie University of Canada and his Ph.D from Niigata University of Japan. His main research is into biomass using fluidized bed combustion and gasification systems. Dr. Winaya was appointed as a Professor in the field of energy conversion systems in 2013 by Indonesia's Ministry of Education and Culture at Udayana University. At presentl, he leads the Renewable Energy Research Laboratory especially in the field of gasification and biogas. He has approved some essential innovations of high volatile matter fuels especially using porous solids as a bed material. A new method has been developed to evaluate the horizontal dispersion of loaded solids at high bed temperatures that resemble those of commercial operations. As the first step to a scaling-up of the fluidized bed (FB) system using carbon loaded solids prepared by capacitant effect, the developed model is considered to be applicable to large scale FBs if the solid dispersion coefficient can be predicted. Prof. Winaya's passion is to transfer research results into industrial practice, having a commitment to spread an advanced knowledge around the globe. Prof. Winaya is a senate member of Udayana University, and a member of the Indonesian Association of Mechanical Engineering, Indonesian Association of Fuel Experts, Indonesian Association of Engineers, American Society of Mechanical Engineering, and Japanese Society of Chemical Engineering.

Journal of Physics: Conference Series **710** (2016) 011003
doi:10.1088/1742-6596/710/1/011003

The 4th International Conference on Science & Engineering in Mathematics, Chemistry and Physics 2016 (ScieTech 2016)

Discovery Room, Discovery Kartika Plaza Hotel, Bali, Indonesia

30–31 January 2016

The 4th International Conference on Science & Engineering
in Mathematics, Chemistry and Physics 2016 (ScieTech 2016)
&
The 3rd International Conferences on Socio-Cultural Relationship
and Education Pedagogy Learning Sciences 2016 (SOCIO-CULTURAL 2016)
The Discovery Kartika Plaza Hotel
Kuta, Bali - Indonesia, 30 - 31,

The 4th International Conference on Science & Engineering
in Mathematics, Chemistry and Physics 2016 (ScieTech 2016)
&
The 3rd International Conferences on Socio-Cultural Relationship
and Education Pedagogy Learning Sciences 2016 (SOCIO-CULTURAL 2016)

Journal of Physics: Conference Series **710** (2016) 011004 doi:10.1088/1742-6596/710/1/011004

Peer review statement

All papers published in this volume of *Journal of Physics: Conference Series* have been peer reviewed through processes administered by the proceedings Editors. Reviews were conducted by expert referees to the professional and scientific standards expected of a proceedings journal published by IOP Publishing.

The approximate solutions of nonlinear Boussinesq equation

Dianhen Lu [1,a], **Jie Shen** [1,b], **Yueling Cheng** [1,c]

[1]Nonlinear Scientific Research Center, Faculty of Science, Jiangsu University, Zhenjiang Jiangsu, 212013,P.R.China

E-mail: [a]dclu@ujs.edu.cn, [b]JDshenjie@163.com, [c]1000003284@ujs.edu.cn

Abstract. The homotopy analysis method (HAM) is introduced to solve the generalized Boussinesq equation. In this work, we establish the new analytical solution of the exponential function form. Applying the homotopy perturbation method to solve the variable coefficient Boussinesq equation. The results indicate that this method is efficient for the nonlinear models with variable coefficients.

1. Introduction

The nonlinear partial differential equation(NLPDE) have become a useful and exact tool to express natural phenomena of science and engineering models. Therefore, the studies of the solutions for NLPDE have great significance in searching the nonlinear natural phenomena. Considerable efforts have been made by many mathematicians and physical scientists to obtain powerful and efficient methods such as the Tanh-function method[1]; the Sine-Cosine method[2]; the Exp-function method[3]; the Adomian decomposition method[4]; the Jacobi elliptic function method[5]; the homotopy perturbation method[6]; the modified simple equation method[7]; the $(\frac{G'}{G})$ expansion method[8]; the F-expansion method[9]; the homotopy analysis method (HAM)[10] and so on. In the paper,through the homotopy analysis method and the homotopy perturbation method,solving the nonlinear Boussinesq equation

$$u_{tt} - \alpha(t)u_{xx} - r(t)u_{xxxx} - \beta(t)(u^2)_{xx} = 0$$

where $\alpha(t), r(t), \beta(t)$ are any function about t . The Boussinesq equation is a classical nonlinear equation, which describes the wave phenomenon of physics, and has been widely studied in many fields of physics. Solving the Boussinesq equation has become a hot topic in the study of nonlinear equations.

2. Basic idea of homotopy analysis method

In this section, we review the main contents of HAM briefly, we consider the nonlinear equation

$$N[u(x,t)] = 0, \tag{2.1}$$

where N is a nonlinear operator, x and t express independent variables, and $u(x,t)$ is an unknown function. According to the basic idea of the traditional homotopy method, we construct a equation in the form

$$(1-q)L[\Phi(x,t;q)-u_0(x,t)] = \hbar q H(x,t) N[\Phi(x,t;q)], \tag{2.2}$$

where $q \in [0,1]$ is the embedding parameter, $\hbar$ is a nonzero auxiliary parameter, L is an auxiliary linear operator, $u_0(x,t)$ is an initial guess of $u(x,t)$; $\Phi(x,t;q)$ is an unknown function , $H(x,t)$ is an auxiliary function. In the above equation(2.2), we can choose L, $H(x,t)$ and $\hbar$ in HAM flexibly. When the parameter q raises from 0 to 1, the solution $\Phi(x,t;q)$ changes from $u_0(x,t)$ to $u(x,t)$. Expanding $\Phi(x,t;q)$ in Taylor series with respect to q, we have

$$\Phi(x,t;q) = u_0(x,t) + \sum_{m=1}^{\infty} u_m(x,t) q^m, \tag{2.3}$$

where

$$u_m(x,t) = \frac{1}{m!}\frac{\partial^m}{\partial q^m}\Phi(x,t;q)\bigg|_{q=0}. \tag{2.4}$$

As pointed by Liao [11], if the auxiliary objects such as L, $H(x,t)$ and $\hbar$ are properly chosen, the series (2.3) will converge at $q=1$ and we has

$$u(x,t) = u_0(x,t) + \sum_{m=1}^{\infty} u_m(x,t). \tag{2.5}$$

The above series solutions generally converge rapidly, and Liao has proved that which must be one solution of the original nonlinear Eq.(2.1). Differentiating Eq.(2.2) m times in the embedding parameter q and then setting $q=0$ and finally dividing by $m!$, we have the equation in the form

$$L[u_m(x,t) - \chi_m u_{m-1}(x,t)] = \hbar H(x,t) R_m(\vec{u}_{m-1}, x, t), \tag{2.6}$$

where

$$\vec{u}_{m-1} = \{u_0(x,t), u_1(x,t), \cdots, u_{m-1}(x,t)\}, \tag{2.7}$$

$$\chi_m = \begin{cases} 0, m \le 1, \\ 1, m \ge 2, \end{cases} \tag{2.8}$$

$$R_m(\vec{u}_{m-1}, x, t) = \frac{1}{m!}\frac{\partial^{m-1}}{\partial q^{m-1}} N[\Phi(x,t;q)]\bigg|_{q=0}. \tag{2.9}$$

Eq.(2.6) is a linear equation, through the iterative process, we can easily get $u_1(x,t), u_2(x,t), \cdots, u_m(x,t) \cdots$.

3. Basic idea of homotopy perturbation method

Assume that a nonlinear differential equation in the form

$$A(u) - f(r) = 0, r \in \Omega \tag{3.1}$$

with the boundary conditions $B(u, \frac{\partial u}{\partial n}) = 0, r \in \Gamma$, where A is a general differential operator, B is a boundary operator, $f(r)$ is known as analytic function, and Γ is the boundary of the domain Ω. Dividing the operator A into two parts linear L and nonlinear N. Thus Eq.(3.1) becomes

$$L(u) + N(u) - f(r) = 0. \tag{3.2}$$

On the basis of the homotopy technique[6], establishing a homotopy

$$H(u,p) = L(u) - L(v) + p[L(v) + N(u) - f(r)], p \in [0,1], r \in \Omega, \tag{3.3}$$

where p is parameter, v is an initial approximation for the solution of Eq.(3.1), which is satisfied with the boundary conditions. Hence, when p increases from 0 to 1, we can obtain

$$H(u,0) = L(u) - L(v) \tag{3.4}$$

$$H(u,1) = A(u) - f(r) = 0. \tag{3.5}$$

Suppose that the solution of $H(u,p) = 0$ can be expressed as a power series with respect to p :

$$u(x,t,p) = \sum_{i=0}^{\infty} u_i(x,t)p^i = u_0 + pu_1 + p^2u_2 + \cdots. \tag{3.6}$$

So when $p = 0$, $u(x,t,0) = u_0(x,t)$ is the solution of $L(u) - L(v) = 0$; when $p \to 1$, the approximate solution of $A(u) - f(r) = 0$ is

$$u(x,t) = u_0 + u_1 + u_2 + \cdots. \tag{3.7}$$

4. The approximate solution of the constant coefficient Boussinesq equation

Now, we study the constant coefficient nonlinear Boussinesq equation:

$$u_{tt} - \alpha u_{xx} - ru_{xxxx} - \beta(u^2)_{xx} = 0 \tag{4.1}$$

where α, r, β are constants.

Firstly, we seek the exponential function solution of Eq.(4.1). Defining $\xi = x + \omega t$, then Eq.(4.1) becomes in the form:

$$\omega^2 u'' - \alpha u'' - ru'''' - \beta(u^2)'' = 0 \tag{4.2}$$

where the prime means the derivative in ξ. Then integrating Eq.(4.2) with respect to ξ, choosing constant of integration to zero, we obtain

$$ru'' - (\omega^2 - \alpha)u + \beta u^2 = 0. \tag{4.3}$$

We choose the solution in the form

$$u(\xi) = \sum_{m=0}^{\infty} a_m e^{-m\xi} \tag{4.4}$$

where a_m is a coefficient, and the linear operator

$$L[\Phi(\xi,q)] = (\frac{\partial^2}{\partial \xi^2} - 1)\Phi(\xi,q), \tag{4.5}$$

meets $L[\alpha_1 e^{-\xi} + \beta_1 e^{\xi}] = 0$, where α_1, β_1 are constants. Based on (4.4), we choose the initial guess $u_0(\xi) = e^{-\xi}$. From (4.3), we definite the nonlinear operator

$$N[\Phi(\xi,q), \omega(q)] = r\Phi'' - (\omega^2 - \alpha)\Phi + \beta\Phi^2. \tag{4.6}$$

Applying the above definition, and defining $H(x,t)=1$,we can get

$$(1-q)L[\Phi(\xi,q)-u_0(\xi)] = q\hbar N[\Phi(\xi,q),\omega(q)], \tag{4.7}$$

$$L[u_m(\xi)-\chi_m u_{m-1}(\xi)] = \hbar R_m(\vec{u}_{m-1},\vec{\omega}_{m-1}), \tag{4.8}$$

where

$$\vec{u}_{m-1} = \{u_0(\xi),u_1(\xi),\cdots,u_{m-1}(\xi)\}, \tag{4.9}$$

$$\vec{\omega}_{m-1} = \{\omega_0(\xi),\omega_1(\xi),\cdots,\omega_{m-1}(\xi)\}, \tag{4.10}$$

and

$$R_m(\xi) = r\frac{\partial^2 u_{m-1}}{\partial \xi^2} - \sum_{i=0}^{m-1} u_{m-1-i}(\sum_{j=0}^{i}\omega_{i-j}\omega_j) + \alpha u_{m-1} + \beta\sum_{i=0}^{m-1} u_i u_{m-1-i}. \tag{4.11}$$

Through the iterative process, we can get the solution of the linear differential equation(4.8) easily .

For example , when $m=0$, Eq.(4.8) becomes

$$L[u_1(\xi)] = \hbar[ru_0'' - \omega_0^2 u_0 + \alpha u_0 + \beta u_0^2]. \tag{4.12}$$

Substituting $u_0(\xi)=e^{-\xi}$ into above equation, that is

$$L[u_1(\xi)] = \hbar[(r-\omega_0^2+\alpha)e^{-\xi} + \beta e^{-2\xi}]. \tag{4.13}$$

The solution of u_1 involves the terms $\xi e^{-\xi}$, if $r-\omega_0^2+\alpha \neq 0$. Thus it must be enforced to zero, i.e. $\omega_0 = \pm\sqrt{r+\alpha}$, and

$$u_1(\xi) = C_1 e^{-\xi} + \frac{\beta\hbar}{3}e^{-2\xi}, \tag{4.14}$$

When $m=2$,

$$L[u_2-u_1] = \hbar[(-\omega_0^2+\alpha)u_1 - 2\omega_0\omega_1 u_0 + ru_1'' + 2\beta u_0 u_1] \tag{4.15}$$

then substituting $u_0=e^{-\xi}$, $u_1 = C_1e^{-\xi} + \frac{\beta\hbar}{3}e^{-2\xi}$ into Eq.(4.15), result in

$$L[u_2-u_1] = \hbar[-2\omega_0\omega_1 e^{-\xi} + (r\beta\hbar + 2\beta C_1)e^{-2\xi} + \frac{2\beta^2\hbar}{3}e^{-3\xi}]. \tag{4.16}$$

Avoiding the term $\xi e^{-\xi}$, let $-2\omega_0\omega_1=0$. Substituting the value of ω_0 into it, we can get $\omega_1=0$. Then solving the linear Eq.(4.16),

$$u_2(\xi) = (C_1+C_2)e^{-\xi} + (\frac{r\beta\hbar^2 + \beta\hbar + 2\beta C_1\hbar}{3})e^{-2\xi} + \frac{\beta^2\hbar^2}{12}e^{-3\xi}, \tag{4.17}$$

where C_1, C_2 are constants. At last the two-degree approximate solution of (4.2) and one-degree approximate solution of ω can be obtained as follows:

$$\omega \approx \omega_0 + \omega_1 = \pm\sqrt{r+\alpha}, \tag{4.18}$$

$$u(\xi) \approx u_0+u_1+u_2 = (2C_1+C_2+1)e^{-\xi} + (\frac{r\beta\hbar^2 + 2\beta\hbar + 2\beta C_1\hbar}{3})e^{-2\xi} + \frac{\beta^2\hbar^2}{12}e^{-3\xi}, \tag{4.19}$$

where $\xi = x + (\pm\sqrt{r+\alpha})t$.

In order to find the reasonable range of $\hbar$, we plot the $\hbar$-curve, which is shown in Fig.1.As proved by Liao [13],the horizontal line segment is a valid region of $\hbar$ and the solution series must converge to the exact solutions. Thus, the valid region of $\hbar$ is $-1.4 \leq \hbar \leq -0.6$. Fig.2 is the exponential style solution under the 10th order of approximation. Error chart is shown in Fig.3.

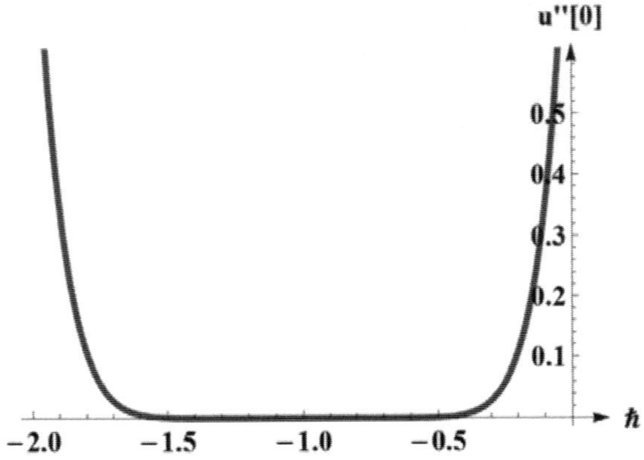

Figure 1. The $\hbar$-curves of $u''_{appr}(0)$ under the 10th order of approximation when $\alpha = 1, \beta = 1, r = 1, u(0) = 1$.

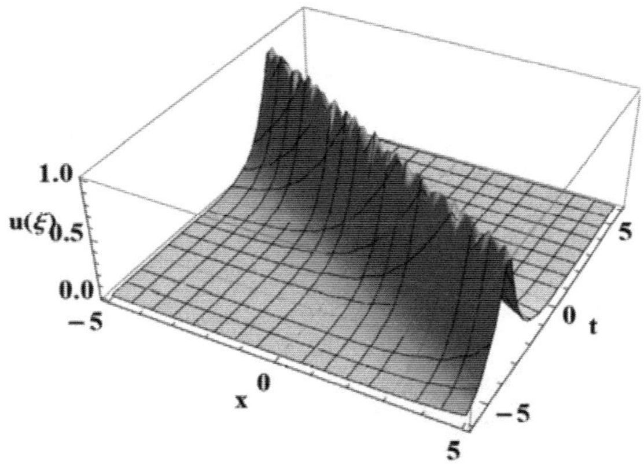

Figure 2. The 10th order of $u_{appr}(\xi)$ with $\hbar = -1, \alpha = 1, \beta = 1, r = 1, u(0) = 1$.

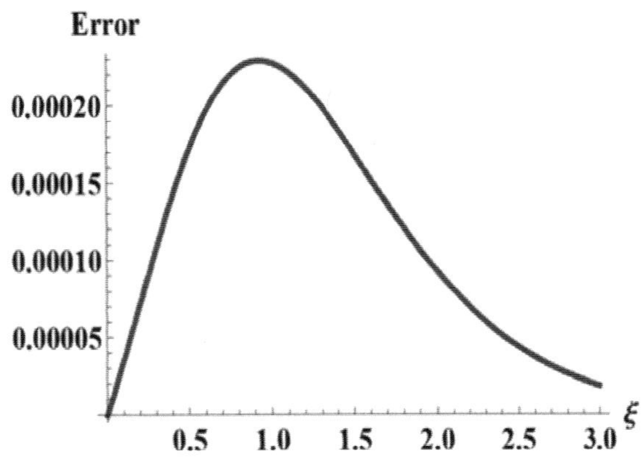

Figure 3. Error chart with $\hbar = -1, \alpha = 1, \beta = 1, r = 1, u(0) = 1$.

5. The approximate solution of the variable coefficient Boussinesq equation

Next,seeking the approximate solution of the variable coefficient Boussinesq equation

$$u_{tt} - \alpha(t)u_{xx} - \beta(t)(u^2)_{xx} - r(t)u_{xxxx} = 0. \tag{5.1}$$

Defining $x_1 = \sqrt{\frac{c_3\alpha(t)}{c_1 r(t)}}x$, $t_1 = \int_0^t \frac{\alpha(\tau)}{c_1}\sqrt{\frac{c_3}{r(\tau)}}d\tau$, $c_1^* = \frac{c_2\alpha(t)}{\beta(t)}$, where $c_1\alpha(t) > 0, c_3 r(t) > 0, c_2 \neq 0$, Eq.(5.1) becomes

$$u_{t_1t_1} - c_1^* u_{x_1x_1} - c_2(u^2)_{x_1x_1} - c_3 u_{x_1x_1x_1x_1} = 0. \tag{5.2}$$

For simplicity, we take $x = x_1$, $t = t_1$. Then (5.2) translates into

$$u_{tt} - c_1^* u_{xx} - c_2(u^2)_{xx} - c_3 u_{xxxx} = 0. \tag{5.3}$$

Next, we applied the homotopy perturbation method to study the approximate solution of Eq.(5.3). Firstly, we set up homotopy mapping $H(u,p) : R \times [0,1] \to R$,

$$H(u,p) = L(u) - L(v) + p[L(v) - c_2(u^2)_{xx} - c_1^* u_{xx}], \tag{5.4}$$

where $R = (-\infty, +\infty)$, v is the auxiliary function, and L is the linear operator in the form: $L(u) = u_{tt} - c_3 u_{xxxx}$.

We can easily get the linear equation

$$v_{tt} - c_3 v_{xxxx} = 0 \tag{5.5}$$

has the following solution:

$$v = \cosh(\omega\sqrt{\frac{1}{c_3}}\xi), c_3 > 0 \tag{5.6}$$

$$v = \cos(\omega\sqrt{\frac{1}{-c_3}}\xi) + \sin(\omega\sqrt{\frac{1}{-c_3}}\xi), c_3 < 0 \tag{5.7}$$

where $\xi = x + \omega t$, ω is any constant. One can easily observe that $H(u,1) = 0$ and Eq.(5.3) is the same, so the solution $u(x,t)$ of Eq.(5.3) is the solution of $H(u,p) = 0$ when under the condition $p \to 1$. Let (3.6) be the solution of $H(u,p) = 0$; this series is uniformly convergent in the $p \in [0,1]$. Thus it yields that

$$u = \sum_{i=0}^{\infty} u_i(x,t) = u_0 + u_1 + u_2 + \cdots. \tag{5.8}$$

Substituting (3.6) into the equation $H(u,p) = 0$, and matching the coefficients of the same power of p, we can get that

$$p^0 : L(u_0) = L(v) \tag{5.9}$$

$$p^1 : L(u_1) = -L(v) + c_2(u_0^2)_{xx} + c_1^* u_{0xx} = c_2(u_0^2)_{xx} + c_1^* u_{0xx} \tag{5.10}$$

$$p^2 : L(u_2) = 2c_2(u_0u_1)_{xx} + c_1^* u_{1xx}. \tag{5.11}$$

From (5.9), we have

$$u_0(x,t) = v(x,t). \tag{5.12}$$

Defining $c_2(u_0^2)_{xx} + c_1^* u_{0xx} = f_1(x,t), 2c_2(u_0u_1)_{xx} + c_1^* u_{1xx} = f_2(x,t)$, and using the Fourier transform method, from Eq.(5.10) (5.11), we have

$$\frac{d^2\overline{u_1}}{dt^2} - \lambda^4 c_3 \overline{u_1} = \overline{f_1}(\lambda, t). \tag{5.13}$$

$$\frac{d^2\overline{u_1}}{dt^2} - \lambda^4 c_3 \overline{u_1} = \overline{f_2}(\lambda, t). \tag{5.14}$$

where $\overline{f_1}, \overline{f_2}$, denotes the Fourier transform of the term $f_1(x,t), f_2(x,t)$, and

$$\overline{u_0}(\lambda,t)=\int_{-\infty}^{\infty}u_0(\xi,t)e^{-i\lambda\xi}d\xi. \tag{5.15}$$

From Eq.(5.13), we assume the initial value is zero, that is

$$\overline{u_1}(\lambda,t)=\int_0^t\frac{\overline{f_1}}{\sqrt{-\lambda^4c_3}}\sin\sqrt{-\lambda^4c_3}(t-t_1)dt_1, c_3<0, \tag{5.16}$$

$$\overline{u_1}(\lambda,t)=\int_0^t\frac{-\overline{f_1}}{\sqrt{\lambda^4c_3}}\sinh\sqrt{\lambda^4c_3}(t-t_1)dt_1, c_3>0. \tag{5.17}$$

Thus we obtain

$$u_1(x,t)=\frac{1}{2\pi}\int_{-\infty}^{\infty}\int_0^t\frac{\overline{f_1}}{\sqrt{-\lambda^4c_3}}\sin\sqrt{-\lambda^4c_3}(t-t_1)e^{i\lambda x}dt_1d\lambda, c_3<0 \tag{5.18}$$

$$u_1(x,t)=\frac{1}{2\pi}\int_{-\infty}^{\infty}\int_0^t\frac{-\overline{f_1}}{\sqrt{\lambda^4c_3}}\sinh\sqrt{\lambda^4c_3}(t-t_1)e^{i\lambda x}dt_1d\lambda, c_3>0. \tag{5.19}$$

From Eq.(5.14), as zero is the initial value, we have

$$u_2(x,t)=\frac{1}{2\pi}\int_{-\infty}^{\infty}\int_0^t\frac{\overline{f_2}}{\sqrt{-\lambda^4c_3}}\sin\sqrt{-\lambda^4c_3}(t-t_1)e^{i\lambda x}dt_1d\lambda, c_3<0 \tag{5.20}$$

$$u_2(x,t)=\frac{1}{2\pi}\int_{-\infty}^{\infty}\int_0^t\frac{-\overline{f_2}}{\sqrt{\lambda^4c_3}}\sinh\sqrt{\lambda^4c_3}(t-t_1)e^{i\lambda x}dt_1d\lambda, c_3>0. \tag{5.21}$$

From (5.18)-(5.21), the two-degree approximate solution of Eq.(5.3) can be obtained as follows:

$$u(x,t)=\cos(\omega\sqrt{\frac{1}{-c_3}}\xi)+\sin(\omega\sqrt{\frac{1}{-c_3}}\xi)+\frac{1}{2\pi}\int_{-\infty}^{\infty}\int_0^t\frac{\overline{f_1}+\overline{f_2}}{\sqrt{-\lambda^4c_3}}\sin\sqrt{-\lambda^4c_3}(t-t_1)e^{i\lambda x}dt_1d\lambda, c_3<0 \tag{5.22}$$

$$u(x,t)=\cosh(\omega\sqrt{\frac{1}{c_3}}\xi)-\frac{1}{2\pi}\int_{-\infty}^{\infty}\int_0^t\frac{\overline{f_1}+\overline{f_2}}{\sqrt{\lambda^4c_3}}\sinh\sqrt{\lambda^4c_3}(t-t_1)e^{i\lambda x}dt_1d\lambda, c_3>0. \tag{5.23}$$

6. Conclusion

This work studies the generalized Boussinesq equation by using the homotopy analysis method and the variable coefficient Boussinesq equation with the homotopy perturbation method. The solution of the generalized Boussinesq equation contains the parameter $\hbar$ and we can adjust the value of $\hbar$ to guarantee the convergence region of the solution series. With the help of some mathematical software, such as MATHEMATICA, MATLAB, we can compile program algorithm to get more accurate approximate solutions. The homotopy perturbation method deforms an nonlinear problem into a set of linear problems which are easier to solve. The two-degree approximate solution of the variable coefficient Boussinesq equation show that the homotopy perturbation method is effective to solve the variable solution equations.

Acknowledgements

The work was supported by the National Nature Science Foundation of China (No. 61070231), and the Natural Science Foundation of Jiangsu Province (BK20140525).

References

[1] M.A. Abdou, The extended tanh-method and its applications for solving nonlinear physical models, Appl. Math. Comput. 190 (2007) 988-996.
[2] Z.Y. Yan, H.Q. Zhang, New explicit and exact traveling wave solutions for a system of variant Boussinesq equations in mathematical physics, Phys. Lett.A 252 (1999) 291-296.
[3] J.H. He, X.H. Wu, Exp-function method for nonlinear wave equations, Chaos, Solitons.Fract. 30 (2006) 700-708.
[4] G.Adomian, A review of the decomposition method and some recent results for non-linear equations,Comput.Math.Appl.21(5)(1991)101-127.
[5] Liu SK, Fu ZT, Liu SD, et al. Jacobi elliptic function expansion method and periodic wave solutions of nonlinear wave equations. Phys Lett A 2001;289:69-74.
[6] J.H.He, Homotopy perturbation technique, Comput. Mathods Appl. Mech. Eng. 178 (3-4) (1999)257-262.
[7] A.J.M. Jawad, M.D. Petkovic, A. Biswas, Modified simple equation method for nonlinear evolution equations, Appl.Math. Comput. 217 (2010) 869-877.
[8] M.Wang, X.Li, J.Zhang, The $(\frac{G'}{G})$-expansion method and travelling wave solutions of nonlinear evolution equations in mathematical physics, Phys. Lett. A,372(2008);417-423.
[9] W.W.Li, Y.Tian, Z.Zhang, F-expansion method and its application for finding new exact solutions to the sine-Gordon and sinh-Gordon equations, Appl. Math. Comput(2012),1135-1143.
[10] S.J. Liao, On the homotopy analysis method for nonlinear problems, Appl. Math. Comput. 147 (2) (2004) 499-513.
[11] S.J. Liao, Beyond Perturbation: Introduction to the Homotopy Analysis Method, Chapman & Hall/CRC, Boca Raton, 2003.
[12] S.J. Liao, An approximate solution technique not depending on small parameters: A special example, Int. J. Non-Linear Mech. 30 (3) (1995) 371-380.
[13] S.J.Liao, Comparison between the homotopy analysis method and homotopy perturbation method, Appl. Math. Comput. 169 (2005) 1186-1194.
[14] D.C.Lu, B.J.Hong, New exact solutions for the (2+1)-dimensional Generalized Broer-Kaup System, Appl. Math. Comput. 199(2)(2008) 572-580.
[15] S.Abbasbandy, The application of homotopy analysis method to solve a generalized Hirota-Satsuma coupled KdV equation, Phys. Lett. A. 361(6) (2007) 478-483.
[16] S.Abbasbandy, Soliton solutions for the Fitzhugh-Nagumo equation with the homotopy analysis method.Appl. Math. Model. 32 (2008) 2706-2714.
[17] J. Singh, D. Kumar, Sushila,Homotopy perturbation sumudu transform method for nonlinear equations. Adv. Theor. Appl. Mech., 4 (2011) 165C175.
[18] Sushila, Jagdev Singh, Y.S. Shihodia, An efficient analytical approach for MHD viscous flow over a stretching sheet via homotopy perturbation sumudu transform method. Ain Shams Eng. J., 4 (2013) 549C555.
[19] J.Wang, Some new and general solutions to the compound KdV-Burgers system with nonlinear terms of any order, Appl. Math. Comput. 217 (2010) 1652-1657.
[20] B.J.Hong, D.C.Lu, New exact solutions for the generalized variable-coefficient Gardner equation with forcing term. Applied Mathematics and Computation, Appl. Math. Comput. 219 (2012) 2732-2738.

Journal of Physics: Conference Series **710** (2016) 012002 doi:10.1088/1742-6596/710/1/012002

Influence of Temperature, Roughness and Washer Shape on Crevice Corrosion for 254 SMO Alloy

SanaaT.Arab[1], Khadijah M. Emran[2], Huda M. Alghamdi[1], Mohammed I. Abdulsalam[3]
[1]Department of Physical Chemistry, King Abdul-Aziz University, Jeddah, Saudi Arabia
[2]Chemistry Department, Faculty of Science, Taibah University, Al-Medina Al-Monawara, Saudi Arabia
[3]Department of Chemical and Materials Engineering, King Abdul-Aziz University, Jeddah, Saudi Arabia
Correspondence: Khadijah M. Emran, Chemistry Department, Faculty of Science, Taibah University, Saudi Arabia. Tel: +966-50-435-5039.
E-mail: kabdalsamad@taibahu.edu.sa

Abstract. Effect of temperature, roughness and washer shape on 254 SMO alloy was applied to study the susceptibility of alloy's crevice corrosion. XPS measurements have been carried out to detect and define the products which formed on the surface of 254 SMO in 22% NaCl at 30oC at applied potential 600 mVSCE. The formation of Fe, Cr and Mo compounds which play a great role to protect the alloy was found.

1.Introduction

254 SMO is considered to be a very high austenitic stainless steel. It was originally developed for use in seawater and other aggressive chloride-bearing environments. Many parameters, such as temperature and crevice geometry, were used to obtain their effect on crevice corrosion.

An increase in temperature tend to stimulate corrosive attack by increasing the rate of electrochemical reactions[1]. For stainless steels and other alloys that are prone to pitting and crevice corrosion, an increase in temperature tends to facilitate initiation of these types of attack [1]. Many parameters that influence corrosion can vary with temperature: dissolved oxygen solubility, solution properties (e.g. viscosity and ion mobility), ferrous oxidation rate and thermodynamic properties of iron scale (leading to formation of different phases or compounds) [2].

Crevice geometry is one of the important factors in crevice corrosion susceptibility. The two dimensions that characterize a crevice, the gap and the depth or length exert a great deal of control on the initiation and propagation of crevice corrosion, in agreement with that in the other studies [3]. The current flow from the crevice to the outer surface is partially supported by the inward migration on anions such as Cl^-, which further lowers the pH. In these respects, the mechanisms of the crevice and pitting corrosion are similar. In view of this sequence, the crevice geometry parameters of gap width and depth become important [4]. The corrosion rate within the crevice increased as the crevice dimension decreases [5].

2. Experimental

In this study, the effect temperature, roughness and washer shape on the crevice corrosion of 254 SMO alloy was tested. XPS technique was used to identifying constituent species of corrosion products.

Table 1. Chemical compositions (%wt) of 254 SMO

C	Cr	Cu	Mn	Mo	Ni	N	P	Si	S	Fe
≤0.020	20	0.7	≤1.00	6.1	18	0.20	≤0.030	≤0.80	≤0.010	53.14

A multiple crevice test assembly made of Teflon from Metal Samples Co. Each washer has 20 teeth. A circular disc of 25 mm diameter was cleaned and degreased by ultrasound using double distilled water. The multiple crevice washers were bolted to both sides of each specimen using screw and nuts (made of Titanium). The screw bolt was inserted in a plastic tube, Fig. 1, to protect it from the solution.

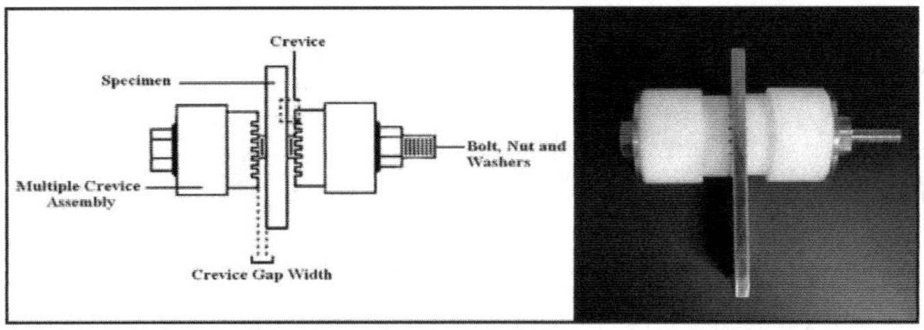

Figure 1. Schematic and photograph of the crevice assembly used in the study.

The corrosion cell contains 1000 ml of aerated 22% NaCl solution. A carbon electrode and a saturated calomel electrode (SCE) were used as a counter and a reference electrodes respectively. Salt bridge with Luggin capillary was used to connect the reference electrode to the corrosion cell. Test solution was used in the salt bridge. The cell was connected to PC14/750 Gamrypotentiostat for the potentiostatic study. Effect of temperature was tested at E_{apply} = 600 mV_{SCE} at room temperature, 30°C, 40°C and 55°C. SCE was put in the Luggin capillary, not in the direct solution and the value of SCE potential doesn't change in the studied range of temperature. After experiments, SCE impedance was tested using Gamry EIS. Effect of roughness was tested at at E_{apply} = 400 mV_{SCE} at final polishing 600 and 1200 SiC. Effect of washers shape was tested using grooved and flat washers at E_{apply} = 400 mV_{SCE}.

Chemical compositions on 254 SMO surface after the study were noticed by XPS using a multi-technique surface analysis system (MAX200, Leybold).

3. Results and Discussion

3.1 Effect of Temperature

The effect of temperature on crevice corrosion behavior for 254 SMO alloy was studied at four temperatures in 22% NaCl at applied potential 600 mV_{SCE} as shown in Table 2. The sample that was tested at room temperature (25°C) did not show any signs of crevice corrosion after three days. Also, it did not show any corrosion activity through current measurements with time (Figs. 2 and 3). The samples tested at 30, 40 and 55oC showed crevice corrosion attacks as shown in Fig.2. It is appeared that the crevice corrosion sets increased with temperature increasing.

Table 2. Crevice Corrosion Susceptibility Results for 254 SMO at Different Temperatures in 22% NaCl at Eappl= 600 mVSCE at Torque Value = 2.5 in lbf

Temp. (°*C*)	Time (*ks*)	Corrosion attack	Induction period(*ks*)
Room temp.(*25°C*)	259.2	No	-
30	259.2	Yes (12/40)	72.13
40	259.2	Yes (34/40)	5.345
55	0.420	Yes (40/40)	0.040

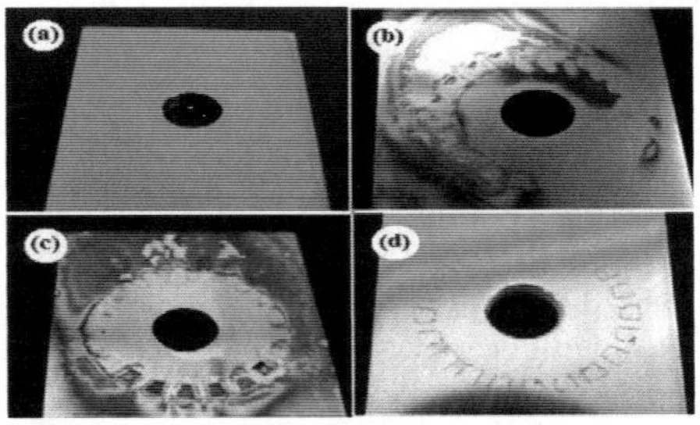

Figure 4. Effect of temperature on crevice corrosion for 254 SMO sample in 22% NaCl at E^{appl}= $600mV_{SCE}$ at torque value = 2.5 in lbf

(a) at room temperature(*25°C*), (b) at 30*°C*, (c) at 40*°C*, (d) at 50*°C*

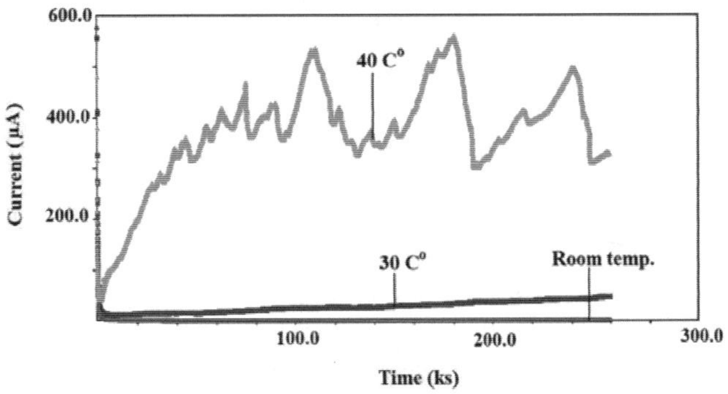

Figure 3. Measured current with time for 254 SMO sample at different temperatures in 22% NaCl at E_{appl}= 600 mV_{SCE} at torque value = 2.5 in lbf

In the test at 30°C, the current continued to increase gradually in very low rate (0.1080 μA/ks) with time (Fig.3). While at 40°C, the current increased significantly with time at the beginning of the experiment then a noticeable fluctuation in current value is found which is suggested to be due to a competition between formation and deformation of oxide passive film. Dissolution and re-built of the protective films were continued until the end of the experiment. The fluctuation in current value may be due to the competitive mechanism between passivating species $[MOOH]_{ads}$, $[MOMOH]_{ads}$ and non-passivating species $[MOH]_{ads}$, $[MOMCl]_{ads}$, $[MOMOHCl]_{ads}$ as suggested by [6].
Figure 4 shows the current resulted from 254 SMO corrosion at 55°C. The current increased sharply to ~ 20.5mA in 7 minutes only. Comparing test at temperatures 30°C, 40°C and 55°C, the induction period decreased with the temperature increasing as shown in Table 2.

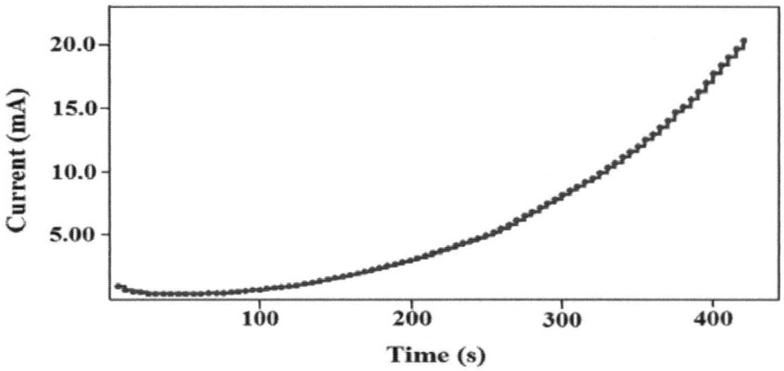

Figure 4. Measured current with time for 254 SMO sample in 22% NaCl at E_{appl}= 600 mV_{SCE} at 55°C at torque value = 2.5 in lbf

The possibility of a temperature induced change in the protective properties of the passive film on alloy surface, resulting in a reduction in its resistance to breakdown as temperature is increased. This change can be rationalized by taking two possibilities into consideration. First, the porosity of the passive film often assumed to increase with temperature [7]. The increase in porosity with temperature is supported by the fact that Cl^- ions have been found to incorporate into the passive film on stainless steel at high temperatures, but not at lower temperatures Wang et al.[8] have proved that at higher temperatures, the passive film on stainless steel becomes thicker but more porous, leading it to be much less protected. So, at higher temperatures, weaker the protective property of the passive film is found. Second, the passive film may undergo an intrinsic modification of its chemical composition and/or physical structure resulting, for example, in a variation of the density of vacancies or voids in the oxide film [9].

In NaCl solution, 254 SMO can form the passive film. Once a film is formed, the reaction rate between a metal and its surrounding environment will be lower several orders of magnitude. The passive film, probably with a bilayer structure: an inner layer rich in Cr_2O_3 and an outer porous layer abundant with $Cr(OH)_3$ [10-12]. According to Qvarfort [13], the two steps that a passive film formed in a solution will be, as follows: first chromium-rich passive film dissolves in the solution, and second the chromium inside the stainless steel diffuses to the metal surface and then forms new passive layers. As the temperature rises, the first step is quicker than the second step. So with the temperature increasing, the passive film becomes thinner and leads the film easier to be ruptured. Both possibilities should be considered [7,14].

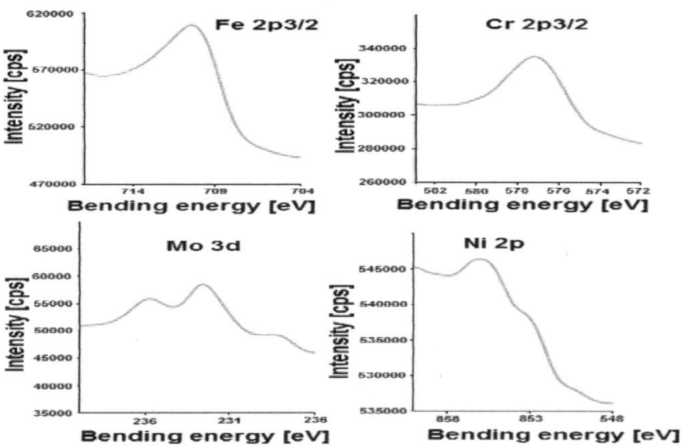

Figure 5. XPS measurement for Cr, Fe, Mo and Ni on 254 SMO alloy surface after corrosion in 22% NaCl at 30oC at 600 Mvsce.

To define the products of the corrosion, XPS has been carried out to 254 SMO surface after corrosion in 22% NaCl at applied potential = 600 mV_{SCE} as shown in Fig. 5. Iron compounds (Fe_2O_3 and $FeCl_2$),

Chromium compounds ($CrCl_3$, Cr_2O_3 and CrOOH), Molybdenum oxide (MoO_3) and Nickel chloride have been detected by XPS measurement. The nature of passive film of stainless steels is a mixture of metal oxides and oxyhydroxides. It has been suggested that Mo6+ ions are present in the outer layer. They may act as cation selective species in the outer layer producing a bipolar passive film, with the inner layer which contains Cr2O3 being anion selective. Thus, the inward passage of cations produced by metallic dissolution at the metal/oxide interface is not favoured [15,16].

3.2 Effect of Surface Roughness

To study the effect of surface roughness, two tests were performed at same experiment conditions, in 22% NaCl at E_{appl}= 400 mV_{SCE}, but they have different final polishing (600 and 1200 SiC). It was found that the number of crevice corrosion sets with final polishing 600 SiC (17/40) is more than the number of crevice corrosion sets with final polishing 1200 SiC (14/40) as shown in Tabale3 and Fig. 6.

Table 3. Crevice Corrosion Susceptibility Results for 254 SMO at Different Surface Roughness in 22% NaCl at E_{appl} = 400 mV_{SCE} at 30°C at Torque Value= 10 in lbf.

Time (ks)	Final polishing	Corrosion attack	Induction period (ks)
172.8	1200	Yes (14/40)	2.681
86.40	600	Yes (17/40)	3.888

The current for 254 SMO corrosion at final polishing 600 SiC takes more time to increase than it at final polishing 1200SiC, Table 4, and it increased slightly to values more than it at final polishing 1200SiC as shown in Fig. 7. The corrosion rates for 254 SMO corrosion in the two tests were almost the same (2.8 μA/ks) but the current in test at final polishing 600SiC continue to increase for longer time.

Table 4. Crevice Corrosion Susceptibility Results for 254 SMO at Different Washer Shape in 22% NaCl at Eappl = 400 mVSCE at 30°C at Torque Value = 10.5 in lbf

Time (*s*)	washers	Corrosion attack
172.8	Grooved washers	Yes (14/40)
86.40	Flat washers	Yes

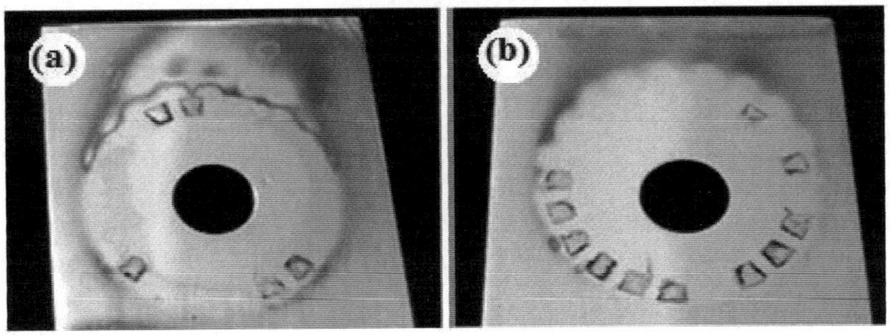

Figure 6. Effect of roughness (final polishing) on crevice corrosion for *254 SMO* sample in 22% *NaCl* at E_{appl}= 400 mV_{SCE} at 30°*C* at torque value= 10 *in lbf.* (a) 1200 *SiC* paper , (b) 600 *SiC* paper

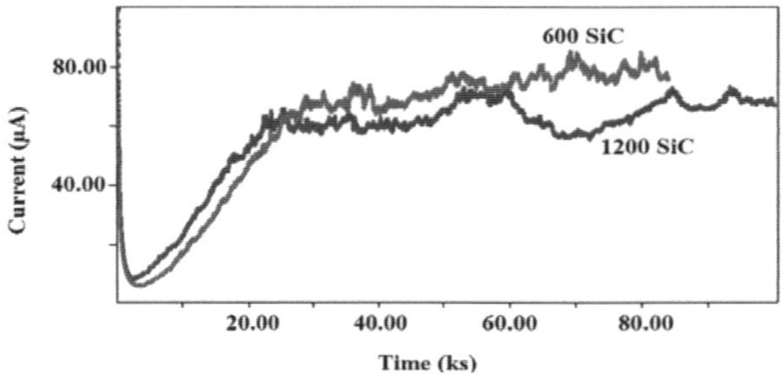

Figure 7. Measured current with time for 254 SMO sample at different final polishing in 22% NaCl at E_{appl}= 400 mV_{SCE} at 30°C at torque value= 10 in lbf

3.3 Effect of Washers Shape

Flat and grooved washers were used for 254 SMO alloy in 22% NaCl at E_{appl}= 400 mV_{SCE} as shown in Table 4 and Fig. 8. By comparing the results between the test using flat washers and the test using grooved washers, the sample using flat takes more time to start corrosion. The value of current with time, Fig. 9, for 254 SMO sample corrosion in the test using flat washers is lower than it in the test using grooved washers.

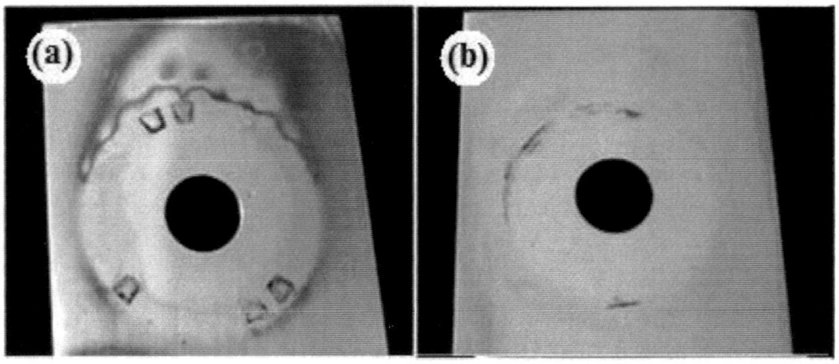

Figure 8. Effect of washers shape on crevice corrosion for 254 SMO sample in 22% NaCl at E_{appl}= 400mV_{SCE} at 30°C at torque value=10 in lbf.
(a)grooved washers,(b) flat washers

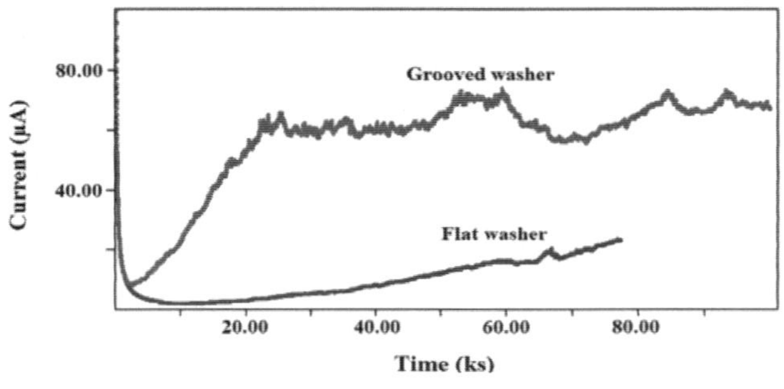

Figure 9. Measured current with time for *254 SMO* sample at different washer shapes in 22% *NaCl* at E_{appl}= 400 mV_{SCE} at 30°*C* at torque value= 10 *in lbf*

In the test using grooved washers, there are 40 locations (40 crevices) where crevice corrosion can initiate during the exposure time in the test. Each crevice may separately initiate crevice corrosion at different times during the study. If crevice corrosion initiated at one location, then stifled may appear in another location, this behavior may give the zigzag shape in the current-time curve .While in the test using flat washers, a gradual increase and smooth shape are found due to the presence of only one crevice.

The grooved washers have smaller crevice than the flat washers which means that the oxygen depletion and chloride-ion influx increased in grooved washer leading to accelerate the crevice corrosion initiation. Oldfield et al. [17], have shown that the gap dimension of the crevice has a major influence on the development of a critical crevice solution within the crevice and initiation/no initiation in any given crevice.

4. Conclusion

1. Increasing temperature leads to increase the number of corrosion sets and decrease the induction period.
2. The effect of polishing on the corrosion results for the sample (from 600 SiC to 1200 SiC) did not show significant differences in current values but the number of corrosion sets decreases with final polishing increasing.
3. Flat washer has delayed the corrosion initiation comparing with grooved washer.

Acknowledgment

This work was financially supported by KACST, Saudi Arabia. The authors wish to thank Prof. Hamad A. Al-Turaif.

References

[1] Roberge P R 2000 *Handbook of Corrosion Engineering*. Inc.USA: The McGraw-Hill Companies.

[2] McNeill L S and Edwards M 2002 . *Environ. Monit. Assess.* **3** 77.

[3] Kelly R G 2003 Crevice, ASM Handbook. *ASM International*. **13A**, 242.

[4]Ahmad Z 2006 . *Principles of Corrosion Engineering* 1st ed. Elssevier Ltd.

[5] Palmer A C and King R A *Subsea Pipeline Engineering*, 2nd ed. Okahoma USA: PennWell Books, 2008, 93.

[6] Pujar M G, Anita T, Shaikh H, Dayal R K and Khatak H S 2007 J. *Mater. Eng. Perform.* **4** 494.

[7] Bogaert W, Van Haute A and Brabers M 1981 *Passivity of Austenitic Stainless Steel in High Temperature – High Pressure Aqueous Solutions International Congress on Metallic Corrosionn.*

[8] Wang J H, Siu C C and Szklarska-Smialowska Z 1988 *Corrosion* **44** 732.

[9] Roy A K, Fleming D L and Lum B Y 1997 . *Fifth International Conference on Nuclear Variables on Localized Corrosion of High-Performance Container Materials.*

[10] Marcus P, and Olefjord I 1988 . *Corr. Sci.* ***28*(6)** 589–602.

[11]. Olefjord I, and L. Wegrelius. 1990. *Corr. Sci.* **31** 89–98.

[12] Dong C, Luo H, Xiao K, Sun T, Liu Q, and Li X 2011 *J. Wuhan University of Technology-Mater. Sci. Ed.* **26(4)** 641.

[13] QvarforR. 1998. *Corr. Sci.* **40 (2-3)** 215.

[14] Manning P E, and Duquette J 1980 *Corr. Sci.,* **20(4)** 597.

[15] Clayton C R and Olejford I 1995 *Corrosion Mechanisms in Theory and Practice*. Edited by J. Marcus, P., Oudar. New York: Marcel Dekker.

[16] McCafferty E 2010 *Introduction to Corrosion Science*. Edited by E. McCafferty. E. McCaffe. New York: Springer New York.

[17] Oldfield J W, Lee T S and Kain R M 1984 Avoiding Crevice Corrosion of Stainless Steels. *Stainless Steel'84, Chalmers University of Technology*, 205–216.

Journal of Physics: Conference Series **710** (2016) 012003 doi:10.1088/1742-6596/710/1/012003

Cryptanalysis of Password Protection of Oracle Database Management System (DBMS)

Timur Koishibayev[1], Zhanat Umarova[2]
[1]Master student, South Kazakhstan State University, Information Systems department Tauke Khan Avenue, 5, Shymkent, Kazakhstan
[2]Associated professor, PhD, South Kazakhstan State University, Information Systems department Tauke Khan Avenue, 5, Shymkent, Kazakhstan

E-mail: koishibayevtimur@gmail.com, zhanat-u@mail.ru

Abstract. This article discusses the currently available encryption algorithms in the Oracle database, also the proposed upgraded encryption algorithm, which consists of 4 steps. In conclusion we make an analysis of password encryption of Oracle Database.

1. Introduction

One of the most characteristic features of the modern level of development of information technology is widespread computer networks. Computers, connected to the network, can significantly improve the effectiveness of the use of the computer system as a whole. Increase of productivity in such case is achieved through a network of information exchange, as well as by using on each computer general network resources [1]. The vulnerability of computer networks is connected with the support of different protocols of information exchange, with the length of communication channels. Therefore, on the first place we put the safety of the transferred information.

The classical scheme of databases' (DB) protection is divided into the following obligatory procedures [2]:

- Access control - each user, including the administrator, has an access only to the necessary for him information according to his position;

- Access protection - the user can obtain an access to the data, who passed the last procedure of identification and authentication;

- Encryption - you must encrypt both data transmitted on the network to protect from eavesdropping and data written to the media, to protect from the theft and unauthorized viewing media / changes without funds of database (DBMS) management system, an access auditing to the database - activities with critical actions should be recorded.

Access to the protocol should not be used by the users on which it is conducted. In the case of applications, used a multi-tier architecture, the security features also take place, with the exception of data protection on the media - this function is left on the database [3].

With all listed security features in one way or another feature are equipped by database and with Oracle applications, which distinguish them from competitors' products. Oracle implements the SSL protocol to encrypt data that is transferred from the customer databases to the database and back. The advantage of SSL is that it is the current Internet standard, and can be used by clients who do not work with protocols of Oracle Net Services [4].

2. The Scheme of Encryption in Oracle Database

In the Oracle database to create a password can be used the following set of symbols [7]:

1. The figures '0123456789'
2. The letters 'abcdefghijklmnopqrstuvwxyzABCDEFGHIJKLMNOPQRSTUVWXYZ'
3. The symbols! "# $% & () + '* - / :; <=>? _.

As well as all symbols are converted to uppercase, then the power of the alphabet passwords is 10 + 26 + 21 = 57 symbols. As a result, the available passwords can be

$$\sum_{i=1}^{30} 57^i \cong 4{,}8 \cdot 10^{52} \quad (1)$$

(all passwords with length of 1 symbol, plus all passwords with length of two symbols and so on up to 30 symbols). However, in many cases this set a little smaller, because some symbols from item 3 are inconvenient to use, e.g., in a script shell.

Experiments have shown that as the password may be used any general symbols [5]

The key moment here is that the syntax of the Oracle server allows a construction with double quotation marks, among which can be any symbol.

On the database – the user's passwords are stored as 8-byte hash in the schedule SYS.USER $.

In [5] this algorithm is described as follows:

1. Concatenate the username and password (this operation is denoted ||).
2. Convert the resulting string to uppercase (UPPER (Username || Password)).
3. If the operating system uses a single-byte encoding, then convert each double-byte symbol by filling the high byte of zeros (0x00).
4. Encrypt the resulting string (supplementing it with zeros to the length of the unit), using the DES (Data Encryption Standard) algorithm in CBC mode with a fixed key, the value of which has 0x0123456789ABCDEF.
5. Encrypt the resulting string again using the DES-CBC, but using the last block of the previous step as the encryption key.

Let X encryption row on the key K as a DES (X, K). Then, in an analytical form of encryption equation looks like this:

$$H = DES\ (UPPER\ (Username||Password),\ DES\ (UPPER\ (Username||Password),\ K)) \quad (2)$$

Where H- hash function.

If we denote UPPER (Username||Password) = x, then the equation will be quite simple:

$$H = DES\ (X,\ DES\ (X,\ K)) \quad (3)$$

The inverse operation (decryption H on the key K) we denote as:

$$X = DES^{-1}(H,K) \quad (4)$$

3. The Encryption Algorithm in the Oracle Database

Let the input of the algorithm receives an input data stream, the hash we should find [6]. Message length may be any (including zero). We write the message's length in L. This number is an integer and non-negative. Multiplicity for any numbers is not required. After receipt of the data – there begin the process of flow's preparation to calculations.

Step 1 Flattening of the stream.

First, we write till the end the single bit stream (bytes 0x80), then the required number of zero bits. Input data are aligned so that their new size L 'was comparable to 448 by module 512 (L' = 512 × N + 448). Alignment occurs even if the length is comparable to 448.

Step 2 Adding the length of the message.

The remaining 64 bits appends a 64-bit representation of the data length (number of bits in the message) to the alignment. First we record lower 4 bytes. If the length exceeds 264 - 1, then appends only the least lower bits. After this the stream's length will be a multiple of 512. The calculations will be based on the presentation of this data stream as an array of words of 512 bits.

Step 3 Initialize the buffer.

The variables with size from 4 to 32 bits are initialized for the calculations, and the initial values are set with hexadecimal numbers (the hexadecimal representation, low byte firstly):

A = 01 23 45 67;

B = 89 AB CD EF;

C = FE DC BA 98;

D = 76 54 32 10. 58

The results of intermediate calculations will be stored in these variables [7]. The initial state of ABCD is called initialization vector.

Let define functions and constants, which we need for calculations.

It will take 4 functions (5) - (8) for four rounds. Let introduce a function of three parameters X, Y, Z - the words; the result will also be a word.

Round 1

$$FunF(X,Y,Z) = (X \wedge Y) \vee (\neg X \wedge Z) \tag{5}$$

Round 2

$$FunG(X,Y,Z) = (X \wedge Z) \vee (\neg Z \wedge Y) \tag{6}$$

Round 3

$$FunH(X,Y,Z) = X \oplus Y \oplus Z \tag{7}$$

Round 4

$$FunI(X,Y,Z) = Y \oplus (\neg Z \vee X) \tag{8}$$

Let define constants array of data T [1...64], built as follows:

$$T[i] = \text{int}(4294967296 \cdot |\sin(i)|) \tag{9}$$

Where $4294967296 = 2^{32}$

The aligned data is divided into units (words) to 32 bits, and each block has 4 rounds of 16 operators. All operators are similar and have the form [abcd k s i], defined by the formula (10):

$$a = b + ((a + Fun(b,c,d) + X[k] + T[i]) <<< s) \tag{10}$$

where X - a block of data;

$$X[k] = M[n \cdot 16 + k] \tag{11}$$

where k - number of 32-bit words from the n-th bit of the message block 512; s - the cyclic shift to the left by s bit received from 32-bit argument.

Step 4 Calculating into cycle.

Let enter n element of the array in the data block. The values of A, B, C & D are stored remaining after operations on the previous blocks (or their initial value if the block is the first).

AA = A

BB = B

CC = C

DD = D

Sum up with the result of the previous cycle: 59

A = AA + A

B = BB+B

C=CC+C

D=DD+D

After the end of the cycle, check whether there are more units for calculations. If so, change the number of array elements (n ++) and go to the beginning of the cycle (figure1).

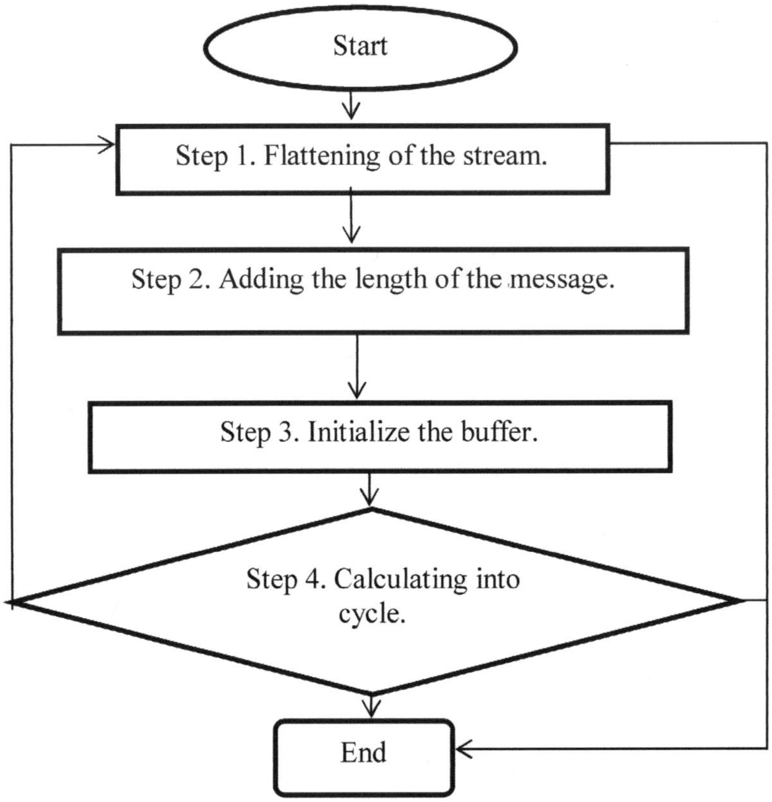

FIGURE 1 Upgraded encryption algorithm in Oracle database.

4. Analysis of the Reliability of Oracle Database

From an algorithmic point of view, during the opening of a modern cipher with a random key, there are two fundamentally different lines of attack: the research and the usage of some vulnerability of the algorithm, and brute force key [8].

Vulnerability - a long and painstaking analytic work, which allows drawing conclusions about the original data, based on the encrypted data and knowledge of the algorithm (cryptosystems).

Bust keys - (the second name "Power Attack», brute force - BF) – a standard method for selecting a password, which generate the next meaning, encrypting it and comparing the result with the existing hash. The main role in brute processing power play the computer powerful intruders, so it is a universal way to crack a variety of ciphers, since this method does not depend on the internal structure of the cipher.

In view of the large amount of time and computing resources, to the sorting we resort in the case, when other methods to solve the problem are impossible. And even in the case of brute force is still trying to narrow the set of sorted out the analysis of peculiarities of a particular algorithm [9].

In the future, we will assume that the attacker knows the user's login, H hash and the encryption key K.

If cryptosystem in Oracle was based on a single encryption, in which H = DES (X, K), then the famous H and K, it would instantly lose a resistance, as X = DES-1 (H, K), and all the parameters of the right are known.

In Oracle, the situation is more difficult, and by our assumptions to calculate the password X, the equation would look like (4).

Note that for the reverse operation, it requires an intermediate key K. So to compute the value of X, in this equation we need the value K = DES (X, K), which is an intermediate key and is not stored anywhere. And calculate the DES (X, K) is not possible because there is X||Password Username and the password for the conditions of the problem is unknown [10].

Written above means that, even knowing the encryption key K, you can only calculate all possible values DES (Username||Password K) for all possible values of a password, and then find in this set a suitable value. And it can be done by trying all possible passwords, i.e. in fact, only by the force of attack!

That is the main focus in the construction of the cryptographic Oracle: to calculate the secret password X, you must first calculate the intermediate key DES (Username||Password K) for all possible values of a password. In other words, the theoretical difficulty of breaking of such cryptosystems with no real vulnerabilities, we use the complexity of power attack on this cipher.

In order to make sure in it perfect, you should explore its characteristics to identify laws that allow it to reduce the cryptographic strength [11]. On the view point of cryptography, used in Oracle a construction has two potentially interesting direction of analysis (attacks) [12].

1. The study of the properties of the intermediate key. Despite the fact that the DES is enough good algorithms, it can be expected that the subset generated values K 'will be small and will have special properties. As a result, from 2^{64} of the theoretically - possible values are K ', only a small part may in fact be realized. From the value of K ', together with the value of X can create a database to continue to K' can be taken from the database and to input the second node DES.

2. Potentially interesting will be an attack with the accumulation DES (X, K) on the one hand, and $DES^{-1}(H,Y)$ on the other, wherein Y - a random value (the probable value of the intermediate key). In this case, it is possible so to parallelize this attack, as reduce its complexity.

5. Conclusion

After the above-mentioned changes - the crypto-scheme of Oracle Corporation should be subjected to certification. This will enhance the credibility of the software of Oracle. For such popular product like Oracle - an official recognition and confirmation of its reliability by the competent institutions' sides is an important moment, both in its exploitation and in promoting the market. In the worst case, the weak link will be found and replaced by a more persistent [13].

As an option for building of a secure database, we could generate a unique key value for each database installation. The fact that for some IS with high security requirement [14]s, it is very desirable to be able to compile the own client with protected in it with a unique key, i.e. the client, who is able to connect only to his/her "native" database, equipped with the same key. It is clear that the ordinary customer with a standard key will never be connected to a database. Despite all the difficulties arise, the proposed construction provides a strong guarantee about the security of information in the database.

6. References

[1] Mikhalev V 2014 *Comparative analysis of compliance of Oracle information security requirements of PCI DSS banking standards* (Molodoy ucheniy) p 103-107
[2] Ochakov A and Orshanskiy A *Information Protection by means of ORACLE database ways.*
[3] Petrov A 2014 *Controlling access to the MySQL and Oracle databases* (Science Publishing Problems)
[4] Bondar A 2008 *Data* encryption in the Oracle database (Cryptographic aspects of information security)
[5] Pudovchenko Yu 2007 *Encrypting passwords in the Oracle database*
[6] Joshua W and Carlos C 2005 *An Assessment of the Oracle Password Hashing Algorithm* 2005

Journal of Physics: Conference Series **710** (2016) 012003 doi:10.1088/1742-6596/710/1/012003

[7] Kulmugariev M 2014 *Construction of information security systems for the software packages used in monopolistic access* (master's thesis) (Almaty)
[8] Dodokhov A.L., Sabanov A.G, Investigation of data privacy protection on Oracle data base use, TUSUR reports, № 2 (24), part 2, December 2011.
[9] Oracle Database Security Guide. http://www.oracle.com/pls/db102/to_pdf?partno=b14266.
[10]Oracle Database Advanced Security Administrator's Guide. http://www.oracle.com/pls/db102/to_pdf?partno=b14268.
[11] Dokhodov A and Sabanov A 2014 *Investigation of the Oracle database for personal data protection / (Reports of the Tomsk State University of Control Systems and Radio Electronics)* №2 (24) p 267-270.
[12] Dokhodov A 2014 *On the question of the protection of personal data using Oracle database* (web journal Fors)
[13] *Ensuring the protection of personal data in the database Oracle, Protection of Personal Data, Data protection, and management of information security risks*
[14] Pishyulin A 2013 *Technology of Oracle Streams (The system administrator)*

Journal of Physics: Conference Series **710** (2016) 012004 doi:10.1088/1742-6596/710/1/012004

Comparison of Catalytic Activities for Sonocatalytic, Photocatalytic and Sonophotocatalytic Degradation of Methylene Blue in the Presence of Magnetic Fe_3O_4/CuO/ZnO Nanocomposites

Ardiansyah Taufik[1,2], Hendry Tju[1,2], and Rosari Saleh[1,2,a]
[1]Departemen Fisika, Fakultas MIPA-Universitas Indonesia, 16424 Depok, Indonesia.
[2]Integrated Laboratory of Energy and Environment, Fakultas MIPA-Universitas Indonesia, 16424 Depok, Indonesia.

[a]Corresponding's author: rosari.saleh@gmail.com ;rosari.saleh@ui.ac.id

Abstract
Fe_3O_4/CuO/ZnO nanocomposites with different molar ratio were synthesized using sol-gel method. The as-synthesized samples were characterized by X-ray diffraction (XRD), UV-visible spectroscopy, BET surface area analyszer, and Electron Spin Resonance (ESR). The sonocatalytic-photocatalytic-sonophotocatalytic activity of as-synthesized samples was investigated based on methylene blue degradation. Moreover, the role of active species was investigated using scavenger technique. The results showed that hole plays an important key role in sono-photo-sonophotocatalytic degradation of methylene blue.

1. Introduction

Methylene blue is widely used in many textile industries. The discharge of methylene blue into wastewater has received a global concern, since it is an environmental hazards. Therefore removal methylene blue from the environment becomes an important issue. Many methods have been investigated for their removal [1-4], however, they are nondestructive methods that only transfer organic pollutant from wastewater to another phase and can easily causing secondary pollution [5].

Sonocatalytic and photocatalytic techniques applied for wastewater treatment have attracted great interest by introducing ultrasonic wave and solar energy without secondary pollution [6-7]. Nanocomposites catalysts by using metal oxide materials have been applied and exhibited some catalytic activity for organic pollutant degradation. It is known that the two most important processes in the degradation of organic pollutants are adsorption and oxidation. Fe_3O_4, a member of the inverse cubic spinel type ferrites, has been found to possess ferromagnetic behavior, and it can be used as an adsorbent. Fe_3O_4, adsorbents engineered with photo- or sonocatalytic ingeredients such as ZnO and CuO can make them multifunctional nanocomposites catalyst [8].

The simultaneous use of sonocatalysis and photocatalysis called sonophotocatalysis has been studied regarding process efficiency to degrade various organics in model solutions [9-14]. The sonophotocatalytic process of oxidation shows interesting advantages at kinetic level, due to the occurance of a synergistic effect between sonolysis and photocatalysis [9-14].

Therefore the aim of this work was to study the degradation of methylene blue by ultrasonic irradiation, UV light irradiations and the combination of ultrasonic as well as UV light irradiations in the presence of Fe_3O_4/CuO/ZnO nanocomposites. The influence of different ZnO loading of Fe_3O_4/CuO/ZnO nanocomposites in the aqueous solution on the sonocatalytic, photocatalytic and sonophotocatalitic efficiency was investigated, especially the mechanism taken place at the degradation processes.

2. Experimental details

CuO and Fe_3O_4 nanoparticles and Fe_3O_4/CuO/ZnO nanocomposites were synthesized using a sol-gel method described in our previous study [8]. The samples were characterized by X-ray diffraction (XRD) measurements using a Philips PW 1710 (CuKα = 1.54 Å). UV-Vis diffuse reflectance spectra were measured using a Shimadzu spectrophotometer with an integrating sphere. Magnetic measurements were performed on Oxford Type 1.2T vibrating sample magnetometer (VSM). These measurements were taken from 0 to 1 T. The nitrogen adsorption/desorption isotherms were measured using Nova Quantachrome 2000. The total surface area and pore size distribution were determined via BET and BJH methods. Electron spin resonance (ESR) was performed using X-band JEOL JES-RE1X at room temperature and X-band spectrometer equipped with a 9.1 GHz field modulation unit. Methylene blue (MB) was selected as models of organic pollutants. Sonocatalytic activity was evaluated by degradation of organic pollutants in the presence of Fe_3O_4/CuO/ZnO nanocomposite in aqueous solutions using ultrasonic bath operated at fixed frequency and power of 40 kHz and 150 W, respectively. In photocatalytic measurements, the experimental setup published previously was used [8]. The experimental setup of sonophotocatalytic system composed of ultrasonic bath operated at 40 kHz frequency with 150 W power and 40 W UV lamps. In all experiments, 0.03 g Fe_3O_4/CuO/ZnO nanocomposite was suspended in 100 mL MB solution with a concentration of 20 mg/L. The suspended solution was placed in the dark for 30 min under magnetic stirring to ensure the establishment of an adsorption-desorption equilibrium in MB solution [15]. After that period of time, the ultrasonic bath was turned on and this was taken as "time zero" for the reaction. The samples were collected for 2h at regular intervals. The solution was analyzed using a Dynamica UV-visible spectrophotometer with a quartz cuvette with an optical length of 10 mm.

3. Results and discussion

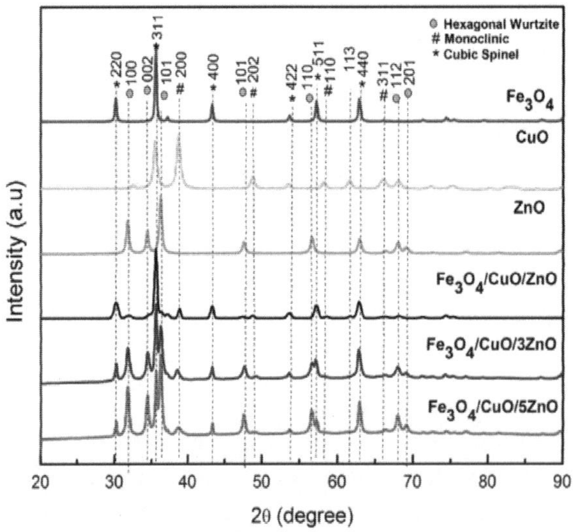

Figure 1. XRD Pattern of pure Fe_3O_4, CuO, ZnO nanoparticle and nanocomposite Fe_3O_4/CuO/ZnO with different molar ratio.

Journal of Physics: Conference Series **710** (2016) 012004 doi:10.1088/1742-6596/710/1/012004

Table 1. Lattice constant and grain size of pure Fe_3O_4, CuO, ZnO nanoparticles and Fe_3O_4/CuO/ZnO nanocomposites with different molar ratio.

Sample	Lattice Parameter (Å)						<D> (nm)		
	ZnO		CuO			Fe_3O_4			
	a=b	c	a	b	c	a=b=c	ZnO	CuO	Fe_3O_4
Fe_3O_4	-	-	-	-	-	8.36	-	-	45
CuO	-	-	4.69	3.42	5.14	-	-	15	-
ZnO	3.25	5.21	-	-	-	-	20	-	-
Fe_3O_4/CuO/ZnO	3.24	5.21	4.69	3.42	5.13	8.37	23	15	38
Fe_3O_4/CuO/3ZnO	3.25	5.21	4.69	3.45	5.13	8.37	25	15	36
Fe_3O_4/CuO/5ZnO	3.25	5.21	4.68	3.45	5.13	8.35	26	15	35

Table 2. Band Gap. Magnetization, and BET analysis of pure Fe_3O_4, CuO, ZnO nanoparticles and Fe_3O_4/CuO/ZnO nanocomposites with different molar ratio.

Sampel	Band Gap (eV)	Magnetization (emu/g)	Surface Area(m^2/g)	Pore Diameter (cc/g)	Meseporous Volume (nm)
Fe_3O_4	-	82	-	-	-
CuO	1.54	-	-	-	-
ZnO	3.37	-	-	-	-
Fe_3O_4/CuO/ZnO	2.83	46	11.99	1.2	0.06
Fe_3O_4/CuO/3ZnO	3.06	23	37.09	2.99	0.13
Fe_3O_4/CuO/5ZnO	3.16	17	39.36	3.17	0.11

The XRD patterns of Fe_3O_4/CuO/ZnO nanocomposites, prepared Fe_3O_4, CuO, ZnO nanoparticles were given in Figure 1. As can be seen in Figure 1, the diffraction angles of $2\theta = 30°$, $2\theta = 30.14°$, 35.49°, 43.28°, 53.76°, 57.20°, and 62.83° correspond well with planes of cubic spinel phase of Fe_3O_4, while the diffraction angle of 32.39°, 35.40°, 38.62°, 48.65°, 53.3°, 58.1°, 61.44°, 65.68°, 66.17°, 67.93°, 72.30° correspond with monoclinic phase of CuO and hexagonal wurtzites phase of ZnO detected at $2\theta =$ 32.12°, 34.48°, 36.6°, 47.76°, 56.84°, 63.02°, 66.95°, 68.23° and 69.24°. The absence of other mixtures of Fe_3O_4, CuO and ZnO diffraction peaks in the spectra confirms the formation of Fe_3O_4/CuO/ZnO nanocomposites. As shown in the figure, obviously, as the molar ratio of Fe_3O_4/CuO/ZnO nanocomposites increased from 1:1:1 to 1:1:5, the diffraction peaks of ZnO become prominent. The average crystallite size for pure nanoparticles and nanocomposites estimated using the well-known Debye–Scherer relation [16] and the values of the lattice parameters obtained using the Rietveld refinement were displayed in Table 1. The result shows that the lattice constant of all samples do not change significantly.

UV-Vis Diffuse reflectance spectroscopy is a powerful tool for obtaining information about the optical absorption properties and the band gap energies of materials.The band gap energies can be calculated from the diffuse reflectance spectra by performing a Kubelka–Munk analysis [17]. The values of the band gaps are tabulated in Table 2. As seen, all of the nanocomposites present a decreased band gap energy compared to that of the pure ZnO nanoparticles, but it is greater than that of the pure CuO nanoparticles.

Pore properties such as surface area, pore volume and size have a close relationship with the catalytic activity efficiency. The surface area the pore diameter are also shown in Table 2. It can be seen from the table that the nanocomposite with a higher ZnO component shows a larger specific BET surface area of approximately 39.36 m^2/g, which decreases as the ZnO content decreases.

Magnetic properties of Fe_3O_4/CuO/ZnO nanocomposites were measured using a vibrating sample magnetometer (VSM) at room temperature in the field range of 0 to 1 T. Table 2 shows magnetization data for all Fe_3O_4/CuO/ZnO nanocomposites with molar ratios of 1:1:1 to 1:1:5. The magnetization of pure Fe_3O_4 nanoparticles were also shown in the table. It is observed that the saturation magnetization in pure Fe_3O_4 nanoparticles (82 emu/g) is lower than that of the reported value (92 emu/g) [18]. These results may be understood on the basis of a core-shell model that predicts a spin canting at the surface of Fe_3O_4 nanoparticles. The saturation magnetization decreased further as CuO and ZnO nanoparticles were introduced into the nanocomposite system.

Figure 2 shows the ESR spectra recorded at room temperature for Fe_3O_4/CuO/ZnO nanocomposites with various molar ratios. The spectra show broad resonance peaks that can be considered as a superposition of several overlapping signals that are too close to be separated with confidence.Therefore, we have carefully studied the spectra and found that the line position, the line width and the line intensity of all signals vary with the molar ratio. The ESR spectra can be fitted satisfactorily to four overlapping signals at (i) a g value range from 2.6591 to 2.5000, (ii) a relatively constant g of 4.4963 to 4.1382, (iii) a g value from and 2.0881 to 2.0194 and (iv) a g value from 1.9976 to 1.9701, indicating the presence of Fe^{2+}, Fe^{3+}, Cu^{2+} [19] and native defects, such as doubly ionized oxygen vacancies, that can act as trapping states for electrons [20], respectively. ESR parameters, such as line width (ΔHpp), g values and integrated areas that reflect the total number of spins taking part in the resonance line, are given in Table 3. As seen from Table 3, with an increase in the molar ratio of Fe_3O_4/CuO/ZnO nanocomposites from 1:1:1 to 1:1:5, a decrease in the integrated areas of Fe^{2+}, Fe^{3+} and Cu^{2+} was observed, indicating that the total number of spins contributing to the ESR spectra decreased, while the integrated area of native defects increased.

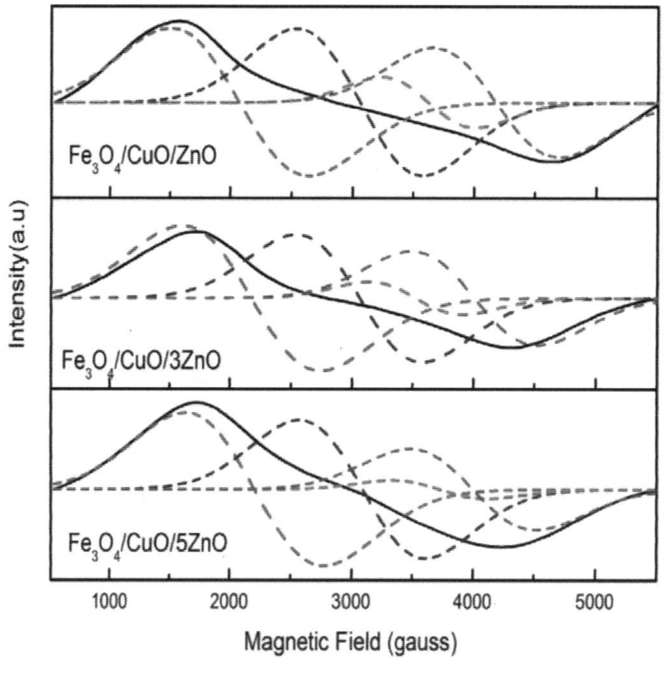

Figure 2. ESR Signal of nanocomposites Fe_3O_4/CuO/ZnO with different molar ratio.

Table 3. g value, Integrated Area, ΔHpp of ESR spectra of Fe_3O_4/CuO/ZnO nanocomposites with different molar ratio.

Sampel	g value	Integrated Area	ΔHpp	Signal Identification
Fe_3O_4/CuO/ZnO	4.4963	9.33 x 10^8	1141.993	Fe^{3+}
	2.6591	7.72 x 10^8	1039.195	Fe^{2+}
	2.0881	1.51 x 10^8	624.8679	Cu^{2+}
	1.9701	4.87 x 10^8	1076.574	Oxygen vacancy
Fe_3O_4/CuO/3ZnO	4.2157	9.04 x 10^8	1201.266	Fe^{3+}
	2.6514	6.41 x 10^8	1039.341	Fe^{2+}
	2.0254	5.34 x 10^8	577.8902	Cu^{2+}
	1.9408	4.92 x 10^8	1034.846	Oxygen vacancy
Fe_3O_4/CuO/5ZnO	4.1382	9.20 x 10^8	1221.993	Fe^{3+}
	2.5015	4.99 x 10^8	919.858	Fe^{2+}
	2.0194	3.30 x 10^8	559.858	Cu^{2+}
	1.9976	5.08 x 10^8	996.574	Oxygen vacancy

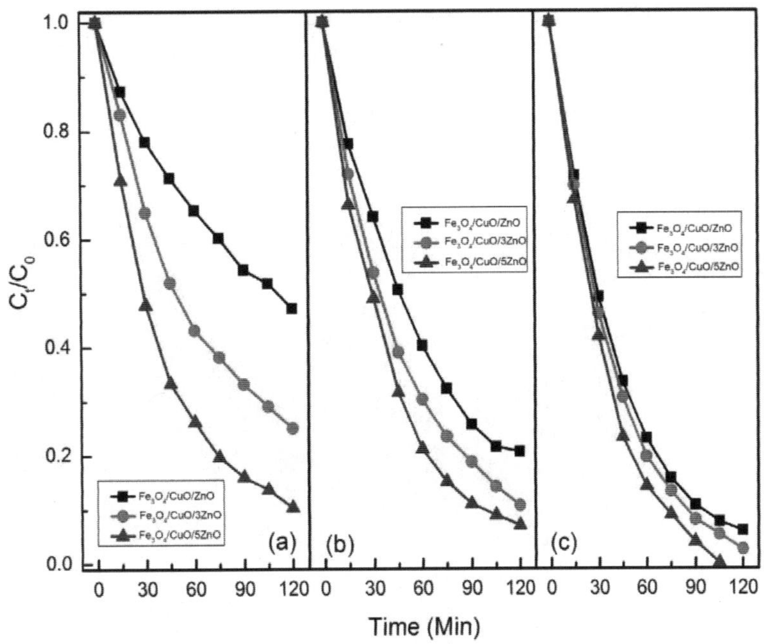

Figure 3. (a) Photocatalytic, (b) sonocatalytic, and (c) Sonophotocatalytic performance of Fe_3O_4/CuO/ZnO nanocomposites with different molar ratio.

The degradation of MB in aqueous solution was investigated in the presence of Fe_3O_4/CuO/ZnO nanocomposites under irradiation of ultrasonic (US) and UV light individually as well as in combination (US+UV). Figure 3 shows under ultrasonic, UV light and combination of both irradiation sources, the degradation of MB conducted at pH values of 13 in the presence of Fe_3O_4/CuO/ZnO nanocomposites. In this study, the initial concentration of MB and the concentrations of nanocomposites as catalysts were fixed at 20 mg/L and 0.3 g/L for all the degradation processes, respectively. As can be seen from the figure 3, $Fe_3O_4/CuO/ZnO$ nanocomposites with molar ratios of 1:1:1, 1:1:3 and 1:1:5 were able to degrade approximately 53%, 75%, 90%, 79%, 89%, 93% and 94%, 97%, 100% of MB, for UV, US and US+UV, respectively, within 2 h. It is clear that $Fe_3O_4/CuO/ZnO$ nanocomposites with molar ratios of 1:1:5 shows the best catalytic performance under all irradiation sources. The degradation is more facile in sonocatalysis compared to photocatalysis. The degradation of 100% was achieved in the presence of UV + US within less than 2h, suggesting the synergy interaction in appropriate manner between UV light and US irradiation plays an important role in enhancing the catalytic activity.

It is known that reusability and stability of catalysts are important parameters of the degradation process and are important issues for practical applications. In this experiment, $Fe_3O_4/CuO/ZnO$ nanocomposites could be easily recovered from the solution using a permanent bar magnet. To evaluate the stability of the catalyst under UV, US and UV+US irradiation, the $Fe_3O_4/CuO/ZnO$ nanocomposites with a molar ratio of 1:1:5 were repeatedly used for four cycles. The stability was evaluated by reusing the catalyst with the same quantity of fresh MB after each run. The results are shown in Figure 4. After four cycles, degradation efficiencies of 75%, 87% and 93% were obtained for UV, US and UV+US, respectively. These results indicated that $Fe_3O_4/CuO/ZnO$ nanocomposites were fairly stable under the condition used in this study.

To understand the active species involved during the degradation of MB in the presence of $Fe_3O_4/CuO/ZnO$ nanocomposites with molar ratio 1:1:5, free radical trapping experiments were carried out by adding various scavengers. Tert-butyl alcohol was used to scavenge hydroxyl radicals, while ammonium oxalate and $Na_2S_2O_8$ were used to scavenge generated holes and electrons, respectively. As shown in Figure 5 after adding scavengers the degradation efficiency of MB exhibited a gradual decrease and a noticeable inhibition was observed in the presence of ammonium oxalate, suggesting the important role of hole in the degradation process of MB. From Figure 5 it is seen that holes were the primary active species in degradation of MB for UV, US and UV+US.

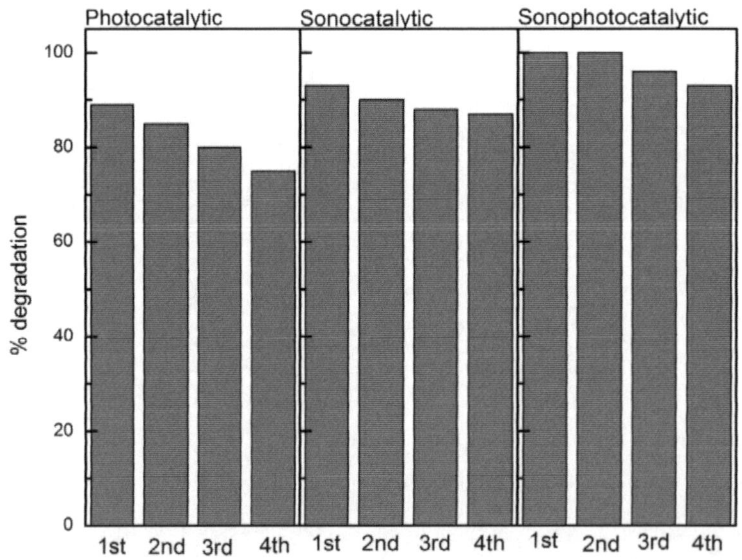

Figure 4. Reusability of Photocatalytic, sonocatalytic, and Sonophotocatalytic. Performance $Fe_3O_4/CuO/5ZnO$ nanocomposites.

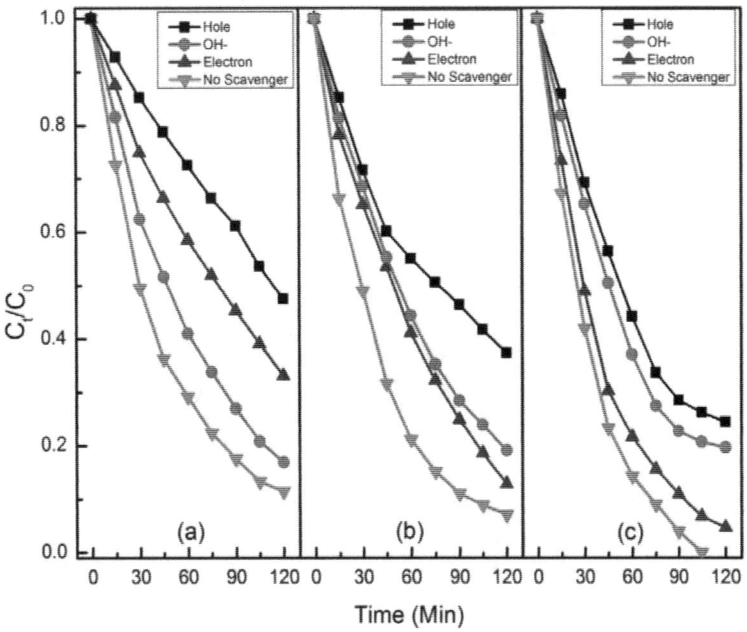

Figure 5. Effect of scavenger on (a) Photocatalytic, (b) sonocatalytic, and (c) Sonophotocatalytic. Performance of Fe_3O_4/CuO/ZnO nanocomposites with molar ratio 1:1:5.

Photocatalytic degradation of methylene blue in aqueous solution by UV irradiation has been reported in our previous papers [8]. However, a mechanism of UV+US processes in the presence of ternary nanocomposites has not been issued yet. The results of this study showed clearly that the oxidation process of methylene blue in all processes was holes dependent. Therefore, the presence of Fe_3O_4/CuO/ZnO nanocomposites should results in the formation of holes in the irradiation solutions. From the diffuse reflectance spectroscopy measurements, the band gap energies of CuO and ZnO were 1.54 and 3.37eV, respectively. The band structure of the above individual components seems unfavorable for the separation of photo- or sonogenerated electrons and holes. However, in the Fe_3O_4/CuO/ZnO nanocomposites, a sufficient contact between the components is a crucial factor in forming effective separation and transfer of electrons and holes. A heterojunction and a new energy band structure will be formed after all components are contacted and may change the charge transfer process. As Fe_3O_4/CuO/ZnO nanocomposites is irradiated by UV light, the ZnO can be activated due to its wide band gap energy [21]. Therefore, it can be seen that the nanocomposites with a molar ratio of 1:1:5 exhibit better photocatalytic performance in the UV.

The degradation of MB is more facile in the US compared to UV. The MB can be rapidly degraded in the presence of Fe_3O_4/CuO/ZnO nanocomposites with lower ZnO loading by US irradiation. It is known that the mechanism of sonocatalytic degradation is related to the acoustic cavitation [22-27]. Besides generating hot spot in the solution, sonoluminescence could also be involved, that excites Fe_3O_4/CuO/ZnO nanocomposites to act as a photocatalyst during sonication to form electron-hole pairs. These pairs can produce OH radicals ($OH^{\bullet}$) and superoxide anions ($O_2^{\bullet -}$) which can be decomposed MB. From scavenger technique, it is found that the degradation process of MB was holes dependent.

According to the predicted band energy position, under irradiation the generated electron in the conduction band of ZnO is prohibited from transferring to the conduction band of CuO by the energy barrier that exists at the interface between ZnO and CuO [21-22]. However, the generated holes in the valence band of ZnO will be readily transferred to the valence band of CuO. The electron could also be transferred to electron trapping states, such as doubly ionized oxygen vacancies formed in the $Fe_3O_4/CuO/ZnO$ nanocomposites. In this process, the electron collected in the trapping state can later react with absorbed molecular oxygen to yield the superoxide radical, $O_2^{\bullet -}$. Simultaneously, the generated holes in the valence band can then react with the hydroxyl group (OH^-) to form a hydroxyl radical ($OH^\bullet$).

4. Conclusion

The purpose of this study was to investigate the degradation of MB under UV, US and UV+US irradiation. The results showed that MB can be effectively removed bu using combination of UV and US irradiation. For individual operation of two modes of irradiation, US has more effeciency compared to UV. For all processes, the catalytic activity of $Fe_3O_4/CuO/ZnO$ nanocomposites with molar ratio of 1:1:5 was the highest, probably due to its higher surface area.

References

[1] M. Ahmad, E. Ahmed, Z. L. Hong , W. Ahmed, A. Elhissi, N. R. Khalid, Ultrasound. Sonochem. 21 (2014) 761–773
[2] A. Houas, H. Lachheb, M. Ksibi, E. Elaloui, C. Guillard, J. M. Herrmann, Appl Catal B: Environ. 31 (2001) 145–157.
[3] M. Saquib, M. Muneer, Desalination 155 (2003) 255–263.
[4] V. Meshko, L. Markovska, M. Mincheva, A.E. Rodrigues, Water Res. 35 (2001) 3357–3366.
[5] W. S. Kuo, P.H. Ho, Chemosphere 45 (2001) 77–83.
[6] J. K. Im, J. Yoon, N. Her, J. Han, K. D. Zoh, Y. Yoon. Sep. Purif. Technol. 141 (2015) 1–9
[7] P. Chowdhury, T. Viraraghavan, Sci. Total. Environ. 407 (2009) 2474–2492.
[8] A. Taufik, I. K. Susanto, R. Saleh. Mater. Sci. Forum. 827 (2015) 37-42
[9] N. Talebian, M. R. Nilforoushan, F. J. Mogaddas, Chem. Eng. J. 39 (2013) 4913-4921
[10] S. Matzusawa, J. Tanaka, S. Sato, T. Ibusuki, J. Photoch. Photobio A. 149 (2002) 183–189.
[11] Y. Kado, M. Atobe, T. Nonaka, Ultrasound. Sonechem. 8 (2001) 69-74.
[12] E. Selli, J. Phys. Chem. 4 (2002) 6123–6128.
[13] L. Davydov, E.P. Reddy, P. France, P. G. Smirniotis, Appl Catal B: Environ. 32 (2001) 95–105.
[14] R. Saleh, N. F. Djaja, Spectrochim. Acta. B. 130 (2014) 581-590
[15] J. Hu, J. Ma, L. Wang, H. Huang, L. Ma, Powder Technol. 254 (2014) 556-562
[16] M. A. Ahmed, A. A. EL-Khawlani, J. Magn. Magn. Mater. 321 (2009) 1959-1963
[17] B. Hapke, Theory of Reflectance and Emittance Spectroscopy, University Press, Cambridge, 1993.
[18] A. Jafari, S. F. Shayesteh, M. Salouti, K. Boustani, J Magn Magn Mater 379 (2015) 305–312
[19] A. V. Kucherov, D. E. Doronkin, A.Y. Stakheev, A. L. Kustov, M. Grill, J Mol Catal A-Chem 325 (2010) 73–78
[20] G. H. Mhlongo, D. E. Motaung, S. S. Nkosi, H.C. Swart, G. F. Malgas, K.T. Hillie, B.W. Mwakikunga, Appl Surf Sci 293 (2014) 62–70
[21] L. Sun, R. Shao, L. Tang, Z. Chen. J. Alloy Compd. 564 (2013) 55–62
[22] A. Roychowdhury, S. P. Pati, A. K. Mishra, S. Kumar, D. Das, J. Phys. Chem. Solids 74 (2013) 811-818.
[23] S. Thangavel, N. Raghavan, G. Kadarkarai, S.J. Kim, G.Venugopal. Ultrasound. Sonochem. 24 (2015) 123–131
[24] A. Khataee, A. Karimi, S. Arefi-Oskui, R. D. C. Soltani, Y. Hanifehpour, B. Soltani, S. W. Joo. Ultrasound. Sonochem. 22 (2015) 371-381
[25] K. P. Jyothi, S. Yesodharan, E. P. Yesodharan .Ultrasound. Sonochem. 21 (2014) 1787–1796

[26] N. Shimizu,C. Ogino, M. F. Dadjour, T. Murat. Ultrasound. Sonochem. 14 (2007) 184–190
[27] R. Saravanan, S. Karthikeyan, V. K. Gupta, G. Sekaran, V. Narayanan, Mater Sci Eng. ;.C 33 (2013) 91-98.
[28] B. Li, Y. Wang, J. Alloy Compd. 47 (2010) 615-623.

Journal of Physics: Conference Series **710** (2016) 012005 doi:10.1088/1742-6596/710/1/012005

Enhanced UV Photocatalytic Performance of Magnetic Fe_3O_4/CuO/ZnO/NGP Nanocomposites

Hendry Tju[1,2], Ardiansyah Taufik[1,2], and Rosari Saleh[1,2,a]
[1]Departemen Fisika, Fakultas MIPA-Universitas Indonesia, 16424 Depok, Indonesia
[2]Integrated Laboratory of Energy and Environment, Fakultas MIPA-Universitas Indonesia, 16424 Depok, Indonesia

[a]Corresponding's author: rosari.saleh@gmail.com, rosari.saleh@ui.ac.id

Abstract
Fe_3O_4/CuO/ZnO/nanographene platelets (Fe_3O_4/CuO/ZnO/NGP) with varied ZnO loadings have been synthesized using a sol-gel method followed by hydrothermal method. X-ray diffraction (XRD) and field emission scanning electron microscopy (FESEM) confirmed the formation of the Fe_3O_4/CuO/ZnO/NGP composites. All of the samples showed the presence of graphene nanoplatelets incorporating Fe_3O_4, CuO and ZnO structures and exhibited ferromagnetic behavior at room temperature. The composites showed photocatalytic activity under UV irradiation, which was used to affect the degradation of methylene blue. The Fe_3O_4/CuO/ZnO/NGP composites showed superior photocatalytic activity than the Fe_3O_4/CuO/ZnO materials.

1. Introduction

Since it was first introduced by Novoselov et al. [1], graphene, a planar sheet of sp^2–bonded carbon atoms with single-atom thickness, has attracted an enormous amount interest within material science and condensed matter physics due to its fascinating physical and chemical properties [2]. It has been recently reported that graphene can compete with carbon nanotubes as a catalyst scaffold, anchoring organic and inorganic groups [3-4]. In addition, through a chemical oxidation process, a wide variety of oxygen-containing functionalities, such as hydroxyl and epoxide groups, can be formed that alter the materials physical and chemical properties [5]. These attractive characteristics have cultivated increased interest in the development of graphene/metal oxide composites with robust photocatalytic properties in environmental remediation applications [6].

A major limitation in achieving the required photocatalytic performance is the quick recombination of photogenerated charge carriers [7]; therefore, slowing this recombination is essential. Recently, many researchers have reported reductions in the charge carrier recombination rate by coupling graphene to two or more metal oxide nanoparticles [8]. Metal oxide nanocomposites, formed by two or more distinct metal oxide nanomaterials, often yield desirable combinations of properties not found from the individual components [9-10]. The similar combination of graphene with several metal oxides photocatalyst may likewise slow the recombination of charge carriers, increasing photocatalytic performance. Therefore, in the present study, Fe_3O_4/CuO/ZnO/NGP composites were developed as photocatalysts using a sol-gel method and employed to remove methylene blue as a model of organic pollutants from water. Fe_3O_4/CuO/ZnO/NGP composites with different ZnO loadings were used in photocatalytic reactions to investigate the role of graphene in the photocatalytic performance. The effect of ZnO loading was also investigated.

2. Experimental

All chemical reagents (Merck) were of analytical grade and used without further purification. Nano graphene platelets (NGP) were purchased from Angstron Materials. All solutions were prepared with distilled water. CuO and Fe_3O_4 nanoparticles and Fe_3O_4/CuO/ZnO nanocomposites were synthesized using sol-gel method as described previously [11]. The Fe_3O_4/CuO/ZnO/NGP nanocomposites were prepared by a simple hydrothermal method based on Khalid's work [12]. Briefly, 200 mg of NGP was dissolved in a solution of water (80 mL) and ethanol (40 mL) through ultrasonic treatment for 2 h, followed by the addition of 2 g of Fe_3O_4/CuO/ZnO to the NGP solution and an additional 2 h of stirring to achieve a homogeneous suspension. The suspension was then heated at 120 °C for 3 h to affect the deposition of Fe_3O_4/CuO/ZnO on the graphene sheets. The resulting nanocomposite was isolated by centrifugation and dried at 70 °C for 12 h.

The samples were characterized by Field Emission Scanning Electron Microscopy (FESEM), X-ray diffraction (XRD). Magnetic measurements were performed by vibrating sample magnetometer (VSM).

Methylene blue (MB) was selected as a model of organic pollutant. The photocatalytic activity was evaluated by degradation of organic pollutant in the presence of Fe_3O_4/CuO/ZnO/NGP nanocomposite in aqueous solution under 40 W UV light irradiation. In this study, organic pollutant solutions of 30 mg/L concentration were prepared and 0.3 g/L of Fe_3O_4/CuO/ZnO/NGP nanocomposite were immersed in these solutions.

3. Results and discussion

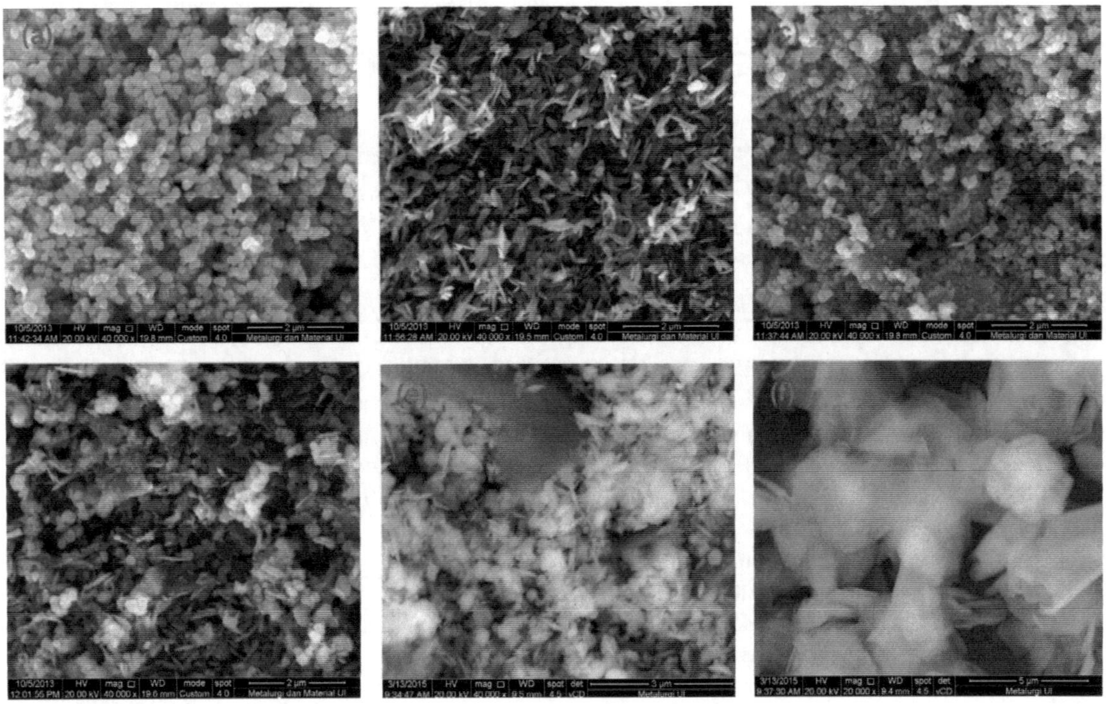

Figure 1. FESEM image of (a) Fe_3O_4, (b) CuO, (c) ZnO nanoparticles, (d) Fe_3O_4/CuO/ZnO nanocomposites and (e) Fe_3O_4/CuO/ZnO/NGP nanocomposites, (f) NGP.

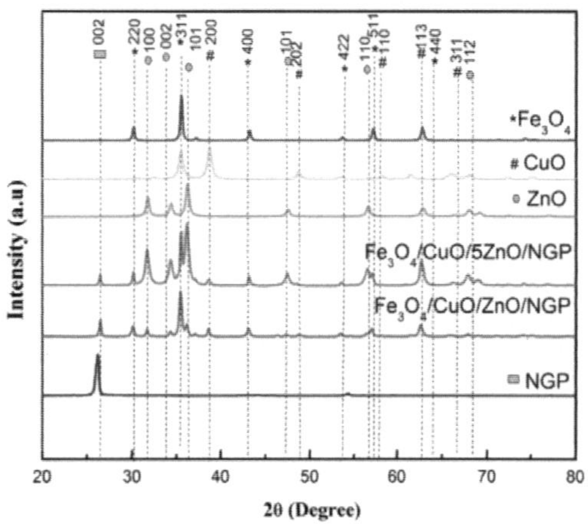

Figure 2. XRD pattern of Fe_3O_4, CuO, ZnO nanoparticles and Fe_3O_4/CuO/ZnO/ NGP nanocomposites.

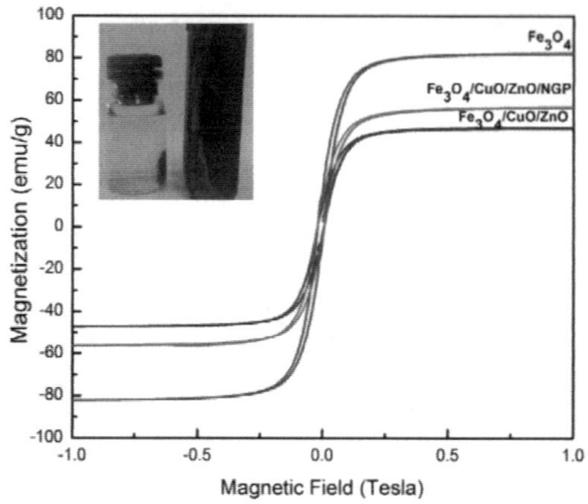

Figure 3. VSM curve of Fe_3O_4 nanoparticles, Fe_3O_4/CuO/ZnO nanocomposites, and Fe_3O_4/CuO/ZnO/NGP nanocomposites.

Table 1. Magnetic Saturation of Fe_3O_4 nanoparticles, Fe_3O_4/CuO/ZnO, Fe_3O_4/CuO/ZnO/NGP, and rate constant of Fe_3O_4, CuO, ZnO nanoparticles, Fe_3O_4/CuO/ZnO nanocomposites and Fe_3O_4/CuO/ZnO/NGP nanocomposites.

Sample	Magnetic Saturation (emu/g)	Rate Constant (k(min^{-1}))
Fe_3O_4	82	0.0029
CuO	-	0.0052
ZnO	-	0.0147
Fe_3O_4/CuO/ZnO	46	0.0059
Fe_3O_4/CuO/ZnO/NGP	57	0.0070
Fe_3O_4/CuO/5ZnO	17	0.0176
Fe_3O_4/CuO/5ZnO/NGP	27	0.0249

Figure 1 depicts the FESEM images of Fe_3O_4/CuO/ZnO/NGP composites. The FESEM images of Fe_3O_4, ZnO, and CuO nanoparticles as well as Fe_3O_4/CuO/ZnO composites and NGP were also observed [11]. It can be seen that the particle morphology of pure Fe_3O_4 and ZnO is generally spherical while the CuO nanoparticles consist of clew-like shapes and the Fe_3O_4/ZnO/CuO hybrid is a combination of spherical and clew-like shapes. Morphology of Fe_3O_4/CuO/ZnO/NGP shows the dispersion of Fe_3O_4/ZnO/CuO in nanographene materials.

Figure 2 shows the XRD patterns of the Fe_3O_4/CuO/ZnO/NGP with different ZnO loadings. The XRD patterns of pure Fe_3O_4, CuO, ZnO nanoparticles and NGP are given as well. The diffraction patterns exhibited a series of peaks characteristic for crystalline cubic spinel Fe_3O_4, monoclinic CuO, hexagonal wurtzite ZnO structures, as well as for NGP. The Fe_3O_4 peaks are very sharp, indicative of their greater crystallinity than found for CuO and ZnO. Moreover, the intensities of the ZnO peaks linearly increased with increasing ZnO loading in the composites. The absence of mixed Fe_3O_4, CuO and ZnO diffraction peaks in the spectra confirmed that the nanocomposites were the desired Fe_3O_4/CuO/ZnO/NGP materials.

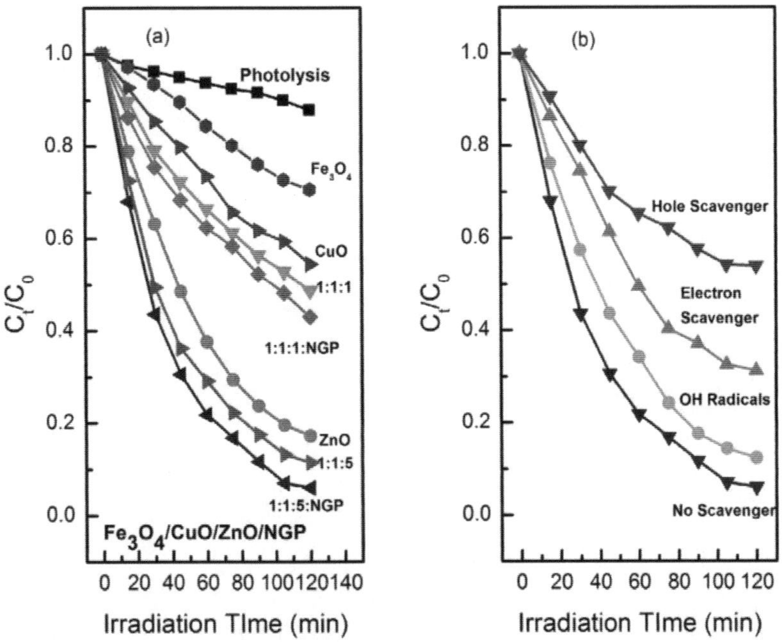

Figure 4. (a) Photodegradation of methylene blue (b) Effect of adding scavenger on photodegradation methylene blue under UV irradiation.

Magnetic hysteresis curves of Fe_3O_4/CuO/ZnO/NGP composites with different ZnO loading are presented in Figure 3. For comparison magnetic hysteresis curves of pure Fe_3O_4 nanoparticles and nanocomposites Fe_3O_4/CuO/ZnO were also measured. The corresponding saturation magnetization values are tabulated in Table 1. All of samples exhibited ferromagnetic behavior. It is found that the value of saturation magnetization decrease as Fe_3O_4 is coupled with CuO and ZnO nanoparticles. However, the incorporation of NGP in nanocomposites Fe_3O_4/CuO/ZnO showed the enhancement of saturation magnetization. According to the literature [13], this might be due to Fe_3O_4 nanoparticles attached to the surface of the NGP materials acting as defects and creating a stable NGP-based magnetic material. To study the magnetic separability of the material, a magnetic bar was placed near the sample holder supporting the Fe_3O_4/CuO/ZnO/NGP composites in water. As seen in the inset of Figure 3, this test demonstrated that the Fe_3O_4/CuO/ZnO/NGP composites can be separated from a suspension by use of a magnetic bar.

The photocatalytic capacity of the Fe_3O_4/CuO/ZnO/NGP composites was tested through the degradation of MB in aqueous solution (Figure 4). Direct photolysis resulted in a ~12 % degradation of MB, while the use of Fe_3O_4/CuO/ZnO composites with molar ratios of 1:1:1 and 1:1:5 removed approximately 52% and 87% of MB, respectively, under UV light irradiation for 2 h. It is indicated that the incorporation of Fe_3O_4/CuO/ZnO nanocomposites could enhance the photoreduction of methylene blue by oxidation process. The removal efficiencies were enhanced to 57% and 93% when NGP was incorporated into the Fe_3O_4/CuO/ZnO composites. It was found that Fe_3O_4/CuO/ZnO composites with molar ratios of 1:1:1 and 1:1:5 showed lower photocatalytic activities than the Fe_3O_4/CuO/ZnO/NGP composites with the same molar ratios. The photocatalytic performances obtained with the Fe_3O_4/CuO/ZnO and Fe_3O_4/CuO/ZnO/NGP composites with different ZnO loadings are plotted in Figure 4a, demonstrating that increased ZnO loading resulted in progressive increases of photocatalytic efficiency. Provided that the photocatalytic reaction follows pseudo-first order kinetics, the apparent reaction rate constants (k) for the degradation of MB could be calculated (Table 1). The values of k for the Fe_3O_4/CuO/ZnO/NGP samples are higher than those of the Fe_3O_4/CuO/ZnO composites, demonstrating that the presence of NGP enhances the photocatalytic performance,

possibly due to

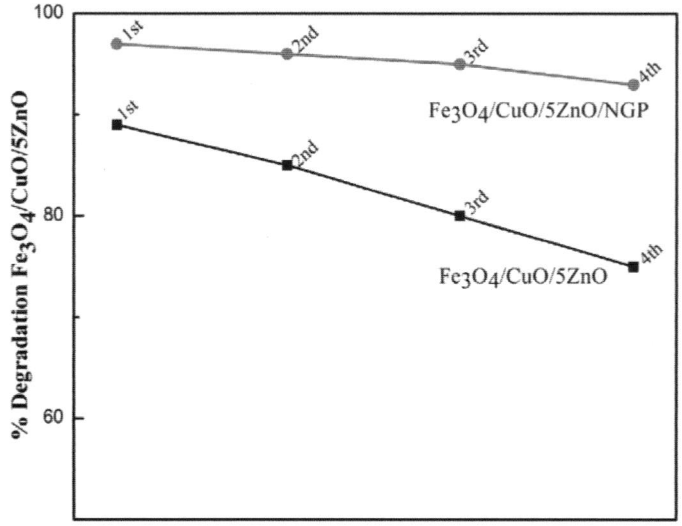

Figure 5. Photocatalytic stability of Fe_3O_4/CuO/5ZnO and Fe_3O_4/CuO/5ZnO/NGP

acceleration of the charge carriers through the conjugated π-structures of the NGP and a slowing of charge recombination [14]. The enhancement of photocatalytic performance in the presence of NGP may be caused by the involvement of radical species in the reaction. To elucidate the primary active oxidative species responsible for the degradation of MB, control experiments employing scavengers of hydroxyl radicals, holes and electrons were carried out. The photodegradation of MB in the presence of tert-butyl alcohol, ammonium oxalate, and $Na_2S_2O_8$, used for scavenging hydroxyl radicals, holes and electrons, respectively, is displayed in Figure 4b. The addition of ammonium oxalate, a hole scavenger, inhibited the photocatalytic removal of MB. The reduced photodegradation percentage suggested that the photocatalytic activity of Fe_3O_4/CuO/ZnO, as well as that of Fe_3O_4/CuO/ZnO/NGP, for the removal of MB is driven by the presence of holes.

To investigate the stability of the photocatalyst, experiments were performed with recycled Fe_3O_4/CuO/ZnO/NGP composites with a molar ratio of 1:1:5 under the same conditions used for freshly prepared composites. Four cycles were examined (Figure 5). For comparison, results for the cycling of a Fe_3O_4/CuO/ZnO composite is also plotted. The results showed that the Fe_3O_4/CuO/ZnO/NGP composites were stable after four cycles of MB photodegradation and are more stable than the composite synthesized without NGP.

4. Conclusions

Magnetically Fe_3O_4/CuO/ZnO/NGP composites have been prepared using sol-gel method for degradation of MB under UV light irradiation. The synthesized composites exhibited obviously higher photocatalytic activity in degradation of MB compared to the same composites synthesized without the presence of NGP. Based on the results of detection active species, it is concluded that the degradation of MB was driven mainly by the participation of holes. Electron and OH radicals to a lesser extent partook in this process. The Fe_3O_4/CuO/ZnO/NGP composites have good photocatalytic stability in the degradation of MB under UV light irradiation after four time cycling process.

References

[1] K. S. Novoselov, M.I. Katsnelson, Graphene: New bridge between condensed matter physics and quantum electrodynamics, Solid State Communications, 143 (2007) 3-13

[2] H. Fan, X. Zhao, J. Yang, X. Shan, L. Yang, Y. Zhang, X. Li, M. Gao, ZnO–graphene composite for photocatalytic degradation of methylene blue dye, 29 (2012) 29-34

[3] S. M. Torres, L. M. P. Martínez, J. L. Figueiredo, J. L. Faria, A. M. T. Silva, Graphene oxide-P25 photocatalysts for degradation of diphenhydramine pharmaceutical and methyl orange dye, Applied Surface Science, 275 (2013) 361-368

[4] H. Wang, X. Yuan, Y. Wu, H. Huang, X. Peng, G. Zeng, H. Zhong, J. Liang, M. Ren, Graphene-based materials: Fabrication, characterization and application for the decontamination of wastewater and wastegas and hydrogen storage/generation, Advances in Colloid and Interface Science, 195 (2013) 19-40

[5] S. Gadipelli, Z. X. Guo,Graphene-based materials: Synthesis and gas sorption, storage and separation, Progress in Materials Science, 69 (2015) 1-60

[6] Y.Gu, M. Xing, J. Zhang, Synthesis and photocatalytic activity of graphene based doped TiO2 nanocomposites, 319 (2014) 8-15

[7] L. Tang, V. H. Nguyen, Y. R. Lee, J. Kim, J.J Shim, Photocatalytic activity of reduced graphene oxide/SnO2 nanocomposites prepared in ionic liquid, Synthetic Metals, 201 (2015) 54-60

[8] S. Cao, C. Chen, T. Liu, B. Zeng, X. Ning, X. Chen, X. Xie, W. Chen, Superfine and closely-packed TiO_2/Bi_2O_3 lamination on graphene nanoplates with high photocatalytic activity, 46 (2014) 61-65

[9] Z. Chen, N. Zhang, Y. J. Xu, Synthesis of graphene–ZnO nanorod nanocomposites with improved photoactivity and anti-photocorrosion, CrystEngComm, 15 (2013) 3022-3030

[10] X. Liu, L. Pan, Q. Zhao, T. Lv, G. Zhu, T. Chen, T. Lu, Z. Sun, C. Sun, UV-assisted photocatalytic synthesis of ZnO–reduced graphene oxide composites with enhanced photocatalytic activity in reduction of Cr(VI), Chemical Engineering Journal, 183 (2012) 238-243

[11] A. Taufik, I. K.Susanto, R. Saleh, Preparation, Characterization and Photocatalytic Activity of Multifunctional Fe_3O_4/ZnO/CuO Hybrid Nanoparticles, Materials Science Forum 827 (2015) 37-42

[12] N. R. Khalid, E. Ahmed, Zhanglian Hong, L. Sana., M. Ahmed, Enhanced photocatalytic activity of graphene–TiO_2 composite under visible light irradiation, Current Applied Physics, 13 (2013) 659-663

[13] E. K. Heidari, A. Ataie, M. H. Sohi, J. K. Kim, $NiFe_2O_4$/graphene nanocomposites with tunable magnetic properties, 379 (2015) 95–101

[14] L. M. P. Martínez, S. M. Torres, V. Likodimos, P. Falaras, J. L. Figueiredo, J.L. Faria, A. M.T. Silva, Role of oxygen functionalities on the synthesis of photocatalytically active graphene–TiO_2 composites, Applied Catalysis B: Environmental 158–159 (2014) 329–340

Journal of Physics: Conference Series **710** (2016) 012006 doi:10.1088/1742-6596/710/1/012006

Sonocatalytic Methylene Blue in The Presence of Fe_3O_4-CuO-TiO_2 Nanocomposites Heterostructure

Malleo Fauzian[1,2], Shofianina Jalaludin[1,2], Ardiansyah Taufik[1,2], Rosari Saleh[1,2,a]
[1]Departemen Fisika, Fakultas MIPA-Universitas Indonesia, 16424 Depok, Indonesia
[2]Integrated Laboratory of Energy and Environment, Fakultas MIPA-Universitas Indonesia, 16424 Depok, Indonesia

[a]Corresponding's author: rosari.saleh@gmail.com; rosari.saleh@ui.ac.id

Abstract. In this work, the emphasis was mainly placed on investigating the sonocatalytic activity of Fe_3O_4-CuO-TiO_2 nanocomposites heterostructure. The prepared samples were characterized by X-ray diffraction (XRD), Vibrating Sample Magnetometer (VSM*),* Brunauer-Emmett-Teller (BET) Surface Area Analysis. Methylene blue dye was selected to examine the sonocatalytic activity of Fe_3O_4-CuO-TiO_2 nanocomposites heterostructure. The degradation reaction processes were monitored by UV–vis spectrophotometer. The influence on the activity of the Fe_3O_4-CuO-TiO_2 nanocomposites heterostructure such as TiO_2 loading was studied. The sonocatalyst Fe_3O_4-CuO-TiO_2 with molar ratio of 1:1:5 showed the highest sonocatalytic activity. At last, the experiment also indicated that holes are the main reactive species in the photodegradation mechanism in methylene blue.

1. Introduction

In the recent years, environmental pollution has been found as one of the biggest problem in the rapid growth in the production processes [1]. Therefore, it is required to develop a new and efficient treatment for the removal of a large number of toxic compounds in the effluents discharged from the industry. As one of advanced oxidation technology, the sonocatalytic degradation has gain more attention for purification of polluted effluents and has good prospect for protecting the environment [2-4]. However, degradation of toxic complex compounds using ultrasound consumes large amounts of energy if one applying sonolysis alone [5]. Therefore, ultrasonic wastewater treatment can be performed in the presence of catalyst. Heterostucture catalysts have received immense interest since they offer a great potential to solve problem of environmental pollution and energy crisis [6]. Among the various heterostructure catalyst, TiO_2–CuO nanocomposites heterostructure is one of the promising materials. Recently, there have been reports of materials having both adsorption and oxidation properties, that are important in degradation of polluted effluents. Fe_3O_4 has been found to possess ferromagnetic behavior, and it can be used as an adsorbent with a large area and small diffusion resistance for removal of chemical species such as organic pollutants [7-9].

The present study aims to investigate the Fe_3O_4-CuO-TiO_2 mediated sonocatalytic degradation of methylene blue as a model of the toxic complex compounds because of its complex aromatic ring structure is difficult to remove from the environtment [10]. The influence of different TiO_2 loading of Fe_3O_4-CuO-TiO_2 nanocomposites heterostructure in the aqueous solution on the sonocatalytic efficiency was evaluated. The presence of radical scavangers was employed to provide insight on the photo-degradation mechanism through different active species.

2. Experimental

A series of Fe_3O_4-CuO-TiO_2 nanocomposites nanoparticles were prepared using sol-gel method described previously with the variation of molar ratio of 1:1:1, 1:1:3 and 1:1:5 [11-13].

The structure and phase of the samples were characterized by X-ray diffraction (XRD) measurements using a Philips PW 1710. Magnetic measurements were performed on Oxford type 1.2 T vibrating sample magnetometer (VSM). The nitrogen ad/desorption isotherms were measured using Nova Quantachrome 2000.

Sonocatalytic activity was evaluated by degradation of Methylene blue (MB) in the presence of Fe_3O_4-CuO-TiO_2 nanocomposite in aqueous solutions using ultrasonic bath operated at fixed frequency of 40 kHz. In the experiment, typically 20 mg Fe_3O_4-CuO-TiO_2 nanocomposite catalyst was suspended in 100 mL MB solution with 20 mg/L concentration of MB. The suspended solution was placed in the dark for 30 min under magnetic stirring to ensure the establishment of an adsorption-desorption equilibrium in MB solution. After that period of time, the ultrasonic bath was turned on and this was taken as 'time zero" for the reaction. During the sonocatalytic reaction, samples were collected for 2 h at regular intervals. The solution was analyzed using a Dynamica UV-visible spectrophotometer with a quartz cuvette with an optical length of 10 mm.

3. Results and discussion

The influence of ultrasonic irradiation on MB degradation by Fe_3O_4-CuO-TiO_2 nanocomposites were investigated and plotted in Figure 1a. The ability of ultrasound to degrade MB in the presence of Fe_3O_4-CuO-TiO_2 nanocomposites with various TiO_2 loading was compared. It is seen, that Fe_3O_4-CuO-TiO_2 nanocomposites with molar ratio of 1:1:5 has degraded the most of MB (92%), whereas nanocomposites with molar ratio of 1:1:1 and 1:1:3 degraded 83 % and 87% of MB, respectively. The sonocatalytic activity of all samples in this study obeys the pseudo-first-order reaction kinetics [5, 8]. The apparent reaction rate constant k of the MB degradation using three different molar ratio of Fe_3O_4-CuO-TiO_2 nanocomposites are plotted in the inset of Figure 2a. It is clear that the degradation efficiency increases with the increase in TiO_2 loading in our nanocomposites.

To get insight of the degradation mechanism, control experiments have been carried out by adding different type of scavengers. The degradation of MB in the presence of tert-butyl alcohol, ammonium oxalate and $Na_2S_2O_8$ scavenger for hydroxyl radicals, holes and electrons, respectively, is displayed in Figure 2b. It is obvious that the order of affecting the rate for degradation of MB follows that of hole > hydroxyl radical > electron. It can be seen that ammonium oxalate provoke a large degree of inhibition

Journal of Physics: Conference Series **710** (2016) 012006 doi:10.1088/1742-6596/710/1/012006

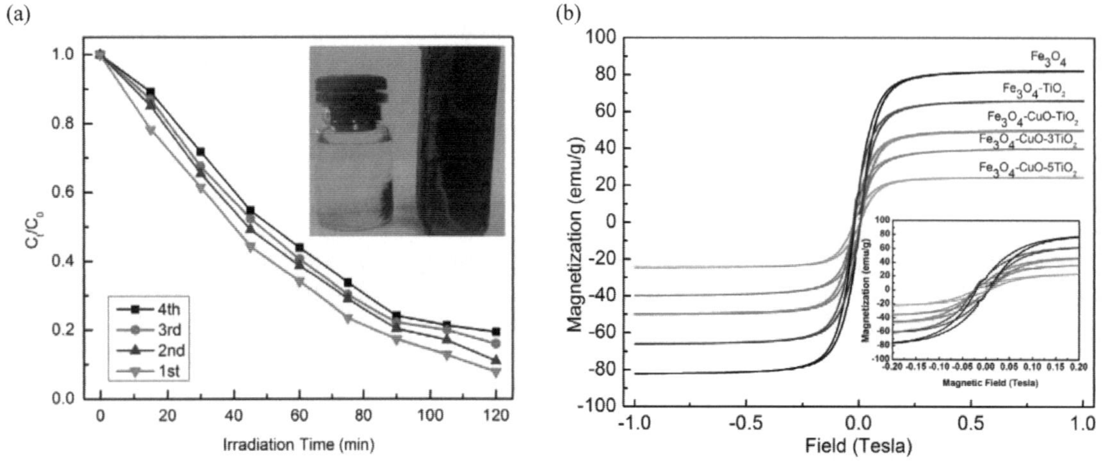

Figure 1. (a) Reusability of Nanocomposites, (b) VSM spectra of Fe_3O_4 and Fe_3O_4-CuO-TiO_2 [13].

Table 1. Magnetic Properties, Grain Size, and Surface Area Analysis.

Sample	M-S (emu/g)	Hc (Tesla)	mr (emu/g)	Grain Size (nm)			Surface Area (m^2/g)	Pore Size (nm)	Pore volume (cc/g)
				Fe_3O_4	CuO	TiO_2			
Fe_3O_4	82	0.0152	13.64	45	-	-	-	-	-
CuO	-	-	-	-	15	-	-	-	-
TiO_2	-	-	-	-	-	33	-	-	-
Fe_3O_4-TiO_2	66	0.0149	12.91	44	-	64	-	-	-
Fe_3O_4-CuO-TiO_2	50	0.0151	7.79	42	18	53	18.36	1.51	0.04
Fe_3O_4-CuO-3TiO_2	40	0.015	6.49	39	24	70	27.12	3.04	0.08
Fe_3O_4-CuO-5TiO_2	25	0.0149	3.66	45	35	84	37.64	1.02	0.16

of MB, which confirmed the indispensable role of holes in the degradation process of MB. Thus, a lower percentage of degradation of MB was obtained, suggesting that the sonocatalytic activity of MB is driven by holes.

It is known that reusability and stability of catalysts are important parameters of the degradation process and are important issues for practical applications. To evaluate the stability of the catalyst

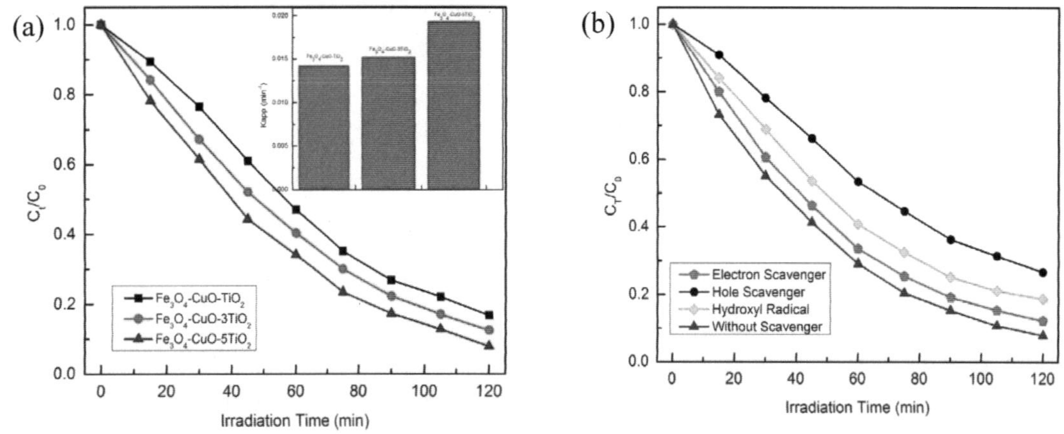

Figure 2. (a) Sonocatalytic activity of Fe_3O_4-CuO-TiO_2 nanocomposites with various molar ratio, (b) Effect of scavenger on sonocatalytic activity of Fe_3O_4-CuO-5TiO_2.

under ultrasonic irradiation, the Fe_3O_4-CuO-TiO_2 nanocomposites with a molar ratio of 1:1:5 were repeatedly used for four cycles. In this experiment, nanocomposites could be easily recovered from the solution using a permanent bar magnet (the inset of Figure 1a). The stability was evaluated by reusing the catalyst with the same quantity of fresh MB. The results displayed in Figure. 1a. It is seen, that after four cycles degradation efficiencies of 80% was obtained.

Figure 1b illustrates the magnetic hysteresis curves of Fe_3O_4 nanoparticles, Fe_3O_4-TiO_2 nanocomposites and Fe_3O_4-CuO-TiO_2 nanocomposites recorded at room temperature. The curve of Fe_3O_4-TiO_2 is also shown in the Figure 1b. The coercivity (Hc), magnetic saturation (M-S) and magnetic remanent (mr) are listed in Table 1. The saturation magnetization decreases gradually with incorporation of TiO_2 nanoparticles and decreased further as CuO incorporated into Fe_3O_4-TiO_2 nanocomposites. The reduction of strength of the super exchange interaction among Fe^{3+} ions with incorporation of diamagnetic TiO_2 in Fe_3O_4-TiO_2 and anti-ferromagnetic CuO in Fe_3O_4-CuO-TiO_2 nanocomposites are responsible for the observed reduction of saturation magnetization [14]. However, the saturation magnetization of Fe_3O_4-CuO-TiO_2 nanocomposites is still large enough for magnetic separation as can be seen in the inset of Figure 2a.

Phase composition and crystallite size of samples are reported to have influence on the sonocatalytic activity. Therefore, XRD measurements are conducted in this study. XRD patterns of nanocomposite Fe_3O_4-CuO-TiO_2 are shown in Figure 3. The characteristic diffraction peaks of all nanocomposites samples clearly identified as individual crystal structure and phase of cubic spinel Fe_3O_4, anatase TiO_2 and monoclinic CuO. Additionally, no diffraction peaks assignable to other segregation of phase was observed in the XRD pattern. The crystallite size calculated according to the Scherrer equation [15-16] and unit cell parameters of all samples are tabulated in Table 1.

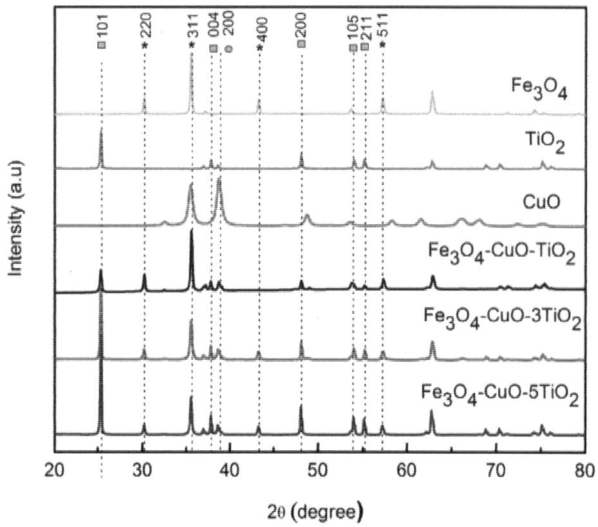

Figure 3. XRD Pattern of Fe_3O_4, CuO, and TiO_2 nanoparticles and Fe_3O_4-CuO-TiO_2 with various molar ratio [13].

The nitrogen adsorption-desorption isotherms of nanocomposites are exhibited in Figure 4. It can be seen clearly in the figure that all samples exhibited a typical type III isotherm with a distinct H3 hysteresis loop according to the IUPAC classification. Table 1 also shows a comparison of the BET surface area, pore volume and pore diameter of all nanocomposite samples were listed in Table 1. This observation suggests that the BET surface area is significantly improved by increasing molar ratio of nanocomposites from 1:1:1 to 1:1:5. It can be seen from the table that the nanocomposite with Fe_3O_4-CuO-TiO_2 molar ratio of 1:1:5 shows a larger specific BET surface area of approximately 37.64 m^2/g. Moreover, the average pore size of Fe_3O_4-CuO-TiO_2 nanocomposites were calculated from desorption branches of the N_2 isotherms via the BJH method.

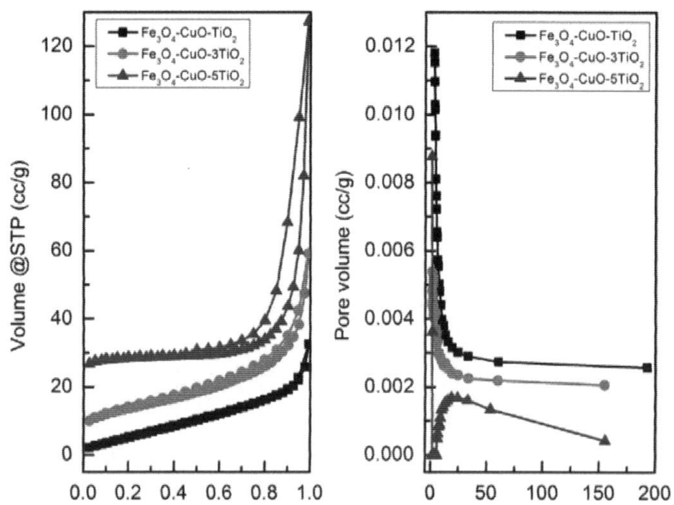

Figure 4. Nitrogen adsorpsion desorpsion and pore size distribution of Fe_3O_4-CuO-TiO_2 with various molar ratio.

It is believed that ultrasonic irradiation can result in the formation of light with comparative band gap energy that could excite electron from valence to conduction band, leaving behind holes in the valence band [17]. Although there have not been satisfying explanation on the sonocatalytic degradation of organic dyes, but the mechanism is often related to the two points of view, namely phenomena of sonoluminescence and “hot spot” [18]. The key steps involved in the mechanism of heterogeneous photocatalyst are charge-carrier generation, charge-carrier trapping, charge-carrier recombination and photocatalytic degradation of organic dye [19]. Moreover there are several parameters that control the photocatalytic activity such as phase purity [20], average crystallite size [21], band gap [22], specific surface area [23] and defect or localized states [24]. It is observed from BET measurements an increase in the specific surface area with increasing TiO_2 loading. However, according to the XRD measurements, the crystallite size did not change significantly. The BET surface areas of samples with molar ratios of 1:1:1 and 1:1:5 are 18.36 and 37.64 cm^2/g, respectively. It has been reported that an increase in the specific surface area will increase the surface active site, which consequently leads to a higher interfacial charge carrier transfer for catalysis, and will enhance the catalytic activity.

It is also reported that the presence of defects or trapping states in catalysts promote the catalytic activity, since defects can act as charge carrier traps for generated electrons and holes under irradiation and can enhance the charge separation, which in turn restrains the loss due to charge recombination. From ESR studies (Figure 5) in our samples confirmed the presence of different kinds of trapping states such as Fe^{3+}, Fe^{2+}, Oxygen vacancy and Cu^{2+} as can be seen in Table 2 [25-28]. It has been reported [26], that these states could trap the generated electrons and later react with the adsorbed O_2 molecules to yield O_2^-. Holes not only could directly decompose the adsorbed dye molecules on the

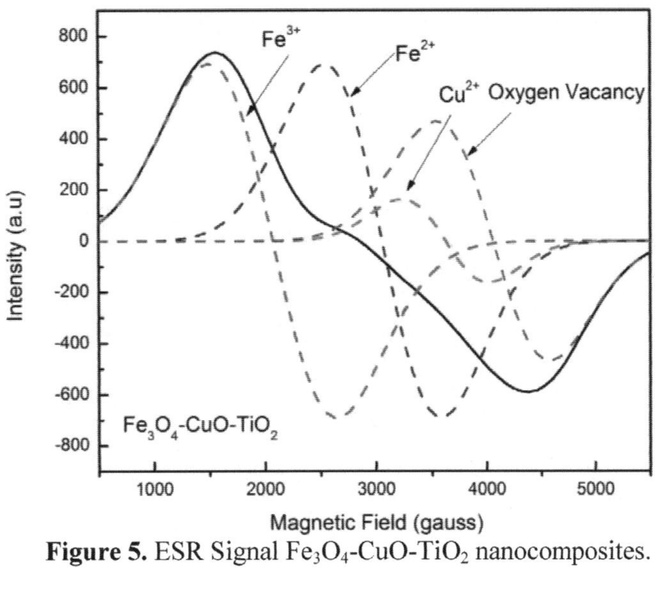

Figure 5. ESR Signal Fe_3O_4-CuO-TiO_2 nanocomposites.

Table 2. ESR Signal Analysis.

Sample	g value	Δ Hpp	Analytical Area	Assign to
	4.50	1141.993	9.33 x 10^8	Fe^{3+}
	2.66	1039.858	7.72 x 10^8	Fe^{2+}
Fe_3O_4-CuO-TiO_2	2.09	779.858	1.02 x 10^8	Cu^{2+}
	2.03	1036.574	5.20 x 10^8	Oxygen Vacancy

surface of the catalyst, it could also accelerate the generation of OH radicals through oxidizing water molecules in the bulk solution. In addition, the "hot spot" produced by ultrasonic cavitation in water medium can produce high temperature that is sufficiently brings about the many holes producing OH radicals on the surface of catalyst. As the molar ratio increased, the probability of electrons collecting in the trapping state increased due to the increase in Ti^{4+}, and more holes can react with the hydroxyl group. Thus, it is concluded that holes play the most important role in the degradation of MB, which is in agreement with the results of experiments using scavenger technique, as shown in Figure 1b.

4. Conclusion

In summary, the present study shows that MB as an organic dye can be removed using ultrasonic irradiation in the presence of Fe_3O_4-CuO-TiO_2 nanocomposites heterostructure. The sonocatalytic activity increased with increasing TiO_2 loading (molar ratio). This might be due to the higher surface area and the enhancement of electron-hole separation. The scavengers such as tert-butyl alcohol, ammonium oxalate and $Na_2S_2O_8$ reduced the MB degradation. Among them, ammonium oxalate was the most effective. These results indicated that holes play an important role in the degradation mechanism of MB.

References

[1] V. M. Correia, T. Stephenson, S.J. Judd, Environmental Technology 15 (1994) 917–924.
[2] E. Sayan, M.E. Edecan, Ultrasonic Sonochememistry 15 (2008) 530–538.
[3] M. Mrowetz, C. Pirola, E. Selli, Ultrasonic Sonochemistry 10 (2003) 247–254.
[4] A. Wang, W. Guo, F. Hao, X. Yue, Y. Leng, Ultrasonic Sonochemistry 21 (2014) 572–575.
[5] J. Wang, Z. Jiang, Z. Zhang, Y. Xie, Y. Lv, J. Li, Y. Deng, X. Zhang, Separation and Purification Technology 67 (2009) 38–43.
[6] B. Li , L. Li , K. Lin , W. Zhang , S.Lu, Q.Luo, Ultrasonics Sonochemistry 20 (2013) 855–863.
[7] N. A. Jamalluddin, A.Z. Abdullah, Ultrasonics Sonochemistry 18 (2011) 669–678.
[8] S. N. R. Inturi, T.Boningari, M. Suidanb, P.G. Smirniotis, Applied Catalysis B: Environmental 144 (2014) 333–342.
[9] A. Z. Abdullah, P.Y. Ling, Journal of Hazardous Materials 173 (2010) 159–167.
[10] Y. Xia, Q. Yao, W. Zhang, Y. Zhang, M. Zhao, Arabian Journal of Chemistry doi:10.1016/j.arabjc.2015.03.010
[11] Z. Xi, C. Li, L. Zhang, M. Xing, J. Zhang, International Journal of Hydrogen Energy 39 (2014) 6345-6353.
[12] S. A. Arifin, S.D.A Jalaludin, R. Saleh, Advanced Materials Research 1123 (2015) 264-269.
[13] S. A. Arifin, S.D.A Jalaludin, R. Saleh, Materials Science Forum 827 (2015) 49-55.
[14] S. D. A Jalaludin, S.A. Arifin, R. Saleh, Materials Science Forum 827 (2015) 13-18.
[15] R.Y. Hong, S.Z. Zhang, G.Q. Di, H.Z. Li, Y. Zheng, J. Ding, D, G. Wei, Preparation, Materials Research Bulletin 43 (2008) 2457-2468.
[16] A. Hamrouni, H. Lachheb, A. Houas, Materials Science and Engineering B 178 (2013) 1371-1379.
[17] A. Cai, Y. Sun, L. Du, X. Wang, Journal of Alloys and Compounds 644 (2015) 344-340.
[18] Y. L. Pang, A. Z. Abdullah, Chemical Engineering Journal 214 (2013) 129–138.
[19] J. Xia, A. Wang, X. Liu, Z. Su, Applied Surface Science 257 (2011) 9724-9732.
[20] C. Hariharan , Applied Catalysis A: General 304 (2006) 55-61.
[21] S. Mukhopadhyay, P. P. Das, S. Maity, P. Ghosh, P.S. Devi, Applied Catalysis B: Environmental 165 (2015) 128-138.
[22] H. Liu, T. Lv, C. Zhu, X. Su, Z. Zhu, Journal of Molecular Catalysis A: Chemical, 396 (2015) 136-142.

[23] J. Huang, S. Wen, J. Liu, G. He, Journal of Natural Gas Chemistry 21 (2012) 302-307.
[24] G. Plantard, V. Goetz, Chemical Engineering Journal 252 (2014) 194-201.
[25] L. Li, H.W. Tian, F. L. Meng, X. Y. Hu, W. T Zheng, C. Q. Sun, Applied Surface Science 317 (2014) 568-572.
[26] A.V. Kucherov, D. E. Doronkin, A.Y. Stakheev, A. L. Kustov, M. Grill, Journal of Molecular Catalysis A: Chemical 325 (2010) 73–78.
[27] J. Zhang, Z. Jin, C.Feng, L. Yu, J Zhang, Z. Zhang, Journal of Solid State Chemistry 184 (2011) 3066-3073.
[28] J. M. Coronado, J. Soria, Catalyst Today 123 (2007) 37-41.

Journal of Physics: Conference Series **710** (2016) 012007 doi:10.1088/1742-6596/710/1/012007

Intelligent QoS routing algorithm based on improved AODV protocol for Ad Hoc networks

Liu Huibin, Zhang Jun

Electronic and Information Engineering Institute, Beihang University

liuhui133134@sina.com, buaazhangjun@vip.sina.com

Abstract. Mobile Ad Hoc Networks were playing an increasingly important part in disaster reliefs, military battlefields and scientific explorations. However, networks routing difficulties are more and more outstanding due to inherent structures. This paper proposed an improved cuckoo searching-based Ad hoc On-Demand Distance Vector Routing protocol (CSAODV). It elaborately designs the calculation methods of optimal routing algorithm used by protocol and transmission mechanism of communication-package. In calculation of optimal routing algorithm by CS Algorithm, by increasing QoS constraint, the found optimal routing algorithm can conform to the requirements of specified bandwidth and time delay, and a certain balance can be obtained among computation spending, bandwidth and time delay. Take advantage of NS2 simulation software to take performance test on protocol in three circumstances and validate the feasibility and validity of CSAODV protocol. In results, CSAODV routing protocol is more adapt to the change of network topological structure than AODV protocol, which improves package delivery fraction of protocol effectively, reduce the transmission time delay of network, reduce the extra burden to network brought by controlling information, and improve the routing efficiency of network.

1. Introduction

Mobile Ad Hoc Networks has self-organizing abilities to dynamically adapt to the changing environments such as in disaster reliefs, military battlefields and scientific explorations. With the continuous expansion of Ad Hoc network application range, more and more people raise different requirements in bandwidth, time delay, computation spending and other aspects. Therefore, as for the specific business requirements, at the same time of providing routing function, Ad Hoc also guarantees to meet with requirements of a certain service quality. However, due to the changing dynamic network structure of Ad Hoc network, it is a really changeling task to offer communication service which meets some certain service quality requirements [1].

In the process of specific application, in order to adapt to different environment, node in Ad Hoc is in moving state at any time, therefore frequent link chain scission and network disconnection can be caused. The current networking protocol can't adapt to this change very well which greatly influences the circulation of network data. Moreover, the continuous expansion of Ad Hoc network application range requires Ad Hoc to retain its advantages, and also guarantee the communication quality of network such as safety requirement and real time requirement. To solve those problems, the key point is that Ad Hoc network should support QoS [2], that is, guarantee that the data transmission of network should conform to a certain requirements of computation spending, bandwidth, time delay and others. The basic method of Ad Hoc network supporting QoS is to provide QoS guarantee system [3], in which, the most core technique is to find a path with enough resources. Because of the dynamic network system, Ad Hoc makes itself different from other normal fixed network. The traditional methods are not unfit for solving QoS problem of Ad Hoc, therefore, new methods should be worked out to make Ad Hoc network satisfy with the requirements of communication quality. To seek for a path which confirms with QoS requirements in the changing topological structure of Ad Hoc network is a multi-constrained NP problem [4]. An effective method of solving multi-constrained NP problem is to find a solution by CS Algorithm [5].

Routing protocol based on QoS is the hot spot of the current research, but the obtained symbolic achievements are not many. Those routing mechanisms brought forward by people are mostly in the phase of theoretical research and hardly can be applied in actual world. To sum up the research thoughts, we can conclude that their routing designs with QoS constraint can start from two directions: (1) improve the traditional routing protocol. The main idea of this thought is to consider QoS constraint in the process of calculating optimal routing algorithm, rather than only take minimum hop as reference. The specific performance in table driven formula routing protocol is to add the information of QoS parameter to routing information which exchange among nodes, the options of optimal-path carry out among originating node; in on-demand routing protocol, QoS constraint is only needed to add in the discovering process of routing, so that the found path can satisfy with the business requirements [6].

2) Design new-pattern routing according to QoS constraint condition. This design thought takes advantage of various routing strategies to realize QoS guarantee. In an ad hoc network, all communication is done over wireless media, typically by radio through the air, without the help of wired base stations. Since direct communication is allowed only between adjacent nodes, distant nodes communicate over multiple hops. The quality-of-service (QoS) routing in an ad hoc network is difficult because the network topology may change constantly, and the available state information for routing is inherently imprecise. In [7], it proposed a distributed QoS routing scheme that selects a network path with sufficient resources to satisfy a certain delay (or bandwidth) requirement in a dynamic multihop mobile environment. The proposed algorithms work with imprecise state information. Multiple paths are searched in parallel to find the most qualified one. Fault-tolerance techniques are brought in for the maintenance of the routing paths when the nodes move, join, or leave the network. In [8], the existing location-based routing approaches, such as flooding-based routing schemes and localized routing schemes, have their limitations. Motivated by ticket-based routing, we propose an on-demand location-aided, ticket-based QoS routing protocol (LTBR). Two special cases of LTBR, LTBR-1 and LTBR-2, are discussed in detail. LTBR-1 uses a single ticket to find a route satisfying a given QoS constraint. LTBR-2 uses multiple tickets to search valid routes in a limited area. All tickets are guided via both location and QoS information. LTBR has lower overhead compared with the original ticket-based routing, because it does not rely on an underlying routing table. On the other hand, LTBR can find routes with better QoS qualities than traditional location-based protocols. In [9], a core-extraction distributed ad hoc routing algorithm for quality-of-service (QoS) routing in ad hoc network environments, CEDAR has three key components: (a) the establishment and maintenance of a self-organizing routing infrastructure called the core for performing route computations; (b) the propagation of the link-state of high bandwidth and stable links in the core through increase/decrease waves; and (c) a QoS-route computation algorithm that is executed at the core nodes using only locally available state.However, this algorithm exist the following defect: the performance of the whole network depends on the division of core regions to a large extent. If the division is inappropriate, the network performance shall be damaged severely, therefore, meticulous core region division of algorithm and maintaining algorithm are necessary.

With the deep research of Ad Hoc On-Demand Distance Vector Routing, facing with the fixed defect of network, people come up with various optimized improvement programs. Starting from the basic function of On-Demand Distance Vector Routing protocol, the ultimate purpose of those improvement methods is reduce time delay and computation spending, and guarantee delivery rate. To achieve this purpose, more information about network state needs to be collected. Moreover, we should discard the dross and select the essence in the process of information transmission, adjust the transmission behaviour according to the current state, reduce the circulation of unnecessary information as much as possible and reduce the network burden. Therefore, self-adaptive design method is a good choice. Swarm intelligence [6] is a kind of bionic intelligence based on assembling cooperative behaviours of simple individual, and can work out complicated distributed problem flexibly. Moreover, it doesn't need being controlled by centre in solving procedure. It only involves some basic mathematic operation and has less constraints of solving problems, therefore, swarm intelligence heuristic routing method is one of ideal solutions for solving Ad Hoc On-Demand Distance Vector Routing problems.

In CS algorithm, though preferential random walks components are beneficial to increase the diversity of species and strengthen the global searching ability of algorithm, as for multidimensional objective functions and solution uses overall update evaluation strategy, it shall influence the convergence rate of algorithm and the quality of solution. The routing protocol research of ant colony algorithm has been mature, but as for routing of such protocol, the parameter controlling process carried out for realizing information exchange is complicated, and the actual operation is much more complex. However, the iterative process mechanism of CS Algorithm is simple, has less parameters and controllable process, therefore in this essay, research is mainly based on Ad Hoc On-Demand Distance Vector Routing protocol of CS Algorithm. To solve the optimization problem of On-Demand Distance Vector Routing with CS Algorithm is a currently new application area. How to take advantage of CS Algorithm self-adaptive distributed optimization characteristic, integrate the abilities of clustering collaboration and multi-objective optimization into solving the routing problems as required, need to fully consider the feasibility and designing difficulty of On-Demand routing protocol technique based on CS Algorithm.

Ad Hoc On-Demand Distance Vector Routing protocol (CSAODV) based on CS Algorithm directs at AODV protocol, with main purpose of getting improvement through AODV protocol routing detecting process and routing calculation strategy, considering bandwidth, consumption and time delay comprehensively when seeking for optimal-path, and further improving the service quality of network. CSAODV protocol obey with the structure of Ad Hoc On-Demand Distance Vector Routing protocol, and comprise of routing detection, routing maintenance and repairing mechanism [10]. Of which, in routing detection, the calculation of optimal-path is completed by improved-CS Algorithm. CS Algorithm is executed in objective node, take bandwidth, time delay and computation spending as optimization objective. The information collecting of bandwidth, time delay and computation spending goes through the whole process of routing detection. The detected network state information in the process of routing detection is all included in routing request package. When objective node collects a certain number of routing packages, according to the detected network state information, it can construct topological structure which reflects the current network state and select one optima-path as per the topological structure to reach a balance among computation spending, bandwidth and time delay. In the process of routing exploration, the routing package submitted by originating node records the information of bandwidth and time delay which passes through node in network transmission, and stores all nodes according to the order of passing through. After the intermediate node receiving routing package, determine whether transmit this package according to ID of request package. This strategy can effectively avoid the production of routing loop. In routing maintenance and repairing mechanism, all intermediate nodes send HELLO package to the former hop and latter hop in stored path periodically and the processing strategy can be determined according to whether reply package can be received in regulated time. As for choosing package transmission path, CSAODV protocol resorts to the same data transmission mechanism as AODV protocol, that is, the current node which has received package only need to know the next hop to the objective nod, instead of storing all nodes in path which package has passed through.

2. Cuckoo Search algorithm

Cuckoo takes a special reproduction strategy of parasitic brooding. She deposits her eggs in other birds' nests in order to make other birds hatch the next generation for her. To reduce the probability of being discovered, some cuckoos shall make up her eggs like those of the selected birds. When other birds find alien eggs in their nests, they will abandon those eggs or put them in their own nests, or build a new nest in other places. Based on the reproduction strategy of cuckoo, Yang and Deb draw out CS-Algorithm which is based on three ideal rules [6]:

Rule 1. Each cuckoo only produces 1 egg for one time, and randomly chooses one bird nest for storage. Rule 2. The bird nest with the best egg shall be reserved for the next generation. Rule 3. The number of available bird nests are fixed, and the probability of being discovered to be the alien eggs is $P_a \in [0,1]$.

Based on the above three rules, the basic procedure of CS algorithm is as shown in algorithm 1.

Algorithm 1. Cuckoo Search.

```
Begin
Initialize the population: n host nests Xi(i=1,2,...,n);
Calculate the fitness value: Fi(i=1,2,...,n).
While (doesn't meet the stop condition)
Take the new solutionXi generated by Lévy flight;
Calculate the fitness value Fiof the new solution Xi;
Choose candidate solution Xj;
If(Fi>Xj)
    Replace candidate solution with the new solution;
End
Discard the bad solutions according to the detection probabilityPa;
Replace the discarded solution with the new solution generated by preferential random walks
Reserve the best solutions.
End
End
```

In CS algorithm, one bird nest represent one candidate solution. Firstly, CS algorithm generates a new solution in Lévy flights random walk way based on current solution, evaluates and reserves better solutions; secondly, discard some solutions according to detection probability P_a; Lastly, regenerates a new solution with the same number as that of the discarded solutions by preferential random walks, complete one-iteration after evaluation and reserving the better solutions.

When takes the new solution X_i generated by Lév y flights random walk, execute the operation of formula (1):

$$X_{g+1,i} = X_{g,i} + \alpha \oplus \text{Lévy}(\beta) \tag{1}$$

Thereinto, $X_{g,i}$ represents the i solution of the g generation; α is step message, used for controlling the range of random searching. In order to gain more useful step messages from the current optimal solutions, reference [8] resorts to formula (2) to calculate step message:

$$\alpha = \alpha_0 (X_{g,i} - X_{best}) \tag{2}$$

Thereinto, α_0 is constant ($\alpha_0 = 0.01$), X_{best} represents the current optimal solution.

In Formula (1), ⊕ is entry-wise multiplications, Lévy(λ) obey Lévy probability distribution:

$$\text{Lévy}(\beta) \sim u = t^{-1-\beta},\ 0<\beta\leq 2 \tag{3}$$

For easy calculation, reference [6] resorts to formula (4) to calculate Lévy random number:

$$\text{Lévy}(\beta) \sim u \frac{\phi\phi \times u}{| v|^{1/\beta}} \tag{4}$$

Thereinto, u,v obey standard normal distribution, β= 1.5.

$$\phi = \left(\frac{\Gamma(1+\beta)\times\sin(\frac{1}{2}\pi\beta)}{\Gamma((\frac{1+\beta}{2})\times\beta\times 2^{\frac{\beta-1}{2}})}\right)^{\frac{1}{\beta}} \quad (5)$$

Obviously, synthesize formula(1) ~formula(5), in Lévy flights random walk components, CS algorithm takes formula(6) to generate the new solution X_i:

$$X_{g+1,i}=X_{g,i}+\alpha_0 \frac{\phi\phi\times u}{|v|^{1/\beta}}(X_{g,i}-X_{g,best}) \quad (6)$$

After discarding some solutions according to a certain probability (detection probability), algorithm takes preferential random walks to regenerate the new solutions with the same number, see formula(7):

$$X_{g+1,i}=X_{g,i}+r(X_{g,j}-X_{g,k}) \quad (7)$$

Thereinto, r is zoom factor, a uniform random number in (0,1) section; $X_{g,j}$and $X_{g,k}$ are two random solutions of the g generation.

3. Routing Detection Strategy of CSAODV protocol

The production of routing request information

When originating node is going to send a package, firstly check if any routing reached objective exists in the current routing table; if exist, send the date directly; if not, cache data and then start routing detection mechanism, and broadcast routing request package RREQ. When originating node has sent the first RREQ package, it shall start a timer and expect to receive RREP sent by objective node before time expiring. If receives RREP package in specified time, get out the data from cache memory and send it; if not, originate the routing request.

3.1. The broadcast and controlling of routing request information

To avoid the unnecessary broadcasting of routing request information in the whole network, the originating node shall expand the searching scope gradually. In searching with bigger and bigger scope, originating node shall set TTL value in IP head of RREQ package as TTL_START, the waiting time of RREP as RING_TRAVERSAL_TIME. If RREP waiting-time out after sending out RREQ, the originating node shall broadcast RREQ again, and TTL value shall be added to TTL_INCREMENT, which shall last until TTL value in RREQ package reaches the pre-set threshold value TTL_THRESHOLD. When reaching the threshold value, TTL shall be set as NET_DIAMENTE each time, the expiration time of waiting RREP is fixed RING_TRACERSAL_TIME. If each routing seeking is carried out in Ad Hoc network scope, TTL_START and TTL_INCREMENT shall set as NET_DIAMENTER to reach the objective.

3.2. The processing and transmission of routing request information

When node receiving a RREQ group, it should firstly check whether the originating node of RREQ routing request package is this node, if it is, discard this package, otherwise, check if this RREQ package is received before. If such RREQ group has been indeed received before, this node shall be discarded to this RREQ group, if not, determine if it is the objective node of this RREQ group. If this current node isn't the objective node in this RREQ group, transmit this RREQ directly, that is to broadcast RREQ package to all neighbouring nodes. Before broadcasting,, calculate the time delay between the current node and the last node, the calculation method is the timestamp of receiving RREQ package minus the timestamp of the last node sending this RREQ, and then fill this time delay into the according position in RREQ package time delay list which shall be transmitted. Except time delay, the currently available bandwidth of node shall be added to the bandwidth list. At last, update the list for nodes which are passed through by RREQ package, add other information required by transmitting communication-package, and then transmit to RREQ package.

When the first routing request package reaches objective node, objective node shall initiate a timer to cache all routing request package before the time expiring. After the time expiring, objective node extract the received path lists and according bandwidth and time delay information in all RREQ groups, and construct topological structure which reflects the current network state. This topological structure includes time delay and bandwidth information of chain. Afterwards, take advantage of network topology constructed by improved-CS Algorithm to find a path with minimum computation spending, the most adequate bandwidth, and minimum time delay. In the operation process of algorithm, each bird nest represent a path between originating node and objective node, the fitness function of determining the superior and inferior of the bird nest as follows:

$$F(r_{s,d}) = \frac{1}{C(r_{s,d})}\left(\prod_{e \in r_{s,d}} \varphi(D(e) - \Delta) + \varphi(B(r_{s,d}) - \delta)\right) \quad (8)$$

Of which, $r_{s,d}$is a routing between originating node r and objecting node d, $C(r_{s,d})$ is computation spending of routing sdr, $D(e)$ is time delay in chain e which is included in routing $r_{s,d}$, $B(r_{s,d})$ is the minimum bandwidth in routing $r_{s,d}$. $\varphi(\cdot)$ is penalty function, when bird nest satisfies with the constraint condition, its value is 1, otherwise, it equals r, $r \in (0,1)$ the extent of penalty. The higher the fitness value of bird nest is, the better the found path performance is. At last, choose the path of bird nest with the highest fitness value as optimal-path. After selecting the optimal-path, objective node constructs routing reply package RREP, fill the optimal-path which is found by CS ALGORITHM into the corresponding area, and then transmit RREP package to the node before reaching objective node in optimal-path.

3.3. Receiving and forwarding routing responses

RREP package returns to RREQ originating node along the reverse direction of the selected optimal-path. After the intermediate node receiving the first RREP package, it shall reach the forward routing of RREQ objective node and backward routing of RREQ originating node according to the record of optimal routing algorithm in RREP before transmitting, and store bandwidth and time delay in responding routing table, and then transmit RREP package to the last node of the current node in optimal-path.

3.4. The processing after originating node receiving the reply information

After originating node receiving the reply package, according to the path information in RREQ package, update the current routing table, extract data from cache memory and sent it out immediately. The flow chart of CSAODV protocol routing detection is shown in Figure 1.

4. Package Transmission Strategy in CSAODV protocol

CSAODV protocol uses forward routing to transmit package. In note routing table, there stores the address of the next hop of reaching objective node and corresponding bandwidth, time delay. When intermediate node receiving the package sent by originating node, it needs to check if any routing has reached objective node in routing table, if exist and bandwidth and time delay conforms to the requirement of package in this routing table, transmit this package to the next hop node in this routing table; if not, initiate routing repairing mechanism.

Originating node receives RREP package sent by objective node means a path which conforms to QoS constraints between originating node and objective node has constructed. At the same time of originating node initiating data transmission, all nodes in this path begin to initiate routing maintenance mechanism. In the expiry date of this path, all nodes in this routing should send HELLO package to the former hop and latter hop nodes periodically and expect to receive the reply of node in regulated time. If no reply is received in regulated time, it shows that chain scission occurs between two nodes, and then initiate routing repairing mechanism.

CSAODV protocol routing repairing, if chain scission occurs between the next hop and this node in routing, check if other path to reach the next node for this node exist; if exist and bandwidth and time delay in path are within the required scope, the next hop can be reached through this node; if not, detect whether any path with conforming bandwidth and time delay exist for this node reaching objective node, if exist, transmit package along this path. If these two paths don't exist, inform that originating node and objective node can't be reached, afterwards originating node initiate a new round of routing detection.

CSAODV protocol and AODV protocol are all Ad Hoc On-Demand Distance Vector Routing protocols. CSAODV protocol is based on AODV protocol. Both of them adapt to On-Demand Distance Vector Routing-network equivalent to node function, and used for data transmission of single objective node. The similarity lies in the basically same routing mechanism used by them, the major difference is the different strategy selected by optimal routing algorithm. AODV protocol updates routing table according to the serial number and the number of hops in RREP package. In AODV protocol, the routing table of node includes the known routing of nodes which has reached the objectives.

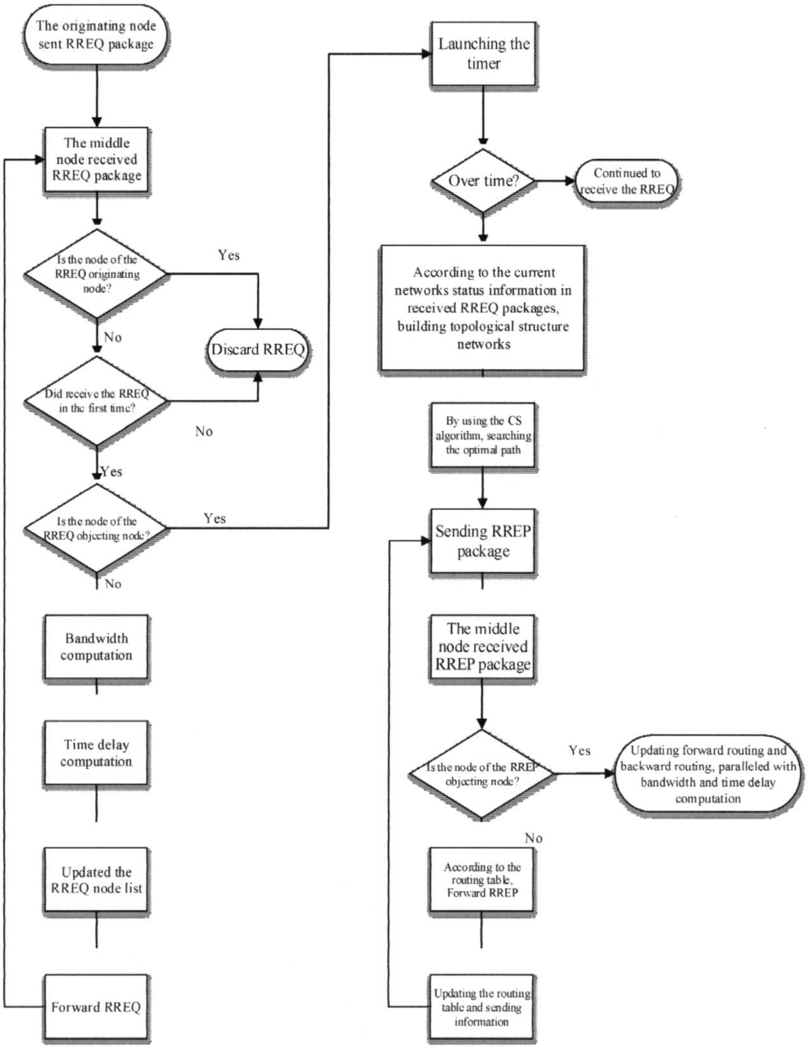

Figure 1.The flow chart of CSAODV protocol routing detection

When originating node receiving a RREP package, extract the serial number and the number of hobs, and compare those two with the corresponding routing items in current routing table. If the seral number of the new package is bigger than the serial number in routing item, then the path represented by the newly received RREP package is the transmission path, update the corresponding routing items; if these two serial numbers equal each other, compare the number of hops, and choose the path with less hops as the transmission path; if the serial number of the newly received RREP package is smaller than the existed serial number in the routing table, then no action and discard RREP package directly. When originating node receiving each RREP package, it will repeat the above procedures. If choose the path represented by RREP package as transmission path, update the corresponding routing items, the following package shall send out the new routing table. CSAODV protocol gets optimal-path by CS ALGORITHM calculation. When choosing optimal-path, comprehensive consider computation spending, bandwidth and time delay, and the optimization process is completed by objective node. First, collect time delay and bandwidth information in path exploration process, and store the order of nodes passed through by RREQ package. When RREQ reaching objective node, the objective node doesn't reply to RREQ package immediately, but initiate a timer to receive all RREQ package before the timer expiring. After the timer expiring, reflect the information of current network state according to the received RREQ package, construct the network topological-structure, and take advantage of CS algorithm to select the most balanced routing among computation spending, bandwidth and time delay as the final routing. Compare the two routings to choose mechanism, CSAODV protocol is added to computation spending, bandwidth and time delay index, then select a routing with certain QoS guarantee, making network more reliable and real-time in data transmission.

5. Simulation

To measure the performance of routing process in CSAODV protocol, this paper chooses AODV protocol before improvement as reference, and designs three network environments to compare packet delivery fraction, time delay and routing spending of CSAODV protocol and AODV protocol in different network environment. To avoid the influence on simulation result from random factors, the simulation of each scene shall repeat five times and then take the average value as the simulation result. To evaluate the performance of CSAODV protocol, this paper carries out simulation test in Window XP+sygwin+NS2-2.8, and compares with AODV protocol in the same setting environment, network environment and state characteristics. The simulation environment set as: simulation area is 1500m*300m, the number of nodes is 10~100, node mobility model is random walk mobility model, the speed of node is 0~30m/s, the protocol type of MAC layer is IEEE802.11, node communicate scope is 250m, bandwidth is 2Mb/s, transmission layer is protocol UDP, data flow CBR, the packet transmitting speed is 4 per second, the data packet capacity is 512bytes,simulation time is 500s. In random walks mobility model, each node is a single individual, the travelling speed of current position of node and objective position is random, its value may be a random value from 0 to the maximum travelling speed. After node reaching an objective position, it will remain for a short while, then select a coordinate in simulation area as the next objective position and move towards the objective position with a constant speed. Node runs repeatedly this process until the simulation ends.

To perform the simulation of routing protocol, evaluate such major indexes of routing protocol as follows: Packet delivery fraction: compare the package received successfully by network application layer and all packages. Packet delivery fraction is the major index of evaluating routing, if it is 1, all packages reach the objective node, which is the best ideal state; if it is 0, which means all packages shall be discarded without reaching objective node, which is the worst ideal state. According to the characteristics of Ad Hoc-network, packet delivery fraction will not less than one appropriate value in different circumstances. How to increase packet delivery fraction as much as possible is the most question for each routing protocol to take account into, and also the ability to exhibit how routing protocol takes advantage of the current network's QoS index.

End-to-End time delay: Within the total simulation time, seek for and maintain the average time of message transmission from originating node to objective node in routing path. End-to End time delay shows the reaction ability of routing protocol, and a kind of index for users to evaluate routing protocol.

Routing expense: in total simulation time, for data transmission, the ratio of the total number of routing controlling message and the number of data message can be shown by percentage.

5.1. The influence on protocol performance of network scale

To compare the adaptive abilities of two protocols to different scaled network, increase the number of network nodes gradually in this simulation. From Table 1, we can see that, with the increasing number of network nodes, the delivery fractions of the two protocols have different growth, but the delivery fraction of CSAODV protocol is always higher than AODV. This is because that with the increasing number of network nodes, the connection probability between two nodes increases accordingly, which makes the selectable number of routing between originating node and objective node increases, the packet delivery fraction increases as well. CSAODV considers the influence of bandwidth factor when routing, the stability of chain is favourable, and therefore packet delivery fraction is always in higher state. Table 1 reflects the changing situation of end-to-end time delay in two protocols when network scale changes, CSAODV protocol has less influence and shorter network time delay in all scales. This is because that CSAODV protocol considers the time delay factor in QoS constraint, and chooses the path with shorter time delay in data transmission, the more network nodes, the bigger the probability of paths with shorter time delay is.

Table 1. Routing performance comparison (Node movement speed: 5m/s; Data originating node number:1)

	Grouped Delivery Rate		Time Delay (S)		Routing Spending	
Network scale (node number)	CSAODV	AODV	CSAODV	AODV	CSAODV	AODV
10	0.928	0.914	0.063	0.070	0.19	0.20
20	0.934	0.921	0.058	0.063	0.21	0.24
30	0.937	0.923	0.057	0.061	0.22	0.25
40	0.941	0.927	0.055	0.059	0.23	0.25
50	0.941	0.930	0.049	0.056	0.23	0.25
60	0.942	0.932	0.041	0.051	0.24	0.27
70	0.944	0.935	0.041	0.045	0.25	0.29
80	0.947	0.937	0.042	0.042	0.25	0.31
90	0.956	0.937	0.034	0.041	0.36	0.32
100	0.965	0.939	0.029	0.041	0.27	0.34

Table 2. Routing performance comparison (Network scale node number: 100; Data originating node nember:1)

	Grouped Delivery Rate		Time Delay (s)		Routing Spending	
	CSAODV	AODV	CSAODV	AODV	CSAODV	AODV
0	1	1	0.024	0.038	0.25	0.29
5m/s	0.965	0.939	0.029	0.041	0.27	0.34
10m/s	0.951	0.937	0.035	0.044	0.31	0.39
15m/s	0.947	0.918	0.039	0.056	0.40	0.50
20m/s	0.922	0.872	0.046	0.062	0.39	0.52
25m/s	0.901	0.855	0.050	0.070	0.50	0.59
30m/s	0.882	0.783	0.064	0.081	0.58	0.66

With the increasing network scale, the connection probability of nodes increases and the network topology becomes complicated; therefore, network needs more controlling information in routing process. Table 1 is the influence on routing expenses affected by network scale. We can see that with the increasing network scale, the gap between CSAODV expense and AODV expense is more and more obvious, and the changing range of CSAODV protocol is small. This is because that stability and durability of the routing selected by CSAODV is higher than AODV, the message for processing chain scission is less, which lessen the proportion taken by controlling information in network flow.

5.2. The influence of moving speed on protocol performance

To compare the influence of network topology change on the two protocols, increase the speed of node continuously in this simulation, the specific experimental parameter settings can refer to the table. The moving speed of node is an important aspect of affecting Ad Hoc network performance, we can see from Table 2, the routing quality of network decrease continuously with the increasing moving speed of node. This is because that the rapid moving of node results in much more frequent change of network topological-structure, chain scission occurs among nodes beyond the communication range of each other, therefore, packet loss rate of network increases, end-to-end time delay increases, the information message for network to judge chain scission processing and controlling message of repairing routing increase accordingly. However, CSAODV protocol considers QoS factor when choosing routing, choose a path with bigger bandwidth for data transmission, which strengthen the reliability and stability of network transmitting data, and reduce influence brought by chain scission.

5.3. The influence of network load on protocol performance

To compare the performance of two protocols in different network load, increase the number of originating nodes gradually in this simulation, the specific experimental parameter settings is shown in the table

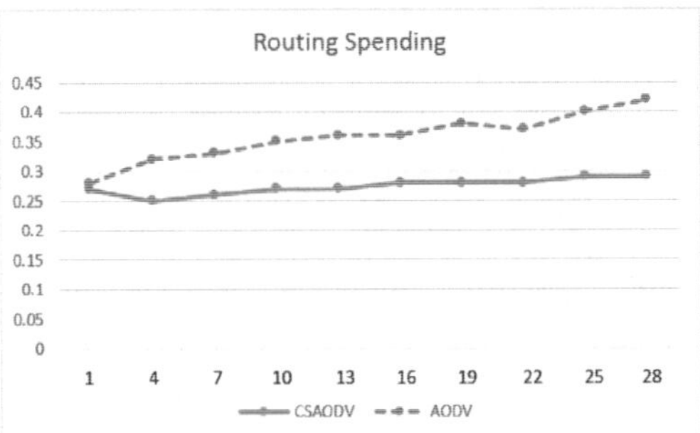

(a)

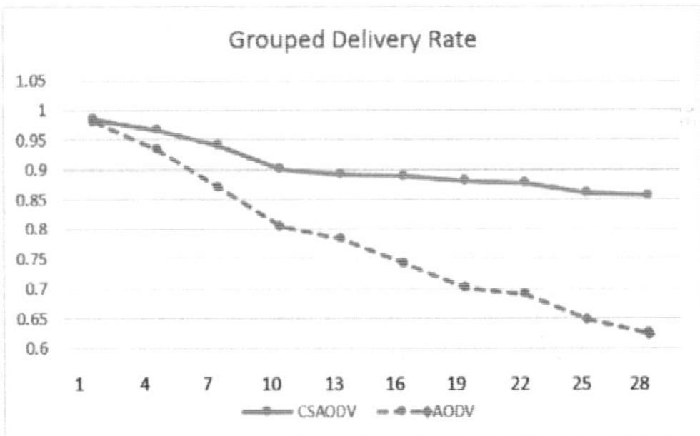

(b)

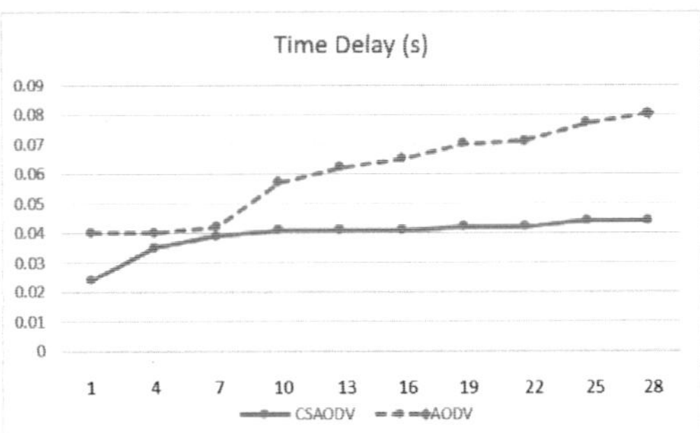

(c)

Fig 2. Table 2. Routing performance comparison (Network scale node number: 100; Node movement speed: 5m/s)

Figure 2(a) shows the influence of network load on packet delivery fraction, we can see that the packet delivery fraction of AODV protocol reduce greatly when network load increases, while the decreasing rapid of the packet delivery fraction in CSAODV protocol is relatively slow. That is because that with the increasing load on network, the probability of network congestion increases accordingly, while constraints are added to bandwidth when CSAODV protocol is routing, which makes packet transmit in routing with larger bandwidth, thereforth guarantee the packet delivery fraction. Figure 2(b) shows that the end-to-end time delay of network increases with the increasing number of originating nodes. This is because that the increasing number of originating nodes shall increase the probability of congestion occurring, and further increases the transmission time delay of network. Compared with AODV protocol, because that constraint is added to time delay when CSAODV selects routing, its time delay performance gets less influence in the increasing process of network load. Figure 2(c) shows the comparison of routing expenses for tow protocols with different number of originating nodes. This is because that it doesn't consider the network flow when routing based on the minimum hopping number. When the number of originating node increases, some nodes shall be selected frequently for routing, the probability of chain scission becomes bigger, therefore the controlling expenses for network to reselect routing shall increase accordingly. Because that bandwidth constraints are added to routing mechanism in CSAODV protocol, the node with better resources can be selected in priority, which to large extent reduce the probability of chain scission and reduce the proportion of controlling information for re-routing.

6. Conclusion

In different scaled network environment, except time delay, the routing quality of CSAODV and AODV protocol reduces with the increasing of scale, but CSAODV protocol is less affected. It is because of QoS system in CSAODV protocol which makes the selected routing relatively stable, and weakens the negative influence brought by the increasing number of nodes. In the environment in which node moving speed increases, though network performance will reduce obviously, the performance of CSAODV protocol is always superior to AODV protocol, which fully manifests the advantage of QoS guarantee to better adapt to the frequently changing network. In the network environment of node changing rapidly, the overall routing quality of network turns out descending trend with the increasing rapid, while the performance of CSAODV protocol is always superior to AODV protocol, which fully manifests the advantage of QoS guarantee to better adapt to the frequent changing of network. In network environment with different loads, CSAODV protocol comprehensively consider the influence of bandwidth and time delay on routing performance when CSAODV protocol is routing, which makes network flow distributed equally in the whole network and avoids bottleneck node in network due to local flow concentration. Overall, the performance of CSAODV protocol is superior to AODV protocol before improvement, and possesses a stronger adaptability and robustness in different-state network environment.

References

[1] Pierobon, M., Jornet, J. M., Akkari, N., Almasri, S., & Akyildiz, I. F. (2014). A routing framework for energy harvesting wireless nanosensor networks in the Terahertz Band. Wireless networks, 20(5), 1169-1183.

[2] Karaboga, D., Okdem, S., & Ozturk, C. (2012). Cluster based wireless sensor network routing using artificial bee colony algorithm. Wireless Networks, 18(7), 847-860.

[3] Lin, C. R., & Liu, J. S. (1999). QoS routing in ad hoc wireless networks.Selected Areas in Communications, IEEE Journal on, 17(8), 1426-1438.

[4] Wang, Z., & Crowcroft, J. (1996). Quality-of-service routing for supporting multimedia

Journal of Physics: Conference Series **710** (2016) 012007 doi:10.1088/1742-6596/710/1/012007

applications. Selected Areas in Communications, IEEE Journal on, 14(7), 1228-1234.

[5] Camilo, T., Carreto, C., Silva, J. S., & Boavida, F. (2006). An energy-efficient ant-based routing algorithm for wireless sensor networks. In Ant Colony Optimization and Swarm Intelligence (pp. 49-59). Springer Berlin Heidelberg.

[6] Wang, J. S., Han, S., Shen, N. N., & Li, S. X. (2014). Features Extraction of Flotation Froth Images and BP Neural Network Soft-Sensor Model of Concentrate Grade Optimized by Shuffled Cuckoo Searching Algorithm. The Scientific World Journal, 2014.

[7] Chen, S., & Nahrstedt, K. (1999). Distributed quality-of-service routing in ad hoc networks. Selected Areas in Communications, IEEE Journal on, 17(8), 1488-1505.

[8] Huang, C., Dai, F., & Wu, J. (2004, August). On-demand location-aided QoS routing in ad hoc networks. In Parallel Processing, 2004. ICPP 2004. International Conference on (pp. 502-509). IEEE.

[9] Sivakumar, R., Sinha, P., & Bharghavan, V. (1999). CEDAR: a core-extraction distributed ad hoc routing algorithm. Selected Areas in Communications, IEEE Journal on, 17(8), 1454-1465.

[10] Jain, J., Gupta, R., & Bandhopadhyay, T. K. (2014). Scalability enhancement of AODV using local link repairing. International Journal of Electronics,101(9), 1230-1243

Acidic Barren Slope Profiling using Electrical Resistivity Imaging (ERI) at Ayer Hitam area Johor, Malaysia

A T S Azhar[1,6], **Z A M Hazreek**[2*,7], **M Aziman**[3,8], **D S Haimi**[4,9] and **Z M Hafiz**[5]
[1,7,8] Research Center for Soft Soil, Universiti Tun Hussein Onn Malaysia, 86400 Batu Pahat Johor, MALAYSIA
[2,3,5,6] Faculty of Civil and Environmental Engineering, Universiti Tun Hussein Onn Malaysia, 86400 Batu Pahat Johor, MALAYSIA
[4] Research Center for Applied Electromagnetic, Universiti Tun Hussein Onn Malaysia, 86400 Batu Pahat Johor, MALAYSIA
[5] College of Engineering, Universiti Tenaga Nasional Malaysia, 43000 Kajang, Selangor, MALAYSIA
[9] Faculty of Electrical and Electronic Engineering, Universiti Tun Hussein Onn Malaysia, 86400 Batu Pahat Johor, MALAYSIA

E-mail: hazreek@uthm.edu.my

Abstract. Recently, non-destructive method such as the electrical resistivity technique has become increasingly popular in engineering, environmental, mining and archeological studies nowadays. This method was popular in subsurface profiling due to its ability to replicate the images of the subsurface indirectly. The soil slope found in Batu Pahat, specifically in Ayer Hitam, is known to be problematic due to its barren condition. This location is believed to contain futile soil due to its difficulty in supporting the growth of vegetations. In the past, acidic barren slope assessment using non-destructive method was rarely being used due to several reasons related to the equipment and knowledge constraints. Hence, this study performed an electrical resistivity imaging using ABEM Terrameter LS in order to investigate the acidic barren slope conditions. Field data acquisition was based on Schlumberger and Wenner arrays while RES2DINV software was used to analyze and generate a 2-D model of the problematic subsurface profile. Based on electrical resistivity results, it was found that the acidic barren slope studied consists of two main zones representing residual soil (electrical resistivity value = 10 – 600 Ωm) and shale (electrical resistivity value = 20 – 2000 Ωm). The results of resistivity value were correlated with the physical mapping and the in situ mackintosh probe test for verification purposes. It was found that the maximum depth of the mackintosh probe test was 1.8 m due to its ground penetration limitation. However, the results of the resistivity section managed to achieve greater depth up to 40 m. Hence, the correlation between electrical resistivity and mackintosh probe results can only be performed at certain depth of the acidic barren slope profile in contrast with the physical mapping which able to define the whole section of the barren soil slope structure. Finally, a good match of electrical resistivity results calibrated with mackintosh and physical mapping data showed that this technique was appropriate to be applied in near-surface acidic barren slope assessment which can further compliment borehole data and other physical mapping data at a lower cost, higher speed, large data coverage and better environmental sustainability.

1. Introduction

Barren slopes along roadways have a high probability to collapse or slide due to heavy rainfall distribution. This catastrophe can cause fatal and heavy damages to our life and properties. Hence, relevant assessment related to field testing is crucial in order to determine the soil profile condition thus minimizing or preventing such disasters. Boring, drilling and excavation are traditional ways of subsurface earth exploration [1]. There are several limitations related to the traditional technique due to it being expensive, time consuming [2,3] as well as limited data coverage [4,5,6,7]. Moreover, conventional site investigation tools such as drilling pose certain difficulties in steep and hilly terrain, swampy areas, coastal regions and complex geomaterial areas which need to be investigated [1]. Furthermore, those conventional techniques are destructive methods which could lead to site damage. This is considered unsustainable in construction industry. As a result, an alternative approach able to reduce cost, save time, ensures larger data coverage and minimal ground destruction needs to be explored.

Recently, geophysical methods such as electrical resistivity, seismic method (refraction and reflection), magnetic, gravity, ground penetrating radar and etc are getting increasingly popular in engineering, environment and archeological studies. Generally, geophysical method applies the principles of physics properties in studying the earth such as resistivity, velocity, density, magnetic susceptibility and etc. Geophysical methods are able to compliment and minimize conventional method limitations due to its efficiency in terms of cost, time, data coverage [5,1,6,8,9] and sustainability [10,11,12]. Field operations require less manpower while data processing and results have become quite easy and quick to obtain compared to the conventional drilling method [12]. In the past, geophysical methods have been used as an alternative tool in many situations related to detection or characterization purposes such as leachate mapping [13,14,15], bedrock mapping [16], overburden materials [10,16,17], cavity and boulder detection [18], groundwater mapping [2,19,20,21], mining exploration [22,23] and archeological studies [24,25,26]. Generally, the process of geophysical methods involved data acquisition, data processing and data interpretation. Geophysical surveys are non-destructive methods thus able to minimize the disruption and damage to the site due to its surface technique [12]. Hence, this technique is considered sustainable since it preserves our environment during the data acquisition stages. However, according to Mauritsch et al. [7], the standard of geophysical methods depends on fundamental physical constraints such as penetration, resolution and signal to noise ratio.

Geophysical method with particular reference to electrical resistivity has been widely used in soil profiling [10,17]. As mentioned previously, geophysical method with particular reference to electrical resistivity method has been used for subsurface profiling related to the soil, boulder, bedrock and groundwater mapping. However, studies related to acidic barren slope profiling based on non-destructive method are still new and need to be explored. Electrical resistivity property was determined by measuring the potential difference at points on the ground surface which caused the propagation of direct current through the subsurface [27]. Basically, the resistivity image was interpreted based on the contrast of geomaterials with verification through previous established references charts, tables, and other related data and information. Finally, a variation of resistivity distribution from the image can be interpreted and the barren soil slope profile was able to be determined successfully.

2. Materials and Methods

This study was located at the Batu Pahat area specifically in Ayer Hitam. Generally, the site of study has mixed topography of undulating hilly terrain and is located near the developing industrial areas, small town and housing area. The general geology of Malaysia has been well documented [28]. The Ayer Hitam area is located on a triassic zone (pale blue colour) which consists of a mixture between interbedded sandstone, siltstone and shale as shown in Figure 1. In general, the presence of this type of rock exhibits geological structures, namely joint, fold, fault and bedding. Those geological discontinuities may possibly be present in the studied area thus accelerating the rate of weathering due to the surface water seepage, groundwater carriage, and water storage in soil and rock formation. During the field observation, it was found that this area consists of two major zones with particular reference to residual soil and weathered sedimentary rock which are confirmed through physical mapping at the surrounding existing rock outcrops. Moreover, during the localized physical mapping, geological structures related to bedding, fold, joint and fault could be spotted as they were exposed on the surface.

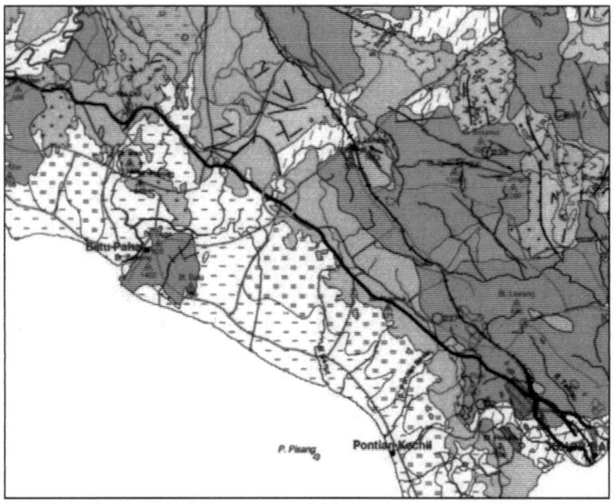

Figure 1. Geology of the study area [28].

Electrical resistivity imaging (2-D electrical resistivity survey) was performed using ABEM Terrameter LS. A single spread line of electrical resistivity survey was performed across the problematic soil slope in this study area. The test configuration was based on Schlumberger and Wenner array using four resistivity land cables, 61 electrodes and 64 jumper cables. Equal electrode spacing of 5.0 m was used for all 61 electrodes producing total electrical resistivity for a survey length of 200 m. Schlumberger array was used during the data acquisition since it is able to provide dense near-surface coverage of resistivity data. As reported by [29], the array provides a good vertical resolution and can give a clear image of groundwater and sand-clay boundaries as horizontal structures. Furthermore, the array is able to provide greater depth of subsurface profiles within limited spaced area during the resistivity data acquisition. Meanwhile, Wenner array was performed at the same location in order to verify the Schlumberger array results. Raw data obtained from data acquisition was processed using commercialized RES2DINV software [30] to provide an inverse model that approximates the actual subsurface structure. The inversion algorithm of RES2DINV was used to process the data, as proposed by [31] in order to replicate the 2-D model of the subsurface profile. An additional test was conducted using X-Ray Fluorescence (XRF) test in order to verify the earth material with the electrical resistivity results obtained.

3. Results and Discussions

Barren soil slope outcrop mapping and electrical resistivity tomography (ERT) results from Schlumberger and Wenner array are presented in Figure 2, 3 and 4 respectively. Moreover, the electrical resistivity value specifically at the center of the ERT was extracted from Figure 4 and 5 and presented in Table 1.

According to physical mapping and Figure 2, the dotted red line indicates the presence of the black soil layer within the slope structure representing the barren geomaterial. According to Figure 3 and 4, the highest resistivity value was located at the middle of the ERT representing the middle of the slope structure. Moreover, the lowest resistivity value was found at both bottom flanks of the ERT. According to Figure 3 and 4, it was found that the Schlumberger array has a greater profiling depth (38 m) compared to the Wenner array result (32 m).

According to Table 1, it was found that the Schlumberger array produced higher electrical resistivity value (ERV) compared to the Wenner array. However, the resistivity contrast and distribution was found to be similar thus able to confirm that the electrical resistivity method was applicable in barren soil slope profiling. In the past, the dissimilarity of resistivity value commonly occurred due to several reasons. Basically, this factor occurred due to the different geometry factor, K, used during the field measurements [32,33,34,35]. The value of apparent ERV (ρa) was greatly influenced by K factor applied in every measurement. Geometry factor, K, describes the geometry of the electrode configuration used in data acquisition. Apparent resistivity (ρa) is ERV estimated based on half-pace geometry assumption which refers to the field ERV. According to [36], apparent resistivity will be equal to the true resistivity provided the current and configuration was applied over the homogeneous isotropic ground. Field ERV was determined using Schlumberger and Wenner array with a geometry factor as given in final Eq. 1 and 2 which is derived from basic Eqs. 3 and 4. Schlumberger and Wenner geometry factor, K, used in this study was derived from basic Eq. 4 based on basic four-electrode system of measurement. The schematic diagram of field array with particular reference to Schlumberger and Wenner resistivity configuration was given in Figure 5 and Figure 6 while the schematic diagram for the four electrode system is given in Figure 7. Hence due to its dissimilarity of ERV, the previous established reference charts and tables related to the resistivity value of earth materials were commonly presented in a certain range of values.

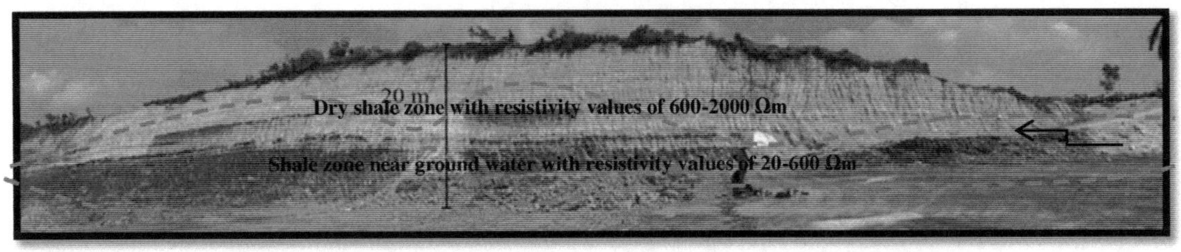

Figure 2. The acidic barren slope studied at Ayer Hitam.

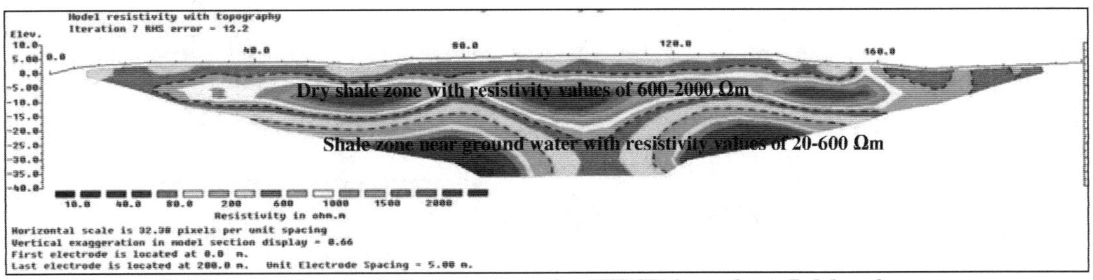

Figure 3. Electrical resistivity tomography (ERT) based on Schlumberger array.

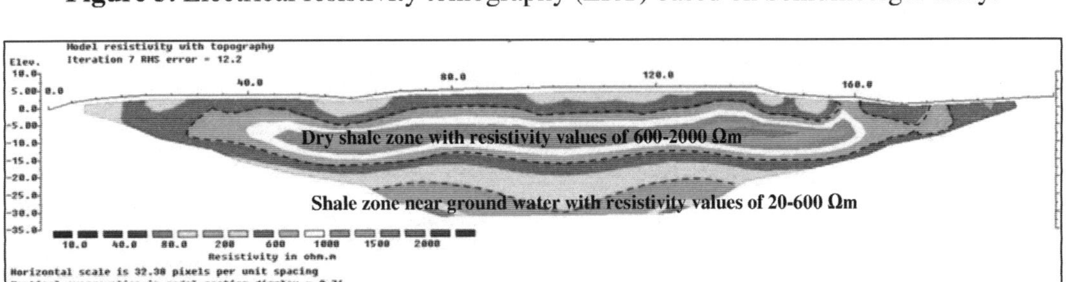

Figure 4. Electrical resistivity tomography (ERT) based on Wenner array.

Table 1. Resistivity data at central of resistivity line (100m from the initial point).

Depth (m)	Resistivity (Ωm)	
	Schlumberger	Wenner
0.94	352.11	343.885
2.91	428.485	390.075
5.07	751.825	546.175
7.45	1441.65	804.30
10.07	3037.35	1303.20
12.96	2705.15	1129.15
16.13	1679.70	748.26
19.62	980.41	471.365
23.45	624.355	312.40
27.67	422.47	227.25
32.31	344.05	186.535
37.42	368.595	

$$\rho_a = 2\pi a * R \tag{1}$$

where R is a resistance term given by R=ΔV/I

$$\rho_a = ((2\pi\Delta V)/I)) * ((1/(1/r1-1/r2)-(1/r3-1/r4)) \tag{2}$$

where r1 = (L – x), r2 = (L + x), r3 = (L – x) and r4 = (L + x) - 1

$$\rho_a = K*(R) \tag{3}$$

where R is a resistance term given by R=ΔV/I, K is geometry factor based on pole-dipole electrode configuration

$$\rho_a = ((2\pi\Delta V)/I))*((1/(1/r1-1/r2)-(1/r3-1/r4)) \tag{4}$$

where $K = ((1/(1/r1-1/r2)-(1/r3-1/r4))$

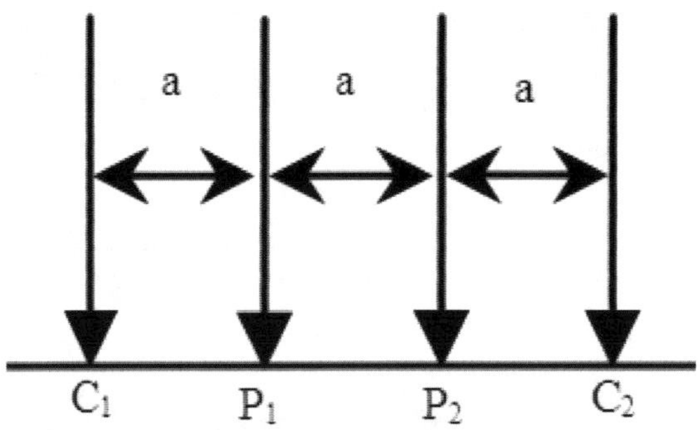

Figure 6. Wenner electrode array arrangement.

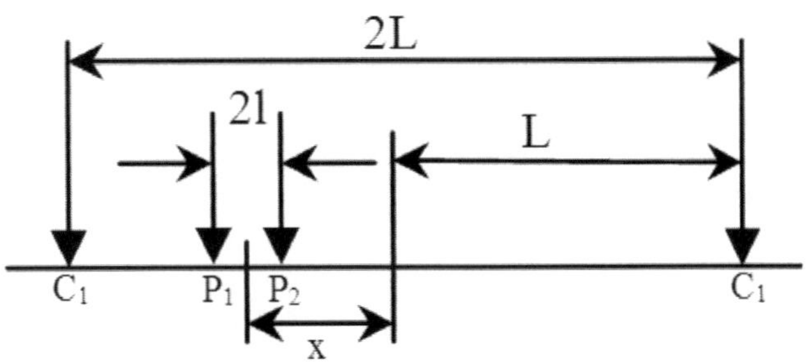

Figure 7. Schlumberger electrode array arrangement.

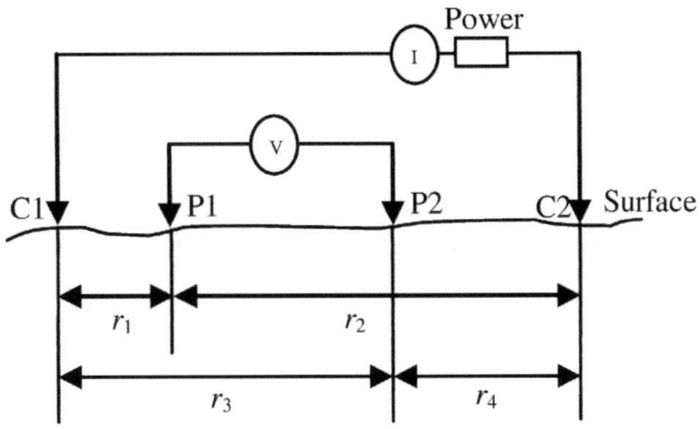

Figure 8. Four-electrode arrangement on the surface of homogeneous isotropic ground of resistivity.

A 2-D resistivity model was compared with the geological map as shown in Figure 1 to display the similarities between both findings. Overall lithology comparison explains that the study area consists of sedimentary rock with particular reference to shale situated at the bottom of the slope structure. According to [37], shale has a resistivity value from 5 – 1000 Ωm. Hence, both ERT from Figure 3 and Figure 4 have reached some good agreement based on physical mapping, geological map and previous reference of ERV. Moreover, it is widely known that the composition of shale consists of a large portion of silica (SiO2). Hence, this study performed an X-Ray Fluorescence (XRF) test which found that that the minerals dominant in Ayer Hitam samples is silica with a concentration of 70.6 % as shown in Table 2.

Table 2. X-Ray Fluorescence data.

Formula	Concentration (%)
CO_2	0.10
SiO_2	70.60
Al_2O_3	21.00
K_2O	4.75
SO_3	1.08
Fe_2O_3	1.01
TiO_2	0.67
Na_2O_3	0.26
MgO	0.20
Zr	0 < LLD

4. Conclusion

The problematic subsurface profile with particular reference to acidic barren slope was successfully investigated using electrical resistivity imaging (ERI). The electrical resistivity value was largely influenced by the variations of sedimentary earth materials beneath the ground surface. The result obtained from the study area consists of two zones, which are residual soil zone with a resistivity of 10-600 Ωm and shale zone with a resistivity of 20-2000 Ωm. This finding has proved that this method is able to predict the slope condition in order to complement conventional borehole data. ERI successfully mapped the acidic barren slope and surface information was extended using the physical mapping technique. The mechanics and physical characteristics of the slope can be easily recognized. The determination of shape, depth and geomaterial stiffness of the problematic slope is relatively easier and cheaper than the conventional borehole method. This geophysical method is suitable for our sustainable ground investigation since it can save time, money and complement other conventional methods especially using its 2D surface technique of investigation. To conclude, the application of ERI in conjunction with geological and borehole information was effectively applied for acidic barren slope assessment.

Acknowledgment

The authors would like to express their deepest appreciation to the Ministry of Higher Education and Universiti Tun Hussein Onn Malaysia for supporting this research under the Multi-disciplinary Research Grant (MDR) Vot. 1320, Incentive Grant Scheme for Publication (IGSP) Vot. U258 and Exploratory Research Grant Scheme (ERGS) Vot. E040.

References

[1] Khatri R, Shrivastava V K and Chandak R 2011 Correlation between vertical electric sounding and conventional methods of geotechnical site investigation. *Int. Journal of Advanced Engineering Sciences and Technologies*. **4** 042-053

[2] Zainal Abidin M H, Baharuddin, M F T, Zawawi M H, Ali N, Madun and Azhar A T S 2015 Groundwater Seepage Mapping using Electrical Resistivity Imaging. *Applied Mechanics and Materials*. **774** 1524–1534

[3] Osman F I S and Osman, S B A S 2012 Integrating geo-electrical and geotechnical data for soil characterization. *International Journal of Applied Physics and Mathematics*. **2** 104-106

[4] Hazreek Z A M, Aziman M, Azhar A T S, Chitral W D, Fauziah A and Saad R 2015 The Behaviour of Laboratory Soil Electrical Resistivity Value under Basic Soil Properties Influences. *IOP Conference Series: Earth and Environmental Science*. **23** 1-9

[5] Abidin M H Z, Ahmad F, Wijeyesekera D C, Saad R and Baharuddin M F T 2013 Soil Resistivity Measurements to Predict Moisture Content and Density in Loose and Dense Soil. *Applied Mechanics and Materials*. **353-356** 911-917

[6] Godio A, Strobbia C and De Bacco G 2006 Geophysical Characterisation of a Rockslide in an Alpine Region. *Engineering Geology*. **83** 273-286

[7] Mauritsch H J, Seiberl W, Arndt R, Römer A, Schneiderbauer K and Sendlhofer G P 2000 Geophysical Investigations of Large Landslides in the Carnic Region of Southern Austria. *Engineering Geology*. **56** 373-388

[8] Liu C and Evett J B 2008 *Soils and Foundation* (New Jersey: Pearson International)

[9] Cosenza P, Marmet E, Rejiba F, Jun Cui Y, Tabbagh A and Charlery Y 2006 Correlations between geotechnical and electrical data: A case study at Garchy in France. *Journal of Applied Geophysics*. **60** 165-178

[10] Hazreek Z A M, Rosli S, Chitral W D, Fauziah A, Azhar A T S, Aziman M and Ismail B 2015 Soil Identification using Field Electrical Resistivity Method. *Journal of Physics: Conference Series*. **622** 1-7

[11] Abidin M H Z, Aziman M, Azhar A T S, Chitral W D, Fauziah A and Saad R 2015 The Influence of Basic Physical Properties of Soil on its Electrical Resistivity Value under Loose and Dense Condition. *Journal of Physics: ConferenceSeries*. **495** 1-13

[12] Abidin M H Z, Saad R, Ahmad F, Wijeyesekera D C and Baharuddin M F T 2012 Seismic Refraction Investigation on Near Surface Landslides at the Kundasang area in Sabah, Malaysia. *Procedia Engineering*. **50** 516 – 531

[13] Abdullahi N And Osazuwa I 2011 Geophysical imaging of municipal solid waste contaminant pathways. *Environmental Earth Sciences*. **62** 1173-1181

[14] Frid V, Liskevich G, Doudkinski, D and Korostishevsky N 2008 Evaluation of landfill disposal boundary by means of electrical resistivity imaging. *Environmental Geology*. **53** 1503-1508

[15] Abu-Zeid N, Bianchini G, Santarato G and Vaccaro C 2004 Geochemical characterisation and geophysical mapping of Landfill leachates: the Marozzo canal case study (NE Italy). *Environmental Geology*. **45** 439-447.

[16] Saad R, Muztaza, N M and Mohamad E T 2011. The 2D Electrical Resistivity Tomography (ERT) Study for Civil and Geotechnical Engineering Purposes. *Electronic Journal of Geotechnical Engineering*. **16** 1537-1545

[17] Yamakawa Y, Kosugi K. I, Masaoka N, Sumida J, Tani M and Mizuyama T 2011 Combined geophysical methods for detecting soil thickness distribution on a weathered granitic hillslope. *Geomorphology*. **145–146** 56-69

[18] Jusoh Z 2010 Application of 2-D resistivity imaging and seismic refraction technique in subsurface investigation for civil engineering, M.S. thesis, Sch. Physics, Science Univ., Penang, Malaysia

[19] Saad R, Nawawi M N M and Mohamad E T 2012 Groundwater Detection in Alluvium Using 2-D Electrical Resistivity Tomography (ERT). *Electronic Journal of Geotechnical Engineering*. **17** 369-376

[20] Juanah M E, Ibrahim S, Sulaiman W and Latif P 2012 Groundwater resources assessment using integrated geophysical techniques in the southwestern region of Peninsular Malaysia. *Arabian Journal of Geosciences*. 1-16

[21] Zawawi M H, Syafalni and Abustan I 2011 Detection of Groundwater Aquifer Using Resistivity Imaging Profiling at Beriah Landfill Site, Perak, Malaysia. *Advanced Materials Research*. **250-253** 1852-1855

[22] Saad R, Adli I and Mohamad A S 2012 The Study of Iron Ore Prospect using 2-D Resistivity and Induced Polarization (IP) Method. *Electronic Journal of Geotechnical Engineering*. **17** 2981-2988

[23] Karimi N S, Hojat A, Kamkar-Rouhani A, Akbari J H and Maknooni S 2011 Successful Use of Geoelectrical Surveys in Area 3 of the Gol-e-Gohar Iron Ore Mine, Iran. *Mine Water and the Environment*. **30** 208-215

[24] Saad R, Saidin M M, Muztaza N M, Ismail N A and Ismail, N E H 2011 Subsurface Study Using 2-D Resistivity Imaging Method for Meteorite Impact at Bukit Bunuh, Perak. *Electronic Journal of Geotechnical Engineering*. **16** 1507-1513

[25] Batayneh A 2010 The use of magnetometry and pole-dipole resistivity for locating Nabataean Hawar archeological site in the SW-Jordan. *Archaeological and Anthropological Sciences*. **2** 151-156

[26] Fabrizio T G 2007 Identifying geological and geotechnical influences that threaten historical sites: A method to evaluate the usefulness of data already available. *Journal of Cultural Heritage*. **9** 302-310

[27] Burger H R, Sheehan A F and Jones C H 2006 *Introduction to Applied Geophysics* (New York: W.W. Norton & Company)

[28] Mineral and Geoscience Department Malaysia, *Geological Map of Peninsular Malaysia* 1985, eighth edn., Ministry of Natural Resources and Environment

[29] Hamzah U, Yaacup R, Samsudin A and Ayub M 2006 Electrical imaging of the groundwater aquifer at Banting, Selangor, Malaysia. *Environmental Geology*. **49** 1156-1162

[30] Loke M H, Acworth I, and Dahlin T 2003 A comparison of smooth and blocky inversion methods 2-D electrical imaging surveys. *Exploration Geophysics*. **34(3)** 182–187

[31] Loke M H and Barker R D 1996 Rapid least squares inversion of apparent resistivity pseudosection using a quasi-Newton method. *Geophysical Prospecting*. **44(3)** 131–152

[32] Hazreek Z A M, Saiful A T S, Aziman M, Rosli S, Wijeyesekera D C and Fauziah A "Integral analysis of laboratory and field electrical resistivity for soil density prediction." *Proc. The 3rd Int. Conf. on Applied Geophysics (Hong Kong 2015)*

[33] Abidin M H Z, Saad R Wijeyesekera D C, Ahmad F and Ismail N A 2014 The Influence of Electrical Resistivity Array on its Soil Electrical Resistivity Value. *Applied Mechanics and Materials*. **510** 185-192

[34] Abidin M H Z, Saad R, Wijeyesekera D C and Ahmad F, 2013 Soil Resistivity Influence due to the Different Utilization of Electrical Resistivity Array. *Electronic Journal of Geotechnical Engineering*. **18** 5643–5654

[35] Abidin M H Z, Saad R, Wijeyesekera D C and Ahmad F "Integral Analysis of Laboratory and Field Electrical Resistivity Value for Soil Moisture Content Prediction," *Proc. The 3rd Int. Conf. on Applied Geophysics (Sarawak, Kuching Malaysia. 2013)*

[36] Telford W M, Geldart L P and Sheriff R E 1990 *Applied Geophysics* (Cambridge: Cambridge University Press)

[37] Griffith D H and King R F 1981 *Applied Geophysics for Geologist and Engineers* (Oxford: Pergamon Press)

Preparation and Application of Chitosan Membranes to Filter Silver from X-ray Film Processing Wastes

N Nyoman Rupiasih[1,2,3,4], Rendra Rustam Purnomo[1] and Made Sumadiyasa[1]
[1] Department of Physics, Faculty of Mathematics and Natural Sciences, Udayana University, Kampus Bukit Jimbaran, Badung, Bali 80362, Indonesia
[2]Group Research Material Sciences and Technology-Polymer and Biomaterial

E-mail: rupiasih@gmail.com

Abstract. Chitosan is a natural polysaccharide biopolymer which has been widely used in different processes and applications. Chitosan based membranes have been used in reverse osmosis, gas separation, dialysis and pervaporation. The object of this research was investigating the possibility of chitosan membrane used as a filter for removing silver (Ag) from X-ray film processing wastes. Several of chitosan membranes such as M1, M2, M3 and M4 have been prepared for the purpose and filtration was done using dead-end filtration method. The filtration experiments were performed on a flat sheet membrane using pure water and X-ray film processing wastes as feeds. The analysis of silver concentration has been done by atomic absorption spectrometers (AAS). The results show that chitosan membrane M2 gave the highest filtration coefficient (R_{coeff}) i.e. 99.9%, with the pure water flux (*PWF*) and product flux (*PF*) are 2972.56 L/m^{2}h and 1761.18 L/m^{2}h respectively. The rejection coefficient of the membranes decreases with increasing the amount of chitosan, while the pure water flux and product flux are increased. The filtration coefficients show that the chitosan membranes are able to filter silver waste from X-ray film processing wastes with performance dependent on their characteristic such as pores size. This suggests that, chitosan membrane can be used as one method that is safe and friendly environment for recovering silver from X-ray film processing waste to improve the quality of treated to an acceptable quality level.

Keywords: Chitosan membrane, X-ray film processing wastes, silver recovery, dead-end filtration

[3]Author to whom any correspondence should be addressed: rupiasih@gmail.com; [4] Grant: RM Udayana University No: 104.41/UN14.2/PNL.01.03.00/2014 date 3 Maret 2014

1. Introduction

Silver halides known as its' photosensitivity, approximately 8.3% of silver is used in photography [1-2]. In the X-ray film processing involve development and fixing of film, some silver-halide crystals in the film that are not exposed to light are removed by thiosulphate from the film into the processing solution [3]. This cause the film processing wastes (effluents) may contain high levels of silver. That silver is primarily present as soluble silver thiosulfate complex, and some smaller amounts present as silver sulfide. As an illustration, waste fixer and rinse solutions commonly contain silver at concentrations of 5 ppm or more causing them to be toxic hazardous wastes [2]. Therefore, X-ray film processing wastes, spent rinse waters, films and negatives may contain high silver concentrations and require management as hazardous wastes [4].

In Indonesia, there are over 2434 public and private hospitals and healthcare centres and, until July 2015, about 5981 radiodiagnostic X-ray machines spread around the country [5]. From these data, it can be estimated that large amount of scrap films and X-ray film processing wastes discarded. Therefore, treatment of these wastes for recovery of silver provides significant economic and environmental benefits.

There are varieties of recovery methods available for silver from film processing wastes. The most common methods are metal replacement, electrolytic recovery, and chemical precipitation [4]. For low-silver wash waters, ion exchange is used. Reverse osmosis, distillation and evaporation are used to concentrate silver in dilute solutions. Every method has their benefit involve efficiency as well as costs.

Recently, chitosan has great attention in many research fields. Chitosan is a biopolymer that contained linear aminopolysaccharide of glucosamine and *N*-acetylglucosamine units [6]. It is obtained by alkaline deacetylation of chitin which extracted from the exoskeleton of crustaceans such as shrimps and crabs, as well from the cell walls of some fungi [7]. Chitosan is also characterized by being cationic, biocompatible, and antibacterial. It can be processed into several products including flakes, fine powders, beads, membranes, sponges, cottons, fibers and gels [6]. These characteristics make chitosan has considerable application in various industrial areas [8-11], such as pharmaceutical and biomedical engineering, food industry, agriculture and wastewater treatment [12]. Such applications involve as drug carriers, wound-healing agents, hemodialysis membranes and surgical dressing materials, film for food packaging, coating of fertilizers and pesticides for their controlled release to soil. Chitosan membranes have been used for active transport of chloride ions in aqueous solution [12].

With the rapid growth of demand in industries and daily life, water has become an increasingly valuable but scarcer resource for human being. Increasing global demand for clean water and increasing environmental concerns make membrane filtration become the technology of choice for industries seeking to reuse their wastewater and reduce their water footprint as well as for potable water production. Membrane filtration is a separation technique that produces a purified water stream (permeate) and a concentrated stream that contains all the separated pollutants. Membrane processes are increasingly used for removal of bacteria, microorganisms, particulates, and natural organic material, which can impart color, tastes, and odors to water and react with disinfectants to form disinfection byproducts. By physically removing the pathogens, membrane filtration can significantly reduce chemical addition, such as chlorination. Disinfection can be performed without chemicals.

In the present study, the recovery of silver from X-ray film processing wastes by using chitosan membranes was investigated. To achieve the purpose, we have used dead-end filtration method by using chitosan membranes as a filter. Further, the influence of various ratios of components of matrix/solvent in membranes preparation also examined. The concentration of silver in the feed and permeate solutions were analyzed using AAS. The feed solution is the wastes of X-ray film processing before filtration. Permeate solution is solution that obtained after filtration. The filtration performance of the membranes was evaluated in term of pure water flux (*PWF*), product flux (*PF*) and rejection value or rejection coefficient (R_{coeff}).

2. Materials and methods

2.1 Materials

The materials used are chitosan powder with characteristics i.e. 87.9% of degree of deacetylation, viscosity of 663,000 cps, solubility in acetic acid is 99.4%, and average molecular weight of 900,000. Acetic acid and sodium hydroxide p.a. were analytical grade and were used without further purification. X-ray film processing wastes was obtained from "Sanglah General Hospital Denpasar", one of the biggest state hospitals in Bali, which was directly used.

2.2 Chitosan membranes preparation

Four sets of chitosan membranes e.g. M1, M2, M3 and M4, have prepared by casting solution method. Those sets obtained by varying the ratios of components of matrix/solvent such as 1%, 2%, 3% and 4% respectively. The preparation involved of dissolving, for example 1% wt chitosan in acetic acid 1% (v/v) solution by stirring at room temperature (30 ^{o}C) for 8 h. The casting solution was filtered to remove not-dissolve chitosan and debris, and was kept for 30 minutes for complete degassing of bubbles. Further, it was cast on a glass plate sized of 18.5 cm x 24.0 cm. At this condition, the membrane was allowed to dry at room temperature for 7 d. The dried membrane was immersed in 1 M NaOH for 12 minutes at room temperature. The membrane was then washed thoroughly using distilled water to remove the excess NaOH and dried at room temperature again. Finally, it obtained dry chitosan membrane M1, which are ready to use as filter. Another three sets of chitosan membranes M2, M3 and M4, were prepared by the same procedure. The average pores sizes of those membranes are 19.13 Å, 25.47 Å, 34.00 Å, and 37.09 Å, respectively.

2.3 Experimental set-up

All experiments were conducted in a dead-end filtration set up. The details of the experimental set up are available elsewhere [13]. The experiments were performed at 50-60 kPa pressure using distilled water and X-ray film processing wastes as feed. The X-ray film processing wastes has been directly used without any treatment. The feed and permeate solutions were analyzed using AAS Varian, SpectrAA-30. The filtration performance of the membranes was evaluated in term of pure water flux (*PWF*), product flux (*PF*) and rejection coefficient (R_{coeff}). Permeate flux is the filtration rate or flow rate through the membrane $\left(\frac{\Delta V}{\Delta t}\right)$ per unit effective surface area of the membrane (A) as expressed by equation 2.1 [13-14].

$$J = \frac{\Delta V}{A\Delta t} \tag{2.1}$$

Rejection coefficient of the membrane is the membrane's ability to hold back or reject a given molecule, in this regard silver wastes, from passing through [12, 13]. The rejection coefficients (R_{coeff}) were calculated by [13-14],

$$R_{coeff} = \left(1 - \frac{C_p}{C_f}\right) x100\% \tag{2.2}$$

where C_p and C_f is concentration of silver in permeate and feed solutions respectively.

3. Results and discussion

3.1 Pure water flux (PWF) and product flux (PF)

Figure 3.1 show the pure water flow rate (volume of water passing through the membrane per unit time) of membranes M2, M3 and M4, at pressure 50-60 kPa. Figure show that membrane M1 does not show the flow rate that means chitosan membrane M1 cannot pass by the distilled water at operating pressure 50-60 kPa. The gradient or slopes of curves are increased with increasing the amount of matrix constituent (chitosan) in membrane preparation i.e. M2, M3 and M4. These suggest that the flow rate of the pure water that passes through the membrane at operating pressure of 50-60 kPa is increased with increasing the amount of chitosan from M2 to M4.

Using equation (2.1), where $\frac{\Delta V}{\Delta t}$ is the pure water flow rate that is equal to the gradient of curve in Fig. 3.1, the effective membrane area (*A*) used in the filtration experiments is 0.000984 m^2, the pure water flux (*PWF*) can be calculated. The results showed in Table 3.1. It shows that the pure water flux increased with increasing the amount of chitosan from M2 to M4. These results are consistent with the pores size of the chitosan membranes which are increased with increasing the amount of chitosan in membrane preparation.

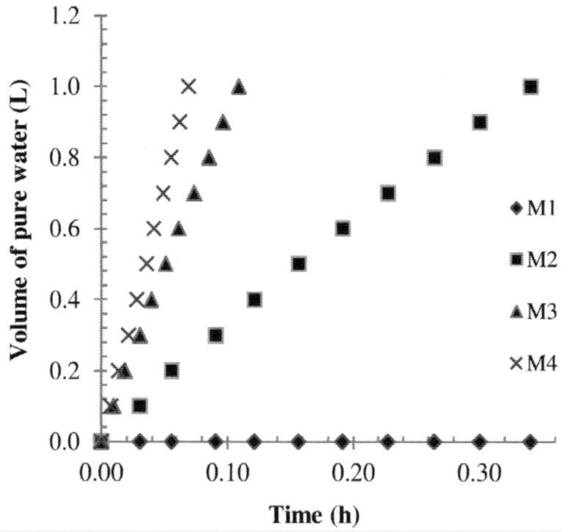

Figure 3.1 The pure water flow rate (volume of pure water passing through the membrane per unit time) of membranes M1, M2, M3 and M4 at pressure 50-60 kPa

Table 3.1 Pure water flux (*PWF*) of chitosan membranes M1, M2, M3 and M4 at operating pressure of 50-60 kPa

Membranes	Gradient (L/h)	*A* (m^2)	*PWF* ($L/m^2.h$)
M1	0	0.000984	0
M2	2.92		2972.56
M3	9.15		9297.76
M4	14.45		14684.96

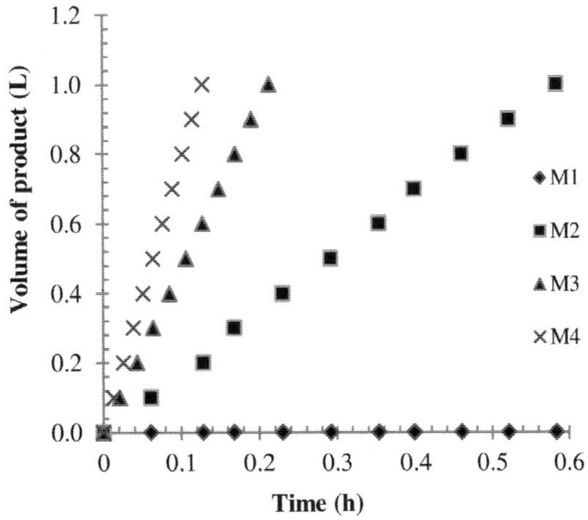

Figure 3.2 The flow rate of product (volume of wastes passing through the membrane per unit time) of membranes M2, M3 and M4 at pressure 50-60 kPa.

Figure 3.2 shows the flow rate of product (volume of X-ray film processing wastes passing through the membrane per unit time) of membranes M1, M2, M3 and M4 at pressure 50-60 kPa. It shows similar pattern with the pure water flux in Fig. 3.1. Figure show that membrane M1 does not show the flow rate of product that means chitosan membrane M1 cannot pass by X-ray film processing wastes at operating pressure 50-60 kPa. The gradient is increased with increasing the amount of matrix (chitosan) from M2 to M4. These suggest that the flow rate of X-ray film processing wastes that passes through the membrane at operating pressure of 50-60 kPa is increased with increasing the amount of chitosan in membranes preparation from M2 to M4.

For the same, using equation (2.1), where $\frac{\Delta V}{\Delta t}$ is product or wastes flow rate, which is equal to the gradient of the curve in Fig. 3.2, the effective membrane area (*A*) used in the filtration experiments is 0.000984 m^2, the product flux of X-ray film processing wastes (*PF*) can be calculated. The results showed in Table 3.2. It shows that the product flux increased with increasing the amount of chitosan from M2 to M4. These results are consistent with the pores size of the chitosan membranes which are increased with increasing the amount of chitosan in membrane preparation.

Table 3.2 Product fluxes of X-ray film processing wastes (*PF*) of chitosan membranes M1, M2, M3 and M4 at operating pressure of 50-60 kPa

Membranes	Gradient (L/h)	*A* (m^2)	*PF* ($L/m^2.h$)
M1	0	0.000984	0
M2	1.73		1761.89
M3	4.70		4778.86
M4	7.82		7950.00

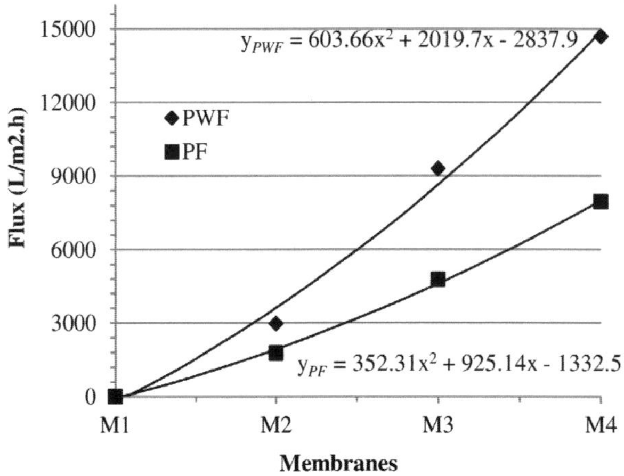

Figure 3.3 Pure water flux (*PWF*) and X-ray film processing wastes flux (*PF*) as a function of chitosan membranes at pressure 50-60 kPa.

Figure 3.3 shows that the flux increases with increasing the amount of matrix (chitosan) making up the membranes. This can be attributed to the pores of the membrane where the pores membrane is getting bigger with increasing the amount of chitosan [14]. The greater the pores size, the easier water pass through the membrane [15]. In laminar flow system, if the membrane consists of straight capillaries, the flux *J* revealed as Hagen-Poiseuille relationship as

$$J = \frac{\varepsilon r^2}{8\eta\tau}\frac{\Delta P}{\Delta x} \tag{3.1}$$

where *r* is the pore radius, Δx is membrane thickness, η is viscosity, ε is porosity, and τ is tortuosity factor (the magnitude is one if the form of the pores are cylindrical). Equation 3.1 shows that flux is proportional to the pressure gradient $\left(\frac{\Delta P}{\Delta x}\right)$ and to the square of the pores radius of the membrane (r^2), which is in accordance with the curve fitting in Fig. 3.3 i.e. the second-order of polynomial (quadratic) with y is the fluxes and x is chitosan membranes M1, M2, M3 and M4. It also observes that the product flux lower than the pure water flux. As revealed by Lojkine 1992, in laminar flow system, the permeate flux (*J*) is formulated by Darcy's law, equation (3.2) [16].

$$J = \frac{\Delta P}{\eta_p R_t} \tag{3.2}$$

Where *J* is the permeate flux, ΔP is Trans membrane pressure (*TMP*), η_p is permeate dynamic viscosity, and R_t is total filtration resistance. R_t is the sum of the resistance of membrane in a clean state (R_m) and in precipitation (fouling) condition (R_f), that mathematically is written as

$$R_t = R_m + R_f \tag{3.3}$$

The physical properties of the feed solution are presented by its' viscosity (η_p) and fouling condition (R_f). From both equations (3.2) and (3.3), it shows that the flux also influenced by the viscosity and total filtration resistance due to the precipitation, which is inversely. Thus, the greater the viscosity and precipitation will give smaller flux. X-ray film processing wastes is comprised of a number of molecules that caused a greater viscosity than the pure water, that allow for the precipitation. As result the flux of X-ray film processing wastes becomes smaller.

3.2 Rejection study

The concentration of silver in the feed and permeate solutions were analyzed using AAS. The results show in Table 3.3. The rejection coefficient (R_{coeff}) of chitosan membranes have been calculated using equation (2.2). The values obtained also were shown in Table 3.3. It shows that membrane M2 gave the highest rejection coefficient compare with M3 and M4. The rejection coefficients decreased as increasing the amount of matrix (chitosan) in making the membranes from M2 to M4. With these results described that the membrane M2 shows the highest ability to hold back or reject silver wastes from passing through compared with M3 and M4. The rejection coefficient is equal to 100% means all the wastes got filtered by the membrane. In sub 2.3 above, it was mentioned that X-ray film processing wastes has been directly used without any treatment. This is a highlight that, membrane filtration using chitosan membrane is one method which is safe and friendly environment for recovering silver from X-ray film processing waste.

Table 3.3 Concentration of silver (Ag) in the feed and permeate solutions, and the rejection coefficients (R_{coeff}) of chitosan membranes

Membranes	Concentration of silver (mg/L)		R_{coeff} (%)
	feed	569	
M2	permeate	0.55	99.90
M3		107	81.20
M4		121	78.73

4. Conclusion

From the experimental data, the study has demonstrated that chitosan membranes prepared by casting method are able to filter silver (Ag) from X-ray film processing wastes. The filtration performance of the membranes was dependent on their characteristics such as the pores size. It suggests that, chitosan membranes can be used as one of method that is safe and friendly environment to recover of silver waste from X-ray film processing wastes to improve the quality of treated to an acceptable quality level.

Acknowledgments

This work was financially supported by the Ministry of National Education, Republic of Indonesia, under Fundamental Research Grant of Udayana University No: 023.04.2.415253/2014, 5 December 2013, MAK 2013.109.011.521119, Year Budget 2014 are highly acknowledged.

References:

[1] Bas A D, Yazici E Y and Deveci H 2012 *Hydrometallurgy* **121–124** pp 22–27
[2] MPCA 2011 *Waste/Hazardous waste* #4.46 (w-hw4-46)
[3] Bober T W, Vacco D, Dagon T J and Fowler H E 2006 *Treatment of photographic wastes* In: Wang L K, Hung Y T, Lo H H, Yapijakis C (Eds.) *Hazardous Industrial Waste Treatment* (CRC Press) pp 361–408
[4] U S EPA 1999 RCRA in Focus: Photo Processing http://www.epa.gov/osw/inforesources/pubs/infocus/photofin.pdf
[5] Nuclear Energy Regulatory Agency of Indonesia (BAPETEN) 2015
[6] Mohamed E I Badawy and Entsar I Rabea 2011 *International Journal of Carbohydrate Chemistry* 460381 p 29
[7] No H K and Meyers S P 1997 *Preparation of chitin and chitosan* in *Chitin Handbook*

Muzzarelli R A A and Peter M G (Eds.) (European Chitin Society, Grottammare, Italy) pp 475–489
[8] Rabea E I, Badawy M E T, Stevens C V, Smagghe G and Steurbaut W 2003 *Biomacromolecules* **4** (6) pp 1457–1465
[9] Hirano S 1996 *Biotechnology Annual Review* **2** pp 237–258
[10] Shigemasa Y and Minami S 1996 *Biotechnology and Genetic Engineering Reviews* **13** pp 413–420
[11] Shahidi F, Arachchi J K V and Jeon Y J 1999 *Trends in Food Science and Technology* **10** (2) pp 37–51
[12] Rupiasih N Nyoman, Yayuk Eka Puspita and Sumadiyasa Made 2015 *J. Phys.: Conf. Series* **622** 012004
[13] Rupiasih Ni Nyoman, Sumadiyasa Made, Ratnawati A A and Hery Suyanto 2011 *Indonesian Jourmal of Materials Scince* **13** (1) pp 54-58
[14] Swapna Rekha Panda, Munmun Mukherjee and Sirshendu De 2015 *Journal of Water Process Engineering* **6** pp 93–104
[15] Mulder M 1996 *Basic Prinsiples of Membrane Tecnology* (Netherlands: Kluwer Academic)
[16] Evenblij, Herman 2006 *Filtration Characteristcs in Membran Bioreactor* (Printed by: Pasmans Offsetdrukkerij Gravenhage)

Analysis of Realized Volatility for Nikkei Stock Average on the Tokyo Stock Exchange

Tetsuya Takaishi[1] **and Toshiaki Watanabe**[2]
[1]Hiroshima University of Economics, Hiroshima 731-0192, Japan
[2]Institute of Economic Research, Hitotsubashi University, Tokyo 186-8603, Japan
E-mail: `tt-taka@hue.ac.jp`

Abstract. We calculate realized volatility of the Nikkei Stock Average (Nikkei225) Index on the Tokyo Stock Exchange and investigate the return dynamics. To avoid the bias on the realized volatility from the non-trading hours issue we calculate realized volatility separately in the two trading sessions, i.e. morning and afternoon, of the Tokyo Stock Exchange and find that the microstructure noise decreases the realized volatility at small sampling frequency. Using realized volatility as a proxy of the integrated volatility we standardize returns in the morning and afternoon sessions and investigate the normality of the standardized returns by calculating variance, kurtosis and 6th moment. We find that variance, kurtosis and 6th moment are consistent with those of the standard normal distribution, which indicates that the return dynamics of the Nikkei Stock Average are well described by a Gaussian random process with time-varying volatility.

1. Introduction

Statistical properties of asset returns have been extensively studied and it is found that asset price returns show some universal properties that are not explained well in the framework of the standard Brownian motion. The universal properties are now classified as the stylized facts of asset price returns which includes: fat-tailed return distributions, volatility clustering, long autocorrelation time in absolute returns and so on[1]. To explain the fat-tailed return distribution Mandelbrot introduced a class of stable processes such as stable Paretian[2]. An alternative idea to explain the asset price dynamics was given by Clark who related the volatility variation to volume and suggested to use the subordinated process to the asset price dynamics[3]. The idea for the asset price dynamics by Clark is also called the mixture of distributions hypothesis (MDH). The return process with the MDH does not conflict with many properties observed in asset returns, e.g. volatility clustering, fat-tailed return distributions. Under the MDH, the asset return at discrete time t can be described by $r_t = \sigma_t \epsilon_t$, where σ_t^2 is a variance of the Gaussian distribution and ϵ_t is a standard normal random variable, and this indicates that the asset return process is viewed as a Gaussian random process with time-varying variance (volatility). Let $p(r|\sigma_t^2)$ be the conditional return distribution with σ_t^2 and $p(\sigma_t^2)$ the probability distribution of volatility. The unconditional return distribution $p(r)$ from this process is obtained by integrating the conditional return distribution and the probability distribution of volatility with respect to volatility, i.e. $p(r) = \int p(r|\sigma_t^2)p(\sigma_t^2)d\sigma_t^2$. Empirical studies suggested that the volatility distributions might be described by the inverse gamma distribution or log-normal distribution[3, 4, 5, 6] with which the unconditional return distributions become fat-tailed distributions.

The verification of the MDH can be made by testing the returns standardized by volatility, r_t/σ_t. Under the MDH, the standardized returns should behave as $r_t/\sigma_t \sim \epsilon_t$, i.e. standard normal random variables. This test has been conducted in the literature[8, 9, 10, 11, 12, 13, 14, 15] and it is shown that the MDH is hold for many cases. For this test it is crucial to use a precise measure of volatility. So far the most precise measure is the realized volatility[7] constructed from high-frequency price data. Under ideal circumstances the realized volatility goes to the integrated volatility in the limit of the infinite sampling frequency. However such circumstances are usually violated and the realized volatilities in empirical cases are biased. There exist two main sources of bias: microstructure noise and non-trading hours. Thus to verify the standard normality of the returns standardized by realized volatility it is important to control such biases. In [14], in order to avoid bias from non-trading hours the MDH is tested separately in morning and afternoon sessions for individual Japanese stocks and it is shown that after removing the finite-sample effect[17] the return dynamics becomes consistent with the MDH[15]. In this paper we focus on the realized volatility of the Nikkei Stock Average index and investigate whether the MDH can also apply for the price dynamics of the Nikkei Stock Average index.

2. Realized Volatility

We assume that the logarithmic price process $\ln p(s)$ follows a continuous time stochastic diffusion,

$$d \ln p(s) = \tilde{\sigma}(s) dW(s), \tag{1}$$

where $W(s)$ stands for a standard Brownian motion and $\tilde{\sigma}(s)$ is a spot volatility at time s. The assumption of a Brownian motion for the logarithmic price process is often used to model stock prices as in the Black-Scholes model[18, 19] and is most widely used to investigate stock price process. Using $\tilde{\sigma}(s)$, the integrated volatility from t to $t+h$ is defined by

$$IV_h(t) = \sigma_h^2(t) = \int_t^{t+h} \tilde{\sigma}(s)^2 ds, \tag{2}$$

where h stands for the interval to be integrated, e.g. $h = 1$ day for the daily integrated volatility.

The realized volatility is designed as a model-free estimate of volatility and constructed as a sum of squared returns[7]. The realized volatility RV_t at time t is given by

$$RV_t = \sum_{i=1}^{n} r_{t+i\Delta}^2, \tag{3}$$

where n is the number of returns at sampling frequency Δ, given by $n = h/\Delta$. The returns sampled at Δ are given by log-price difference,

$$r_{t+i\Delta} = \ln P_{t+i\Delta} - \ln P_{t+(i-1)\Delta}, \tag{4}$$

for $i = 1, \ldots, n$. Eq.(3) goes to the integrated volatility defined by eq.(2) in the limit of $n \to \infty$. However what we observe in the real financial markets is not the returns given by eq.(4). The asset prices observed in the real financial markets are contaminated with the microstructure noise originating from discrete trading, bid-ask spread and so on. Following Zhou[20] let us assume that the microstructure noise introduces independent noises and the log-price observed in financial markets is given by

$$\ln P_t^* = \ln P_t + \xi_t, \tag{5}$$

where $\ln P_t^*$ is the observed log-price in the markets which consists of the true log-price $\ln P_t$ and noise $\xi_t \sim N(0, \omega^2)$. Under this assumption the observed return r_t^* is given by

$$r_t^* = r_t + \eta_t, \tag{6}$$

where $\eta_t = \xi_t - \xi_{t-\Delta}$. The realized volatility RV_t^* actually observed at the financial market is obtained as a sum of the squared returns r_t^*,

$$RV_t^* = \sum_{i=1}^{n} (r_{t+i\Delta}^*)^2, \tag{7}$$

$$= RV_t + 2\sum_{i=1}^{n} r_{t+i\Delta}\eta_{t+i\Delta} + \sum_{i=1}^{n} \eta_{t+i\Delta}^2. \tag{8}$$

After averaging RV_t^*, we find that the bias in RV_t^* appears as $\sum_{i=1}^{n} \eta_{t+i\Delta}^2$ which corresponds to $\sim 2n\omega^2$. Thus with independent noise the RV_t^* diverges as $n \to \infty$. Another bias appeared in the realized volatility is due to the existence of the non-trading hours on the real financial markets. At the Tokyo Stock Exchange domestic stocks are traded in the two trading sessions: (1) morning trading session (MS) from 9:00 to 11:00. (2) afternoon trading session (AS) from 12:30 to 15:00. If we calculate the daily realized volatility without including returns during the non-traded periods it can be underestimated.

In order to avoid the non-trading hours issue we consider two realized volatilities:(i) RV_{MS}, realized volatility in the morning session and (ii) RV_{AS}, realized volatility in the afternoon session. Since these realized volatilities are calculated separately each volatility does not have bias due to the non-trading hours issue.

3. Normality of the Standardized Return

In this study we analyze the high-frequency data of Nikkei Stock Average (Nikkei225) index from May 1, 2006 to December 30, 2009, that corresponds to 900 working days. Figure 1 shows return time series of the Nikkei Stock Average in the different time zones of the Tokyo Stock Exchange.

In each zone the return is calculated by the log-price difference between the beginning and end prices in the corresponding zone. For instance let $R_{MS,t}$ be the return in the morning session at day t. $R_{MS,t}$ is given by $\ln P_{MS,t}^o - \ln P_{MS,t}^c$, where $P_{MS,t}^o$ is the opening price of the morning session and $P_{MS,t}^c$ the closing price of the morning session. Similarly the return in the afternoon session $R_{AS,t}$ is given by $\ln P_{AS,t}^o - \ln P_{AS,t}^c$. Moreover the return in the lunch break (LB), $R_{LB,t}$ is given by $\ln P_{MS,t}^c - \ln P_{AS,t}^o$ and the return in the overnight break (ON), $R_{ON,t}$ is given by $\ln P_{MS,t}^o - \ln P_{AS,t-1}^c$. As seen in the figure returns in the lunch break and overnight break also vary. Variation of returns in the lunch break is small. This is because the lunch break duration takes only 90min. On the other hand considerable variation of returns is seen in the overnight break which has 18 hrs duration. Since returns in both breaks can vary the (daily) realized volatility without including the data in both breaks can not be accurate.

In order to avoid the non-trading hours issue we calculate two realized volatilities in the morning and afternoon session separately. Let $RV_{MS,t}(RV_{AS,t})$ be the realized volatility in the morning (afternoon) session. Then we analyze standardized returns separately in each session. $RV_{MS,t}$ is given by

$$RV_{MS,t} = \sum_{i=1}^{n} r_{MS,t,\Delta,i}^2, \tag{9}$$

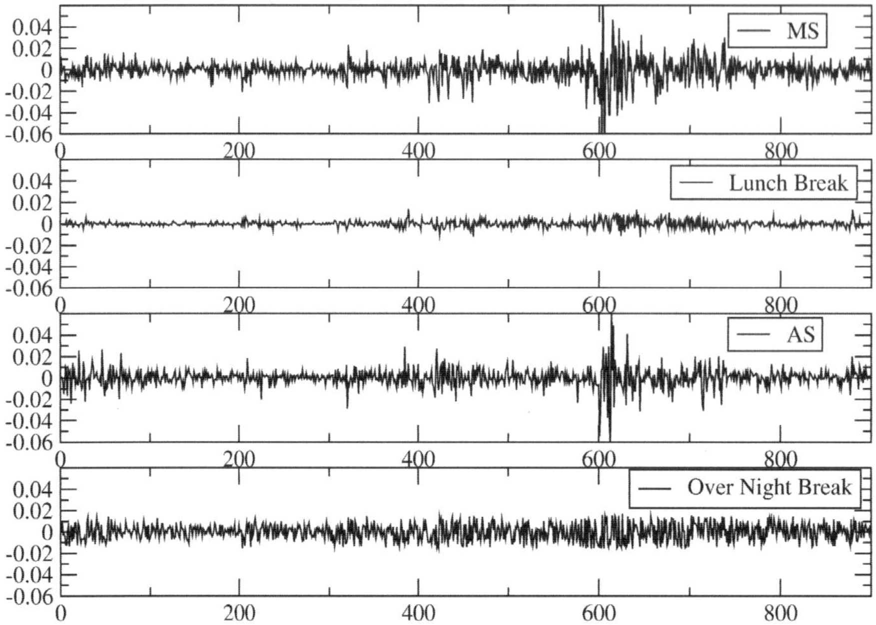

Figure 1. Return time series in different time zones from May 1, 2006 to Dec. 30, 2009.

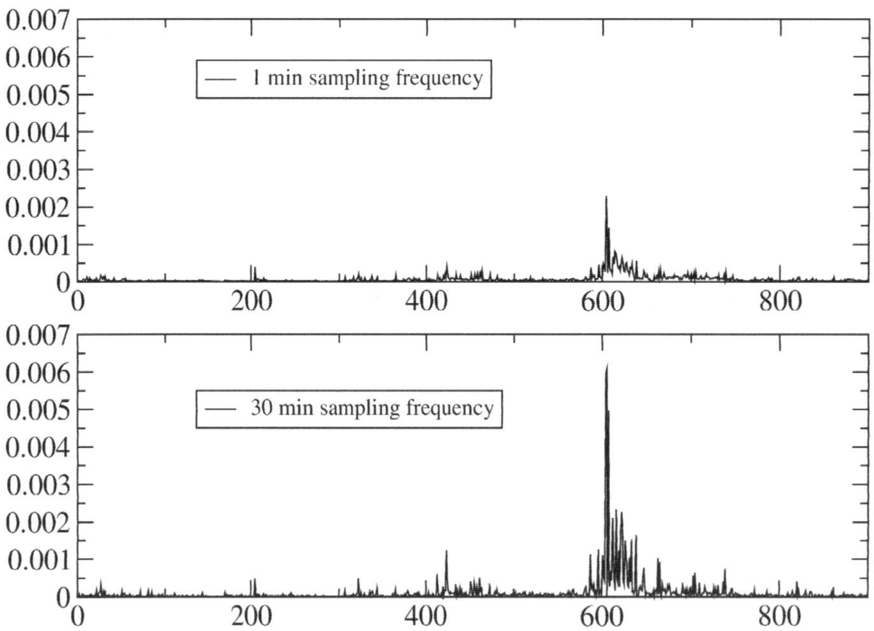

Figure 2. Time series of the realized volatility in the MS from May 1, 2006 to Dec. 30, 2009.

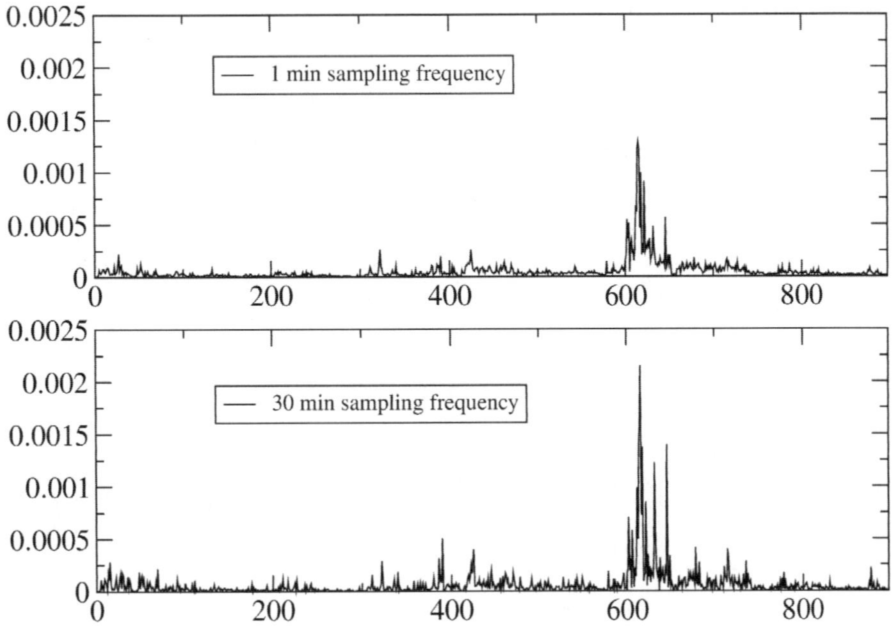

Figure 3. Time series of the realized volatility in the AS from May 1, 2006 to Dec. 30, 2009.

where $r^2_{MS,t,\Delta,i}$ is the i-th intraday return in the MS on day t, sampled at Δ min. Similarly, $RV_{AS,t}$ is given by

$$RV_{AS,t} = \sum_{i=1}^{n} r^2_{AS,t,\Delta,i}. \tag{10}$$

Here n is the number of returns generated at sampling frequency Δ during trading sessions. We calculate the realized volatility for $\Delta = (1, 2, ..., 40)$. Figure 2 and 3 show the realized volatility in each trading session at $\Delta = 1$ and 30 as representative ones.

In figure 4 we show the average of realized volatility as a function of the sampling frequency Δ. Such plot is also called volatility signature plot[16] and visualizes the effect of the microstructure noise. It seems that the average realized volatility decreases with the sampling frequency Δ. This is the opposite result to the expectation of eq.(8) and empirical results of individual Japanese stocks[14] where the realized volatility diverges as the sampling frequency Δ decreases. Thus for the Nikkei Stock Average the assumption of the independent noise of eq.(5) does not apply.

4. Normality of Standardized Returns

According to the MDH, let us assume that $R_{MS,t}$ and $R_{MS,t}$ are described by

$$R_{MS,t} = \sigma_{MS,t}\epsilon_t, \tag{11}$$

$$R_{AS,t} = \sigma_{AS,t}\epsilon'_t, \tag{12}$$

respectively. $\sigma^2_{MS,t}(\sigma^2_{AS,t})$ is an integrated volatility in the morning (afternoon) and ϵ_t and ϵ'_t are independent standard normal random variables $\sim N(0,1)$. Substituting the realized volatility for the integrated volatility we standardize $R_{MS,t}$ and $R_{MS,t}$ as $R_{MS,t}/RV^{1/2}_{MS,t}$ and $R_{AS,t}/RV^{1/2}_{AS,t}$.

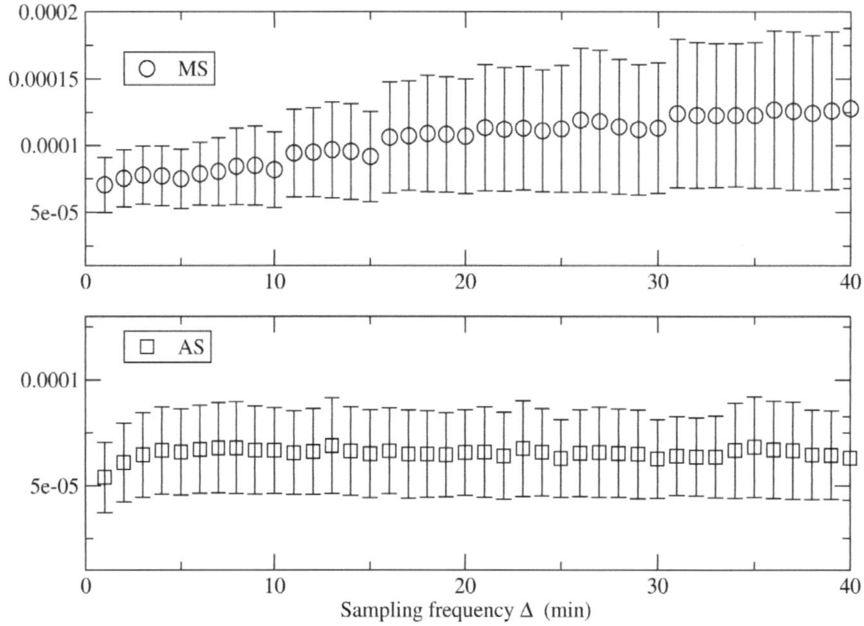

Figure 4. Average of the realized volatility as a function of sampling frequency Δ.

These standardized returns are expected to be standard normal variables provided that eqs.(11)-(12) are hold. However for the realized volatility constructed from the finite number of intraday returns the standardized return receives the finite-sample effect. Let R_s be a standardized return. The distribution of the standardized return R_s is given by[17]

$$f(R_s) = \frac{\Gamma(n/2)}{\sqrt{\pi n}\Gamma((n-1)/2)}\left(1 - \frac{R_s^2}{n}\right)^{(n-3)/2} \times I(\sqrt{n} \le R_s \le \sqrt{n}), \tag{13}$$

where n is the number of returns used to construct the realized volatility. The indicator function $I(X)$ means that $I(X) = 1$ if X is true and otherwise $I(X) = 0$. Under this distribution the even moments of R_s are also calculated to be[17]

$$m^{2k} = \frac{n^k(2k-1)(2k-3)\dots 1}{(n+2k-2)(n+2k-4)\dots n}. \tag{14}$$

Actually this finite-sample effect on the standardize return has been observed in empirical results[11, 12, 15] and also in the spin model simulation[21].

Figure 5 shows the variance of the standardized returns as a function of sampling frequency Δ. Here note that from eq.(14) m^2 always takes one, i.e. the variance has no finite-sample effect. As seen in the figure the variance is very close to one through every sampling frequency. Only we see an increase at small Δ, which may indicate the effect of the microstructure noise.

Figure 6 shows the kurtosis as a function of the sampling frequency Δ. It is seen that the kurtosis decreases as the sampling frequency Δ increases as expected from the functional form of the finite-sample effect in eq.(14). This behavior is also observed for individual stocks on the Tokyo Stock Exchange[15]. The solid lines in the figure are results fitted to a fitting function

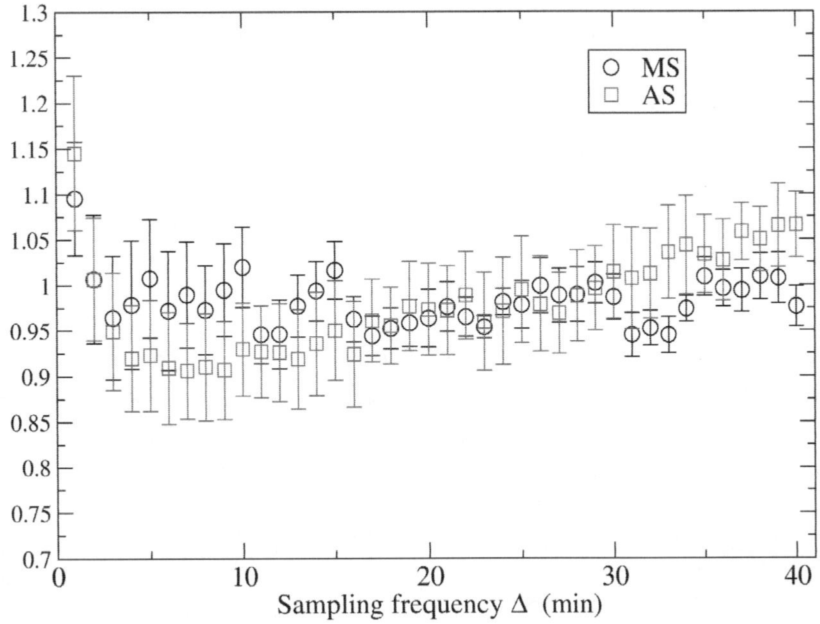

Figure 5. Variance of the standardized returns as a function of sampling frequency Δ.

Table 1. Fitting results. The values in parentheses show the theoretical values expected from the standard normal distribution. The moment m^{2k} of the standard normal distribution is given by $m^{2k} = \int x^{2k}P(x)dx$ where $P(x)$ is the standard normal distribution. Then we obtain $m^2 = 1, m^4 = 3$ and $m^6 = 15$. Kurtosis is $m^4/(m^2)^2 = 3$. It is found that K and M_6 extracted at $\Delta = 0$ come close to their theoretical values, which indicates that the results are consistent with the MDH.

	Kurtosis (3)		6th moment (15)	
	K	B_4	M_6	B_6
MS	2.42	216.5	9.17	176.2
AS	2.86	216.7	11.6	219.7

of $K(1 - \frac{2}{B_4/\Delta+2})$ where K and B_4 are fitting parameters. K corresponds to the kurtosis in the limit of $\Delta \to 0$. The fitting results are listed in Table 1. We obtain $K = 2.42(2.86)$ for the MS(AS). These values are close to 3 expected for the standard normal distributions although the value of the MS deviates slightly from 3.

We further analyze the 6th moment of the standardized returns. The higher moments of the standardized returns have been investigated in the spin financial model where no microstructure noise exists and it has been shown that the behavior of the higher moments with varying Δ is consistent with the finite-sample effect described by eq.(14)[21].

Figure 7 shows the 6th moment of the standardized returns as a function of Δ. The 6th moment increases as the sampling frequency Δ decreases and it seems that the 6th moment approaches the theoretical value, i.e. 15 as Δ goes to zero. We assume that the 6th moment is fitted by a function of $\frac{M_6L^2}{(L+4)(L+2)}$ where $L = B_6/\Delta$, and M_6 and B_6 are fitting parameters. For this fitting function the value of the 6th moment at $\Delta = 0$ is obtained by M_6. The fitting results are listed in Table 1 and we obtain $M_6 = 9.17(11.6)$ for the MS(AS).

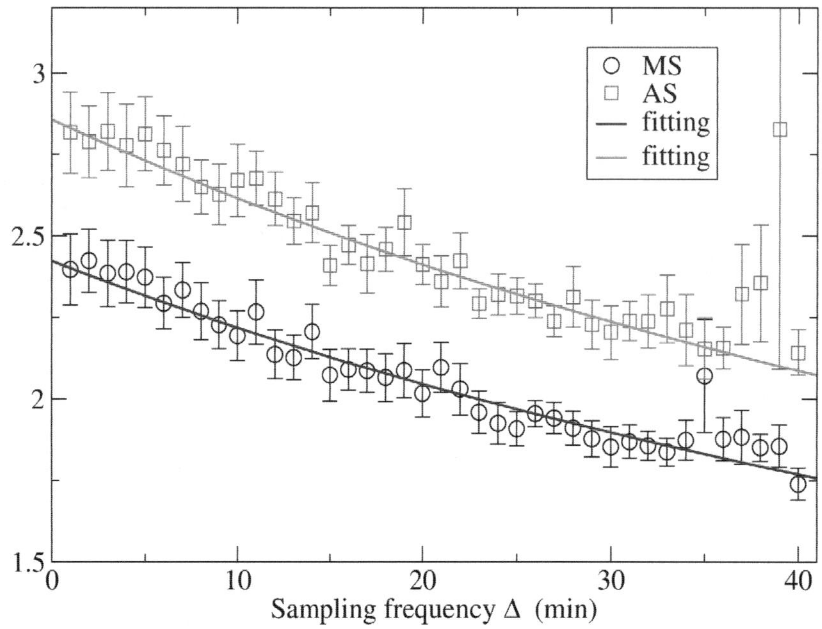

Figure 6. Kurtosis of the standardized returns as a function of sampling frequency Δ.

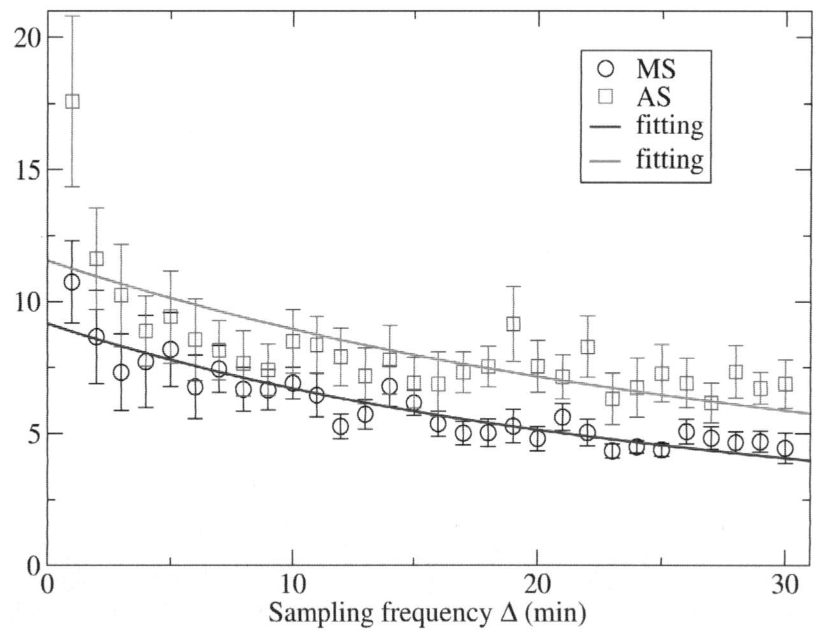

Figure 7. 6th moment of the standardized returns as a function of sampling frequency Δ.

5. Conclusion

We calculated realized volatility using high-frequency data of the Nikkei Stock Average (Nikkei225) Index on the Tokyo Stock Exchange. To avoid the bias from the non-trading hours issue realized volatility was calculated separately in the morning and afternoon trading sessions of the Tokyo Stock Exchange. As seen in figure 4 it is found that for the Nikkei Stock Average the microstructure noise decreases the realized volatility at small sampling frequency. This finding is the opposite result to the microstructure noise appeared on individual Japanese stocks[14]. Using realized volatility as a proxy of the true volatility we standardized returns in the morning and afternoon sessions and investigated the normality of the standardized returns by calculating variance, kurtosis and 6th moment. It is found that variance, kurtosis and 6th moment are consistent with those of the standard normal distribution, which indicates that the return dynamics of the Nikkei Stock Average are well described by Gaussian random process with time-varying volatility, expected from the mixture of distributions hypothesis.

Acknowledgements

Numerical calculations in this work were carried out at the Yukawa Institute Computer Facility and the facilities of the Institute of Statistical Mathematics. This work was supported by JSPS KAKENHI Grant Number 25330047.

References

[1] Cont R 2001 *Quantitative Finance* **1** 223
[2] Mandelbrot B 1963 *J. of Business* **36** 394
[3] Clark P K 1973 *Econometrica* **41** 135
[4] Praetz P D 1972 *J. of Business* **45** 49
[5] Van der Straeten E and Beck C 2009 *Phys. Rev. E* **80** 036108
[6] Takaishi T 2010 *Evolutionary and Institutional Economics Review* **7** 89
[7] Andersen T G and Bollerslev T 1998 *International Economic Review* **39** 885
[8] Andersen T G, Bollerslev T, Diebold F X and Labys P 2000 *Multinational Finance Journal* **4** 159
[9] Andersen T G, Bollerslev T, Diebold F X and Labys P 2001 *J. of the American Statistical Association* **96** 42
[10] Andersen T G, Bollerslev T, Diebold F X and Ebens H 2001 *Journal of Financial Economics* **61** 43
[11] Andersen T G, Bollerslev T, and Dobrev D 2007 *Journal of Econometrics* **138** 125
[12] Andersen T G, Bollerslev T, Frederiksen P, and Nielsen M Ø2010 *Journal of Applied Econometrics* **25** 233
[13] Fleming J and Paye B S 2011 *Journal of Econometrics* **160** 119
[14] Takaishi T, Chen T T and Zheng Z 2012 *Prog. Theor. Phys. Supplement* **194** 43
[15] Takaishi T 2012 *Procedia - Social and Behavioral Sciences* **65** 968
[16] Andersen T G, Bollerslev T, Diebold F X and Labys P 2000 *Risk Magazine* **13** 105
[17] Peter R G and De Vilder R G 2006 *Journal of Business & Economic Statistics* **24** 444
[18] Black F and Scholes M 1973 *Journal of Political Economy* **81** 637
[19] Merton R C 1973 *Bell Journal of Economics and Management Science* **4** 141
[20] Zhou B 1996 *J. of Business & Economics Statistics* **14** 45
[21] Takaishi T 2014 *JPS Conf. Proc.* **1** 019007

Compressive and Shear Wave Velocity Profiles using Seismic Refraction Technique

M Aziman[1*,5], **Z A M Hazreek**[2,6], **A T S Azhar**[3,7] and **D S Haimi**[4,8]
[1,2,7] Faculty of Civil and Environmental Engineering, Universiti Tun Hussein Onn Malaysia, 86400 Batu Pahat Johor, MALAYSIA
[3,5,6] Research Center for Soft Soil, Universiti Tun Hussein Onn Malaysia, 86400 Batu Pahat Johor, MALAYSIA
[4] Research Center for Applied Electromagnetic, Universiti Tun Hussein Onn Malaysia, 86400 Batu Pahat Johor, MALAYSIA
[8] Faculty of Electrical and Electronic Engineering, Universiti Tun Hussein Onn Malaysia, 86400 Batu Pahat Johor, MALAYSIA

E-mail: aziman@uthm.edu.my

Abstract. Seismic refraction measurement is one of the geophysics exploration techniques to determine soil profile. Meanwhile, the borehole technique is an established way to identify the changes of soil layer based on number of blows penetrating the soil. Both techniques are commonly adopted for subsurface investigation. The seismic refraction test is a non-destructive and relatively fast assessment compared to borehole technique. The soil velocities of compressive wave and shear wave derived from the seismic refraction measurements can be directly utilised to calculate soil parameters such as soil modulus and Poisson's ratio. This study investigates the seismic refraction techniques to obtain compressive and shear wave velocity profile. Using the vertical and horizontal geophones as well as vertical and horizontal strike directions of the transient seismic source, the propagation of compressive wave and shear wave can be examined, respectively. The study was conducted at Sejagung Sri Medan. The seismic velocity profile was obtained at a depth of 20 m. The velocity of the shear wave is about half of the velocity of the compression wave. The soil profiles of compressive and shear wave velocities were verified using the borehole data and showed good agreement with the borehole data.

1. Introduction

Soil velocity profile is often used as subsurface characterization by using seismic-based techniques [1,2,3,4]. Seismic-based techniques are the most sensitive to the physical properties of geo-materials and relatively insensitive to the chemistry of geo-materials and their fluids [5]. The seismic-based techniques shake the ground and produces very small strains. Thus, the soil velocities derived from the seismic-based measurements are related to soil modulus. Therefore, the seismic-based techniques can be used to directly derive the geotechnical properties related to the strain including maximum shear modulus, bulk modulus (B), Young's modulus (E), and Poisson's ratio (v) [5,6,7,8,9].

There are four types of elastic seismic waves, produced by impulses and all of which travel at different velocities. The four are: compression wave, called the P-wave, shear wave called the S-wave, Rayleigh wave and Love wave. P and S waves are known as body waves; Rayleigh and Love waves are surface waves as shown in Figure 1 [10]. A seismic wave transmits energy by the vibration of soil particles in different directions. For P-waves, the soil particles vibrate in the direction of the wave propagation and S-waves vibrate in the direction perpendicular to the direction of wave propagation. Both waves are propagated along a hemispherical wave front; thus wave amplitude is attenuated in proportion to $1/r^2$ where r is the distance from seismic source.

The P-wave has the greatest velocity of the elastic seismic waves. The P-wave travels through all media: liquids, gases and solids. On the other hand, the S-wave travels slightly slower than the P-wave in solids and does not propagate through liquids and gases, as these media have no shear strength. Table 1 [6] shows the typical values of P- and S-wave for different earth materials which illustrate key differences. The P-wave causes volumetric strains and the velocity depends on bulk density, bulk modulus and any deformation which are considered to be undrained. Meanwhile, S-waves induce shear distortion in the soil without volumetric changes. The velocity of a P-wave in water is 1450 m/s. Therefore, the undrained bulk modulus of the ground is usually a combination of stiffness in both the soil and the water. In contrast, S-wave velocity in water is 0 m/s. Thus, the measured modulus is represented by the soil only. The use of S-waves is preferred when studying saturated soils because these are independent of water content [11,12].

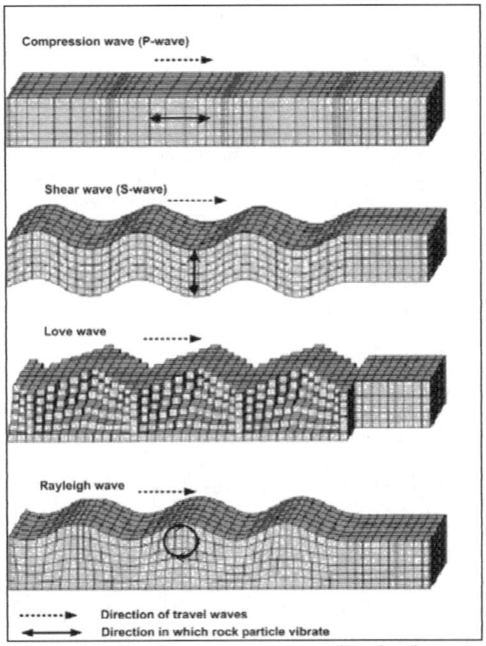

Figure 1. Schematic elastic wave propagation in the ground [10].

Table 1. P- and S-wave typical values for different earth materials [6].

Material	P-wave velocity, m/s	S-wave velocity, m/s
Air	330	0
Water	1450	0
Sands and clays	300-1900	100-500
Glacial till	1500-2700	600-1300
Chalk	1700-3000	600-1500
Strong limestone	3000-6500	1500-3500
Weathered granite	100-3000	500-1500
Fresh granite	3000-6000	1500-3000
Slate	5000-7000	2500-3800

There are two common types of seismic survey using body waves: refraction and reflection (see Figure 2). The seismic refraction is fundamentally based on Snell's Law which states that at the critical incident angle, a wave is refracted along the soil layer boundary before it returns to the surface. Snell's Law is;

$$Sin.i_c = {}^{V_1}\!/_{V_2} \tag{1}$$

where i_c is the critical incident angle (degree), V_1 is the velocity of the upper layer (m/s) and V_2 is the velocity of the lower layer (m/s).

The seismic refraction method requires the soil layers to increase in density with depth. The reflection method requires density contrast to reflect waves back to surface [13]. The seismic refraction method involves recording the travelling time of either the P-wave or S-wave energy. Thus, interpretation of the data provides layer thicknesses and seismic velocities [6]. Figure 2 shows the difference between refraction and reflection.

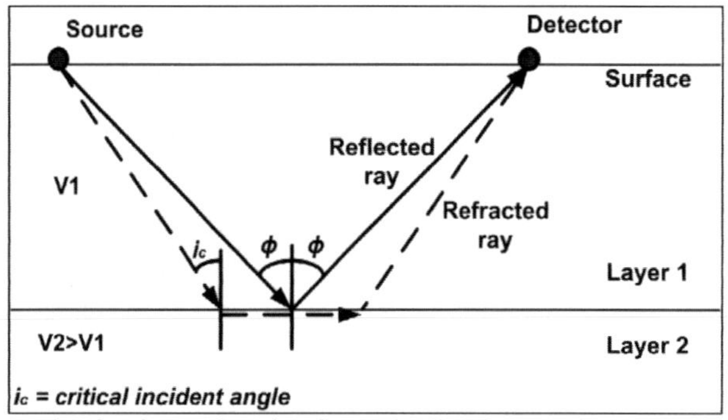

Figure 2. The path of refracted and reflected seismic rays in a two-layer system [6].

The way in which S-waves and P-waves are generated in refraction surveying differ. Generating P-wave energy is relatively easy. It can be done by either dropping any mass vertically onto the ground surface or by using explosives to generate more P-wave energy. The generation of S-waves is more difficult. S-wave can be generated by putting energy into the ground in a direction perpendicular to the row of receivers; thus the soil particle motion will be in a perpendicular direction to the direction of wave propagation [14]. The perpendicular direction energy generation technique enhances the S-wave amplitude and at the same time decreases P-wave amplitude. Thus, the S-waves are easier to recognize in the seismic records [13].

Seismic refraction surveys can be used for these two body waves, with appropriate wave sources and receivers. The arrival times of waves are measured and converted to wave velocities. The seismic waves are directly utilised to calculate engineering properties. From the P-wave and S-wave velocities input, the maximum shear modulus, G_{max}, the constrained modulus, M_{max} or bulk modulus, B, Young's modulus or dynamic elasticity modulus, E, and Poisson's ratio, v, at very small strains can be calculated from the following relationships;

$$G_{max} = \rho v_s^2 \tag{2}$$

$$B = M_{max} = \rho v_p^2 = \frac{E}{3(1-2\upsilon)} \tag{3}$$

$$E = 2\rho v_s^2(1+\upsilon) = \frac{\rho v_p^2(1-2\upsilon)(1+\upsilon)}{(1-\upsilon)} \tag{4}$$

$$\upsilon = \frac{\left[\left(\frac{v_s}{v_p}\right)^2 - 2\right]}{2\left[\left(\frac{v_p}{v_s}\right)^2 - 1\right]} \tag{5}$$

Where ρ is the bulk density of the soil (kg/m³), v_p is the P-wave velocity (m/s) and v_s is the S-wave velocity (m/s) [15,16,17]. G_{max} can be obtained from measurements of v_s and bulk density using equation 2 [17]. Geo-materials have values of Poisson's ratio in the range of 0.05 for very hard rocks and nearly 0.5 for saturated unconsolidated clays [18]. According to the theory of elasticity, Young's modulus, E, and constrained modulus, M_{max}, are related to shear modulus, G_{max}, by:

$$G_{max} = \frac{E}{2(1+\upsilon)} \tag{6}$$

$$G_{max} = \frac{M_{max}(1-2\upsilon)}{(1-\upsilon)} \tag{7}$$

Where v is the Poisson's ratio, E is the Young's modulus (N/m²) and M_{max} is the constrained modulus (N/m²).

Seismic refraction measuring the travel times of the seismic body waves, the nature and depth of the surface can be computed and later reveal the velocity of soil layers. The velocity of seismic wave that travels through the subsurface varies with material composition and stiffness [19,20]. The layers of soil in p and s wave velocities are determined using intercept-time formula or crossover distance formula as shown in Equation 8 and 9 respectively [18].

$$Z = \frac{t_1}{2}\frac{v_2 . v_1}{\sqrt{v_2^2 - v_1^2}} \tag{8}$$

$$Z = \frac{x_c}{2}\sqrt{\frac{v_2 - v_1}{v_2 + v_1}} \tag{9}$$

Where t_i is the intercept time (s), Z is the depth to refractor two (m), x_c is the crossover distance (m), v_1 is the seismic velocity in layer 1 (m/s) and v_2 is the seismic velocity in layer 2 (m/s).

The aim of this paper is to investigate the P-wave and S-wave velocities obtained from the seismic refraction technique. The location of this study is shown in Figure 3, at Sejagung Sri Medan which is located at the geological boundary between quaternary marine deposit and volcanic rock.

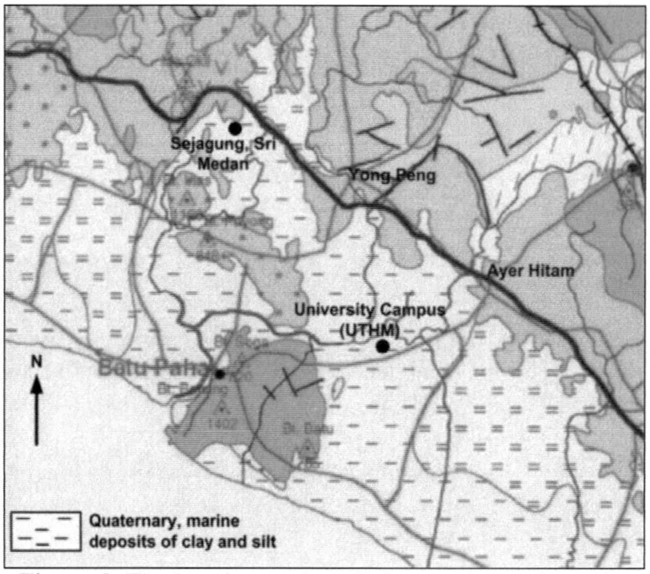

Figure 3. The location study at Sejagung Sri Medan [21].

2. Seismic Refraction Method

In this study, the seismic refraction equipment consists of source, detector and recorder. The source of this seismic survey is a sledge hammer weighing 7 kg that is used to strike on an impact plate. The hammer strikes should be in the vertical direction of the steel plate to produce the compressive wave as shown in Figure 4. Meanwhile, the hammer strikes in the horizontal direction to the side of the steel plate to produce horizontal shear wave energy. For detectors, a 24-unit 10 Hz vertical geophone and a 24-unit 14 Hz horizontal geophone were used to detect the compressive wave and the shear wave, respectively. The ABEM Terraloc MK-8 seismograph was used as the recorder for the seismic raw data. For data acquisition, firstly, the compressive wave refraction was recorded followed by shear wave refraction. There were two reels of geophone cables and each reel consisted of 12 geophone connector points. The geophone spacing was set at 3 m and 20 m offset from the first and last geophones. For seismograph settings, the record length used 1 second recording length where the sampling interval was 1000 μs and the number of samples was 1024. Seven shot points were taken at offset and intervals of 1st and 2nd, 6th and 7th, 12th and 13th, 18th and 19th, and 23rd and 24th geophones as shown in Figure 4.

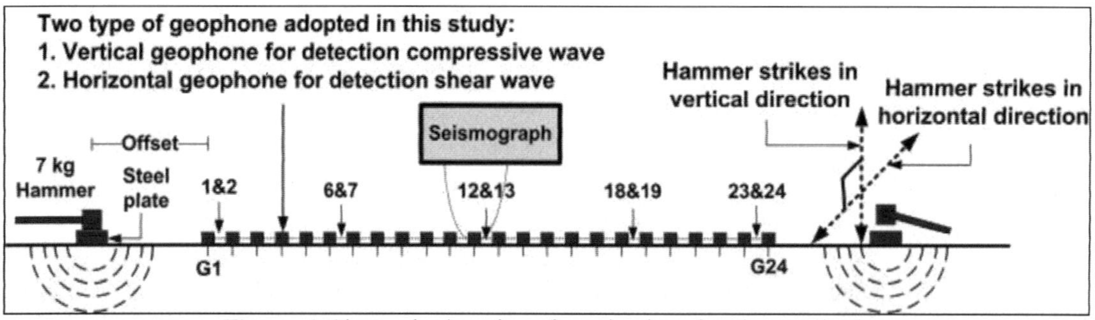

Figure 4. Shot point locations for seismic refraction method.

3. Results and Discussions

All result was presented at section 3.1 – 3.3 while the discussion of the results was discussed in section 3.4.

3.1. P-Wave Velocity Soil Profile Result

The first arrival of P wave was selected from the raw seismic data using SeisOptPicker software. Then the data were processed using SeisOpt@2D software to generate a tomography plot along the spread line. The soil profile was divided into several layers depending on the colour contrast which showed the difference of its soil velocity. Figure 5 shows that the P-wave velocity soil profile was divided into 4 layers. The layer 1 boundary varied between 5 m and 10 m while the P-wave velocities were in the range of 400 m/s to 1100 m/s. Layer 2 had a boundary between 7 m and 17 m and the velocities were between 1600 m/s and 2200 m/s. Layer 3 boundary was between 11 m and 19 m and the velocities were between 2200 m/s and 3700 m/s. The final layer indicated a velocity higher than 3700 m/s.

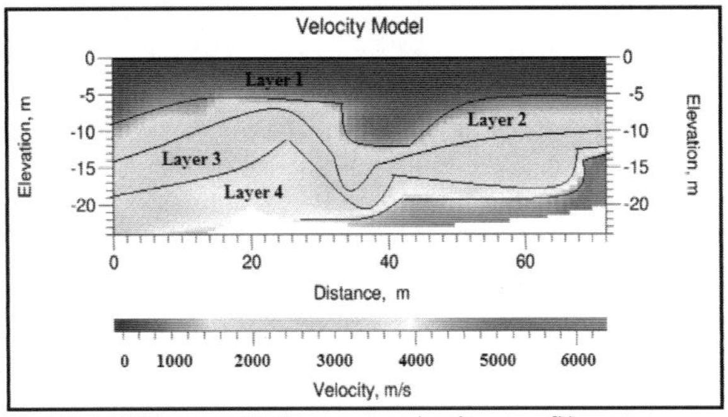

Figure 5. P-wave velocity profile.

3.2. S-Wave Velocity Soil Profile Result

The first arrival of S-wave was selected from the raw seismic data using SeisOptPicker software. Then the data were processed using SeisOpt@2D software. Figure 6 shows the shear wave velocity in horizontal direction (SH-wave) tomography image across the spreadline. The shear wave velocity profile was divided into 4 layers based on its velocity. The layer 1 boundary varied between 4 m and 14 m and the SH-wave velocities were between 190 m/s and 650 m/s. Layer 2 had a boundary between 7 and 20 m and the velocities were between 750 m/s and 1300 m/s. Layer 3 boundary was between 9 m and 19 m and the velocities were between 1500 m/s and 1700 m/s. The final layer indicated a velocity higher than 1700 m/s.

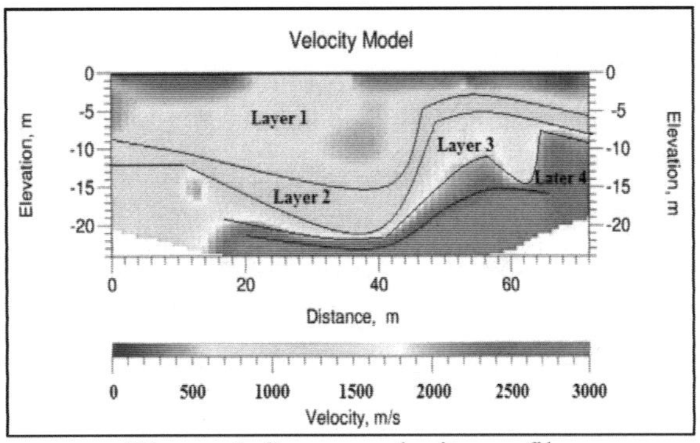

Figure 6. S-wave velocity profile.

3.3. Borehole Result

The seismic refraction test investigation was conducted at the borehole location in Sejagong, Sri Medan. The borehole indicated that the soil layer consists of silt, sand and gravelly sand and shallow bedrock at 9 m deep. Soil was categorized based on SPT N-value, where the first layer was classified as loose layer up to 6 m deep, followed by a dense layer up to 9 m and a rock core up to 15 m. Figure 7 shows the plot of SPT-N value with depth.

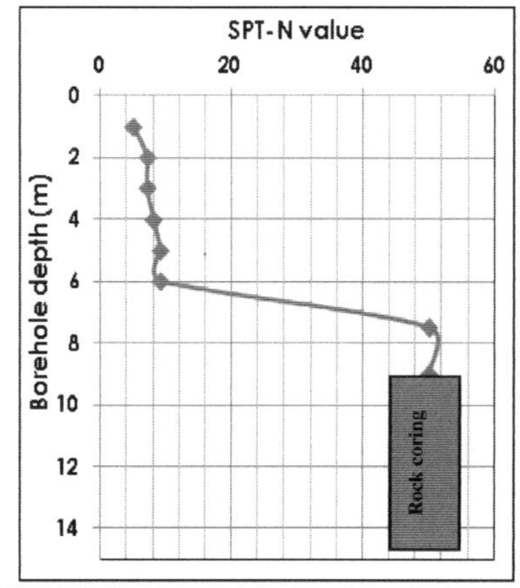

Figure 7. Plot of SPT–N value with depth at Sejagong, Sri Medan site.

3.4. Discussion

The velocity of shear wave is about half the velocity of compression wave. At a depth of 9 m where the rock head was found, the P- and S-wave velocities were above 3000 m/s and 1500 m/s respectively. The seismic velocities show good agreement with the borehole data. In addition, the velocity of igneous rock reported in Table 1 shows good agreement with the obtained P- and S-wave velocities. Furthermore, the seismic velocity profile was able to provide a visual representation of the rock head across the seismic test spread line. By referring to the geological map in Figure 3, it can be seen that Sejagung is located at the boundary of volcanic formation and quaternary marine clay. Thus, it was expected that the rock head is uneven and drastically changed. The shear modulus, Young's modulus and Poisson's ratio for each layer can be calculated when the P- and S-wave velocities are obtained. For example, the Poisson's ratios were calculated using the velocity of P- and S-wave in equation 5. The Poisson's ratio of the first layer was between 0.33 and 0.35, between 0.32 to 0.33 for the second layer and between 0.33 to 0.34 for the third layer and final layer. On the other hand, the bulk density of material is required to obtain the shear modulus and Young's modulus.

4. Conclusion

The soil velocity profiles of P- and S-wave velocities at Sejagung Sri Medan were obtained using Seismic Refraction. The velocity of shear wave is about half the velocity of compression wave. The soil profiles of compressive and shear wave velocities were verified using the borehole data and showed good agreement with the borehole data.

Acknowledgment

The authors would like to express their deepest appreciation to the Ministry of Higher Education and Universiti Tun Hussein Onn Malaysia for supporting this research under the Exploratory Research Grant Scheme (ERGS) Vot. E040, Multi-disciplinary Research Grant (MDR) Vot. 1320 and Incentive Grant Scheme for Publication (IGSP) Vot. U258.

References

[1] Zuraidah Z S, Aziman M, Ariffuddin J and Faiz L A M 2015 Seismic Surface Wave Testing for Investigating the Shallow Soil Profile. *Applied Mechanics and Materials*. **773-774** 1565-1568

[2] Jazlan M S M, Adnan Z and Aziman M 2015 Soil Velocity Profile on Soft Soil Using Seismic Refraction. *Applied Mechanics and Materials*. **773-774** 1549-1554

[3] Fahmy K A, Aziman M and Jazlan M S M, "Integrated Methods of Geophysical and Geotechnical Study on Saturated Ground Profile," *Proceeding of Soft Soil Engineering International Conference (Sarawak, Malaysia 2013)*
[4] Hazreek Z A M, Rosli S, Fauziah A, Wijeyesekera, D C and Faizal T B M 2012 Seismic Refraction Investigation on Near Surface Landslides at the Kundasang area in Sabah, Malaysia. *Procedia Engineering*. **50** 516-531
[5] Steeples D W and Miller R D 1990 Seismic reflection methods applied to engineering, environmental and groundwater problems. Inv*estigations in Geophysics: Review and Tutorial, Society of Exploration Geophysicists*. **1(5)** 1-30
[6] McDowell P W, Barker R D and Butcher A P et al 2002 Geophysics in engineering investigations. London: Construction Industry Research and Information Association, CIRIA C562
[7] Luna R and Jadi H Determination of Dynamic Soil Properties Using Geophysical Methods. "*Proceedings of the First International Conference on the Application of Geophysical and NDT Methodologies to Transportation Facilities and Infrastructure," (St. Louis, MO, December 2000)*
[8] Lankston R W 1990 High-resolution refraction seismic data acquisition andinterpretation." *Investigations in Geophysics: Review and Tutorial, Society of Exploration Geophysicists*. **1(5)** 1-30
[9] Charles J A and K.S. Watts K S 2002 Treated ground engineering properties and performance. Construction Industry Research and Information Association, CIRIA C572, London.
[10] Crice D 2005 MASW: the wave of the future. *Journal of Environmental and Engineering Geophysics*. **10(2)** 77–79
[11] Clayton C R I 2011 Stiffness at small strain: research and practice. *Géotechnique*. **61(1)** 5-37
[12] Fam M and Santamarina J C 1997 A study of consolidation using mechanical and electromagnetic wave. *Géotechnique*. **47(2)** 203-219
[13] Matthews M C, Clayton C R I and Own Y "The use of field geophysical techniques to determine geotechnical stiffness parameter," *Proceedings of the Institution of Civil Engineers Geotechnical Engineering (143 31-42, 2000)*
[14] Menzies B K 2001 *Near-surface site characterisation by ground stiffness profiling using surface wave geophysics. Instrumentation in Geotechnical Engineering*, (H.C. Verma Commemorative).
[15] Clayton C R I Matthews M C and Simons N E 1995 *Site Investigation* (London: Blackwell Science Ltd.)
[16] Menzies B and Matthews M "The continuous surface-wave system: A modern technique for site investigation," *In Special lecture: Indian Geotechnical Conference (Madras, December 11-14th 1996)*
[17] Massarsch K R, "Deformation properties of stabilized soil columns," *Proceedings of International Conference on Deep Mixing (Stockhom, 23 - 25 May 2005)*
[18] Telford W M, Geldart L P and Sheriff R E 1990 *Applied Geophysics* (Cambridge: Cambridge University Press)
[19] Abidin M H Z, Rosli S, Fauziah A and Wijeyesekera D C, "Application of Geophysical Methods in Civil Engineering," *Proceeding of Malaysian Technical Universities International Conference on Engineering & Technology (Batu Pahat, Johor Malaysia, 2011)*
[20] Madun A, Jefferson I, Foo K Y, Chapman D N, Culshaw M G and Atkins P R 2012 Characterization and quality control of stone columns using surface waves testing *Canadian Geotechnical Journal*. **49(12)** 1357-1368
[21] Mineral and Geoscience Department Malaysia, *Geological Map of Peninsular Malaysia* 1985, eighth edn., Ministry of Natural Resources and Environment

Production of Starch Based Bioplastic from Cassava Peel Reinforced with Microcrystalline Celllulose Avicel PH101 Using Sorbitol as Plasticizer

Maulida[1], M Siagian[1], P Tarigan[1]

[1]Department of Chemical Engineering, University of North Sumatera, Medan, Indonesia

E-mail: lock_piranha@yahoo.com

Abstract. The production of starch based bioplastics from cassava peel reeinforced with microcrystalline cellulose using sorbitol as plasticizer were investigated. Physical properties of bioplastics were determined by density, water uptake, tensile strength and Fourier Transform Infrared Spectroscopy. Bioplastics were prepared from cassava peel starch plasticized using sorbitol with variation of 20; 25; 30% (wt/v of sorbitol to starch) reinforced with microcrystalline celllulose (MCC) Avicel PH101fillers with range of 0 to 6% (wt/wt of MCC to starch). The results showed improvement in tensile strength with higher MCC content up to 9,12 mpa compared to non-reinforced bioplastics. This could be mainly attributed to the strong hydrogen bonds between MCC and starch. On the contrary, the addition of MCC decreased the elongation at break, density and water uptake. Fourier Transform Infrared Spectroscopy showed the functional groups of bioplastics, which the majority of O-H groups were found at the bioplastics with reinforcing filler MCC that represented substantial hydrogen bonds. The highest tensile strength value was obtained for bioplastic with MCC content 6% and sorbitol content 20%. With good adhesion between MCC and starch, the production of bioplastics could be widely used as a substitute for conventional plastics with more benefits to the environment.

1. Introduction

Packaging plays a significant part in recent years as over 67 millions tonnes of packaging waste resulting in environmental concerns. This provokes many different processing, for example the using of additives such as fillers, colourants and plasticizers to produce polymers as packaging applications [1]. Environmental, economic and safety challenges have caused many scientists to invent biodegradable polymers as a substitute for petrochemical based polymers. This biodegradable polymers or bioplastics indicate more eco-friendly environment than the conventional plastics which lead to environmental issues [2]. But there is still a lack in the development of bioplastics as its high cost of production leads to certain limitations. An alternative for a low cost and renewable substrate has been proposed by using agriculture waste (AW) [3]. In addition to performance and abundant sources, starch from agriculture waste is the solution for an alternative. Starch as biodegradable polymer becomes reasonable material for the production of bioplastics beacause of its low cost [4]. Therefore, production of bioplastics is the breakthrough innovation to solve the environmental issues by using renewable and degradable natural resources and to provide more cost effective bioplastics.

Raw materials for bioplastics originate from natural constituents such as polysaccharides (e.g. starch, cellulose, chitin and lignin), proteins (e.g. gelatine, casein and wheat gluten) and lipids (e.g. plant oils and animals fats) [1]. In Indonesia, development of starch based bioplastics has a big potential, because starch can be easily obtained by majority of Indonesia's plants. One of the most abundant source for starch production in Indonesia is from cassava. Cassava processing results in organic waste such as cassava peel that can be used as bioplastic matrix for its high starch content. However, bioplastic based on starch still has many disadvantages like poor mechanical properties and high moisture adsorption, therefore an alternative is used to improve its properties [5]. Reinforcing fillers such as cellulose has proven to be the most promising material. Microcrystalline cellulose used as reinforcement filler for starch based edible films was analyzed by Psomiadou et al. (1996). The results showed higher strength and elongation and lower water vapour transmission rate (WVTR) of bioplastics. Microcrystalline cellulose Avicel PH 101 is used as reinforcing filler because it offers higher density of hydroxyl groups on its surface that is available for hydrogen bonding [5].

Utilization of organic waste such as cassava peel for production of starch based bioplastic can help reducing the environmental damages that are caused by conventional plastics. Higher value bioplastics can be obtained by improving their properties with the most abundant and biodegradable reinforcing filler like cellulose. The goal of this work is to study the properties of bioplastics from agriculture waste. Effect of reinforcement fillers and plasticizers on bioplastics are also examined.

2. Materials and Methods

2.1. Materials

Starch derived from cassava peel was obtained from Traditional Market Medan, microcrystalline cellulose (MCC) Avicel PH101 was obtained from Pusat Penelitian Kelapa Sawit Medan with particle size 50 μm, and sorbitol 99% was obtained from Rudang Jaya Medan.

2.2. Preparation of Cassava Peel Starch

Cassava peel (100 gram) obtained from waste container at local traditional market was washed with clean water before shredded to small pieces. The shredded cassava peel was later peeled and placed in mixing blender and soaked in water for about 100 ml. After mixing process, starch slurry was filtered and later placed in tank for settling that took at least 30 minutes. Starch sediment was separated from the slurry and then washed again with distilled water. After the second settling, starch sediment was dried using oven with temperature 70 °C for removal of free water. Starch was sieved with strainer 100 mesh/inch for better homogeneous size.

2.3. Bioplastic Preparation

Solution containing 0, 2, 4 and 6% wt/wt of fillers to starch was prepared by dispersing 200 ml distilled water and sorbitol with concentration varied from 20, 25 and 30% wt/v of plasticizer to starch. Solution was placed into ultrasonic homogenizer KUDOS tank and processed for about 50 minutes. After ultrasonication, solution was removed from the tank. Starch (10 gram) was added to the solution and heated using hot plate while stirred until it gelatinized (70 °C). After mixing, the solution was cast onto flat and dried with temperature 60 °C for 24 hour. Once set, bioplastic was cooled to ambient temperature before peeled off the flat.

2.4. Bioplastic Characterizations

2.4.1. *Tensile Strength and Elongation at Break.* Tensile strength was measured with GoTech Universal Testing Machine using the standard of ASTM D882-91. Tensile strength value was obtained from the observed data. Elongation at break is an indication of bioplastics flexibility and is expressed as a percentage. Elongation at break was calculated as follows :

$$\text{Elongation at break } (\varepsilon) = \frac{\text{extension x 100}}{\text{gauge}} \quad (1)$$

2.4.2. *Density.* Density of bioplastic was investigated by the standard of ASTM D792-91 on film with size approximately 5x5 cm. The film's mass was measured using analytical balance. Density was calculated as follows :

$$\text{Density} = \frac{\text{mass (gr)}}{\text{volume (cm}^3\text{)}} \quad (2)$$

2.4.3. *Water Uptake.* Water uptake was investigated by cutting film with size approximately 2x2 cm and then weighed the mass. Film was put into a container filled with distilled water for 24 hour. After immersion in water, film was removed from the water and weighed to measure the wet weight. Water uptake was calculated as follows :

$$\text{Water uptake} = \frac{\text{wet weight} - \text{dry weight}}{\text{wet weight}} \text{ x } 100\% \quad (3)$$

2.4.4. *Fourier Transform Infrared Spectroscopy (FT-IR).* Functional groups of bioplastics were analyzed by using IR Prestige-21 Shimadzu. The analysis using FT-IR represented spectrum data in graphic and wave numbers of each data that provided functional groups of bioplastics.

3. Results

3.1. Tensile Strength and Elongation at Break

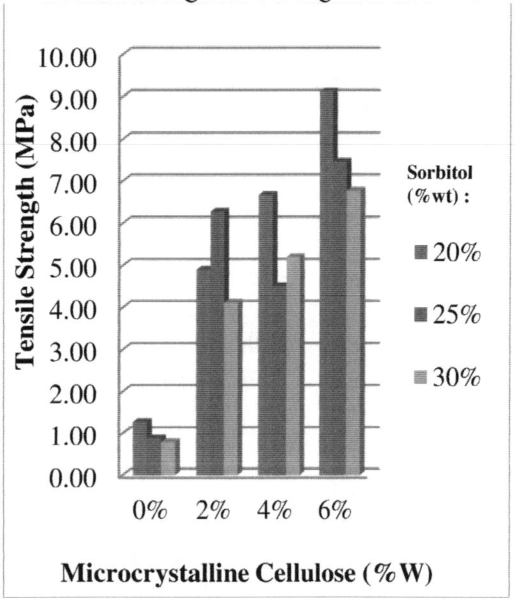

Figure 1. Tensile strength of starch based bioplastic reinforced with MCC Avicel PH101 using sorbitol as plasticizer.

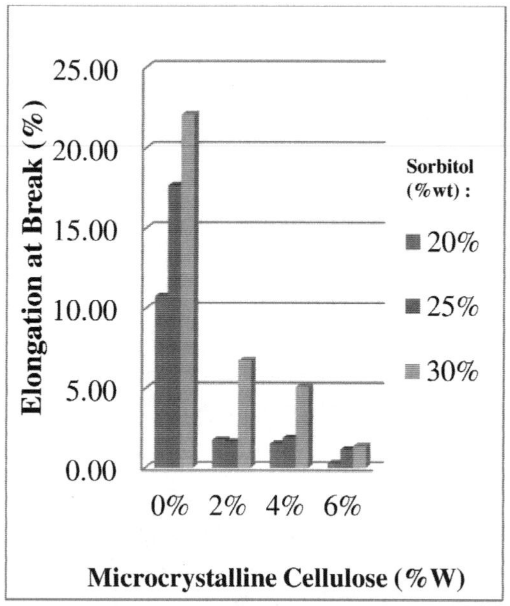

Figure 2. Elongation at break of starch based bioplastic reinforced with MCC Avicel PH101 using sorbitol as plasticizer.

Figure 1 shows the results of tensile strength and elongation at break as a function of MCC content for plasticized starch based bioplastic. It indicated that tensile strength increased with the increasing of MCC content. Results showed an increase of tensile strength from 0 to 6% MCC content for bioplastics with 20 and 30% sorbitol. The improvement of bioplastic properties can be attributed to good interfacial adhesion that can form strong 3D hydrogen bond network that is occured between starch matrix and MCC fillers [5]. Thus, bioplastic with MCC content 6% and 20% sorbitol provided the maximum tensile strength for 9,12 MPa. A drop in tensile strength value is shown for bioplastic with MCC content 4% and sorbitol content 25%. It is possibly due to higher content of MCC contributed to retarding the intermolecular interaction thus resulting in aggregates and heterogeneous film structure [5]. Plasticizer is necessary for production of bioplastics to increase its flexibility and workability, on the other hand addition of plasticizer results in lower hardness [8]. Thus higher sorbitol content resulted in lower tensile strength. It should be noted that increasing the sorbitol content up to 30% resulted in lower tensile strength. However, for bioplastics with sorbitol content below 30%, a primary concern that there is sometimes not enough plasticizing effect and starch molecules become brittle [9]. This could be seen on bioplastics with MCC content 2 and 4% for 25% sorbitol.

Contrary to tensile strength value, figure 2 shows the results of elongation at break as the addition of MCC content decreased the value. Results showed a decrease in elongation at break with the increasing MCC content from 0 to 6%. It is possibly due to hydrogen bond [5] between hydroxyl group (O-H) from starch and hydroxyl and carboxyl group (COOH) of cellulose. This bonding resulted in a higher strength and lowered the elongation at break. Meanwhile, addition of plasticizers to bioplastics have the opposite effect on elongation at break. Plasticizers give elastic properties to plastic and have the ability to weaken the intermolecular hydrogen bond and increase spacing [10]. An increase in elongation at break with higher sorbitol content was linked to the addition of plasticizers that act as the compatibilizer to improve the interaction between matrix and fillers. However, bioplastic with MCC content 2% and 25% sorbitol provided an increase in elongation at break. This could be attributed to an antiplasticisation effect. Myllarinen et al. (2002) reported that film with sorbitol content below 27% has the tendency to strong interaction between starch and sorbitol which resulted in decreasing mobility of system. Antiplasticisation effect also results in rigid and brittle plastic. The highest elongation at break value was observed at bioplastic with MCC content 0% and 30% sorbitol.

Base on the values obtained in figure 1 and figure 2, the results show the relationship between the addition of reinforcement fillers and plasticizers to the mechanical propeties of bioplastics. While the addition of MCC as reinforcement fillers represented an improvement in tensile strength with increasing value, elongation at break showed otherwise. This indicates that MCC is used to strengthen materials but results in more brittle plastics. On the other hand, plasticizers improved the brittleness by increasing sorbitol's content but resulted in lower tensile strength.

3.2. Density

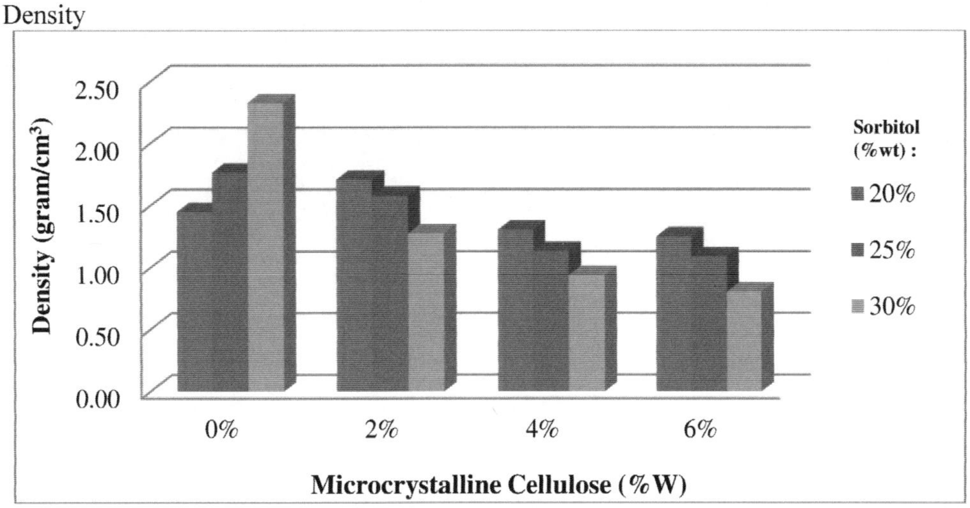

Figure 3. Density of starch based bioplastic reinforced with MCC Avicel PH101 using sorbitol as plasticizer.

The effect of MCC content on density of plasticized starch based bioplastic is shown in Figure 3. Density increased for bioplastics with 0% MCC, on the other hand density of bioplastics decreased as the content of MCC increased from 2 to 6%. Bierley et al. (1988) conveyed that plastics with lower density tended to have open structure which can be penetrated by fluids, such as H_2O, O_2 or CO_2. The results showed that addition of MCC trigerring in lower density. This could be attributed to decrease in crystallinity index of MCC after ultrasonic treatment as reported by Zhang et al. (2013). Decrease in crystallinity leads to an increase in amorphous fraction. Higher amorphous fraction attributes to lower polymer's mass because of the un-uniformity and less dense molecules that results in lower density. Plasticizers also effect the density of bioplastics as increasing plasticizers content causes network to swell and results in the decrease of network density [14]. The highest density value was found at bioplastic without reinforcing MCC and 30% sorbitol content.

3.3. Water Uptake

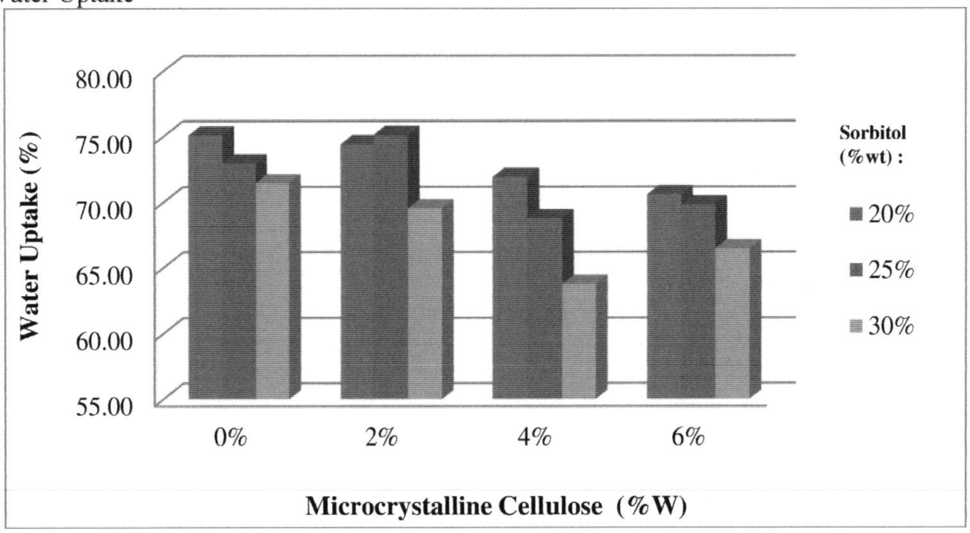

Figure 4. Water uptake of starch based bioplastic reinforced with MCC Avicel PH101 using sorbitol as plasticizer.

The effect of MCC content on water uptake of plasticized starch based bioplastic is shown in Figure 4. Water uptake decreased as the content of MCC increased for bioplastics with MCC content 0, 4 and 6%. Furthermore, starch was more sensitive to water because of its hydrophilic nature than cellulose [15]. On the other hand, an increase in water uptake was found for bioplastic with MCC content 2% and 25% sorbitol. Cellulose has strong hydrogen bond and characteristic that has the difficulty to bond with water. However, the excess addition of cellulose has the capability to increase water absorption because the intramolecular hydrogen bonding. An increase in water uptake could not fully homogeneous and resulted incorporated particles. This could be attributed to the less handled ultrasonication process.

3.4. Fourier Transform Infrared Spectroscopy (FT-IR)

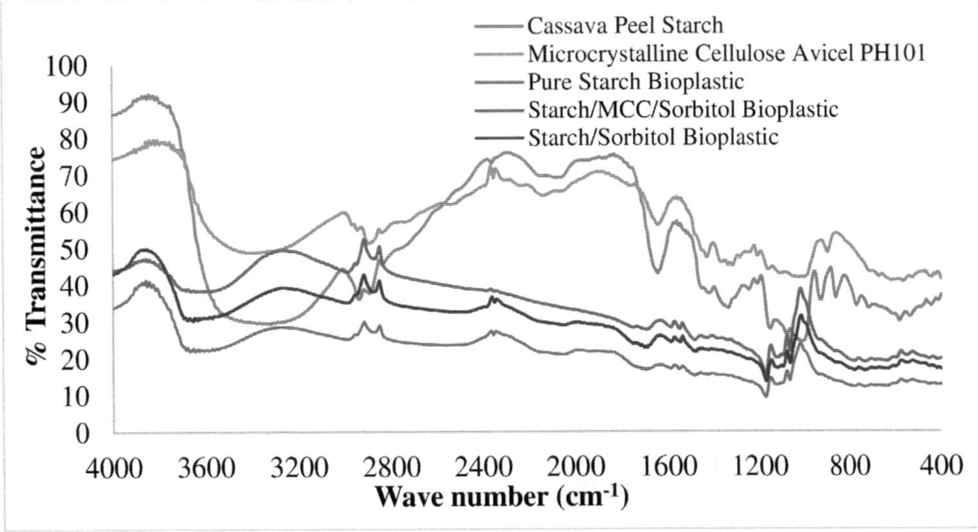

Figure 5. FTIR specs of cassava peel starch, MCC, pure starch bioplastic, strach/MCC/sorbitol bioplastic and starch/sorbitol bioplastic

The analysis of FT-IR shows the characteristic peaks from bioplastics. FT-IR is used for its indirect analytical method to study molecular structures [16] and has the capability to measured functional groups in greater speed and greater sensitivity [17]. The interaction between bioplastic's components can be identified from FT-IR spectra. If components form immiscible blends, then no appreciable changes can be found in FT-IR spectra. However, if the components are compatible, chemical interaction exists between their chains culminating in the change of FT-IR spectra [18]. Figure 5 shows the results of characteristic peaks of cassava peel starch, MCC, pure starch bioplastic, starch/MCC/sorbitol bioplastic and starch/sorbitol bioplastics. The rose colored line represented characteristic peaks for cassava peel starch. It could be observed that the hydrogen bond –OH stretching at 2835 cm^{-1} and the C-H stretching at 709 cm^{-1}. Characteristic peak at 1242 cm^{-1} indicated C-O bending of C-O-H group [9]. The green colored line represented characteristic peaks for MCC with –OH bending at 3383 cm^{-1}. The peaks at 1002 cm^{-1}, 1103 cm^{-1} and 1157 cm^{-1} were attributed to C-O stretching. The bending peak of C=C bond appeared at 1639 cm^{-1}. The blue colored line represented pure starch bioplastic's characteristic peaks. For –OH stretching and bonding appeared at 2877 cm^{-1} and 3649 cm^{-1} repectively. Characteristic peaks at 1111 cm^{-1} and 1176 cm^{-1} indicated C-O strecthing. The maroon colored line represented starch/MCC/sorbitol bioplastic's characteristic peaks. It could be observed that the C-O stretching at 1064 cm^{-1}, 1118 cm^{-1} and 1168 cm^{-1}. The peaks at 2870 cm^{-1} and 2989 cm^{-1} indicated –OH stretching while peak at 3633 cm^{-1} indicated –OH bonding. The purple colored line represented starch/sorbitol bioplastic's characteristic peaks. For –OH stretching and bonding appeared at 2877 cm^{-1} and 3668 cm^{-1} repectively. The peaks at 1107 cm^{-1} and 1172 cm^{-1} indicated C-O stretching.

The addition of MCC and sorbitol aims to modify starch matrix. Lower wave number results in stronger hydrogen bond [9] as C-O peak of starch/MCC/sorbitol bioplastic is wider than the pure starch bioplastic. This means that the strong hydrogen bonding exist between C-O of starch and –OH of MCC. Furthermore, the addition of sorbitol can improve the intermolecular interaction.

4. Conclusion

The cassava peel starch/MCC bioplastics increased in tensile strength from 0,88 MPa for pure starch bioplastic to 9,12 MPa. Contrary to tensile strength, a decrease in elongation at break was reported. The improvement in bioplastics with reinforcing MCC could be attributed to the strong hydrogen bond between hydroxyl groups of the interface of both MCC fillers and starch matrix. This formation is also influenced by the percolation mechanism. Density of bioplastics decreased as MCC content increased. Furthermore, addition of MCC content decreased water uptake of bioplastics. The incorporation of MCC particles into starch matrix resulted in agglomerates that caused deflections in bioplastics. FT-IR spectra provided information about hydrogen bond through its characteristic peak. Bioplastic with the highest tensile strentgh assumed to have better characteristics than the other bioplastic was found at MCC content 6% and 20% sorbitol.

References

[1] Song J H, Murphy R J, Narayan R, Davies G B H 2009 Biodegradable and compostable alternatives to conventional plastics. *Phil. Trans. R. Soc. B* **364** 2127–2139

[2] Reddy L, Reddy S, Gupta A 2013 Study of bio-plastics as green and sustainable alternative to plastics *IJETAE* **3** 76–81

[3] Jain R, Tiwari A 2015 Biosynthesis of planet friendly bioplastics using renewable carbon source *J Environ Health Sci Eng* **13**

[4] Mose B R, Maranga S M 2011 A review on starch based nanocomposites for bioplastic materials *J Mat Sci Eng* **1** 239-245

[5] Wittaya T 2009 Microcomposites of rice starch film reinforced with microcrystalline cellulose from palm pressed fiber. *Int. Food. Res. J* **16** 493–500

[6] Psomiadou E, Arvanitoyannis I, Yamamoto N 1997 Edible films made from natural resources; microcrystalline cellulose (MCC), methylcellulose (MC) and corn starch and polyols. Part 2. *Carbohydr Polym* **31** 193–204

[7] Jonat S, Albers P, Gray A, Schmidt P C 2006 Investigation of the glidant properties of compacted colloidal silicon dioxide by angle of repose and X-ray photoelectron spectroscopy *Eur J Pharm Biopharm.* **63** 356–9

[8] Vieira M G A, Silva M A D, Santos L O D Beppu M M 2011 Natural-based plasticizers and biopolymer films: A review *Eur. Polym. J.* **47** 254-263

[9] Ahmad Z, Anuar H, Yusof Y 2011 The Study of Biodegradable Thermoplastics Sago Starch *Key Engineering Materials* **471-472** 397-402

[10] Cao N, Yang X, Fu Y 2009 *Food Hydrocolloids* **23** 729–735

[11] Myllarinen P, Partanen R, Seppala J, Forssell P 2002 Effect of glycerol on behaviour of amylose and amylopectin films *Carbohydr Polym.* **50** 355-361

[12] Bierley A W, Heat R J, Scott M J 1988 *Plastic Materials Properties and Aplications* (New York : Chapman and Hall Publishing)

[13] Zhang Q, Benoit M, Vigier K D O, Barrault J, Jégou G, Philippe M, Jérôme F 2013 Pretreatment of microcrystalline cellulose by ultrasounds: effect of particle size in the heterogeneously-catalyzed hydrolysis of cellulose to glucose *Green Chem.* **15** 963–969

[14] Wypych G 2004 *Handbook of Plasticizer* (Toronto : ChemTec Publishing)

[15] Dufresne A, Vignon M R 1998 *Macromolecules* **31** 2693-2696

[16] Porras M A, Cubitto M A, Villar M A 2015 A new way of quantifying the production of poly(hydroxyalkanoate)s using FTIR *J Chem Tech Biotechnol*

[17] Pavia D L, Lampman G M, Kriz G S 2001 *Introduction to Spectroscopy* (Philadelphia : Thomson Brooks/Cole) p 23

[18] Qu P, Gao Y, Wu G F, Zhang L P 2010 *BioResouces* **5** 1811-1823

Mobile Attendance Checking System on Android Platform for Kazakhstani University

N Saparkhojayev[1], E Shakhov[2], Y Mailybayev[3]
[1]PhD, Head of "IT in Education" Scientific Lab, Almaty Management University, 227 Rozybakiev str, Almaty, Kazakhstan
[2] Almaty Management University, 227 Rozybakiev str, Almaty, Kazakhstan
[3] MSc in IS, Head of IT Center, Kazakh University Ways of Communication

E-mail: nursp81@gmail.com

Abstract. In the 21st century- the century of Information Technology, it is difficult to imagine life without any gadgets: phones, tablets, computers, laptops, and so on. Today, smartphones and tablets are becoming popular, and thus their operating systems become popular too. Android is designed for low-power devices that run on battery power at full capacity, using all of its services, such as cameras, lights, GPS navigation, Wi-Fi, etc. In Kazakhstani universities, the process of checking students' attendance is one of the important issues, because final grade evaluation of students is based on their total number of appearances and their grades during the whole semester. This raises the question of having some tool to control students' attendance. There are many possible ways of controlling attendance: there are many examples when universities prefer to control attendance by the use of paper sheet, and some universities prefer to use two-stage way of controlling attendance: firstly, teachers and professors use paper sheet for checking students' attendance and after this, they fill out these information into a system manually. However, this is not an efficient way since there will be spent much of time for calling students names and putting marks like "presence" or "absence" if the class is a lecture class, and in this class at least 5 groups are presented. Furthermore, some students may call his/her friend as "presence" nevertheless to the fact that this student is currently absent. After taking into consideration all these issues and the fact that many gadgets use Android platform, authors of the following research paper decided to create a mobile system that makes easier to check students' attendance automatically, and this system is implemented in Almaty Management University, Kazakhstan. The system is based on Android platform, and in this paper, details of this system are presented.

1. Introduction

OS Android - an operating system that is based on the Linux kernel and is used in many different gadgets: cell phones, tablets, computers, laptops, televisions, watches and many other devices. Initially, the operating system was developed by the same company Android Inc., which later was purchased by Google, and then created an alliance within the company Open Handset Alliance (OHA), which is responsible for supporting and further development of the platform. The first version of the operating system Android coming from Google was released in September 2008. By the end of 2010 Android has become the best-selling operating system for mobile devices [1]. Android supports building applications that use phone features and at the same time, Android protects users by minimizing the consequences of bugs and malicious software. Android's process isolation precludes the need for difficult policy configuration files for sandboxes. This gives applications the flexibility to utilize native code without compromising security of Android or granting the application additional rights [2]. Over one million Android applications were published by the Google Play store in July 2013 and over 50 billion applications were downloaded, and 71 % applications for Android were created which was founded by survey of mobile application developers in spring of 2013. In 2015, survey found that 40 % of full time professional developers see Android as the target platform which is more than iOS [3]. Figure 1 depicts the architecture of Android operating system. The operating system is based on Linux kernel version 2.6.x that is a monolithic kernel. The kernel consists of drivers for different mobile device hardware, like screen, keyboard, camera, USB, Bluetooth etc. Kernel provides interface hardware and memory management, processes and other resources.

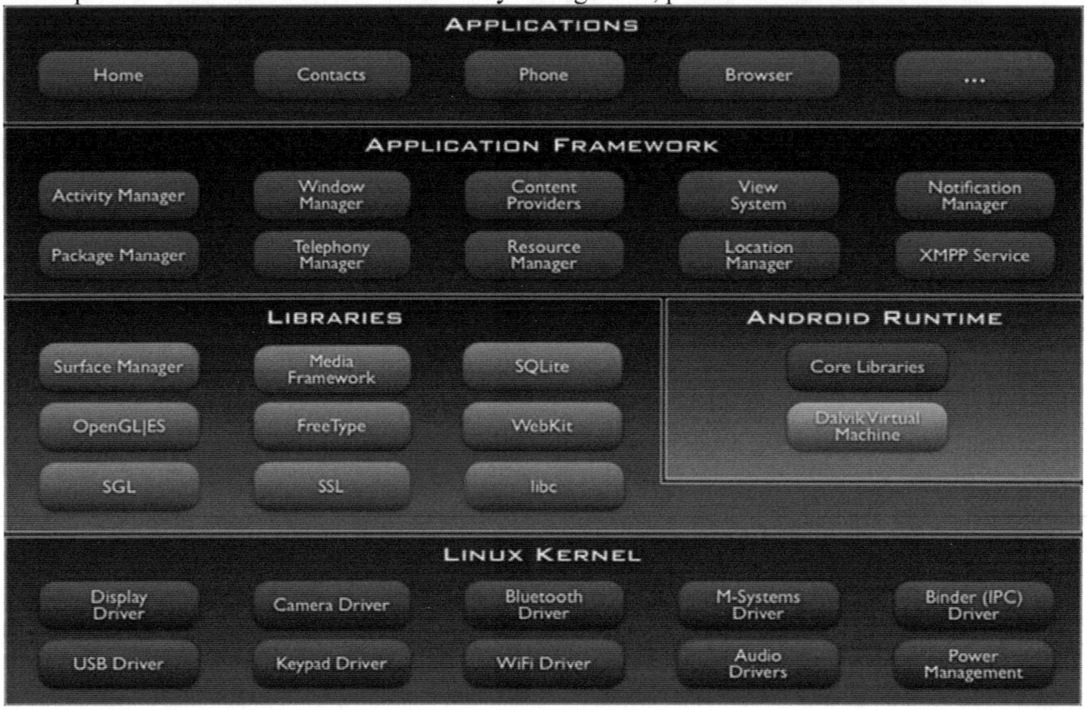

Figure 1. Android Architecture.

Native libraries on the next level are dependent on hardware architecture of the mobile device. These libraries include support for 2D and 3D graphics (Single, OpenGL ES), multimedia, security, storage, browsing (WebKit) and standard C library [4]. Android applications are developed by using Java programming language, and applications require an environment to manage their lifecycle. This includes a Java virtual machine (named Dalvik virtual machine) and Java class libraries that provide basic support for applications (collections, input/output operations management etc.). Android applications are not compatible with Java ME or Java SE. The applications are optimized for mobile devices constraints. The application programming interface allows accessing a framework that includes components used by all Android applications. The application framework includes components for Android application management (installation, execution), windows management and user interface graphical, event handling etc. Application level includes pre-installed applications (contact management, phone, calendar, Internet browser) and user applications. Applications are based on Java technologies and use classes provided through application programming interface. In addition, there is the possibility to use native functions in programs written in C/C++ programming language through NDK (Native Development Kit). Widgets unlike applications, occupies only part of the main

display screen and associated (Home). Android operating system is multitasking, each application running in a separate thread [5].
Android applications are developed using one or more basic components [6], [7]:

- activities (Activity base class);
- services (base class Services);
- content providers (ContentProvider base class);
- components that receive and act on messages sent to all applications (the base class BroadcastReceiver);
- messages (class Intent).

A particular importance in application development is the resources that enable separation of interface code. Activities represent the screen associated to an application. An application can have one or more activities. Services are routines that run in parallel with the main thread and do not have GUI. They allow the development of actions in the background without blocking the main thread execution and interaction with such application. Content providers are used for sharing data between applications. Data sharing is done through files, databases or other means. An alternative to content providers is the use of communication between processes. Applications can respond to the occurrence of events in the system by using classes derived from BroadcastReceiver. In order to activate components like activities, asynchronous messages encapsulated in objects of Intent type are used. Android applications are developed mainly using Eclipse IDE with Android Development Tools (ADT) plug-in. Android SDK and emulators are necessary for application development. According to [8], designing an Android app needs to be a constant collaboration between design and development. Additionally, app creators should target specific phones. There is too much trial and error involved in asset creation to be left squarely in the hands of a designer and this can lead to resource problems.

2. Motivation for implementing mobile application

In some Kazakhstani universities, professors and lectures take attendance by calling out the names and surnames of students, and then marking them, while, in others, teachers pass around a sheet of paper, asking students to sign in attendance sheet just next to their surnames. Both practices have their disadvantages. In the first case, if numerous groups attend the same lesson, checking all of these students by name and surname might take more than 10 minutes out of each lesson; in the second case, friends of absent students may sign for them next to their names and surnames. These practices place university professors, lectures and their institutions at considerable disadvantages when it comes to taking attendance. To rectify these systematic failings, we have decided to use mobile cell phones on Android platform for upgrading this service. We have taken into consideration the popularity of Samsung cell phones and smartphones here in Kazakhstan. We have found that many professors and lectures in ALMU use gadgets from this company. Each professor has its own cell phone, and each professor downloads this mobile application into his/her cell phone and adds specific data related to classes and courses he/she teaches. We wanted to show to our Administration that this mobile application has bright future, and that we need to get rid of paperwork and do everything by using innovative technologies. Our system allows our professors and lecturers to do the following:

- View the list of students;
- Add and delete groups;
- View the list of students in specific group;
- Add and remove students;
- Grading and checking attendance;
- Creating reports about attendance and students achievement;
- Viewing and changing reports about attendance and achievement;
- Ability to select the type of assessment for different classes (credit, 5 - point system, 10 - point system, a 100 point system);
- View information about a student and/or a group.

Journal of Physics: Conference Series **710** (2016) 012013 doi:10.1088/1742-6596/710/1/012013

As, it was mentioned above, this was beta version, so we decided not to cover all the functionalities, instead, we wanted to have some application, which will perform basic, but important operations regarding the checking Student's attendance.

3. System architecture and its description

To begin with, we would like to go through some applications that exist on the market and do similar operations. There are many applications, but the most popular ones are: application titled "Educate", which was created for Apple iOS [9], and some other applications oriented for OS Android. This mobile app is very functional, and it allows users to store and edit schedule, information about students, students' attendance and performance. The program has a synchronization feature with Google Docs and the ability to send e-mails and calls to students. The screenshot of the application is shown in Fig. 2.

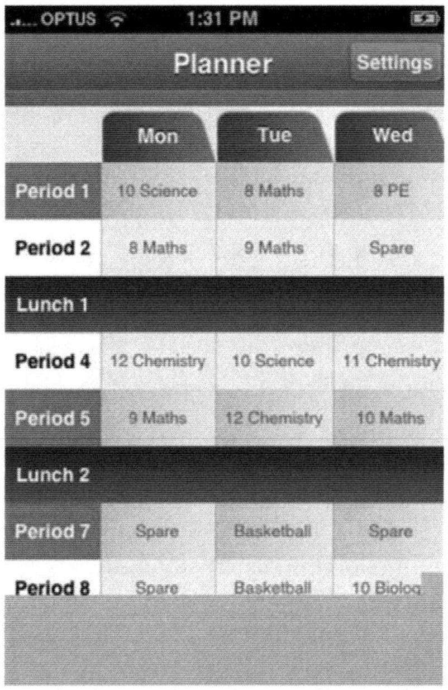

Figure 2. The "Educate" mobile application.

The most popular, similar mobile applications for OS Android are «Grade Book for Professors», «Attendance», and «GradeA». Applications «Grade Book for Professors» and «Attendance» are developed by Academy Geeks Inc[10-11]. These applications allow users to keep record of progress and attendance of students, respectively, and also they support synchronization with Google Docs. Screenshots of the application shown in Fig. 3 and Fig. 4 respectively.

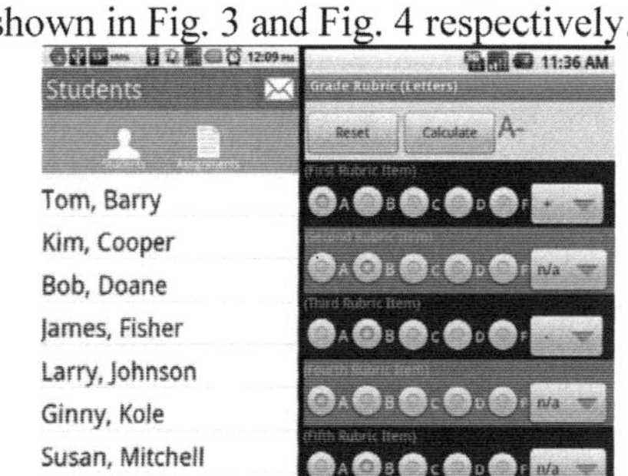

Figure 3. The screenshot of «Grade Book for Professors» application.

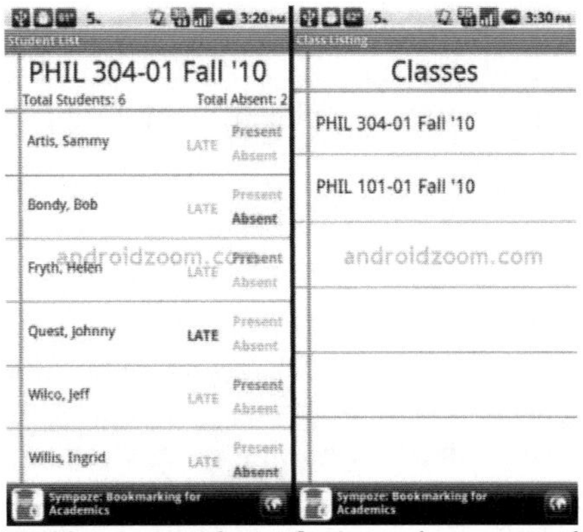

Figure 4. The screenshot of «Attendance» application.

Applications «Grade Book for Professors», «Attendance» and «GradeA» compared to «Educate» application are less convenient to use and do not allow you to create and store course schedule; moreover, in these applications instead of tables, lists are used. Also, they use American grading system, which is not the same as Kazakhstani grading system. The program, developed as part of this research paper in terms of functionality is closer to the program «Educate» and allows users to store and edit training schedule, information about students, their attendance and performance, but in our application compared to «Educate» application timetables are associated with students groups and journals. After learning and analysing all these applications we decided to create our own mobile application that allows professors and teaching staff of ALMU perform attendance checking automatically by the means of their smart phones. As a starting point for creating the architecture for this mobile application, we used previous work in [12], which was using RFID technology and cards instead of mobile smart phones. In Fig. 5, the diagram of how teachers use this mobile application is shown.

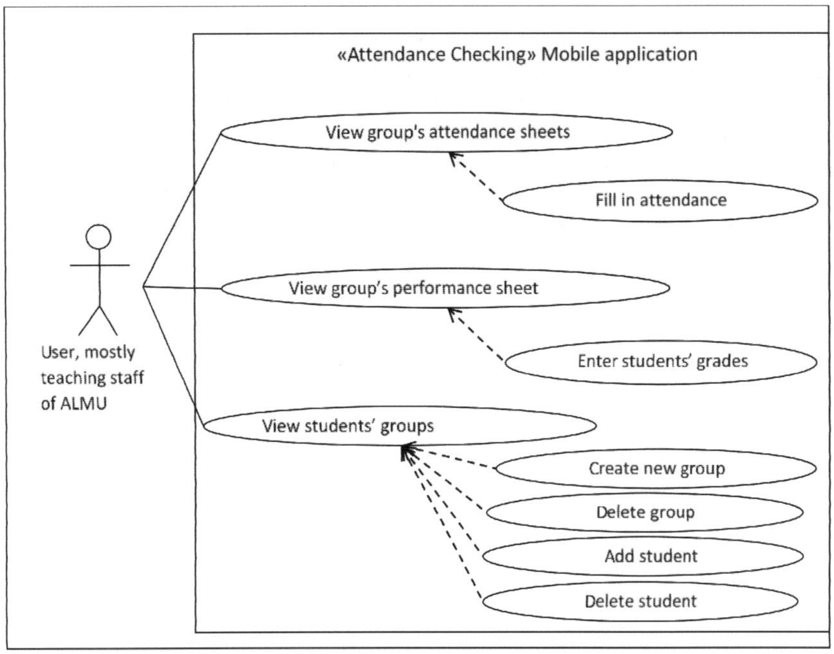

Figure 5. The diagram of the use of mobile application.

From the Fig. 5, we may see that this mobile application covers the most important operations that teaching staff of our University need, the operations of Attendance checking, Entrance of students' grade and so on. In the next section, we will describe this mobile application in more detailed form.

4. The description of mobile application

As programming language for implementing this mobile application, we have chosen Java. Java is an object-oriented programming language developed by Sun Microsystems. Java applications are usually convert into a special byte code, so they can run on any Java-virtual machine regardless of computer architecture. Java programs are translated into bytecode executed by the virtual machine Java (JVM), a program of processing the byte code and transmitting instructions to the hardware as an interpreter [13]. As the Development environment the newest modern IDE called AndroidStudio was chosen. Its main characteristics are as follows:

- In this IDE, the ability to layout in real-time is implemented;
- Many options of sizes and screen resolutions are added;
- Help section has been developed;
- Built-in-tools enabling improvement of the quality of applications and applications monetization are added;
- Tools for tracking the performance of ads, as well as tool for interaction with beta-testers are included in this IDE [14].

For DB, the SQLite was chosen due to its some characteristics that we liked. SQLite - a lightweight, embedded database. The word "embedded" means that SQLite does not use the client-server paradigm, meaning that SQLite engine is not a separate working process that program interacts with, instead of this, it provides a library, with which the program is assembled and the engine becomes an integral part of the program. Thus, as the exchange protocol function calls (API) of SQLite library are used. This approach reduces overhead costs, response time and simplifies the program. SQLite stores all database (including definitions, tables, indexes, and data) in a single standard file on the computer on which the program is executed. Ease of implementation is achieved due to the fact that prior to the execution of a transaction record the entire file that stores the database is locked; ACID-functions are achieved, including by creating a log file [15]. The structure of Database is described in Figure 6.

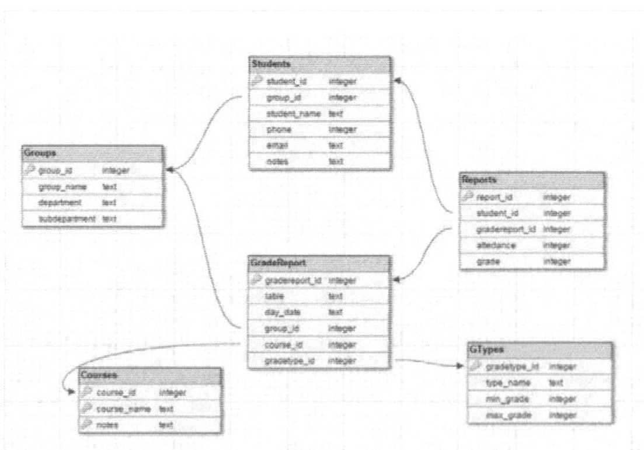

Figure 6. The structure of Database for Mobile system.

We have six tables, and each of these tables are used and due to the fact that our teachers may have difficulties we decided not to overload the process of attendance checking and for this reason, we minimized the number of tables.

After the successful compilation, our program will launch and at the screen of the smart phone the picture like in Fig. 7 will appear.

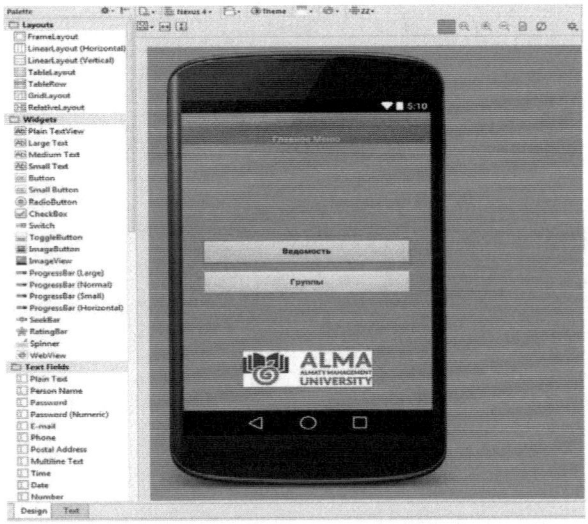

Figure 7. The start of our mobile application.

After this, user needs to choose appropriate group, and checks students' attendance by checking boxes near to students' last names. We have tested this application and used this application for one specific course, on which teacher used this application as attendance checking tool. During this period, application performed correctly and stable.

In general, after presenting this app, teaching staff of ALMU in general accepted this app with a joy, since they all are tired of manual checking and in their opinion, this is good replacement of paperwork.

5. Conclusion

In this research paper, the idea of implementing the mobile system checking students' attendance based on OS Android is discussed. This system is flexible and this is beta-version, which means that it may be extended by adding more functionalities. For the future work, this research should be extended by adding more modules and making some updates or changes. We are planning to extend this mobile system by adding some new modules, specifically "Library module", "Doors Access Control module", "Payment module", "Parking lot module", and so on. There was research done in [16], which showed how to build and implement Doors Access Control system based on NFC technology. Simultaneously, other versions for other Mobile OS should be implemented so that we can cover all possible smartphone models that our teaching staff use. Also, we plan to start doing research on secure wireless data transmission between the mobile smart phones and our server, meaning that we need to think about situations when students may try to attack our system for replacing their grades, putting Attended Mark near their surnames, and so on. We have to add some secure encryption algorithm that will exclude above-mentioned scenarios.

6. References

[1] Google's Android OS: Past, Present, and Future. Retrieved from: http://www.phonearena.com/news/Googles-Android-OS-Past-Present-and-Future_id21273.

[2] Jesse Burns, "Developing Secure Mobile Applications For Android: An introduction to making secure Android applications", October, 2008, Version 1.0. Retrieved from: https://www.isecpartners.com/.

[3] Need To Know Facts About Android Version. Posted on August 21, 2015. Retrieved from: http://androidnios.com/need-to-know-facts-about-android-version/.

[4] The Developer's Guide. Android Developers [Online]. Retrieved from: http://developer.android.com/guide/index.html.

[5] Paul Pocatilu, "Developing Mobile Learning Applications for Android using Web Services". Informatica Economică vol. 14, no. 3/2010.

[6] E. Burnette, "Hello, Android: Introducing Google's Mobile Development Platform". 2nd Edition, The Pragmatic Bookshelf, 2009.

[7] R. Meier, "Professional Android 2 Application Development". Wiley Publishing, Inc., 2010.

[8] Adam Beckley. "Android design guidelines. Version 1". Retrieved from: https://www.som.soton.ac.uk/learn/elearning/lab/mobile/images/MMAndroidDesignGuidelines.pdf.

[9] Official website: https://itunes.apple.com/en.

[10] Official website for application: http://androidforacademics.com/grade-book/.

[11] Official website for application: http://androidforacademics.com/attendance/.

[12] Nurbek Saparkhojayev and Selim Guvercin, "Attendance Control System based on RFID-technology", IJCSI International Journal of Computer Science Issues, Vol. 9, Issue 3, No 1, May 2012.

[13] Java. Official website: http://www.oracle.com/technetwork/java.

[14] AndroidStudio. Official website: http://developer.android.com/sdk/index.html.

[15] SQLite. Official website: https://www.sqlite.org/.

[16] Nurbek Saparkhojayev, Aigul Dauitbayeva, Aybek Nurtayev, Gulnaz Baimenshina. "NFC-enabled access control and management system". 2014 International Conference on Web and Open Access to Learning (ICWOAL), pp.1-4, Nov. 25-27 2014.

The Synergize effect of Chain extender to Phosporic acid catalyst to the ultimate property of Soy-Polyurethane

Flora Elvistia Firdaus
Director of Research Jayabaya University
Dept. of Chem Eng, Jayabaya University-Jl Pulomas Kav 21 Jakarta 13201, Indonesia

E-mail: flora_elvistia@yahoo.com

Abstract. The polyurethanes (PUs) foam were made from vegetable oil; a soybean based polyol. The foams were categorized into flexible and semi rigid. This research is manufacturally designed polyurethane foams by a process requiring the reaction of mixture of 2,4- and 2,6-Toluene di Isocyanate isomers, soy polyol in the presence of other ingredients. The objective of this work was to functionalized soy-polyol using phosporic acid catalyst and chain extender, study their collaborative reaction in producing ultimate property of PU foam. Correlates the foam morphology images in accordance to mechanical properties of foams.

1. Introduction

Polyurethanes are an important group of polymers that submit chances to obtain the designed properties by proper selection of different composition [1]. The blends of polyol, isocyanate, and chain extender from low molecular gives rise to a multitude of forms which suitable for extremely different practical applications [2].

In the current years a fast growing interest in the development of bio-based products can reduce the widespread dependence on non renewable resources. Petroleum is non renewable, with inevitable depletion resources coupled with high cost has prompted researchers to develop alternatives to renewable based. Using vegetable oils and its derivatives for polyurethane are well known [3]-[4]. The nature of easy to degrade renewables and their derivatives, a symbol of environmental friendly has made them as an attractive candidates for development of sustainable technologies [5].

The properties of polyurethane (PU) foam can be modified within wide limits depends on raw material used. The major raw materials of polyurethane synthesis are polyols and isocyanates, which are largely influence to properties of foams and mainly contributes to bursting process of foam.The are requisite of the process that can decides the rigidity or flexibility of foams which are; chemical structure, equivalent weight, and functionality, and combination of polyols [6].

The preparation of polyols from vegetable oils for polyurethane and its application has been the subject of many areas of research. The very attractive of vegetable oils based polyols are its multiple hydroxyl functionality [7].

Vegetable oil-based polyols are traditionally prepared from triglyceride molecules. Polyols are successfully prepared using different methods. The most common methods are carbon-carbon double bonds which further oxirane ring-opening with alcohols or other nucleophiles [8]- [12].

Presenting phosporic acid as an appropriate catalyst in epoxy compounds, will react to reactive isocyanates forming oxazolidone during foaming process of polyurethane [13]. Most of rigid PU's are prepared from judicially formulated mixtures of polyol. The mixtures may contain a mixed polyols to the accomplishment ultimate properties. The benefit of using polyol mixtures for PU foam synthesis, it has minimum tendency to crystallize which further benefit to the better physical properties. The difunctional such as 1,2 ethanediol or ethylene glycol are essential as chain extenders; in general is low molecular weight reactants produces hard segments; an increased intermolecular association or bonding effect [14].

The soft and hard part of multi-element array influenced the products obtained. The research was conducted to determine the synergize effects of phosporic acid catalyst in soy-epoxide hydroxylation reaction and chain extender to property of polyurethanes product. However, the reports did not explored the optimization amount of chain extenders; an ethylene glycol. Moreover, had studied the effect of blending of chain extender and polyol to properties of polyurethane. Chemical blowing agents and distilled water are extensively allow for the prior in PU foams synthesis, the responses to isocyanate has generated carbon dioxides and polyureas. The carbon dioxide causes of foaming which forming sectional structures. The practice of mixture of chain extender and polyol in making polyurethanes are to improve the properties and the foaming process which seldom studies.

The objective of the work is to determine the effect of phosporic acid to the effectiveness of chain extender in the synthesis to mechanical and the subtle changes of foams cellular images.

1. Experimental Procedures

2.1.Materials

The soybean oil was manufactured from local grocery. Hydrogen Peroxide and phosporic acid are obtained from Brataco Chemika, acetic acid from Merck Germany, and sulfuric acid from Tedia Company, Inc. All solvents and reagents were of laboratory grade and there were no treatments before used.

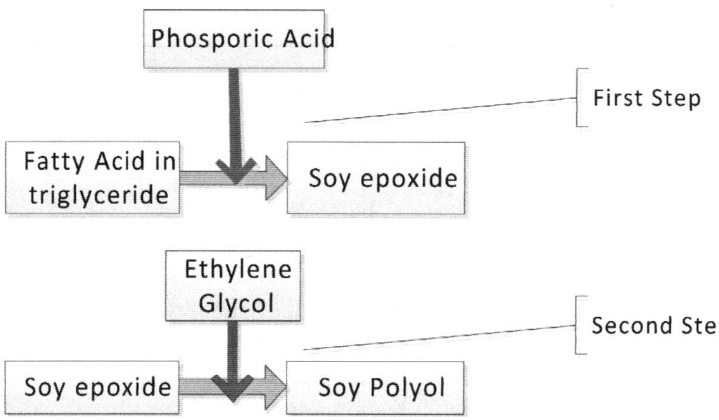

Figure 1. The Flowchart of the Process

For the first step, the unsaturated fatty acids in triglycerides chain of soybean oil are converted to soy-epoxide using phosphoric acid; having an oxirane rings. In the second step, the soy-epoxide with ethylene glycol are hydroxylated to soy-polyol.

Four blended of soy-polyols (SP); SP 1, SP 2, SP 3, and SP 4 are prepared from individual soy-polyol to achieve certain PU products, which then categorized. The catalyst were applied in the synthesis; phosporic acid with concentration 0,5% (v/v), 1,0% (v/v), and 1,5% (v/v) using three variables of temperature; 50^{0}C, 60^{0}C, and 70^{0}C. Another catalyst; sulfuric acid were used as for comparative. The soy-polyol were characterized before proceed to polyurethane synthesis, as can be seen in table 1.

Table 1. The Property of Material

Material	Properties	Value	Unit
Soybean Oil	1. Iodine Value	53,89	gr Iod/100 gr
	2. Viscosity	65,53	cps
	3. Acid Value	0,024	mgr KOH/ gr sample
Soy Polyol (SP)			
Temp 50 C			
SP 1	Oxirane Number	4,7	mgr KOH/ gr sample
SP 2		5,1	mgr KOH/ gr sample
SP 3		5,1	mgr KOH/ gr sample
SP 4		4,7	mgr KOH/ gr sample
Temp 60C			
SP 1	Oxirane Number	4,6	mgr KOH/ gr sample
SP 2		4,7	mgr KOH/ gr sample
SP 3		5,1	mgr KOH/ gr sample
SP 4		4,7	mgr KOH/ gr sample
Temp 70 C			
SP 1	Oxirane Number	4,7	mgr KOH/ gr sample
SP 2		4,7	mgr KOH/ gr sample
SP 3		5,1	mgr KOH/ gr sample
SP 4		4,7	mgr KOH/ gr sample

3.Methods

The property of polyol especially the physicochemicals have direct effect to the performances of polyurethane, as can be described below as important criterias for characterizing polyols.

3.1. Equivalent weight of Polyol

Equivalent weight more focused on polyols which has a weight distribution , the average equivalent weight can be calculated using the hydroxyl and acid number.

Equivalent weight of Blend $= \frac{56{,}1\, x\, 1000}{Hydroxyl number + Acid\ number}$ (1)

Acid number is calculated using = % weight of polyol A + % weight of polyol B (2)

So that

Equivalent weight of Blend $= \frac{100}{\frac{\%\,weight\ equivalent\ polyol\ A}{weight\ equivalent\ of\ polyol\ A} + \frac{\%\,weight\ equivalent\ B}{weight\ equivalent\ of\ polyol\ B}}$ (3)

3.2.Impact Resilience

The impact resilience of foam measures bouncing, elasticity, or springiness which describe % resilience or as % of return. The method has explained, percentage (%) of resilience of using two metal balls weighed 46 gr and 2,7 gr. The balls were dropped from 50 cm onto foam which every 5%

of return calibration is marked. Three drops were executed, and every three readings are averages counted.

3.3.The Obvious Density
The obvious density of foams are measured as per ASTMD 1622-03. The sample are prepared in scale 1cm^3. The obvious density of of PU 1, PU 2, PU 3, and PU 4 were measured.

3.4.The Foam Morphology
The PU cellular images were studied by using Scanning electron microscope.

The synthesis of PU foam was done in a 400 mL glass reactor at normal pressure. The prepolymer product is a soy polyol was reacted to TDI isomers. It was designed for 2 hours of reaction. The result of the synthesized material was poured into a glass mold at room temperature for 26 hours.

3.5.Foam Preparation and Evaluation
The preparation of polyurethane foam were marked by its free rise method with formula which is shown in table 2.

Table 2. The formula of Polyurethane

Step 1	Step 2	Code of
Epoxide: EG (mol/mol)	Soy-polyol/H2O /TDI Isomers (70:30) (mol/mol)	PU (Foam)
1:01	2,5:1:2,5	PU 1
1:05	2,5:1:2,5	PU 2
1:07	2,5:1:2,5	PU 3
1:09	2,5:1:2,5	PU 4

[a]The epoxide temp. synthesis are respectively 50,60, and 70^{0}C
[b]The elasticity of foams are influenced by polyol series
[c] Concentration of isocyanates are constant and stoichiometric.

4.Results and Discussion

4.1.The Effect of Chain Extender, catalyst, and temperature to Density
The synergize effect of phosporic acid catalyst to polyol synthesis using three different of temperatures; 50^{0}C, 60^{0}C, and 70^{0}C for PU synthesis, as can be seen in figure 2. The polyol was synthesized using four different ratios of epoxide to ethylene glycol; (1:1), (1:5), (1:7), and (1:9), using phosporic acid catalyst; 0,5%, 1,0%, and 1,5% (v/v).

The reaction is more pronounced at 50^{0}C to phosporic 0,5% (v/v) with the ratio of epoxide to ethylene glycol 1:1 (mol/mol), where the density of PU 1 is existed the highest among others, eventhough in the form of soy-polyol product the SP 3 has achieved high oxirane number compared to SP 1. But if the synthesis is decide at 70^{0}C, the best H_3PO_4 catalyst amount is 1,0% (v/v) for all ratio concentration of epoxide-glycol.

Journal of Physics: Conference Series **710** (2016) 012014 doi:10.1088/1742-6596/710/1/012014

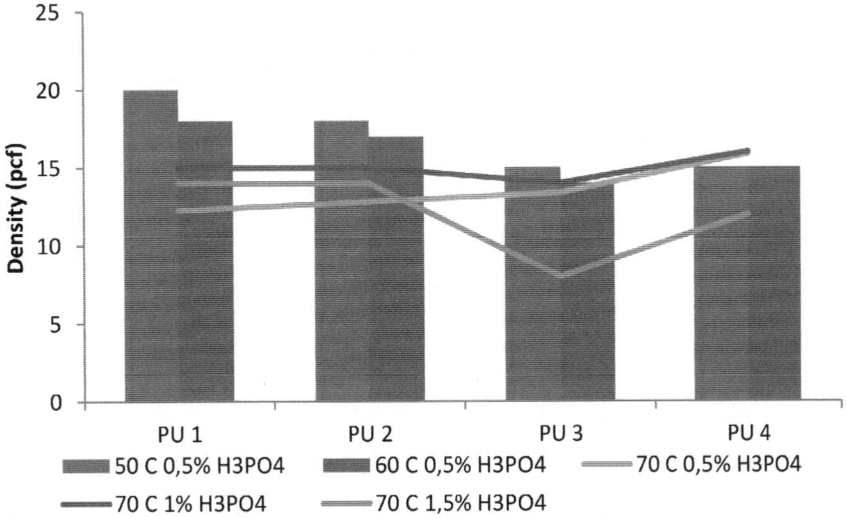

Figure 2. Catalyst concentration and temperature to density

4.2. Effect of Polyol and its blends on the properties of PU

Mechanical and physical properties PU foams are depends on functionality, equivalent weight, and hydroxyl value of polyols. The equivalent weight of polyol in common is 13,5 grek as compared to polyol 1 was 208 grek; polyol 2 was 181 grek, and polyol 3 was 122 grek. The equivalent weight is one among others parameter should be consider for the preparation of PU. The value are correlates to PU density which declined as value increased. The polyurethane from soy-polyol 4 labeled as PU 4 is found more rigid than PU 1, PU 2, and PU 3.

Aligned to the research founding, the foam manufacture using temperature 70^0C has the optimum density compared to 50^0C and 60^0C. Among temperature 70^0C, the optimized density has occured either using 0,5%, 1,0%, and 1,5% (v/v), they are categorized as flexible foam specially in the form of PU 2.

4.3.The Effect of Chain extender on PUF Impact Resilience

Water reacts with isocyanates forming carbon dioxide as chemical blowing agent, which in most cases chemical blowing agents such water are used in lower quantities compared to physical blowing agents. The ball used in this experiment was weighted 46 gr for ball 1 and 27 gram for ball 2. The addition of EG in PU's formula (4,26; 21,3; 29,8, and 38,34 mL) are respectively for PU 1, PU 2, PU 3, and PU 4 which made different responses to foam elasticity, as summarized in table 3. The resilience impact of PU are viewed as function of chain extender, phosporic concentration, and temperature of reaction to the bouncing responses. Phosporic react directly with oxirane group. Reduction oxirane content was observed with 1,5% (v/v) of phosporic. The result of PU 2 is classified as flexible while for PU 1 and PU 4 are less flexible. The fact for PU 1, the ethylene glycol addition of approximately 4,26 mL to either 1,0 % and 1,5% (v/v) of H_3PO_4 showed no bouncing effect during the test. But the compression set test using 1,5% (v/v) H_3PO_4 as for an advanced test has proven is the best amongst.

Table 3. The Impact Resilience of PU

Type of PU	Elasticity to Ball Bouncing (%)			
	H3PO4 1% (v/v)		H3PO4 1,5% (v/v)	
	Ball 1	Ball 2	Ball 1	Ball 2
PU 1	nb	nb	70,42	nb
PU 2	7,04	9,39	70,42	3,36
PU 3	3,36	nb	8,39	3,36
PU 4	1,3	3,91	70,42	3,91

[a]nb is indicated as the ball is not bouncing. Ball 1:46 gr, ball 2: 27 gr

[b]PU Elasticity (%) = $\{100 - \frac{(initial\ height - foam\ height\ of\ bouncing\ impact)}{initial\ hegiht}\}$x 100%

4.4.The Effect of Formula to Cellular Morphology

The mechanical property of a cellular material are mainly depend on density. The cellular images

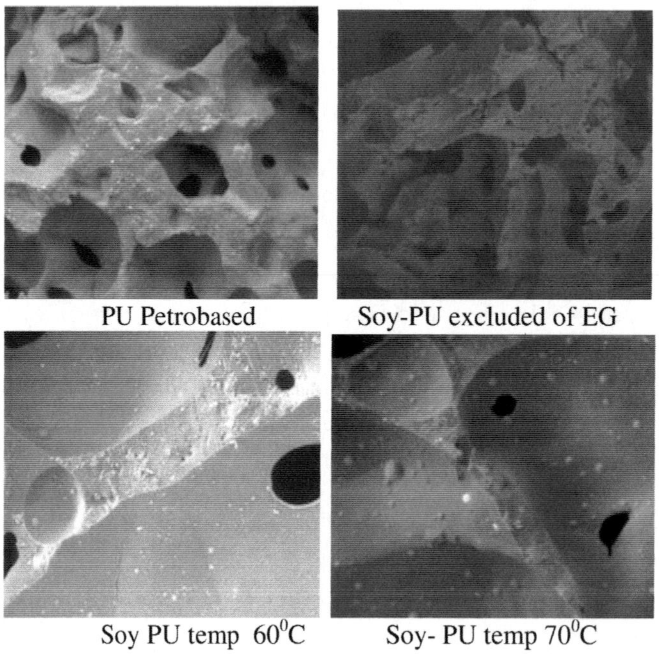

Figure 3. The SEM of PU

using SEM reveals the foam cells are slightly narrow with the enhancing temperatures. In this reaction, water molecules are as chemical blowing agent, it generates CO_2 on reaction with TDI isomers, resulted the evolution of heat. The temperature is effected to the volume of blowing gas in mixtures which exceeds its solubility limits and bubble nucleation. The SEM result has illustrated differentiate images of temperature effect during epoxide processing as previously designed was 50^0C, 60^0C and 70^0C. SEM images like those shown in figure 3 were examined and the average cell diameter are 0,1 mm for 50^0C; 0,5 mm for 60^0C; and 0,5 mm for 70^0C. The size difference between the foams is not statistically significant. The cellular images of PU without EG was vagued where at the initial state most of the cell faces are broken. Thus the SEM study and cell size, will only change slightly suggested to the pathway from partial SP 1 to PU 1 with temperature 70^{0}C, and thus the observed foam mechanical properties are unlikely to be due to cell sizes.

Quite obvious and comparable using petroleum based polyol to soy based polyol with the same composition, it slightly higher mechanical strength is due to higher crosslink density,which results in a higher hard segment content. The cell size are reveals more larger this could be effected to the exceeds water content [15].

4.5.Polyurethane Foam Water absorption behavior

The water absorption behavior of PU prepared from soy-polyol using phosporic catalyst, and sulfuric acid was used as for comparative. The hydrophobic nature of long carbon chain as well as ester linkage of PU in the synthesis, are much comparable of using sulfuric acid to phosporic acid. The amount of H_3PO_4 1,5% (v/v) is more water absorbance than 1,0% (v/v). In general H_3PO_4 catalyst in PU synthesis resulted more water absorbance than sulfuric catalyst.

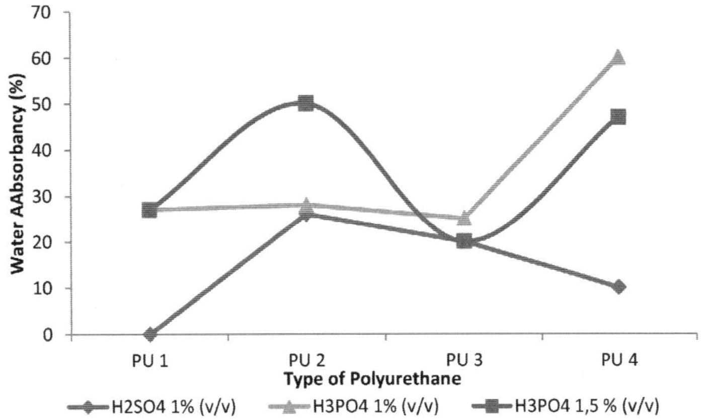

Figure 4. The polyurethane water absorbancy

[a] The foam is classified to the differentiation concentration of acid catalysis of H_3PO_4, H_2SO_4 was used as for the comparative

[b] H_2SO_4 1 % (v/v) is failed to react because of gel formation

4.6.Compression set of Polyurethane Foam

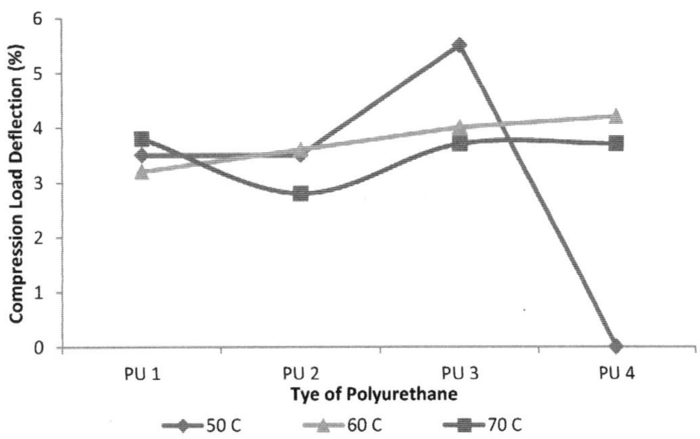

Figure 5. Polyurethane Foam Compression result

The foam characterization to percentage of compression set test has justified the foam flexibility value are recovered after compressed between two metal plates at controlled conditions: time and room temperature. The foam PU synthesized at 70^0C were discovered as the best among other foams because of its lowest value to recover to its original thickness.

Conclusion

On this study we found formula of making PU with optimum mechanical properties. Using phosporic acid catalyst, ratio of chain extender to epoxide, temperature in the PU synthesis has made responses to foams during characterization which then categorized into flexible and less flexible. The formula has much related to density, impact resilience, compression set, water absorbency behavior, and cellular morphology.

ACKNOWLEDGMENT

The author gratefully acknowledge for the financial assistance from Indonesia Ministry for Higher Education. Thanks are due to Department of Chemical Engineering Jayabaya University for providing laboratory and facilities.

REFERENCES

[1] E. Yamazaki *et. al.*, Polym J, 29, (1997), p.811
[2] T.L. Wang T.H Hsiech, Polym J, 28, (1996), p.839
[3] N. Jiratumnukul, Van de Mark MR, Preparation of glycol esters of soybean oil fatty acids and their potential as coalescent aids in paint formulations. J Am Oil Chem Soc 77, 2000, pp.691-697
[4] Thirumail et.al., Mechanical, morphological and thermal properties of rigid polyurethane foam: effect of chain extender, polyol and blowing agent, Cellular Polymers, vol. 28, no 2, (2008), pp. 145-158
[5] R.E. Harry-O'kuru, R.A. Holser, T.P Abbott, D. Weisleder D. Synthesis and Characterization of Polyhydroxy Triglycerides from Milkweed Oil. Ind. Crop. Prod., 15, (2002), pp.51–58
[6] R.E. Harry-O'kuru, C.J. Carriere, Synthesis, Rheological Characterization, and Constitutive Modeling of Polyhydroxy Triglycerides derived from Milkweed Oil. J. Agric. Food Chem.,50, (2002), pp. 3214–3221
[7] M. Desroches *et.al*, From Vegetable Oils to Polyurethanes: Synthetic Routes to Polyols and Main Industrial Products, Polymer Reviews, DOI: 10.1080/15583724.2011.640443, Taylor Francis Group, 52; (2012), pp.38-79
[8] Y. Guo, J.H. Hardesty, V.M. Mannari, J.J.L Massingill, Hydrolysis of Epoxidized Soybean Oil in the Presence of Phosphoric Acid. J. Am. Oil Chem. Soc., 84; (2007), pp. 929–935
[9] B.K.Sharma, A. Adhvaryu, S.Z. Erhnan, Synthesis of Hydroxy Thio ether Derivatives of Vegetable Oil. J. Agric. Food Chem., 54, (2006), pp.9866–9872
[10] Korshak et.al., Reaction of cyanurates with the epoxide ring. Russ chem. Bull 32, 1983, pp. 2135-2138
[11] N.J. Mills, Polymer foam Handbook, Elsevier Ltd, New York, 2007.
[12] F.E Firdaus, Silicone on Blending Vegetal Petrochemical Based Polyurethane, Journal of World Academy Science Engineering and Technology (WASET), p ISSN 2010-376x, e ISSN 2010-3778, WASET, Mei, 65; (2010), pp. 349-352
[13] W.J. Seo W.J. et. al., Properties of water blown rigid polyurethane foams with reactivity of raw materials, J. Apply Polymer Science, 93; (2004), p.2334
[14] M.Thirumal, Effect of foam density on the properties of water blown rigid polyurethane foam, J.Appl. Polymer Science, 108; (2008), pp.1810
[15] F.E, Firdaus, Dachyar M., Silicone on Blending Vegetal Petrochemical Based Polyurethane, WASET, 65:2010, pp. 152-156

Study on online community user motif using web usage mining

Meera Alphy[1] and Dr Ajay Sharma[2]
[1] PhD Scholar, Department of Computer Science and Engineering, SRM University, Haryana, India
[2] Assistant Professor, Department of Computer Science and Engineering, SRM University, Haryana, India

[1]Corressponding author e-mail: meeraalphy.urumbath@gmail.com

Abstract. The Web usage mining is the application of data mining, which is used to extract useful information from the online community. The World Wide Web contains at least 4.73 billion pages according to Indexed Web and it contains at least 228.52 million pages according Dutch Indexed web on 6th august 2015, Thursday. It's difficult to get needed data from these billions of web pages in World Wide Web. Here is the importance of web usage mining. Personalizing the search engine helps the web user to identify the most used data in an easy way. It reduces the time consumption; automatic site search and automatic restore the useful sites. This study represents the old techniques to latest techniques used in pattern discovery and analysis in web usage mining from 1996 to 2015. Analyzing user motif helps in the improvement of business, e-commerce, personalisation and improvement of websites.
Keywords. Log file, Pattern analysis, Pattern discovery, Pre-processing

1.Introduction

1.1Initial Period

As everyone believed Oren Etzioni mentioned web mining in his paper "The World Wide Web: quagmire or gold mine?" According to his Web mining [1] may be broken down into the following subtasks' Resource Discovery, Information Extraction and Generalization. Resource discovery was unearthing records and favors on the Web. Information Extraction automatically took out particular data from newly revealed network resources. Generalization was revealing motif at individual online community with other online community.

1.2Later Period

Web mining [21] is the use of data mining techniques to take out information from the online repository. Web mining consists of web content mining, web structure mining and web usage mining. Web content mining is the process of taking out useful data from the collective data of web document. Web structure mining was the process of finding already available structure information from the online community. It can be divided into two types' hyperlinks and document structure. Web usage mining is the use of data mining techniques to find a useful pattern of data by analyzing web users browsing patterns in the online community.

2. Web usage mining (wum) process

Web usage mining [29] is the process of discovering the patterns used by the web user from the web sites. Web usage having mainly three phases such as pre-processing, pattern discovery and pattern analysis. Collecting the necessary data's from web server logs, proxy server logs or browser logs. Log files are automatically stored by the web server which contains the information about the web pages accessed by the web user during browsing the websites. Pre-processing [19] is the first phase in WUM which having sub tasks' data cleansing, user identification and user session identification. Data cleansing coined unwanted entities to be removed. User identification can be done by internet protocol address, cookies and registration of users. User session identification can be done by analyzing a web user's browsing activity in a particular time period. Pattern discovery is the main factor in the web usage mining. The pattern can be discovered using statistical analysis, association rules, clustering, classification, sequential pattern, dependency modeling and etc. Pattern analysis is the third phase in web usage mining which gives refine the output patterns from the pattern discovery which can be done using Knowledge query mechanism or visualisation technique.

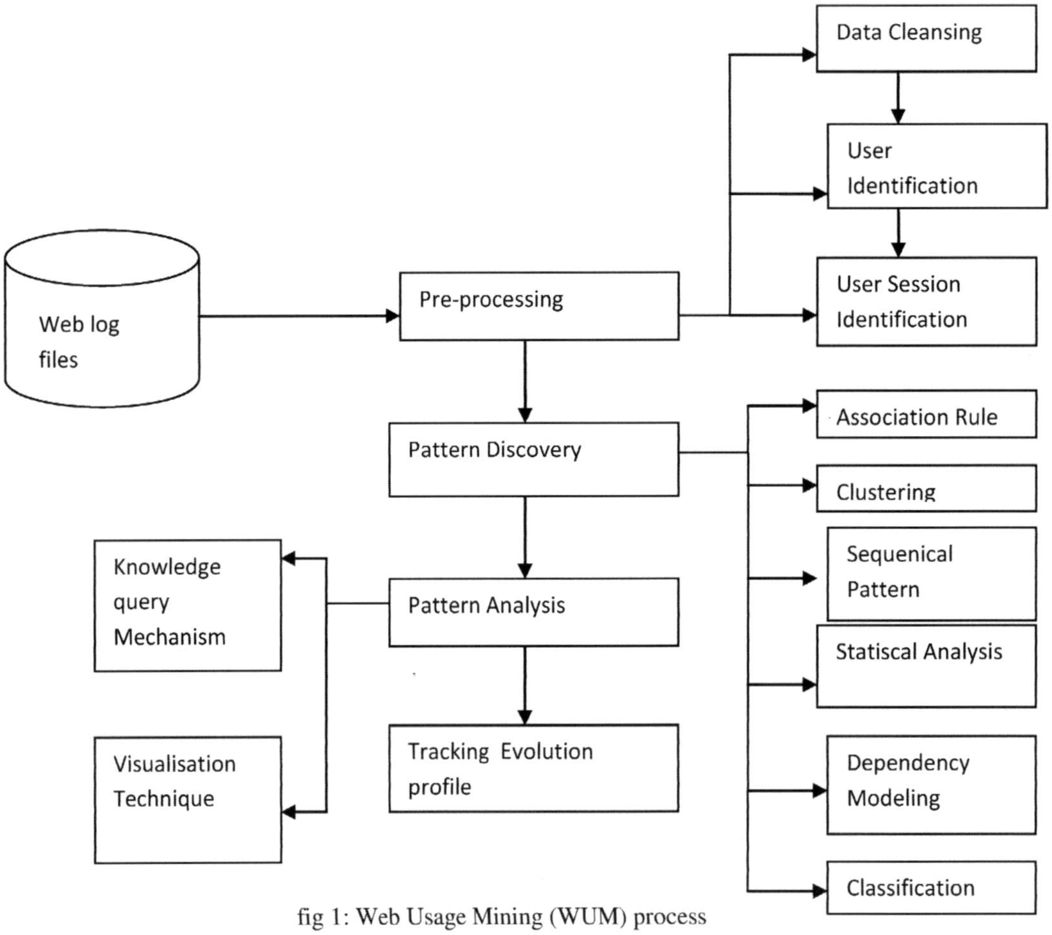

fig 1: Web Usage Mining (WUM) process

3. SURVEY ON WEB USAGE MINING

3.1 Knowledge and motif discovery on world wide web

R Cooley [2] *et.al* proposed web mining can be used for two distinct ways' web content mining and web usage mining. Web content mining is defined as extracting knowledge from different resources of World Wide Web. Two approaches are mainly used for web content mining; they are agent based approach and database approach. Web usage mining is defined as extracting user access patterns from the browser. Web usage mining has four process' Pre-processing, transaction identification, pattern discovery tools and pattern analysis tools. In pre-processing, the first task was data cleaning, which removes irreverent information and second task was transaction identification which gets all the information about the user browsing for a particular visit. Path analysis, association rules, clustering and classification, one among these can be used for discovering the web transactions. Path analysis could be used to occur in a particular way most regularly look in on track in a Web site. Association rule discovery techniques are in most cases applied to a structured set of data held in a computer, especially one that is accessible in various ways of proceedings where each proceeding was made up of a collection of individual data. Discovering classification rules allow one to start to exist in the collection of individual data to be a member of a specific set according to their common characteristics. Clustering analysis precede one to set of clients together or facts that have similar attributes. Web miner tool is mainly used tool for pattern discovery and analysis.

3.2. Interesting navigational patterns can be discovered and utilized to improve the services provided on the site

Georgios Paliours [3] *et.al* proposed the group of the people who used an online site into set of people who used similar navigational way of behaving in the online site. The database used in these experiments was collected from the Web site of the Advanced Course on Artificial Intelligence 1999 log files. The unverified machine learning algorithm COBWEB is used to arrange the consumer of the online community, who go after relative paths, into a small group of people having particular characteristics in common. Author given more importance to the elucidation of the common behaviour group that was developed through this procedure. And also COBWEB was an application of clustering which was suitable for symbolic training data such as access sessions. The resulting communities can be used to improve the services provided in the Web site. Author used each trouble-free metric for different navigational behaviour community. This data were used by the site developer to re-arrange it in a useful way that is customized to the importance of each community. Limitation was the representation of user sessions for the learning algorithm. Advantage was reducing the overlap and increasing the coverage.

3.3. Heuristic method for real time web usage mining

Florent Masseglia [4] *et.al* proposed a heuristic method for real time web usage mining. In view of the rising number of online community' caller, and the rate at which operates the web site at which consumer' activities may change from one particular time period to another time period, the output obtained by examining an access online repository may lose its importance, if it is exploited too late. Author proposed a Heuristic based Distributed Miner (HDM) designed to avoid such obstacles and to get regular activities motif. This method allows discovering regular activities motif in real time, which doesn't depend on connected users. In view of how quick the regular activities motif can vary since the final examination of the access online repository, this output thus provides completely tailored direction-finding schemas for user activity prophecy. Based on a distributed heuristic, author says this method also answers quite a few unfolded issues within the data mining framework are recognizing most popular zone and find out very long chronological motif and collaborative data mining. Corroborate HDM activity and output, the author had verified it with local databases, where each input, arranged in a particular order was allocated a thread (imitating in washer machine) on the evaluating in the machine. The advantage was recognizing a new kind of motif, collaborative data mining and mining vast regular chronological patterns. Disadvantage was limited neighbourhood operators used.

3.4 Ant clustering algorithm and linear genetic programming for web usage mining

Ajith Abraham [5] *et.al* proposed an ant clustering algorithm to find out web usage mining and a linear genetic programming approach to examine the visitor inclination. First-hand outcome allow that ant colony clustering perform well when compared to a self organizing feature map for clustering Web usage motifs, even though the performance, quality of state is not that performance efficient when compared to evolutionary-fuzzy clustering approach. An ant colony clustering was proposed to set apart from other visitors and thereafter a linear genetic programming approach to analyze the visitor motif. Data was collected from web access log data at the Monash University's Web site. The inputs were obtained from the statistical text data generated by the online repository from 1st January 2002 to 7th July 2002. Isolating the useful data from the raw file was an important part in the data pre-processing phase. After some preparatory analysis, the author selected the statistical data be made up of domain byte requests, hourly page requests and daily page requests as a focal point of the cluster structures for taking out Web consumer's usage motifs. The maximum currently used data were recorded with higher index while the minimum currently used data were recorded with a lower index. Disadvantage for these approaches where each of the datasets in the approach was run twice (for f= my to 10, 00,000) in order to test if somehow the output were alike. Hence time consumption is high. The advantages of this approach were classification space was always 2D, non-parametric and toroidal (Toroidal_describes something which resembles or relates to torus or toroid. Toroid was a doughnut-shaped object which resembles a torus).

3.5 USIA approach for users and session identification

Zhang Huiying[6], *et.al* the USIA was used for online users' recognition and session recognition. There are a lot of online users following a sequence of records come from the same Internet Protocol address in online repository files. Time consumption is higher when checking every record in an online repository. Hence, record efficiency, reduced. If two record internet protocol address is same, then it's assumed that the two records used by the same web user. The premise of USIA approach was based on two prerequisites as follows by acceptable user's identification precision and algorithm efficiency. Algorithm's advantage was good precision, high efficiency and favourable structure of the store. Algorithm's disadvantage was algorithm need check several elements when it examines whether the current online user is the same online user or not, this results time complexity. The data were collected from the online repository file of Tianjin University's Web site (http://www.tju.edu.cni) and time period of 1st march 2003 to 7th March 2003. The size of the total online repository file was 105M bytes and after pre-processing the size of the file was 378,747 records.

3.6 An individual approach for person name bipolarization

Chien Chin Chen [7], *et al.* an individual approach that founds out the polarity of persons' names. Author gave the indication of data in the principal eign vector of PCA could separate person name to bipolar sets fast. Two techniques' weighted correlation coefficient and off-topic block elimination were introduced to solve sparseness problem. Data were collected from 33 news documents related to the opening game from Google news and also 166 news documents related to the 2010 World Cup Final. The advantage of the approach was applied to different languages and diverse topic, identify bipolar person names in topic documents correctly without using any external knowledge source and context-oriented. Disadvantage of the approach were person names possessed neutral orientations.

3.7 Motif-based clustering approach to grouping customer transactions

Yinghui Yang [8] *et.al* using a motif-based clustering approach to assembling consumer transactions. The main idea was to cluster consumer transactions such that motifs generated from each cluster, while similar patterns to be within the cluster, are very different patterns generated from other clusters. The objective of the function was that maximize in order to goal a good clustering of consumer transactions. Advantage was user-centric Web usage data that illustrate that GHIC shows a highly effective clustering of transactions. For a good clustering algorithm had two measures of cluster qualities' internal quality and external quality. The internal quality measure that used was the average M measure across all set of two clusters generated by GHIC. The data was collected from market data vendor web browsing data. Disadvantage was algorithm should haven't considered global objective functions to develop better heuristic functions.

3.8Semantic web usage mining approach for consumer emotions and behaviours through self-reporting and behavioural tracking

A.C.M. Fong [9] *et al.* proposed a method a semantic web usage mining approach for consumer emotions and behaviours through self – reporting and behavioural tracking. Thoughts and mindset of the consumer have been found to influence a person's activity of moving from site to site on the online repository. The proposed approach has four different steps of Personal Web Usage Ontology (PWUO). First step, Personal Web Usage Lattice (PWUL) using input as relative online community records which consists of a group of attributes of a specific time period and a group of attributes on specific subjects necessary for the development of consumer emotions periodic patterns influencing the web surfing activities. A set of attributes at a specific period coined on different time period of a day (e.g., morning, afternoon, etc.), the days of a week (e.g., Monday, Tuesday etc.) and other concepts (e.g., weekdays, weekend, etc..). A set of attributes specific subject defines the useful domain ontological concepts. Second step, Global Web Usage Lattice (GWUL) to entitle all specific time period pattern-based worldwide online repository access activities and relative arranged in order of rank between these activities which can be very large. Third step, Global Web Usage Ontology (GWUO) derived from GWUL step by making a map of World Wide Web access activities and their relative arrangement in order of rank into a category of tasks having common property or attribute. The fourth step, PWUO is the based on information that can subsequently provide meeting the consumer's needs more effectively and efficiently, making interactions faster and easier and, consequently, increasing consumer satisfaction and the likelihood of repeat visits. Here the inputs were obtained from online repository file Nanyang Technological University, Singapore. The main advantage of this approach was a fast response (low latency) and other were good aid, relevant results. The main limitation of this method was not ease of use.

3.9New method to identify navigation related usability problems

Ruli Geng [10] *et al.* proposed a new method to identify navigation related usability problems. The inputs were obtained from Furniture Giveaway (FG) 2009 website. The method includes three modules where usage pattern extraction, Ideal User Interactive Path (IUIP) modelling and usability problem identification. Usage pattern extraction was done by extract real activity of accurately ascertaining the position and planning and following a route from server repository and finds a repeated design for some typical work data. Usage pattern extraction can be done by process' Data Preparation and Pre-processing, Transaction Identification and Trail Tree Construction. Data Preparation and Pre-processing consist of data cleaning, User identification, User session identification and path completion. Transaction Identification was sequence of the pages seen by the consumer to purchase the product. Trail Tree Construction used the same path followed by the different consumers make a tree structure which developed by trie algorithm and its frequency taken as trail tree. Meanwhile design IUIP models for same work. IUIP models are based on the perception of consumer action and can be noted by predicting tracks for particular consumer-based on a series of action. IUIP model specifies the maximum time visited by the consumer in a particular page and sequence of pages followed (path). This idea is obtained by reducing the complexity of Adaptive Control of Thought-Rational (ACT-R) model. Usability problem identification helps to find the main cause of web design problems using logical deviation calculation and temporal deviation calculation. Logical deviation was calculated by paths participated by IUIP model, but not select, it can be only count as one deviation and total transaction is equal to the deviation counted by total page visited by a consumer. Temporal deviation calculated by the consumer visited the page than the maximum time allowed for each page can count as one deviation and total deviation calculated by total deviation counted from total pages visited by a consumer. The main advantage of this model was usability improvement and reducing the complexity in the larger system. The limitation of this method was user satisfaction and presentation aspects of usability.

3.10 The request dependency graph representation of the relationships among HTTP Requests
Jun Liu [11] *et.al* introduced the request dependency graph was a graph representation of the relationships among HTTP requests. A directed link from A to B in the graph indicates the accessing of online community object B is caused by the accessing of A; hence B depends on A. Methodology to develop such a graph by mining the of time and connecting information among aggregated HTTP requests. The main objective of design and implement algorithm for primary request identification which was a significant task of web usage mining. Advantage was algorithm applied to large-scale real world, where the web access online repository shows that the request dependency graph is a useful tool for web usage mining.

TITLE	**AUTHOR**	**PROPOSED METHOD**	**LOG FILE**	**ADVANTAGE**	**DISADVANTAGE**
Knowledge and motif discovery on world wide web	R.Cooley, B.Mobasher, J.Srivastava [2]	Web Mining can be used for two distinct ways'web content mining and web usage mining			
Interesting navigational patterns can be discovered and utilized to improve the services provided on the site	Georgious Paliours, Christos Papathedorou, Vangelis Karkaletsis [3]	the group of the people who used an online site into set of people who used similar navigational way of behaving in the online site	Web site of the Advanced Course on Artificial Intelligence	representation of user sessions for the learning algorithm	reducing the overlap and increasing the coverage
Heuristic method for real time web usage mining	Florent Masseglia, Maguelonne Teisseire, Pascal Poncelet [4]	heuristic method for real time web usage mining	local databases	recognizing a new kind of motif, collaborative data mining and mining vast regular chronological patterns	limited neighbourhood operators used
Ant clustering algorithm and linear genetic programming for web usage mining	Ajith Abraham, Vitorino Ramos [5]	ant clustering algorithm to find out web usage mining and a linear genetic programming approach to examine the visitor inclination	Monash University's Web site.	Classification space was always 2D, non-parametric and toroidal	time consumption is high
USIA approach for users and session	Zhang Huiying, Liang Wei [6]	USIA was used for online users' recognition and session recognition	Tianjin University's Web site	good precision, high efficiency	algorithm need check several elements when it examines

identification		describes the online user navigational patterns		and favourable structure of the store	whether the current online user is the same online user or not, this results time complexity
An individual approach for person name bipolarization	Chien Chin Chen, Zhong-Yong Chen, Chen –Yuan Wu [7]	individual approach that founds out the polarity of persons' names	33 news documents related to the opening game from Google news and also 166 news documents related to the 2010 World Cup Final	approach was applied to different languages and diverse topic, identify bipolar person names in topic documents correctly without using any external knowledge source and context-oriented.	approach were person names possessed neutral orientations.
Motif-based clustering approach to grouping customer transactions	Yinghui Yang, Balaji Padmanabha n [8]	cluster consumer transactions such that motifs generated from each cluster, while similar patterns to be within the cluster, are very different patterns generated from other clusters	market data vendor web browsing data	user-centric Web usage data that illustrate that GHIC shows a highly effective clustering of transaction s	global objective functions haven't considered to give better huristic function
Semantic web usage mining approach for consumer emotions and behaviours through self-reporting and behavioural tracking	A.C.M. Fong, Baoyao Zhou, Siu C.Hui, Jie Tang, Guan Y.Hong [9]	semantic web usage mining approach for consumer thoughts and mindset of have been found to influence a person's activity of moving from site to site on the online repository	Nanyang Technologi cal University website	fast response (low latency), good aid, relevant results	Not ease of use.
New method to identify	Ruli Geng, Jeff Tian [10]	new method to identify navigation related usability	Furniture Giveaway (FG)	usability improveme nt and	user satisfaction and

navigation related usability problems		problems and method includes three modules such as usage pattern extraction, Ideal User Interactive Path (IUIP) modelling and usability problem identification	website	reducing the complexity in the larger system	presentation aspects of usability
The request dependency graph representation of the relationships among HTTP Requests	Jun Liu, Cheng Fang, Nirwan Ansari [11]	introduced the request dependency graph was a graph representation of the relationships among HTTP requests and primary request identification which was a significant task.	Local database	applied to large-scale real world	Less accurancy

Table1 : table of review

4. Areas were web usage mining applied

E-Learning Personalization [12] recommendation system put forward the skills through study resources were reckon based on the current learner's recent activity of accurately ascertaining his/her position and planning and following a route history, as well as utilizing likes and dislikes among learners' predilection and educational content. Intelligent agents to find missing and exploited children [13] provide a user interface system were customizable. Data mining used intelligent agents to find missing and exploited children over the World Wide Web, chat rooms and intranets. Data mining in World Wide Web is known as web mining. Predictive Web usage visualizations [14] help to understand usage information and identify major traffic motif in e-commerce sites. Web customized index combination in commercial web sites [15] helps for self customization and web site usually contains vast amounts of information across hundreds of pages. Without proper direction for the visitor often wanders pointlessly without visiting significant pages, loses attention, and exit the site sooner than expected. Web ad's effectiveness [17] making it easier for companies to conduct business and transfer information to customers. Online Handwritten Script Recognition [18] helps automatic identification of handwritten script facilitates many important applications such as automatic transcription of multilingual documents and search for documents on the Web containing a particular script. Strong regularities in Web surfing helps to characterize empirical Web usage regularities [20]. Web personalization [22] [26] techniques have three major categories were decision rule-based filtering, content-based filtering, and collaborative filtering [16]. Decision rule-based filtering surveys users to obtain user demographics or static profiles, then lets Web sites manually specify rules based on them. Content-based filtering relies on items being similar to what a user has liked previously. Collaborative filtering uses explicit user ratings of products or preferences to sort user profile information into peer groups. It then tells users what products they might want to buy by combining their personal preferences with those of like-minded individuals E-science in astronomy [23] helps in tracking the web and SQL queries in E-science usage. Unsupervised feature selection for social media data [24] helps to find a tough task due to nonappearance of label information based on which feature relevances

5.WEB USAGE MINING TOOLS

SpeedTracer [27] a Web usage mining and analysis tool, was developed study user surfing pattern by checking out the online repository files with data mining techniques. Rapid Miner 4.6 [28] supports automatic web page visitor session extraction and clustering of the visit. Weka [29] helps in collecting visualization tools and algorithms for data analysis and predictive modeling. Orange [30] is a data mining tool which can be used for web usage mining analyzing.

CONCLUSION

Tim Berners-Lee invented the World Wide Web in 1989. In 2009 Tim burners lee coined the term Linked Data in his semantic web project. Linked data describe the relation of a web page to another web page. This made World Wide Web semi structured and easy to access related data. Limited numbers of data's are accessed under this linked data in online communities. Hence unlimited data's in online community gives Web usage mining importance in day to day life. Web usage mining can be applied to every area in the online community to extract useful information. Vast areas of research available in web usage mining to develop user friendly online community and can be generated more structured data's in world wide web.

ACKNOWLEDGEMENT

Firstly, I would like to thank almighty for making me motivated. Secondly, Dr Ajit Kumar and Dr jawahar Lal, SRM university, Haryana, Delhi-NCR, Sonepat for their kind help.

REFERENCES

[1] Oren Etzioni "*The World Wide Web: quagmire or gold mine?*" Appears in Comm. of ACM, Nov. '96

[2] R. Cooley, B. Mobasher, and J. Srivastava "*Web Mining: Information and Pattern Discovery on the World Wide Web*" IEEE 1997

[3] Georgios Paliouras, Christos Papatheodorou, Vangelis Karkaletsis, Costantine Spyropoulos, Panayiotis Tzitziras "*From Web Usage Statistics to Web Usage Analysis*" *volume : 2* Pages: 159 - 164 IEEE 1999

[4] Florent Masseglia, Maguelonne Teisseire, Pascal Poncelet *" Real Time Web Usage Mining: a Heuristic based Distributed Miner"* pages: 288 - 297 *volume:1* IEEE 2002

[5] Ajith Abraham, Vitorino Ramos "*Web Usage Mining Using Artificial Ant Colony Clustering and Linear Genetic Programming*"pages: 1384 - 1391 *Volume:2* IEEE 2003

[6] Zhang Huiying, Liang Wei "*An Intelligent Algorithm of Data Pre-processing in Web Usage Mining* ", Proceedings of the 5Ih World Congress an Intelligent Control and Automation, Pages: 3119 - 3123 *Vol.4* June 15-19, 2004, Hangzhou, P.R. China

[7] Chien Chin Chen,Zhong-Yong Chen, and Chen –Yuan Wu " *An Unsupervised Approach for Person Bipolarization using Principal Component Analysis*", IEEE Transactions on Knowledge and Data Engineering,*Volume:24*, Issue: 11 Pages: 1963 - 1976 ,November 2012

[8] Yinghui Yang and Balaji Padmanabhan" *GHIC: A Hierarchical Pattern-Based Clustering Algorithm for Grouping Web Transactions"* IEEE Transactions On Knowledge And Data Engineering, *Vol. 17*, No. 9, September 2005

[9] A. C. M. Fong, Baoyao Zhou, Siu C. Hui, Jie Tang and Guan Y. Hong "*Generation of Personalized Ontology based on Consumer Emotion and Behavior Analysis*", IEEE transaction on Affective Computing,*Volume:3*, Issue: 2,Pages: 152 - 164, IEEE 2012

[10] Ruili Geng and Jeff Tian "*Improving Web Navigation Usability by Comparing Actual and Anticipated Usage"*, IEEE Transactions on Human-Machine Systems, *Volume: 45*, Issue: 1,Pages: 84 – 94, IEEE 2014.

[11] Jun Liu, Member, IEEE, Cheng Fang, and Nirwan Ansari, Fellow, IEEE "Request Dependency Graph: A Model for Web Usage Mining in Large-scale Web of Things" IEEE Internet of Things Journal, *Volume: PP*, Issue: 99,Pages: 1 – 1, IEEE 2015

[12] Mohamed Koutheaïr Khribi, Mohamed Jemni and Olfa Nasraoui *"Automatic Recommendations for E-Learning Personalization Based on Web Usage Mining Techniques and Information Retrieval"* Advanced Learning Technologies, 2008, ICALT '08,Eighth IEEE International Conference 2008,Pages: 241 – 245.

[13] Steve G. Romaniuk *"Using intelligent agents to identify missing and exploited children"* IEEE Intelligent Agent, *Volume: 15*,Issue: 2,Pages: 27 - 30,March/April 2000

[14] Ed H. Chi *"Improving Web Usability through Visualization"* IEEE Internet Computing, *Volume: 6*, Issue: 2,Pages: 64 - 71, April 2002

[15] Mamata Jenamani, Pratap K.J. Mohapatra, and Sujoy Ghose "*Online Customized Index Synthesis in Commercial Web Sites*" IEEE Intelligent Systems, *Volume: 17*, Issue: 6, Pages: 20 -26, November/December 2002

[16] Sung Ho Ha *"Helping Online Customers Decide through Web Personalization"* IEEE Intelligent Systems, *Volume: 17*, Issue: 6, Pages: 34 - 43, November/December 2002

[17] Sung Min Bae , Sang Chan Park and Sung Ho Ha "*Fuzzy Web Ad Selector Based on Web Usage Mining"* IEEE Intelligent Systems, *Volume: 18*, Issue: 6,Pages: 62 - 69, November/December 2003

[18] Anoop M. Namboodiri and Anil K. Jain "*Online Handwritten Script Recognition*" IEEE Transactions On Pattern Analysis And Machine Intelligence, *Volume: 26,* Issue: 1,Pages: 124 - 130, January 2004

[19] Doru Tanasa and Brigitte Trousse "*Advanced Data Preprocessing for Intersites Web Usage Mining*" IEEE Intelligent Systems, *Volume: 19*, Issue: 2,Pages: 59 - 65, March/April 2004

[20] Jiming Liu, Shiwu Zhang and Jie Yang *"Characterizing Web Usage Regularities with Information Foraging Agents"* IEEE Transactions on Knowledge and Data Engineering,*Volume:16*, Issue: 5,Pages: 566 – 584,May 2004

[21] Pranam Kolari And Anupam Joshi *"Web mining: Research and Practice*" the IEEE CS and AIP, *Volume:6*, Issue: 4, Pages: 49 - 53, July/August 2004

[22] Sung Ho Ha" *Helping Online Customers Decide through Web Personalization"* IEEE Intelligent Systems , *Volume : 17*, Issue: 6, Pages: 34 - 43, November/December 2002

[23] M. Jordan Raddick, Ani R. Thakar, and Alexander S. Szalay "*Ten Years of SkyServer I: Tracking Web and SQL e-Science Usage"* IEEE CS and the AIP, *Volume:16,* Issue: 4,Pages: 22 - 31, July/August 2014

[24] Jiliang Tang and Huan Liu *"An Unsupervised Feature Selection Framework for Social Media Data"* IEEE Transactions On Knowledge And Data Engineering, *Volume: 26*, Issue: 12, Pages: 2914 – 2927, December 2014

[25] Mehdi Adda and Petko Valtchev" *Toward Recommendation Based on Ontology Powered Web-Usage Mining*" IEEE Computer Society, *Volume : 11*, Issue: 4, Pages: 45 - 52, July/August 2007

[26] Dimitrios Pierrakos, Member, IEEE, and Georgios Paliouras "*Personalizing Web Directories with the Aid of Web Usage Data*" IEEE Transactions On Knowledge And Data Engineering, *Volume: 22*, Issue: 9, Pages: 1331 - 1344, September 2010

[27] Wu,K.-L.; Yu,P.S.; " *SpeedTracer : A Web usage mining and analysis tool*" IBM system journal, *Volume: 37*, Issue: 1, Pages: 89 – 105,IEEE1998.

[28] https://rapidminer.com/

[29] http://www.cs.waikato.ac.nz/ml/weka/

[30] http://orange.biolab.si/

Journal of Physics: Conference Series **710** (2016) 012016 doi:10.1088/1742-6596/710/1/012016

Application of the Augmented Reality in prototyping the educational simulator in sport – the example of judo

1. Wojciech B. Cieśliński – University School of Physical Education in Wrocław, Faculty of Sport Sciences, Department of Communication and Management in Sport, Division of Organization and Management

2. Janusz Sobecki – Wrocław University of Technology, Faculty of Computer Science and Management, Departament of Computer Science, Institute of Informatics

3. Paweł A. Piepiora – University School of Physical Education in Wrocław, Faculty of Sport Sciences, Department of Sport Teaching, Division of Individual Sports

4. Zbigniew N. Piepiora – Wrocław University of Environmental and Life Sciences, Faculty of Environmental Engineering and Geodesy, Department of Spatial Economy, Division of Spatial Planning and Rural Shaping

5. Kazimierz Witkowski – University School of Physical Education in Wrocław, Faculty of Sport Sciences, Department of Sport Teaching, Division of Combat Sports

Abstract. The mental training (Galloway, 2011) is one of the measures of the psychological preparation in sport. Especially such as the judo discipline requires the mental training, due to the fact that the judo is a combat sport, the direct, physical confrontation of two opponents. Hence the mental preparation should be an essential element of preparing for the sports fight. In the article are described the basics of the AR systems and presents selected elements of the AR systems: sight glasses Vuzix glasses, Kinect sensor and an interactive floor Multitap. Next, there are proposed the scenarios for using the AR in the mental training which are based on using both Vuzix glasses type head as well as the interactive floor Multitap. All options, except for the last, are provided for using the Kinect sensor. In addition, these variants differ as to the primary user of the system. It can be an competitor, his coach, the competitor and the coach at the same time. In the end of the article are presented methods of exploring, both, the effectiveness and usefulness, and/or the User Experience of the proposed prototypes. There are presented three educational training simulator prototype models in sport (judo) describing their functionality based on the theory of sports training (the cyclical nature of sports training) and the theory of subtle interactions, enabling an explanation of the effects of sports training using the augmented reality technology.

1. Introduction

The mental training (Galloway, 2011) is one of the measures of the psychological preparation in sport. Especially such as the judo discipline requires the mental training, due to the fact that the judo is a combat sport, the direct, physical confrontation of two opponents. Hence the mental preparation should be an essential element of preparing for the sports fight.

1. The aim of the research is to develop three types of models of the prototype simulator dedicated the sports education in these studies for the competitors practicing the combat sports, in particular the judo. It is, however, pointed to the need to develop three models of the educational training simulator in judo, such as:
2. Developing a prototype of the educational training simulator for practicing judo competitors in recruitment groups and youth (the ideal model);
3. Developing a prototype of the educational simulator for the competitors of the championship class (the reference model);
4. Developing a prototype of the educational simulator for all ages and experienced competitors, namely developing a software model, capable of supplying an optimal state of emotional arousal (the concentration) before the sports fight. This model simulator, can be called 'the model simulator of the emotional arousal'. In a word, the simulator used to capture the low-signal and subtle magnetic interactions of the emotional states of the competitor to stimulate optimal psycho-emotional and neuromuscular disorders;

In connection with these objectives, it can be formulated the research questions which are a key aspect defined in the research problems.

It should be considered the following paradigm of the scientific research concerning the defined issues, namely: is the theory of 'low-signal interactions' [Cieśliński, Sczuka, 2015] in the extent to which the theory of 'low-signal interactions' [Perechuda, 2010] can explain the impact of the proposed simulator on the behaviour of the competitors in sports training and combat sports? Proceeding from the above formulated paradigm assumed in this study, it have been indicated that it is possible to formulate the following problems defined in the form of questions. The research problem is as follows:
in what period of the sports preparation is best to use the mental training with the AR?

The hypothesis: the mental training using the AR the can be used as a training in:
• the period of the general preparation in teaching new techniques,
• the period of the special preparation for improving the selected individual techniques,
• in the period of the immediate starting preparation in stimulating the mental structures of movement (the imaginary training),
• during the takeoff and direct sportsmanship to build optimal arousal (concentration) before a fight sport,
• during detraining to 'unlearning' the movement techniques and structures which are not effective in the sports fight.

Summarizing the presuppositions, it is pointed to the need to assume a pattern of conduct leading research (paradigm) associated with the theory of subtle interactions. There is a need to locate a cycle of sports training before using the simulator AR. Finally, it is pointed out that it can be built three models of the prototype of the educational simulator in sport: for beginners, a model for master and the model of the prototype whose function is primarily to optimize the emotional arousal and the ability to concentrate in the field of the low-level impacts of the stimulus training and the incentives related to the sports fight.

In recent times, it has been noticed a significant increase in both the interest in and the applications of the augmented reality systems (called. augmented reality – AR), among others in art, museums, exhibiting, architecture, infrastructure maintenance, exchange of information on human teams, railways, education, editorial work, maintaining security, psychology, supporting browse and searching the information. The augmented reality (AR) can be defined as (Stephanidis, 2015), direct or indirect view of the real world viewed in real time enriched with the virtual elements generated by the computer. This definition points to the following elements: real and virtual objects embedded in a real environment; their mutual arrangement that allows the observer to view them naturally with different views; It works interactively, in real time and in three dimensions.

Ultimately, it has been presented the idea to develop and test a prototype of the simulator dedicated to education for the sport. Preliminary studies indicate that the prototype of a dedicated simulator will include two models. The first relates to the development program enabling exploration model of a perfect execution of the individual techniques. It is a prototype dedicated the coaches involved in coaching recruited groups. Proper performance of throws in judo at the beginning of science is the basis for learning the normal motor structures. However, for the experienced competitors, it cannot be designed the ideal model of a particular throw, because every competitor, especially the champion acquires the specific structure of the movement, reference to the type of personality, physique and mentality of the competitor. Thus, in this study, it is pointed to the construction of an expanded real-life technology in two models; the ideal model dedicated to an opponent's quick recruited groups and the reference model, dedicated individually different for each competitor.

In the article, besides the basics of the augmented reality systems, the selected devices that enable the implementation of the AR systems, will be presented the idea of using the AR in the training mental of the judo and an outline of the method of examining the implemented prototype of the AR system supporting the judo training for increasing the effectiveness of training and/or the prototype of the User Experience.

This article was prepared on the basis of the results of the earlier stages of the grant NCN called 'Application of the augmented reality in perfecting the judo techniques Tachi Waza (the technique in the high position)', and in particular, the implementation of the project 'Development of the idea of using the augmented reality in teaching the judo techniques' (Cieśliński, Witkowski, Maslinski, Kalina, 2015). The project has been hypothesized that the use of the AR enables to: increase the attractiveness of teaching, deepen the effect of the mental and imaginary training; and it is applicable in practice without an opponent, with the opponent, the imaginary combat the and combats in the competition.

2. The augmented reality systems

The augmented reality is a technology that allows to simulate and improve the technique of the throw using what and how it is performed by the competitor in the reality. It allows also to the imposition of a given physical structure of the computer-generated sequences of movements. The imposition of these two images on each other (overlapping in the image of the real and computer image), is called 'imaging' of reality. The consequence of imaging in this technology is the ability to conduct the mental training using the so-called sighted glasses which indicate the examined competitors considered how they should do the throw in the best way.

For many years one of the most technologically advanced forms of human interaction with the computer system is the Virtual Reality (Virtual Reality - VR) which was originally used to describe the sensory imersive experiences with the artificially generated worlds (N. & J. Chapman, 2004). The VR is defined as a measure of visualization, manipulation and human-computer interaction and complex data. The VR utilizes a set of technologies, such as the helmet HMD (Head Mounted Display), the gloves or overalls entire interactive devices that generate sound. Even today the AR is defined as a variant of the VR (Kipper, Rampolla, 2012) in which the digital information, in the form of images, audio, video or haptics experience, is applied to the real world surrounding the recipient. Although the AR can be used to impact all five senses, usually the AR allows users to see the real world overlaid by virtual graphical objects. One of the first applications of the AR is to use a special head-up displays (HUD heads-up display) for the presentation of information on the state of avionics and weapons systems for fighter pilots. For the most popular hardware platforms used by the AR systems are considered to the tablets and the mobile phones where on the screen of these devices is screened an image from a camera mounted in the device (usually, the rear camera) enriched with some virtual graphics elements. In the LSOA-HCI lab of the Wrocław University of Technology, to implement a judo training simulator currently are used:

• the sighted glasses Vuzix Star 1200 XLD are the AR system that supports audio and 2D and 3D video for most devices with HDMI output (eg. desktops, laptops, tablets, smartphones or DVD or 3D Blu-ray players). The glasses are also equipped with the HD camera with 1080p mounted between displays. The camera uses a separate USB connector through which it can be connected to the control computer as a standard web camera in MS Windows,

• the Kinect sensor enables the implementation of the new ways of human interaction with the computer systems (Miles, 2012). It's a combination camera, microphone and directional sensor in the integrated product. It enables the software developers to create the interactive systems of all forms of kinetic interactions, for example the complex gestures made with hands. Currently, the following versions are used: Kinect for Xbox 360, Kinect for Windows and Kinect for Xbox One,

• the interactive MultiTap floor is an integrated device combining a PC and a high power video projector in a single chassis. The mirror surface is used by the projector to display the image on a substrate and a Kinect sensor (on the extended arm). The system allows creating the modern multimedia systems that displaying the image on the floor. It also allows the natural interaction of the user with the system using body movements or its individual parts, for example the feet and/or the hands (Multitap, 2015). The AR is used in implementing the information systems currently applicable in many areas already mentioned in the introduction, and such as advertising, industry, games, entertainment and translations.

It is worth to notice that the research is free of conflict of interest.

3. The theory of subtle interactions, the sports training efficiency and using the identified models of the educational training simulator AR

The theory suggests the subtle interactions in the context of the theory of sports training and 'the cyclical sports training' that there is a need for accumulation of effects which consist of the direct and prolonged effects. To obtain the cumulative effect (the increased state of the readiness to exercise – fitness, achieving the effect of improving strength, speed, stamina and/or what in the object of the study is essential, the effective execution of kata with and without a partner) each of direct and prolonged effects has to have a synergistic impact to another. Therefore, it is important what measures and in what point of the training cycle have to be used [Cieśliński, 2008].

The model for stimulating by the training measures with respect to the training cycles is presented in the table 1. The training cycles should produce results in the form of fitness and / or improving the elements of technical, tactical, motor and psychological preparation. As it can be seen, an important element of the application of a specific model of simulator is:

• to which the training effect it refers: current, prolonged
cumulative.

• in which phase of the preparing cycle it occurs:

- the general preparation (it refers to recruited groups in general and master level as part of the overall preparing mesocycyle),

- the targeted preparation (for recruited groups it is the core of the curriculum where the main objective is the efficiency in carrying throws in the form of the kata and during the imaginary combat and in the relation to the championship level it should be noted to increase the efficiency and effectiveness of the throws in the qualifying events)

- the special preparation (it relates only to the champions and it points to the need to obtain the exercise capacity whose function is to earn the highest championship degrees during the sports events).

Table 1. The model of the relations between the sports efficiency, the training cycle and the selected model of using the AR. Source: own study.

Type of cycle / effects of training	General preparation	Targeted preparation	Special preparation	Overtraining
Current	Primary education	Improving individual technique	Preparing for combat	Indicating other techniques
prolonged		Improving individual technique	Improving technique before the imaginary combat	Changing the pattern of movement
cumulative		Effectiveness of own throws	Optimal level of concentration	Unlearning the structure of movement
Model of simulator	**ideal**	**referral**	**To the mental training**	**Maintaining fitness at a level slightly below the previous cycle**

Initially, to define and identify major research problems related to the application of the various models of simulator and in general, functionality of the judo educational simulator, it should built the scenarios of the development which are possible, in order to answer the question: how to program the model to meet the expectations for the performance of his particular function? Recalling, these include:

a. the educational model (ideal) – its feature is the quest for the perfect physical structure enabling the coach for effective teaching throws among the recruited groups.

b. The model for the championship level (reference) - the feature is the search for perfectly chosen the structure of movement for the individual competitor for a particular throwing technique under the static conditions and during the imaginary combats, and especially and above all during the sports fight.

c. The model of the simulator used for mental training which allows among the recruited groups the mental playing (imaging) of the correctly throwing technique and for the master, the funcion of the AR simulator will be optimizing the psycho-emotional states (the concentration) before the sports fight.

4. Using the AR in the mental training of the judo

This section of the article will present the selected scenarios for the implementation of the prototype of the judo simulator. The study assumed that the sensor tracking the movement of the competitor performing the throws will be the Kinect sensor for the Xbox One, but for the purpose of presenting the information is used a tablet/smartphone running on Android, the sighted glasses Vuzix Star 1200 XLD or interactive floor. Each of these displays enables to visualize the virtual models applied to real image of an athlete performing a throw. They show the reference (suiting to the learner, the weight, the height, the level of adjustment to the technique and others) motor activity. At the same time it may be presented the adequate for on-going throws, the feedback of the coach, the psychologist, the biomechanics or the teacher, discussing the proper performance of the individual sequences of movements.

4.1. The option 1 – learning and improving the kata by the competitor

In this scenario, they are used the desktop computer, the Kinect sensor for the Xbox One, the glasses Vuzix Star 1200 XLD or alternatively the interactive floor MultiTap. The scenario is as follows:

1. With the Kinect sensor, to which is set frontally the athlete which is located in the filed of view of the sensor (0,4-4,5 m), is recorded the standard judo kata;
2. On the basis of the position information for the highlighted joints calculated for each frame of the video recording the kata, there is created a mathematical model of the standard judo kata. This model takes into account the position of each joint in certain quanta of time (1/30 second). In addition, this model can be created not on the basis of one of the standard execution of the kata but rather on the basis of the selected performances of the kata;
3. Exercising the kata by an athlete in a field of view of the Kinect sensor with the additional use of the glasses Vuzix Star 1200 XLD or the MultiTap interactive floor that allows the presentation of the feedback. During executing the kata it can be optionally used the displayed icon (the eyeglasses) or the background color of the ground (the floor) to display the information whether in the moment of executing the kata the athlete performs it properly (good – the green colour) or incorrectly (wrong – the red colour), or other intermediate colour (between the green and red), depending on the deviation from the ideal;
4. After executing the kata the athlete receives depending on the used device (at glasses or floor) the following feedback:

- the video of the executed kata,
- the video of the executed kata with the imposed competitor's image as a skeleton;
- the information whether the kata was performed correctly,
- in the case of deviations are displayed differences from ideal (eg. an animated perfect skeleton and the skeleton of the practitioner)
- in addition, the deviations from the standard are explained by the 'voice of the coach' or using the display on a suitable text,
- in the same way are given the advices on how to improve executing the kata (in the version for less advanced athletes can be presented the detailed instructions for performing the kata with the accuracy to display on the floor the places of putting feet and taking a specific position in particular highlighted the key elements of executing the kata). Using the proper position on the one hand can be presented in the form of picture and in the form of an interactive manner that is verified by the skeleton read by the Kinect.

5. After the sequence of the throw the athlete receives the following feedback:

- the information that all the kata in the sequence are made correctly,
- the indication using the video or the animated skeleton of these parts of kata in which the athlete committed the highest number of errors (deviations from the standard).

were committed the most errors (deviations from standard) - is presented in the proper execution along with the information that in some parts the kata should be improved.

4.2. The option 2 – assisting the coach's work

In this scenario, they are used the desktop computer, the Kinect sensor for the Xbox One, the tablet/smartphone or the sighted glasses Vuzix Star. This scenario is executed in the initial portion as described in section 4.1 as follows:

1. As in Option 1;

2. As in Option 1;
3. The athlete performs the certain kata in the field of view of the Kinect, practicing kata is observed by the coach or biomechanic which uses a tablet/smartphone or Vuzix glasses. At the time, the coach is observing the kata by a suitable device on which he obtains (an option) the following information: displays the skeleton against the real picture of the competitor (this requires the recognition of the competitor's position and to overlay the skeleton at a certain angle), the display of the parallel the animation model of the executing kata in the form of a skeleton, the display of the information about errors during executing the kata (live) displayed eg. by changing the color of the skeleton;
4. After executing the kata or its sequence the coach receives the following information:
• the video of the kata executed by the competitor;
• the video of the executed kata with the imposed competitor's image as a skeleton;
• the information whether the kata was properly executed;
• in the case of deviations are displayed the differences from the ideal (eg. an animated perfect skeleton and the skeleton of the practitioner);
• in addition, there is presented the text list of deviations from the pattern;
• in the same way they are given suggestions on how to improve the executed of kata;
• in the case of the sequences, the kata are presented in the form of the video or the animation of the skeleton of these parts of the kata in which the athlete committed the highest number of errors (deviations from the standard).

4.3. The option 3 – collaborative, assisting the competitor and the coach's work with the AR
This option is a connection of the options 1 i 2.

4.4. The option 4 – the motivational training
Using the sight glasses Vuzix (with dimming) the competitor have presented before the fight the multimedia content that result reaching by the competitor the optimum level of motivation before the fight, for example :
• the video of the competitor's won fights;
• the video awards for his victory;
• the customized multimedia content affecting the respectively agitation, calm or motivate of a competitor.

5. Conclusion
The effectiveness of supporting through the mental training of the judo by the techniques using the AR systems can only be determined experimentally. Therefore, it has to be implemented the certain options and then they have to be subjected to verification with the athletes and, possibly, their coaches and biomechanics. Some difficulty in this task is the fact that the applications executing each scenario will be the prototypes which in many ways will not be fully functional and free of at least some disadvantages associated with the utility. It may, in more or less, influence the perception of the suitability of the proposed solution. The proposed procedure for testing the effectiveness of the mental training support using the AR may, on the one hand, consist of examining the usefulness and/or the User Experience of the proposed solution, on the other, consists of verifying whether using of such solutions in some way on how they influence the improvement achieved by a competitor's results. The idea of the usability is defined in many ways. According to the PN-EN ISO 9241-11 the usability is defined as 'the level to which the product can be used by the specified users to achieve the specified objectives to obtain: the effectiveness, the efficiency and the satisfaction in a specified context of use". And the ISO / IEC 9126-1 defines the usability as a model for the external and internal quality and is defined as 'the ability of the software product to be affordable, operational and attractive when it is used in the certain conditions.' In the literature are currently more than 100 different methods

providing or otherwise testing the usability. In the case of testing the usability of the prototypes of the systems supporting the mental training of the judo using the AR, it can be proposed the following methods:

- the focus survey which involves performing the discussion moderated by an expert, in a group of users – the result of that session are the observations, the comments and the evaluation related to the investigated product;
- the tests of the individual users which are based on the observation of the user during the implementation of the tasks set by the experimenter when using the system under test – they allow users to see how they deal with their performance but also their intentions and interpretation of the encountered problems;
- the eye-tracking, a technique for registration of the user's eyesight focal point while working with the product which is usually used during testing of individual users – it allows for a better understanding of the process of the implementation of the tasks and establishing the design errors in the graphical interfaces;
- the satisfaction survey, is carried out usually through the use of the specially developed survey questionnaires that are filled both before the use and after controlled using the system – they enable to determine the level of user's satisfaction with using the system.

6. Acknowledgement
It is worth to notice that the research were funded from the measures for the science in the years 2013-2016 and 2015-2017, by the Ministry of Science and Higher Education under direction of dr hab. Wojciech B. Cieśliński, prof. AWF Wrocław, grant No.; 0014/RS2/2013/52,. „The e-AZS platform" of the AZS-AWF Wrocław sport club and the applications connected with the Augmentem Reality, primary investigator: dr hab. Kazimierz Witkowski and Wojciech B. Cieśliński (prototyping educational training simulator) – grant No. 0011/RS3/2015/53.

7. References

[1] Chapman N., Chapman J., (2004) Digital Media. John Willey & Sons, Chichester.
[2] Cieśliński W. B., Witkowski K., Maśliński J., Kalina R. M. (2015) Augmented realisty (AR) In teaching and developing judo techniques – project assumptions [in:] Kalina R. M. (ed.) Proceedings of the 1st Word Congress on Health and Martial Arts in Interdisciplinary Approach. HMA, 17-19 September 2015, Częstochowa, Poland. Warsaw: Archives of Budo, 2015, p. 185.
[3] Cieśliński W. B. (2008) Model trenadżeryzmu w doskonaleniu i rozwoju kadr przedsiębiorstw [in:] Potocki A. (ed.) Organizacja i Zarządzanie. Politechnika Śląska, Gliwice.
[4] Cieśliński W. (1987) Trening heurystyczny [in:] Lachowicz Z. (ed.) Letnia Szkoła Zarządzania w Sporcie. AWF Wrocław, Olejnica.
[5] Galloway J. (2011) Mental training for runners: how to stay motivated. Meyer & Meyer Sport, UK.
[6] Kipper G., Rampolla J. (2012) Augmented Reality: An Emerging Technologies Guide to AR. Syngress, USA.
[7] Miles R. (2012) Start Here! Learn the Kinect API. Microsoft, USA.
[8] Multitap (2015) Podłoga interaktywna Multitap, pobrano październik 2015 z adresu http://multitouch.pl/produkty/podloga-interaktywna
[9] Stephanidis C. (2015) Constantine Stephanidis and Eija Kaasinen, Augmented reality. ERCIM News nr 103, October 2015, p. 10.
10. Vuzix (2015) Vuzix Star 1200 XLD, October 2015: http://www.vuzix.com/augmented-reality/products_star1200xld.

Casing strength degradation in thermal environment of steam injection wells

M I P Hidayat[1, 2], **S Irawan**[1] and **Mohamad Zaki Abdullah**[1]

[1] Department of Petroleum Engineering, Universiti Teknologi PETRONAS, Bandar Seri Iskandar, 31750 Tronoh, Perak Darul Ridzuan, Malaysia

[2] Department of Materials and Metallurgical Engineering, Institut Teknologi Sepuluh Nopember, Kampus ITS Keputih Sukolilo 60111 Surabaya, Indonesia

E-mail: irfan@mat-eng.its.ac.id, drsonny_irawan@petronas.com.my, zaki_abdullah@petronas.com.my

Abstract. Degradation of the casing strength in relation with thermal cycles of steam injection process is still less explored in literature. In this paper, three-dimensional finite element (FE) analysis of casing strength degradation in thermal environment of steam injection wells is presented. 3D FE models consisting of casing-cement-formation system are developed in this study. Grade N80 casing is employed with the casing length of 3.048 m. In the analysis, cyclic thermal stresses induced on the casing in thermal environment of steam injection wells from 25 °C to 360 °C are first examined to verify the feasibility of the 3D FE models. Degradation of the casing strength in the thermal environment is subsequently investigated by applying an external pressure that represents formation pressure to the casing-cement system. The results show that the casing capability to resist the pressure is lowering as the number of thermal cycles extends, thus causing casing strength degradation in the thermal application. It is also shown that the casing may fail under external pressure below its specified collapse strength i.e. 10 % lower than the reference casing strength obtained at 360 °C.

Keywords*:* steam injection, casing strength degradation, transient 3D finite element analysis, collapsing pressure

1. Introduction

Casing plays one of the most important functions in a well. The structure maintains the capacity of oil production by isolating the wellbore fluids from the subsurface and formation fluids and providing a high-strength flow passage to the surface. Its function also allows safe control of formation pressure and prevents collapse of the borehole during drilling (Rodriguez, 2003). Due to its functional role that determines operational life of a well, reliability and integrity aspects of casing always become main concerns in oil and gas industries.

In order to provide sound design basis for the thermal recovery well casing and gain greater and deeper understanding in its reliability aspects, intensive investigations have been carried out to study casing strength degradation under thermal environment of the wells. Besides experimental investigations, numerical studies also play an important role in casing and well integrity assessment. The main reasons are: numerical studies allow quick assessment in accordance with technology advancement of casing materials. Another reason is that there is limitation in experimentations. Due to cost and time constraints, combined experimental and numerical methodology are often the options in gaining complete casing and well information.

Maruyama et al. (1990) performed an experimental study of casing performance under thermal cycling conditions. Thermal stress behavior, biaxial collapse resistance of casing and leak resistance of pipe and connections under cyclic steam injection conditions up to 354 °C were investigated in the study. Referring to the experimental study of Maruyama et al. (1990), Placido et al. (1997) presented experimental and numerical stress analyses of casing string submitted to cyclic steam injection. It is observed that residual stress increased for the casing grades of H40, K55 and N80 examined in the study, from which collapse ratio needs to be considered in casing design. As casings of K55 and N80 grades all become hot-yielded, Wu et al. (2005) suggested the use of high strength casing (HSC) of P110 and Q125 as they are not yielded under the cyclic thermal stresses due to higher yield strength values. The HCS can be also used for deep and sour oil and gas fields due to its high resistance to tension load, internal pressure and collapse as well as high resistant to sulfide stress corrosion cracking. Lin et al. (2013) noted however that the available API Bulletin 5C3 and new ISO 10,400 collapse models are in appropriate to predict the collapse strength of the high collapse casings. New HSC collapse equations for improvement in the casing design have been further proposed by the authors and validated through a series of collapse tests. Moreover, Liu et al. (2014) proposed a collapse strength formula to calculate the collapse strength of defective casings in steam injection wells. The formula has taken into account the effect of temperature and pit parameter which are still not considered in the previous standards and formulations (ISO/TR 10400, 2007).

It is noted from the previous studies that some research works have been focusing on stress development and mechanism affecting the casing performance, while the others focused on impact of mechanical properties and imperfection factors on casing strength integrity. However, the present authors notice that transient, three-dimensional fully coupled thermal-mechanical analysis describing degradation of the casing strength in relation with thermal cycles of steam injection process is still less explored in literature. It has been observed that casings in the thermal recovery wells fail after a period of steam injection (Li, 2008). Therefore, such a study is important in gaining complete information of casing strength degradation during the periods of steam injection process, in which detailed numerical analysis on such aspect has not been presented so far in the previous studies.

In this paper, three-dimensional finite element (FE) analysis of casing strength degradation in thermal environment of steam injection wells is presented. 3D FE models consisting of casing-cement-formation system are developed in this study. In the analysis, cyclic thermal stresses induced on the casing in thermal environment of steam injection wells from 25 °C to 360 °C are first examined to verify the feasibility of the 3D finite element models. The effect of temperature on the casing yield strength is taken into account. Degradation of the casing strength in the thermal environment is subsequently investigated by applying an external pressure that represents formation pressure to the casing-cement system.

2. Thermo-mechanical finite element analysis

Problem of casing strength degradation in thermal environment of steam injection wells is represented by coupled thermo-mechanical analysis. Transient thermal analysis needs to be carried out in order to obtain histories of thermal stresses generated in the casing as well as load carrying capability of the casing during the steam injection cycles.

2.1. Governing equations

Governing equation for heat conduction in a homogeneous isotropic solid body occupying a closed domain Ω bounded by Γ is given by (Hidayat et al., 2014, Hidayat et al., 2015):

$$\rho(\mathbf{x})c(\mathbf{x})\frac{\partial T(\mathbf{x},t)}{\partial t} = k(\mathbf{x})\nabla^2 T(\mathbf{x},t) + Q(\mathbf{x},t) \text{ in } \Omega \tag{1}$$

The boundary conditions are:

$$T(\mathbf{x},t) = \bar{T}(\mathbf{x},t) \qquad \text{on } \Gamma_1 \tag{2a}$$

$$\mathbf{n}.k\nabla T(\mathbf{x},t) = \bar{q}(\mathbf{x},t) \qquad \text{on } \Gamma_2 \tag{2b}$$

$$\mathbf{n}.k\nabla T(\mathbf{x},t)=h_f\left(T_f(\mathbf{x},t)-T(\mathbf{x},t)\right) \quad \text{on } \Gamma_3 \tag{2c}$$

The initial condition is:

$$T(\mathbf{x},t)\big|_{t=0}=T_0 \qquad \text{in } \Omega \tag{3}$$

Here, $T(\mathbf{x},t)$ represents the temperature field, T_0 denotes the initial temperatures, t denotes time, $\rho(\mathbf{x})$ and $c(\mathbf{x})$ are mass density and specific heat at constant pressure, respectively, $k(\mathbf{x})$ is the thermal conductivity, $Q(\mathbf{x},t)$ is the heat generation/source per unit volume at the internal point $\mathbf{x}$ and Γ_1, Γ_2 and Γ_3 denote the three kinds of boundary conditions of first kind (specified temperature), second kind (specified heat flux) and third kind (convection heat transfer), respectively. Moreover, $\bar{T}$ and $\bar{q}$ are the prescribed temperature and the prescribed heat flux on the corresponding boundaries, respectively, h_f is the convection heat transfer coefficient, T_f is the environmental temperature, and $\mathbf{n}$ is the unit normal vector outward to the boundary.

Furthermore, generic thermo-stress elastic equations that couple with the heat transfer effect can be expressed in terms of displacements u and v and temperature T as:

$$\frac{\partial^2 T}{\partial x^2}+\frac{\partial^2 T}{\partial y^2}=Q \tag{4a}$$

$$\frac{\partial^2 u}{\partial x^2}+\frac{1-\mu}{2}\frac{\partial^2 u}{\partial y^2}+\frac{1+\mu}{2}\frac{\partial^2 v}{\partial x \partial y}-(1+\mu)\alpha\frac{\partial\left(T-T_{ref}\right)}{\partial x}=0 \tag{4b}$$

$$\frac{\partial^2 v}{\partial y^2}+\frac{1-\mu}{2}\frac{\partial^2 v}{\partial x^2}+\frac{1+\mu}{2}\frac{\partial^2 u}{\partial x \partial y}-(1+\mu)\alpha\frac{\partial\left(T-T_{ref}\right)}{\partial y}=0 \tag{4c}$$

α is the thermal expansion coefficient, μ is Poisson ration and T_{ref} refers to the reference temperature.

2.2. Finite element equations

Finite element equations of the coupled thermal and mechanical analysis are obtained by choosing a suitable form of interpolation function representing the variation of temperature and displacements as field variables within the FE of the geometrical model and applying subsequently weighted-residual or variational argument to the governing equations along with boundary and initial conditions.

The thermal FE equations including boundary conditions can be written as:

$$[\mathrm{C}]\left\{\dot{T}\right\}+[\mathrm{K}]\{T\}=\{F_T\} \tag{5}$$

in which:

$$[\mathrm{C}]=\int_V \rho c[\mathrm{N}]^T[\mathrm{N}]dV \tag{6}$$

$$[\mathrm{K}]=\int_V k[\mathrm{B}]^T[\mathrm{B}]dV+\int_S h_f[\mathrm{N}]^T[\mathrm{N}]dS \tag{7}$$

$$\{F_T\}=\int_V \mathrm{Q}[\mathrm{N}]^T dV+\int_S h_f T_{ref}[\mathrm{N}]^T dS \tag{8}$$

[N] is the matrix of element shape functions, [B] is the matrix of shape functions derivative and $\{T\}$ is the vector of nodal temperature.

Temperature distribution and history results from (1) are further inserted as thermal loads into the mechanical model. Incorporating the elasto-plasticity material model, the mechanical FE equations are written in the following incremental form:

$$^{i+1}[K_1]\{\Delta U\} - {}^{i+1}[K_2]\{\Delta T\} = {}^{i+1}\{R\} - {}^{i}\{R\} \tag{9}$$

in which:

$$[K_1] = \int_V [B]^T [D^{ep}][B] dV \tag{10}$$

$$[K_2] = \int_V [B]^T [C^{th}][M] dV \tag{11}$$

$$\{R\} = \int_S [N]^T \{p\} dS + \int_V [N]^T \{f\} dV \tag{12}$$

$$[D^{ep}] = [D^e] + [D^p] \tag{13}$$

$\{\Delta U\}$ and $\{\Delta T\}$ are the incremental of nodal displacements and temperature, respectively, [B] is the matrix of strain-displacement, $[D^e]$ is the matrix of elastic stiffness, $[D^p]$ is the matrix of plastic stiffness, [C] is the matrix of thermal stiffness, [M] is the temperature shape function, $\{p\}$ is the vector of traction, $\{f\}$ is the vector of body force and i is the current step of analysis.

The vector of nodal displacement at the next step of analysis, $^{i+1}\{U\}$ proceeds as:

$$^{i+1}\{U\} = {}^{i}\{U\} + \{\Delta U\} \tag{14}$$

The updated stress conditions in the structure are obtained from the following stress-strain relations:

$$^{i+1}\{\sigma\} = {}^{i}\{\sigma\} + \{\Delta\sigma\} \tag{15}$$

$$\{\Delta\sigma\} = [D^{ep}][B]\{\Delta U\} + [C][M]\{\Delta T\} \tag{16}$$

The iterative method of Newton-Raphson is commonly employed in the FE solver to solve the nonlinear equations (Bathe, 1996).

3. Casing strength degradation and collapse analysis

Casing collapse is a failure state in which casing losses its capability or integrity against the applied loads. The casing-cement system is therefore to withstand thermal stresses and uneven external pressure from the interaction with surrounding formation. For instances, non-uniform initial in-situ stresses in the rock salt formation as well as sanding induced cavities that causes the formation more likely to fail may induce larger deformations to the casing and cement (Peng et al., 2007).

By assuming that casing-cement-formation system is in a consolidation state, casing and cement may be viewed as ideal and concentric cylinders interacting with the surrounding formation. 3D FE models consisting of casing-cement-formation system to assess the casing strength degradation are developed in this study, as shown in Figure 1. Contact elements are employed to model interaction between casing and cement as well as cement and formation, in which the Coulomb friction model is used to describe friction between the contacting surfaces. ANSYS software is used in the present numerical study of casing strength degradation.

A common grade N80 casing is used in this study with the casing length of 3.048 m. The material properties are taken from Placido et al. (1997) in which the effect of temperature on the casing yield strength is taken into account. Geometrical data and material properties for the casing, cement and formation are given in Table 1.

Table 1. Geometrical data and material properties (at 25 °C) used in the numerical calculations

Parameters and properties	Casing	Cement sheath	Rock
Dimension or OD (m)	0.1778	0.2238	1.572
Wall thickness (m)	0.0103	0.023	-
Young's modulus (E, GPa)	209	2.8	100
Poisson ratio (ν)	0.33	0.2	0.31
Density (ρ, kg/m^3)	7850	2400	2650
Thermal expansion (α, 1/°C)	1.69E-6	9E-6	5.4E-6

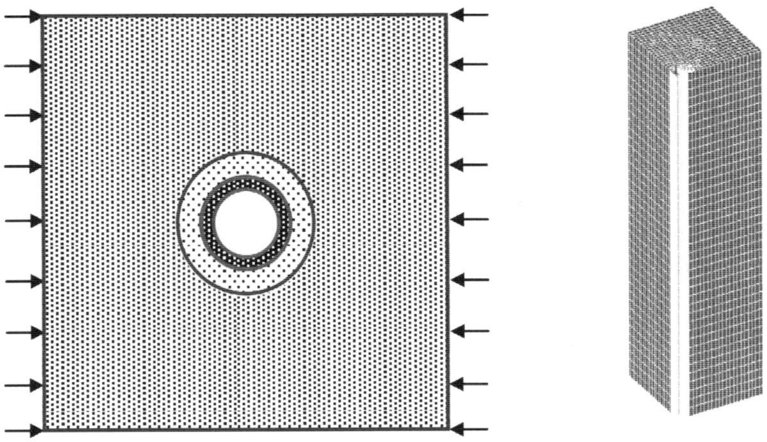

(a) (b)

Figure 1. (a) Schematic of casing-cement-formation system under collapse pressure and (b) three-dimensional FE mesh.

4. Results and Discussion

In this study, firstly the following aspects are simulated and examined to ensure the present 3D FE models suitability for casing strength degradation analysis: (i) axial thermal stress in the casing due to cyclic steam injection process. The process is simulated up to three thermal cycles, and (ii) temperature-dependent collapse pressure analysis for the casing in the temperature range between 25 °C and 360 °C. Further, degradation of the casing strength in thermal environment of steam injection wells is subsequently investigated.

Figure 2 shows axial thermal stress in casing for three cycles of steam injection. In addition, three different thermal cycles at respectively 260 °C, 300 °C and 360 °C are shown in Figure 3. It is noted that the casing become hot-yielded after the heating phase. The simulated axial stress in the casing is 639 MPa, while the casing yield strength at 360 °C is 550 MPa. It is obvious that the casing is prone to high failure rate in steam injection wells. Next examination is casing collapse strength with respect to temperature of steam injection well. Figure 4 depicts simulated casing collapse strength with the increase of steam injection well temperature. The values of D/t = 17.26 and 20.09 are examined. It is observed that the casing capability for collapse pressure is reduced as the well temperature increases. It is noted here that the simulation results obtained are similar with those obtained by Placido et al. (1997) and Liu et al. (2014), thus showing the suitability of the present 3D FE models.

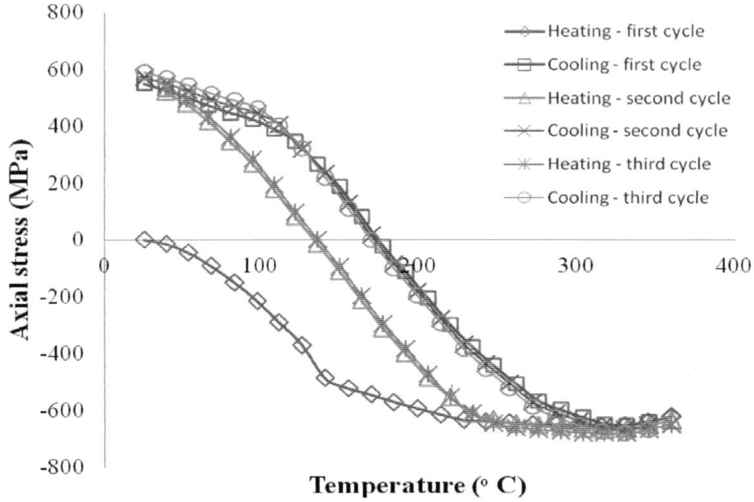

Figure 2. Axial thermal stress in casing for three cycles of steam injection.

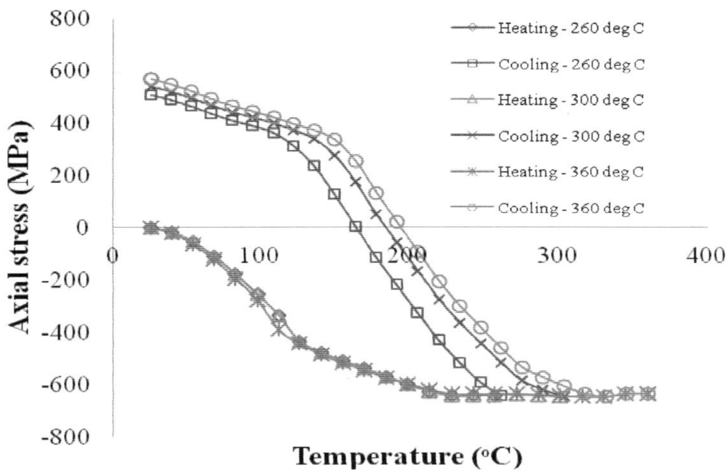

Figure 3. Three different thermal cycles of steam injection.

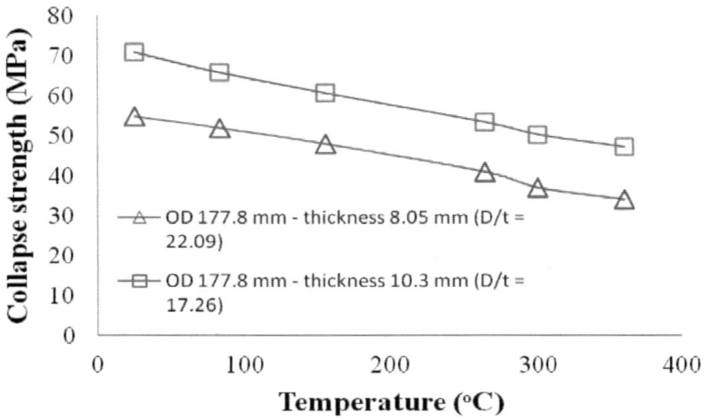

Figure 4. Casing collapse strength with respect to well temperature.

It is worthy noting that the results obtained by Placido et al. (1997) and Liu et al. (2014) still did not describe the casing behaviour i.e. degradation of casing strength along with the increased cycles of steam injection process. On the other hand, it is important to gain complete information of casing strength degradation during the periods of steam injection process for improved design as well as predictive maintenance. This is the contribution of the present research.

To investigate degradation of the casing strength, the casing integrity against the thermal load is examined as the thermal cyclic increases. Figure 5 describes the development of axial thermal stress in casing during six cycles of steam injection process.

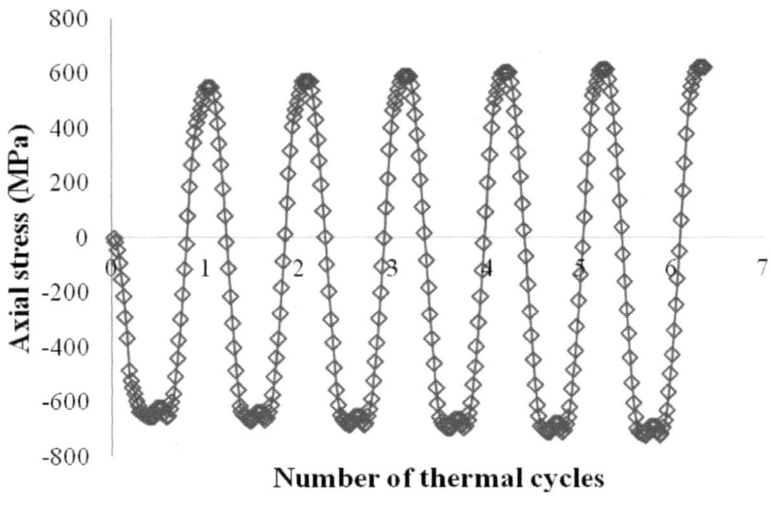

Figure 5. Axial thermal stress in casing during six cycles of steam injection process.

It is seen that the integrity of casing is reduced along with the increase of steam injection cycles. The casing becomes more hot-yielded and the tensile stress of casing also becomes greater as the number of thermal cycles increases. Obviously, this indicates casing strength degradation during the periods of steam injection process. The results also show that the casing is in severe condition due to the thermal stress even though no external pressure is applied.

Moreover, effect of external pressure to the casing strength degradation is further investigated by applying a pressure representing formation pressure to the casing-cement system. Figure 6 shows casing deformations in term of vertical displacement under the applied external pressures of 17.2, 30, 42 and 47 MPa, respectively. The deformation is shown for casing cross-sectional area under the influence of formation pressure exerted to the casing-cement system. The results show that the casing capability to withstand formation pressure is lowering as the number of thermal cycles extends, thus causing casing strength degradation in the thermal application. In particular, the casing strength degradation is severe in the first and second periods of steam injection.

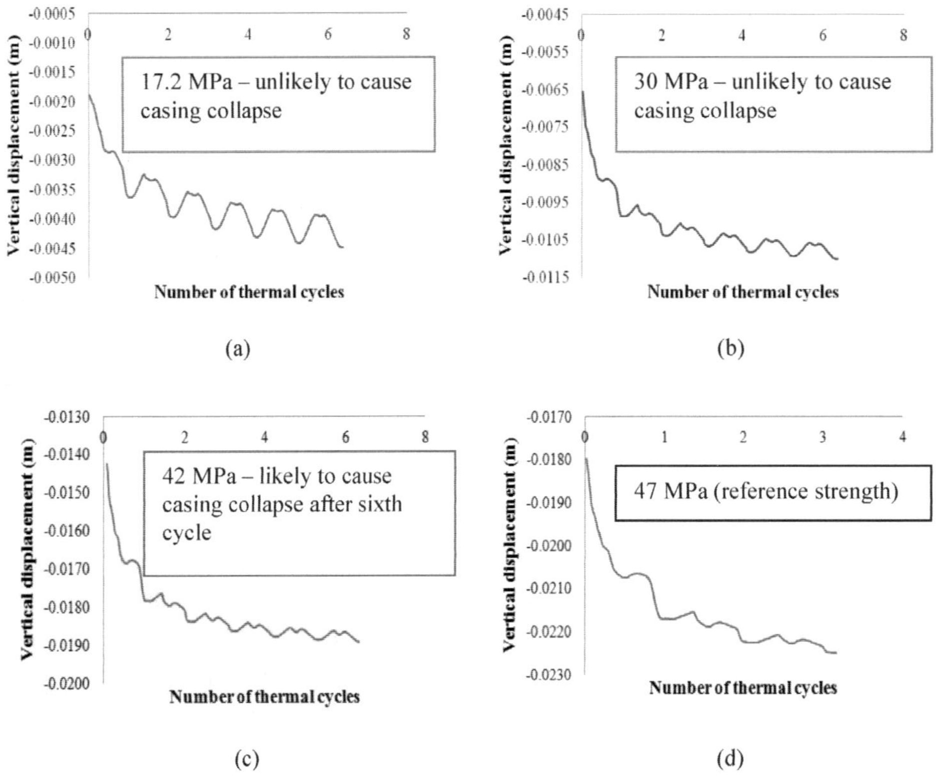

Figure 6. Casing vertical displacements under the influence of external pressures of: (a) 17.2 MPa, (b) 30 MPa, (c) 42 MPa and (d) 47 MPa (reference collapse strength).

Journal of Physics: Conference Series **710** (2016) 012017 doi:10.1088/1742-6596/710/1/012017

In Fig. 7, the corresponding casing deformations under external pressures of 17.2 and 30 MPa are respectively shown. It is noted here that the formation pressure of 17.2 MPa was reported for the Cat Canyon well (Maruyama et al., 1990). Note also that the deformation results are obtained at the end of sixth thermal cycle. It is observed that formation pressure can severely deform the cemented casing and leading to collapse failure of casing although the formation pressure is lower than collapse pressure specified for the casing. For an instance, the simulated external pressure of 42 MPa will not cause collapse failure of casing at 360 °C, where at the temperature the casing collapse strength is 47 MPa. However, from Figure 6 it is seen that the casing deformation after 6 thermal cycles under the external pressure of 42 MPa is 19 mm, which is closer to the casing deformation under the pressure of 47 MPa after one cycle of steam injection process i.e. 20.75 mm. Hence, as the cycle of steam injection process extends, the formation pressure of 42 MPa would constitute collapse failure of casing.

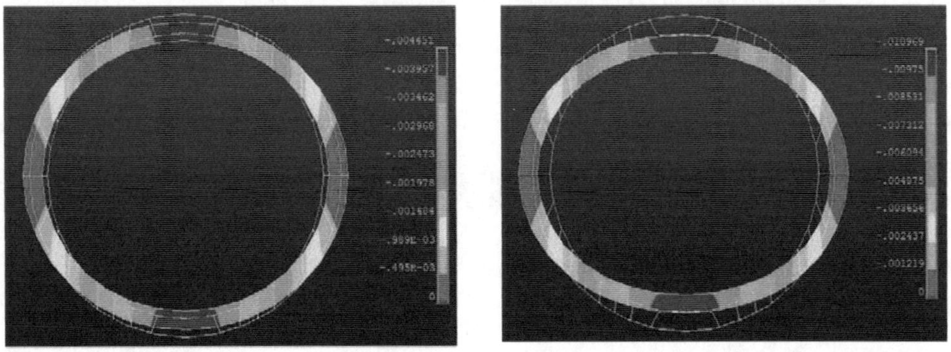

(a) (b)

Figure 7. Casing deformation under external pressure of:
(a) 17.2 MPa and (b) 30 MPa. The undeformed casing is shown for clarity.

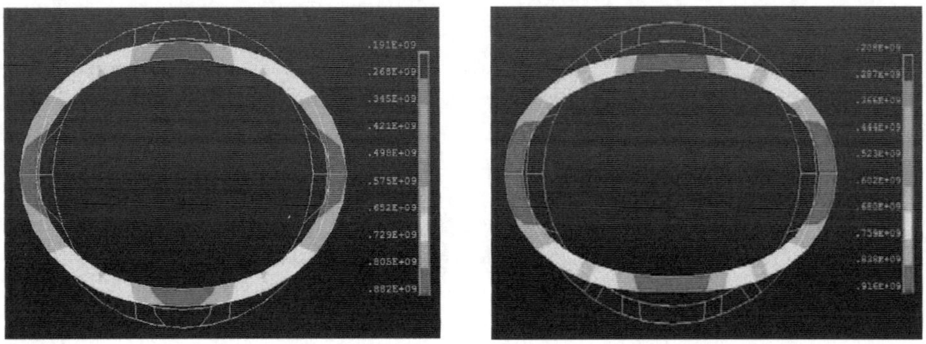

(a) (b)

Figure 8. Casing stress of von Mises under external pressure of:
(a) 33 MPa and (b) 42 MPa.

Moreover, casing stresses of von Mises under external pressures of respectively 33 and 42 MPa are depicted in Figure 8. Note that the external pressure of 42 MPa is 10.6 % lower than the reference casing collapse strength at 360 °C (47 MPa). It can be observed clearly that the cross-sectional area of casing is highly deformed, denoting the casing deforms plastically with high stress. It is hence noted that casing may fail under external pressure below its specified collapse strength as the cycle of steam injection process extends. Along with the effect of thermal stress, the severity of casing condition is clearly increased along with the steam injection cycles.

5. Conclusions

3D FE analysis of casing strength degradation in thermal environment of steam injection wells has been presented in this paper. The simulation results show that the casing capability to withstand formation pressure is lowering as the number of thermal cycles extends, thus causing casing strength degradation in the thermal application. It is also shown that the casing may fail under external pressure below its specified collapse strength i.e. 10 % lower than the reference casing strength obtained at 360 °C. As recommendation in practical operation, it is suggested that the maximum working temperature of steam injection well is better to be below 360 °C, such as for example 300 °C. At this temperature, it is shown that the casing has lower thermal stress as well as better capability for collapsing pressure.

Acknowledgment
The present authors gratefully acknowledge grant provided by Universiti Teknologi PETRONAS.

References

[1] "Petroleum and Natural Gas Industries-Formula and Calculation for Casing, Tubing, Drill Pipe and Line Pipe Properties," ISO/TR 10400, 2007.
[2] Bathe, K. J. (1996). Finite Element Procedures, Prentice Hall, Inc. , New Jersey.
[3] Hidayat, M. I. P., Ariwahjoedi, B. and Parman, S. (2015), "A new meshless local B-spline basis functions-FD method for two-dimensional heat conduction problems", *International Journal of Numerical Methods for Heat and Fluid Flow*, Vol. 25, pp. 225-251.
[4] Hidayat, M.I.P., Wahjoedi, B.A., Parman, S. and Megat Yusoff, P.S.M. (2014), "Meshless local B-spline-FD method and its application for 2D heat conduction problems with spatially varying thermal conductivity", Applied Mathematics and Computation, Vol. 242, pp. 236-54.
[5] Li, Z. (2008), "Casing cementing with half warm-up for steam injection wells", *Journal of Petroleum Science and Engineering*, Vol. 61, pp. 94–8.
[6] Lin, Y., Sun, Y., Shi, T., Deng, K., Han, L., Sun, H., Zeng., D and Zhu, H. (2013), "Formula to calculate collapse strength for high collapse casing", *Journal of Pressure Vessel Technology*, Vol. 135(4), pp. 041202-1 - 041202-5.
[7] Liu, S., Zheng, H., Zhu, X. and Tong, H. (2014), "Equations to calculate collapse strength of defective casing for steam injection wells", *Engineering Failure Analysis*, Vol. 42, pp. 240–251.
[8] Maruyama, K., Tsuru, E., Ogasawara, M., Inoue, Y. and Peters, E.J. (1990), "An experimental study of casing performance under thermal cycling conditions", *SPE Drilling Engineering*, Vol. 5(2), pp. 156–64.
[9] Peng, S., Fu, J. and Zhang, J. (2007), "Borehole casing failure analysis in unconsolidated formations: a case study", *Journal of Petroleum Science and Engineering*, Vol. 59, pp. 226–38.
[10] Placido, J.C.R., Ademar P. Jr., Paulo, L.A.B.F., Pasqualino, I.P. and Estefen, S.F., *Stress analysis of casing string submitted to cyclic steam injection*, The Fifth Latin American and Caribbean Petroleum Engineering Conference and Exhibition, 30 August-3 September 1997, Rio de Janeiro, Brazil.
[11] Rodriguez, W.J., Fleckenstein, W.W. and Eustes, A.W., *Simulation of collapse loads on cemented casing using finite element analysis*, SPE Annual Technical Conference and Exhibition, 5-8 October 2003, Denver, Colorado.
[12] Wu, J., Gonzales, M.E. and Hosn, N., *Steam injection casing design*, SPE Western Regional Meeting, 30 March-1 April 2005, Irvine, CA, USA.

Logistics Management: New trends in the Reverse Logistics

A Antonyová[1], P Antony[2] and B Soewito[3]
[1]University of Prešov in Prešov, Faculty of Management,
Konštantínova 16, 08001 Prešov, Slovak Republic
[2]APmikro, Pavlovičovo námestie 19, 08001 Prešov, Slovak Republic
[3]Bina Nusantara University, Jakarta 11530, Indonesia

E-mail: antonyova@gmail.com

Abstract. Present level and quality of the environment are directly dependent on our access to natural resources, as well as their sustainability. In particular production activities and phenomena associated with it have a direct impact on the future of our planet. Recycling process, which in large enterprises often becomes an important and integral part of the production program, is usually in small and medium-sized enterprises problematic. We can specify a few factors, which have direct impact on the development and successful application of the effective reverse logistics system. Find the ways to economically acceptable model of reverse logistics, focusing on converting waste materials for renewable energy, is the task in progress.

1. Introduction

There are a few example definitions of the reverse logistics, which are more or less general or stressing some of its function. The European Working Group on Reverse Logistics (REVLOG) has analyzed the concept of reverse logistics and specified the definition as follows: "The process of planning, implementing and controlling backward flows of raw material, in process inventory, packaging and finished goods, from manufacturing, distribution or use point, to a point of recovery or point of proper disposal." [23]

More general definition was formulated by the Reverse Logistics Executive Council (RLEC): "Reverse logistics is a movement of materials from a typical final consumption in an opposite direction in order to regain value or to dispose of wastes. This reverse activity includes take back of damaged products, renewal and enlargement of inventories through product take back remanufacturing of packaging materials, reuse of containers, and renovation of products, and handling of obsolete appliances." [24]

The Council of Logistics Management (CLM) defined reverse logistics using the definition of logistics: "The process of planning, implementing, and controlling the efficient, cost effective flow of raw materials, in-process inventory, finished goods and related information from the point of origin to the point of consumption for the purpose of conforming to customer requirements. Reverse logistics

[1] Anna Antonyová, Department of Mathematical Methods and Managerial Informatics

Journal of Physics: Conference Series **710** (2016) 012018 doi:10.1088/1742-6596/710/1/012018

includes all of the activities that are mentioned in the definition above. The difference is that reverse logistics encompasses all of these activities as they operate in reverse." [25]

2. Overview of the literature

The individual aspects of the chain process that is characteristic for the reverse logistics, particularly how it is manifested in the various sectors of the economy, are focusing in the scientific works of certain researchers.

Alok Agrawal and Vijay Choudhary [1] examined reverse logistics in the light of product lifecycle. Reverse logistics activities are specified as follows:

- Gatekeeping as the activity of controlling,
- Sorting and storing,
- Asset recovery,
- Transportation.

They also determined strategies as recommendations for companies to keep the reverse logistics as successful process:

- Customer satisfaction,
- New technology implementation,
- Eco-compatibility,
- Strategic alliances
- Knowledge management,
- Value recovery.

Turkish researcher Aras constructed model with his colleagues Boyaci and Verter [3] for products intended to manufacturing, which are keeping in the serviceable inventory for categorization in the chain of the reverse logistics. Robert Bucki and Bronislav Chramcov pay attention to information support with simulation the manufacturing tasks to build the proper logistics channels [5]. It is important to recognize also the proportions as firms across economic sectors contribute to the overall reverse logistics of the country's economy [9], particularly the following sectors: industry, magazine publishing, book publishers, book distributors, greeting cards, catalog retailers, electronic distributors, computer manufacturers, CD-ROMs, printers, mail order computer manufacturers, mass merchandisers, auto industry, consumer electronics and household chemicals.

Nowadays the interest is devoted especially to e-waste and its place in the chain in the reverse logistics as e-waste increases equally to the development of computer industry. American researchers K. K. Dhanda,a and A. A. Peters [8] specify various methods to alleviate the e-waste amount from the resource reduction in the input process through reuse, computer refurbishing as a form of remanufacturing or recycling. Disposal results in toxic waste as well as the groundwater contamination. Extended producer responsibility can also help to cope with the increasing e-waste in its amount. The reason for avoiding to disposal of e-waste in China [20] is mainly prevention to restrain of considerable loss of valuable secondary raw materials. From this point of view the sustainable management of waste electrical and electronic equipment (WEEE) involves organizing small companies, groups, or even individuals for collecting, trading and processing of already used electronic material leading to increasing also the economic viability.

Civil engineering is part of the industry which is typical with its problems how to cope with the continuous waste production. The Waste Management Plan (WMP) is a strategy how to minimize the construction waste. The causes of waste generation are set in the work of Malaysian researcher Sasitharan Nagapan with his colleagues [11], [16]. The practice proved the lack of waste management plan (WPM) as well as its implementation as the most significant factor for the increasing of the construction waste in its amount. This directly costs the following problems such as: mistakes in quantity surveys, ordering errors and wrong material storage. Poor supervision can costs poor materials handling as well as poor attitudes of workers. The managers sometimes can not influence the quality of construction materials. Natural materials to compose the building envelopes can not only increase the energy efficiency in buildings but also cope with the problem of artificial waste that is harmful for the environment. Scientists in Vienna [15] tried to composite new materials made of jute, flax, and hemp. However there is still problem with the moisture that influences the insulation properties.

Some countries are specific in the development of their system of the reverse logistics that is under the direct influence with the government legislation. Ismam [12] with his colleagues specifies new strategies for construction waste management in developed countries as a plan for government implementation in policy and regulations.

Customers' involvement into the chain of the reverse logistics system is a necessity that can be realized not only through the government legislative but also through the supermarkets. Brazilians Renata Borini Marcondes e Santos and Sergio Silva Braga Junior [18] focus on processing the goods through the recycling of materials as important part of the reverse logistics. They also collected data through the questionnaire regarding implementing the reverse logistics in supermarkets. The researcher Patricia Oom do Valle with her coauthors [17] applies multivariate statistical methods such as discriminant analysis, exploratory factor analysis and confirmatory factor analysis to examine the level of customer service in the reverse logistics chain in Portugal. The obtained findings enable to improve the complex logistics service in the recycling program to stress the role of the customers at the collection stage.

Wright, Richey, Tokman and Palmer deal with building the recycling channel for products [21]. Reverse logistics activities for products they specify as: return to supplier, resell, sell via outlet, salvage, recondition, refurbish, remanufacture, reclaim materials, recycle, and landfill. Reverse logistics activities for packaging: reuse, refurbish, reclaim materials, recycle, and salvage.

3. Designing the waste logistics system

Assessment the optimal costs to manage the chain for the general logistics system with taking into account the specific qualities of the possible agents, which may appear in the production process, requires using the appropriate quantitative methods. Both empirical as well as modelling analyses play an important role in modelling the waste logistics system. Strategic and operational factors can be set for the specific chain of the waste logistics, when the mathematical model of the logistics system is already created. It is important to observe the principles of mathematical modelling as algorithm for creating the system model, which is applied to the specific conditions of the waste logistics system.

General methodology for designing the waste logistics system can be described as following:

- main agents of the waste in the system are identified,
- analyzing the economic behavior of the waste agents in the system,
- decision process that is based on the quantitative methods and empirical analysis,
- mathematical modelling with setting the main variables,

- identifying the dependency and interrelationships between the variables,
- assigning the coordination mechanisms where appropriate,
- economic behavior is applied using the mathematical apparatus,
- identifying special cases for the certain agents,
- completing the details of the system model,
- testing the system model.

We can specify a few factors, which have direct impact on the development and successful application of the effective reverse logistics system. We recognize especially strategic and operational factors. Strategic factors:

- environmental concerns,
- customer service,
- overall quality,
- strategic costs,
- legislative concerns.

Operational factors:

- recycling,
- supply management,
- remanufacturing,
- warehousing,
- cost-benefit analysis,
- packaging,
- transportation.

Both strategic and operational factors are identified for every enterprise according to the entrance materials to the production process, production specifics and packaging used for both purposes: immediate packaging to protect products and to keep products safety during the transportation.

4. Specification the key trends in the reverse logistics development

Dynamic progress of society in unity with the technology development brings also some specific trends in both a new material development as well as key trends for the waste logistics. Creating the sustainable production cycles becomes a necessity for the further society development.

The trends in the waste logistics development are motivated economically as well as according to the government regulations. We can specify them as follows:

- research in setting the suitable material with the recycling possibility before the production program starts,
- reusage of the components that are used already in the production process,
- new product development process with advanced technology that is based on usage the recycled materials,

- creation the funds for development and building the reverse distribution channels,
- organizing the logistics programs for sustainable development,
- increasing responsibility of producers for the whole product life cycle,
- a „green" image as important marketing element,
- remanufacturing of machine parts,
- renovation in technology of packaging materials,
- invention activities aiming to reusable materials,
- collaboration with the retail chains in the collection of materials for recycling,
- ensure the increased safety in transport of dangerous materials,
- decision-making processes based on advanced information technology applications,
- specific attention to the E-waste logistics.

5. Modelling of reverse logistic processes

Reverse logistics system is influenced with not only economic reasons or more less environmental feelings of the company owner but also with the legislation. Henrique Luiz Corrêa and Lucia Helena Xavier consider the establishment of legal environment-related limits such as maximum emissions for certain modes of transportation in the United States in the 90s [7] as a milestone in the logistics – environment interaction. The same way the public policy in many countries can influence through the rules included in the law the regulation of reverse logistics. However some modern companies build their corporate brand directly on their relationship to the environment.

There several approaches to building the model of the reverse logistics processes. The optimization model under the stochastic environment was applied for the chains of the reverse logistics, using genetic algorithm in the work of Mostafa Hosseinzadeh with Emad Roghanian [12] from Iran. They proposed also strategy with the minimum shipping cost as well as the fixed opening cost, taking into account the locations with possible transportation links and capacities of the recovery facilities. The selected items were used to measure the performance outcome attributed to the adoption of reverse logistics activities in the research of Kuan Siew Khor and Zulkifli Mohamed Udin [14] from Malaysia. The obtained results were used to apply Chi-Square Test and build the model. Yu Xie Y, George Bruce and John Mangan [22] present model that is typical for the Chinese industry with respect to the role of profession in the reverse logistics system.

Quantitative model of Fleischmann with his co-researchers was constructed [10] with respect to the type of items, which are recovered. Finally there were appointed the following categories:

- packages such as bottles or pallets,
- routable spare parts, for instance TV tubes and machine parts,
- consumer goods such as refrigerators or copiers.

Reverse channel was appointed according to the flow from the producer through distributor to the user. Another model expresses the causal relationships between energy consumption, economic growth and pollutant emissions for Malaysia [2] over the period of the years 1970-2010.

While in the large factories reverse logistics system is a direct part of the production process in small and medium sized enterprises logistics channels are self-built, often not directly dependent on the manufacturing process. The example model of the channels in the reverse logistics system for small and medium sized enterprise devoted to paper box production in Slovakia is expressed in Figure 1. The production program of the company influences also the specific waste materials in the reverse logistic system. Our approach in constructing the model was to stress the activities, which are necessary for running directly in the enterprise and which are more suitable in collaboration with the external company:

- main agents of the waste in the system: paper, water, plastic, metal and electronics,
- economic behavior of the waste agents in the system:

  transportation of pellets of paper is managed with the company using the railways (additional costs for the company),

  transportation of electronics means additional costs for the company,

  repair and reuse of electronics mean a profit for the company,

  transportation of metal and plastic is managed with the external company (no additional cost for the company),
- decision process that is based on the quantitative methods and empirical analysis is a subject matter of our future research as well as mathematical modelling with setting the main variables and their dependences using the mathematical apparatus,

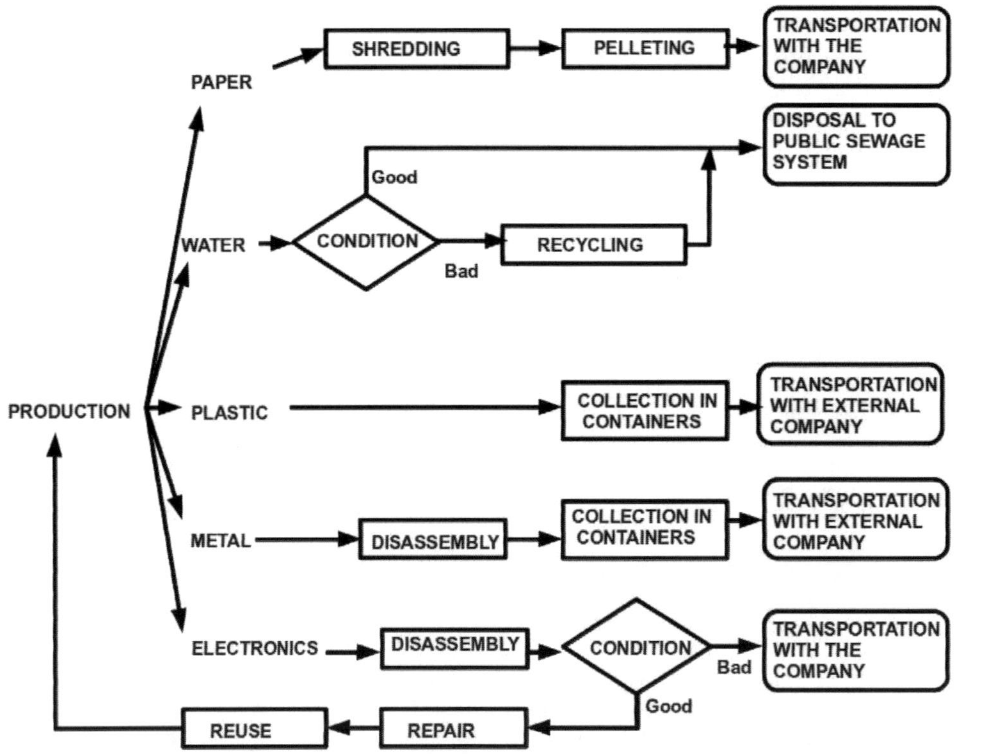

Figure 1. Reverse logistics channels for small and medium sized enterprise.

- special cases for the certain agents: water is contaminated mostly with the dispersion colorants,
- the system model will be tested after completing the details of the system model.

Some details of the specific channels including our research activities are subject matter also in our previous research [4], [6], such as for instance details of the shredding machine construction (Figure 2) for the waste paper material as well as the chamber with the pressing apparatus that is expressed in Figure 3 together with the pelleting process.

Suggested way for the water recycling is for instance the flotation technology. The example of the water cleaner working on the base of electro-flotation process is expressed in Figure 4. The water cleaner with the control system that is placed in the control box was designed in collaboration with a researcher and constructor Peter Antony from APmikro company. The concentrated dust is collected as foam from the surface of contaminated water during its recycling.

Figure 2. Shredding machine.

The container as a part of the water cleaner comprises 1000 liters of dirty water. The flotation tank is the place for the direct process of the water treatment. The water recycling is necessary part of every production where the water contamination is part of the manufacturing process as legislation in European Union does not allow discharging water directly into the public sewer network when there is occurrence of the specific contamination.

Figure 3. Pelleting.

6. Conclusion

The production process is the main source in contamination of our environment therefore consistency in building the channels of the reverse logistics is an important part of management in every company. The paper deals with the general trends in reverse logistics development. It specifies also some waste materials occurring in the production process as well as the ways how companies cope with the decision between the recycling or collection the waste materials for transportation to the external company. The channels of the reverse logistics system, which are typical for small and medium sized enterprises, are expressed using the schematic representation.

Model specification with its benefits as well as the cost of running the process the reverse logistics system for small and medium sized enterprise is the subject also for our future research. The costs optimization together with specification the proper paths in the decision making processes can make the reverse logistics more affordable for the companies.

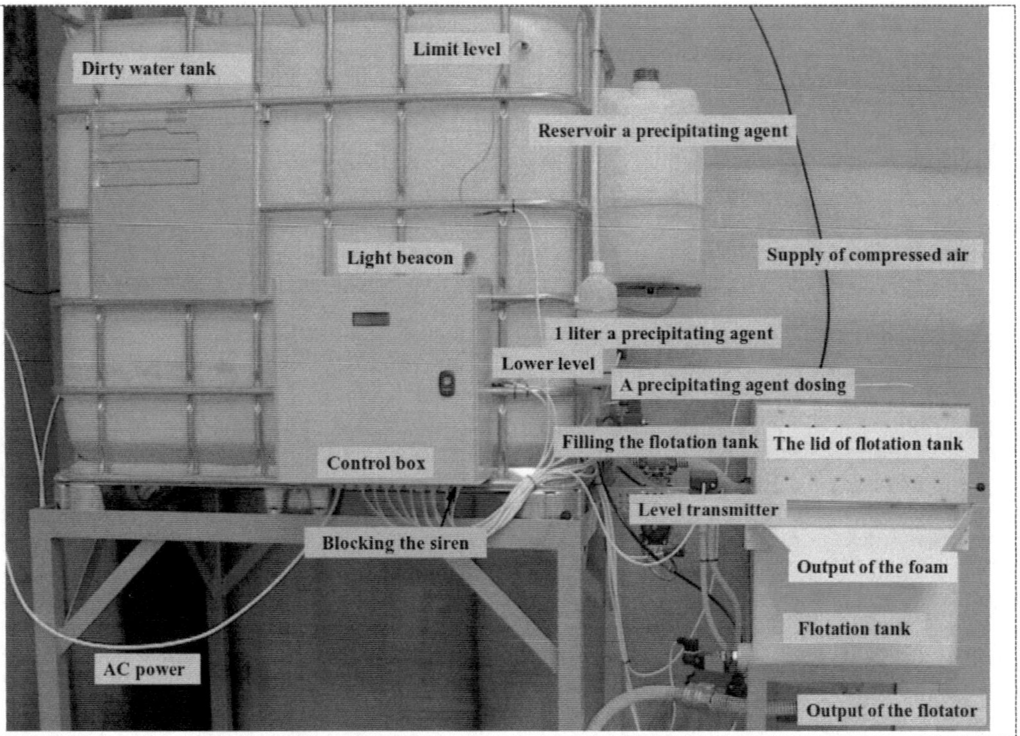

Figure 4. Cleaner for the contaminated water working on the flotation principle.

References

[1] Agrawal A and Choudhary V 2014 Reverse Logistics: Performance Measures and their effect in product lifecycle *International Journal of Core Engineering & Management (IJCEM)* **1**(2) 14-22

[2] Azlina A A and Mustapha N H N 2012 Energy, Economic Growth and Pollutant Emissions Nexus: The case of Malaysia *Procedia - Social and Behavioral Sciences* **65** 1-7

[3] Aras N, Boyaci T and Verter V 2004 The effect of categorizing returned products in remanufacturing *IIE Transactions* **36(4)** 319–331

[4] Antonyová A, Antony P and Soewito B 2015 Analytical Approach to Automation Control of Electro-flotation Process for Treatment the Water Contaminated with Dispersion Colorants *Advanced Science Letters* **21(1)** 24-28

[5] Bucki R and Chramcov B 2013 Information Support for Logistics of Manufacturing Tasks *International Journal of Mathematical Models and Methods in Applied Sciences* **7(3)** 193-203

[6] Antonyová A, Antony P and Soewito B 2015 Analytical Approach to Automation Control of Electro-flotation Process for Treatment the Water Contaminated with Dispersion Colorants *Advanced Science Letters* **21(1)** 24-28

[7] Corrêa H L and Xavier L H 2013 Concepts, design and implementation of Reverse Logistics Systems for sustainable supply chains in Brazil *Journal of Operations and Supply Chain Management* **6(1)** 1-25

[8] Dhanda K K and Peters A A 2005 Reverse Logistics in the Computer Industry *International Journal of Computers, Systems and Signals* **6(2)** 57-67

[9] Elmas G and Erdoğmuş F 2011 The Importance of Reverse Logistics *International Journal of Business and Management Studies* **3**(1) 161-171
[10] Fleischmann M, Bloemhof-Ruwaard J M, Dekker R, van der Laan E, van Nunen J A E E and Van Wassenhove L N 1997 Quantitative Models for Reverse Logistics: A review *European Journal of Operational Research* **103** 1-17
[11] Foo L Ch, Rahman I A, Asmi A, Nagapan S and Khalid K I 2013 Classification and Quantification of Construction Waste at Housing Project Site *International Journal of Zero Waste Generation* **1(1)** 1-4
[12] Hosseinzadeh M and Roghanian E 2012 An Optimization Model for Reverse Logistics Network under Stochastic Environment Using Genetic Algorithm *International Journal of Business and Social Science* **3(12)** 249-264
[13] Ismam J N, Ismail Z and Hashim H 2014 Construction Waste Management Successful Implementation Framework in Developed Countries *IEEE Colloquium on Humanities, Science and Engineering* 97-102
[14] Khor K S and Udin Z M 2012 Impact of Reverse Logistics Product Disposition towards Business Performance in Malaysian E&E Companies *Journal of Supply Chain and Customer Relationship Management* **2012** 1-19
[15] Korjenic A, Petránek V, Zach J and Hroudová J 2011 Development and performance evaluation of natural thermal-insulation materials composed of renewable resources *Energy and Buildings* **43(9)** 2518–2523
[16] Nagapan S, Rahman I A, Asmi A and Adnan N F 2013 Study of Site's Construction Waste in Batu Pahat, Johor *Procedia Engineering* **53** 99-103
[17] Oom do Valle P, Menezes J, Reis E and Rebelo E 2009 Reverse logistics for recycling: The customer service determinants International Journal of Business Science and Applied Management 4(1) 1-17
[18] Santos R B M, Braga Junior S S, da Silva D and Satolo E G 2014 Analysis of the economic and environmental benefits through the reverse logistics for retail *American Journal of Environmental Protection* **3(3)** 138-143
[19] Senthil S and Sridharan R 2014 Reverse Logistics: A Review of Literature *International Journal of Research in Engineering and Technology (IJRET)* **3(11)** 140-144
[20] Wolfer S, Sander H and Gogoll F 2011 Reverse Logistics for Waste Electrical and Electronic Equipment (WEEE) in China: Application of Linear Programming to Eco-Innovation in Industry *Globus Working Paper* **2011-4** 1-31
[21] Wright R E, Richey R G, Tokman M and Palmer J C 2011 Recycling and Reverse Logistics *Journal of Applied Business and Economics* **12(5)** 9-20
[22] Xie Y, Bruce G and Mangan J 2010 Role of professional reverse logistics in china's shipping sector and shipyards: a case study off shipyards in Zhejiang China *Logistics & Sustainable Transport* **2(1)** 1-13
[23] Dolgui A, Morel G and Pereira C E (Ed) 2006 Information Control Problems in Manufacturing 2006, A Proceedings Volume from the 12th IFAC Conference, Saint-Etienne, France, 17-19 May 2006, Elsevier IFAC publications
[24] Ritha W and Antonitte Vinoline I 2014 Environmentally Responsible Repair and Waste Disposal Inventory Models. *Aryabhatta Journal of Mathematics & Informatics* **6(1)** 1-9
[25] Wu Y (Ed.) 2012 Software Engineering and Knowledge Engineering: Theory and Practice. Advances in Intelligent and Soft Computing, Springer

Application of The Full-Sweep AOR Iteration Concept for Space-Fractional Diffusion Equation

A. Sunarto[1], **J. Sulaiman**[2] **and A. Saudi**[3]
[1,2] Faculty of Science and Natural Resources,
Universiti Malaysia Sabah, Malaysia.
[3]Faculty of Computing and Informatics,
Universiti Malaysia Sabah, Malaysia.

E-mail: [1]andang99@gmail.com, [2]jumat@ums.edu.my, [3]azali60@gmail.com,

Abstract. The aim of this paper is to investigate the effectiveness of the Full-Sweep AOR Iterative method by using Full-Sweep Caputo's approximation equation to solve space-fractional diffusion equations. The governing space-fractional diffusion equations were discretized by using Full-Sweep Caputo's implicit finite difference scheme to generate a system of linear equations. Then, the Full-Sweep AOR iterative method is applied to solve the generated linear system. To examine the application of FSAOR method two numerical tests are conducted to show that the FSAOR method is superior to the FSSOR and FSGS methods.

1. Introduction

Many problems in science, industry and engineering can be formulated as mathematical model in a form of fractional partial differential equations (FPDE's) [1,2]. Some different numerical methods have been proposed for solving fractional partial differential equations (FPDE's): Neamaty [3] solved fractional partial differential equations by using wavelet operational method, method of line tranform of the space-fractional Fokker-Plank equations [4] and modified decomposition method for the analytical solution of space fractional diffusion equation [5].

Space-fractional diffusion equations are a type of fractional partial differential equations. Therefore many reaserchers solved the problems numerically. For instance Saadatmandi and Dehghan [6] used a tau approach for solution space-fractional diffusion equation, Feng et al [7] developed second-order approximation for space-fractional diffusion equation with variable coefficient and Sousa [8] applied Spline for solving space-fractional diffusion.

To solve the proposed problem, let us describe some basic definitions and mathematical preliminaries of the fractional derivative theories which are required for our subsequent development.

Definition 1.[9] The Riemann-Liouville fractional integral operator , J^{β} of order-β is defined as

$$J^{\beta}f(x)=\frac{1}{\Gamma(\beta)}\int_0^x (x-t)^{\beta-1}f(t)dt,\ \beta>0,x>0 \qquad (1)$$

Definition 2.[10] The Caputo's fractional partial derivative operator, D^{β} of order -β is defined as

$$D^{\beta}f(x)=\frac{1}{\Gamma(m-\beta)}\int_0^x \frac{f^{(m)}(t)}{(x-t)^{\beta-m+1}}dt,\ \beta>0 \tag{2}$$

with $m-1<\beta\le m, m\in N,\ x>0$.We have following properties when $m-1<\beta\le m, m\in N,\ x>0$:

$D^{\beta}{}_{k}=0$, (k is constant)

$$D^{\beta}x^{n}=\begin{cases}0, & \text{for } n\in N_0 \text{ and } n<[\beta]\\ \dfrac{\Gamma(n+1)}{\Gamma(n+1-\beta)}x^{n-\beta}, & \text{for } n\in N_0 \text{ and } n\ge[\beta]\end{cases}$$

where function $[\beta]\le\beta$, $N_0=\{0,1,2,...\}$and $\Gamma(.)=$ Gamma function.

To construct the approximation equation of problem.(1), this work considers implicit finite difference scheme with Full-Sweep Caputo's operator. Both scheme and operator applied over Eq.(3) will derive Full-Sweep Caputo's implicit finite difference approximation equation. Actually this approximation equation has been discussed in constructing a linear system for time-fractional diffusion equations and space-fractional diffusion equations [11,12,13]. Since the linear system will be generated via this approximation equation, this paper consider the Full-Sweep Accelerated Over-Relaxation (AOR) method as linear solver.

According to development of standard AOR iterative method, this methods has been introduced by Hadjidimos [14]. This concept of this method is based on the point iteration concept together with two accelerated parameters. Actually this iterative method is one of the efficient point iterative methods. Due to the advantages of the AOR method for solving linear systems, the aim of this paper is to construct and examine the effectiveness of the FSAOR iterative method for solving space-fractional diffusion equations (SFDE's) based on the Caputo's implicit finite difference approximation equation. To examine the effectiveness of the FSAOR method, we also implement the FSSOR and FSGS iterative methods being used control methods.

Now let us consider the space-fractional diffusion equation being given as :

$$\frac{\partial W(x,t)}{\partial t}=a(x)\frac{\partial^{\beta}W(x,t)}{\partial x^{\beta}}+b(x)\frac{\partial W(x,t)}{\partial x}+c(x)W(x,t)+f(x,t) \tag{3}$$

subject to the initial condition $W(x,0)=f(x),\ 0\le x\le\ell$, and the Dirichlet boundary conditions $W(0,t)=g_0(t),\quad W(\ell,t)=g_1(t),\ 0<t\le T$.

2. Caputo's Impliciti Finite Difference Approximation for SFDE's

In the section, the process of solving the space-fractional diffusion equations is described in Section 3. Before that let $h=\frac{\ell}{k}$, where k is positive integer. By applying the operator (2), we get for *m=2* that

$$\frac{\partial^{\beta}W(x_i,t_n)}{\partial x^{\beta}}=\sigma_{\beta,h}\sum_{j=0}^{i-1}g_j^{\beta}\left(W_{i-j+1,n}-2W_{i-j,n}+W_{i-j-1,n}\right) \tag{4}$$

where $\sigma_{\beta,h}=\frac{(2h)^{-\beta}}{\Gamma(3-\beta)}$ and $g_j^{\beta}=(j+1)^{2-\beta}-j^{2-\beta}$.

Then by imposing the implicit finite difference scheme and the Caputo's operator in Eq.(4) over Eq.(3), the Full-Sweep Caputo's implicit finite difference scheme can be written as

$$\lambda\left(W_{i,n}-W_{i,n-1}\right)=a_i\sigma_{\beta,h}\sum_{j=0}^{i-1}g_j^{\beta}\left(W_{i-j+1,n}-2W_{i-j,n}+W_{i-j-1,n}\right)+b_i\frac{\left(W_{i+1,n}-W_{i-1,n}\right)}{2h}+C_iW_{i,n}+f_{i,n} \tag{5}$$

for *i = 1,2,4,...m-1*. Then the approximation equation (5) can be simplified again as

$$\lambda W_{i,n-1}=-a_i\sigma_{\beta,h}\sum_{j=0}^{i-1}g_j^{\beta}\left(W_{i-j+1,n}-2W_{i-j,n}+W_{i-j-1,n}\right)-\frac{b_i}{2h}\left(W_{i+1,n}-W_{i-1,n}\right)-C_iW_{i,n}+\lambda W_{i,n}-f_{i,n}$$

(6) After simplifying from Eq.(6), we have

$$b_i^* W_{i-1,n} + (\lambda - c_i^*) W_{i,n} - b_i^* W_{i+1,n_i} - a_i^* \sum_{j=0}^{i-1} g_j^\beta (W_{i-j+1,n} - 2W_{i-j,n} + W_{i-j-1,n}) = f_i \quad (7)$$

where $a_i^* = a_i \sigma_{\beta,2h}$, $b_i^* = \frac{b_i}{2h}$, $c_i^* = c_i$, $F_i^* = f_{i,n}$, $f_i = \lambda(W_{i,n-2}) + F_i^*$.

By referring to Eq.(7), be expanded to get the following approximation equation

$$R_i + \alpha_i W_{i-3,n} + s_i W_{i-2,n} + p_i W_{i-1,n} + q_i W_{i,n} + r_i W_{i+1,n} = f_i \quad (8)$$

where

$$R_i = a_i^* \sum_{j=3}^{i-1} g_j^\beta (W_{i-j+1} - 2W_{i-j,n} + W_{i-j-1,n}),\ \alpha_i = (-a_i^* g_2^\beta),\ s_i = (-a_i^* g_1^\beta + 2a_i^* g_2^\beta),$$

$$p_i = (b_i^* - a_i^* g_2^\beta + 2a_i^* g_1^\beta - a_i^*),\ q_i = (-a_i^* g_1^\beta + 2a_i^* + (\lambda - c_i^*)),\ r_i = (-a_i^* - b_i^*)$$

Thus by referring Eq.(8), we can develop to a linear system in matrix form as

$$A\underset{\sim}{W} = \underset{\sim}{f} \quad (9)$$

3. Full-Sweep AOR Iterative Method

By considering linear system in Eq.(9), it is clear that the characteristics of its coefficient matrix is large scale and sparse. In previous studies, a lot of works have been done to establish various iterative methods such as Young [15], Hackbusch [16], Saad [17]. To solve the linear system which is generated by the Caputo's implicit finite difference approximation equation, we consider the FSAOR method which is known as the Standard Accelerated Over-Relaxation (AOR) iterative method [11]. This iterative method is the most known and widely using for solving any linear systems [12,13]. To derive the formulation of FSAOR method in matrix form, we consider the coefficient matrix *A* (9) being expressed as summation of the three matrices for FSAOR method

$$A = D - L - U \quad (10)$$

where *D*, *L* and *U* are diagonal, lower triangular and upper triangular matrices respectively. From Eq.(10), the general scheme for FSAOR iterative method can be written as [14,18]

$$\underset{\sim}{W}^{(k+1)} = (D - \omega L)^{-1}[\beta U + (\beta - \omega)D + (1 - \beta)D]\underset{\sim}{W}^{(k)} + \beta(D - \omega L)^{-1} f \quad (11)$$

where $\underset{\sim}{V}^{(k)}$ represents an unknown vector at k^{th} iteration. As $\beta = \omega$, this iterative method (11) will be named as the FSSOR [14,15], while $\beta = \omega = 1$ we get the FSGS method. Based on Eq.(11), the general algorithm for FSAOR iterative method to solve linear system (9) would be generally described in Algorithm 1.[11]

Algorithm 1: FSAOR method

i. Initialize $\tilde{V} \leftarrow 0$ and $\varepsilon \leftarrow 10^{-10}$.
ii. For $j = 1,2,\ldots,n-1$ implement

   For $i = 1,2,\ldots,m-1$ calculate $\underset{\sim}{W}^{(k+1)} = (D - \omega L)^{-1}[\beta U + (\beta - \omega)D + (1 - \beta)D]\underset{\sim}{W}^{(k)} + \beta(D - \omega L)^{-1} f$
iii Convergence test. If the convergence criterion i.e $\left\| \underset{\sim}{W}^{(k+1)} - \underset{\sim}{W}^{(k)} \right\| \leq \varepsilon = 10^{-10}$ is satisfied, go to Step (iii). Otherwise go back to Step (ii).
iv Display approximate solutions.

Journal of Physics: Conference Series **710** (2016) 012019 doi:10.1088/1742-6596/710/1/012019

4. Numerical Experiment

In this section , two numerical examples are illustrated to show acccuray and effectiveness of the proposed method. Three criteria will be considered in comparison for FSGS, FSSOR and FSAOR such as K (number of iterations), Time (execution time in seconds) and Max Error (maximum error) at $\beta=1.2$, $\beta=1.5$ and $\beta=1.8$ and several mesh sizes as 128, 256, 512,1024 and 2048. During the implementation of numerical experiments, the convergence test considered the tolerance error $\varepsilon=10^{-10}$. The results of numerical simulations which were obtained from applications of the FSGS, FSSOR and FSAOR iterative methods for the following examples 1 and 2 have been recorded in Tables 1 and 2 respectively.

Example 1 : [10]

$$\frac{\partial w(x,t)}{\partial t} = d(x)\frac{\partial^{\beta} w(x,t)}{\partial x^{\beta}} + p(x,t), \tag{12}$$

The exact solution of problem (12) is $w(x,t) = (x^2+1)\sin(t+1)$.

Examples 2 : [10]

$$\frac{\partial w(x,t)}{\partial t} = \Gamma(1.2)x^{\beta}\frac{\partial^{\beta} w(x,t)}{\partial x^{\beta}} + 3x^2(2x-1)e^{-t}, \tag{13}$$

The exact solution of this problem is $w(x,t) = x^2(1-x)e^{-t}$.

TABLE 1. Comparison between K (number of iterations), TIME (the execution time in seconds) and Max Error (maximum errors) for the iterative methods using Example 1 at $\beta=1.2, 1.5, 1.8$

M	Method	β = 1.2			β = 1.5			β = 1.8		
		K	Time (Seconds)	Max Error	K	Time (Seconds)	Max Error	K	Time (Seconds)	Max Error
128	**FSGS**	74	1.48	2.37e-02	251	4.95	6.20e-04	930	18.29	3.99e-02
	FSSOR	66	1.36	2.37e-02	205	4.08	6.21e-04	733	14.47	2.42e-02
	FSAOR	65	1.32	2.37e-02	188	3.88	6.21e-04	269	5.35	3.99e-02
256	**FSGS**	152	11.64	2.44e-02	666	51.01	5.69e-04	3029	233.01	3.97e-02
	FSSOR	129	10.13	2.44e-02	545	42.29	6.69e-04	1361	107.33	2.39e-02
	FSAOR	128	10.00	2.44e-02	370	28.88	5.69e-04	756	58.90	3.97e-02
512	**FSGS**	352	99.64	2.47e-02	1780	550.52	5.36e-04	9840	755.31	3.96e-02
	FSSOR	278	85.9	2.47e-02	1459	144.73	5.35e-04	3472	725.25	2.37e-02
	FSAOR	270	84.05	2.47e-02	983	104	5.35e-04	2497	703	3.96e-02
1024	**FSGS**	709	672.27	2.49e-02	4750	1870.68	5.13e-04	21847	5259.97	3.95e-02
	FSSOR	607	140	2.49e-02	3906	756.12	5.13e-04	5539	1259.97	2.36e-02
	FSAOR	577	125	2.49e-02	3640	689	5.13e-04	5220	1119	2.36e-02
2048	**FSGS**	1547	1227.21	2.50e-02	8320	4348.68	5.02e-04	47322	8979.18	3.93e-02
	FSSOR	1230	577.00	2.52e-02	6320	3348.68	5.09e-04	13643	3979.18	2.30e-02
	FSAOR	1150	540	2.52e-02	5950	3102	5.09e-04	13203	3920	2.30e-02

TABLE 2. Comparison between K (number of iterations), TIME (the execution time in seconds) and Max Error (maximum errors) for the iterative methods using Example 2 at $\beta = 1.2, 1.5, 1.8$

M	Method	β = 1.2			β = 1.5			β = 1.8		
		K	Time (Seconds)	Max Error	K	Time (Seconds)	Max Error	K	Time (Seconds)	Max Error
128	**FSGS**	57	1.42	1.80e-01	182	4.41	5.44e-02	569	13.70	8.88e-04
	FSSOR	49	1.19	1.80e-01	156	3.77	5.44e-02	332	3.24	1.25e-04
	FSAOR	48	0.93	1.80e-01	133	1.41	5.44e-02	148	1.52	1.25e-04
256	**FSGS**	117	10.95	1.84e-01	481	45.32	5.58e-02	931	174.77	4.09e-04
	FSSOR	103	5.45	1.84e-01	225	14.80	5.58e-02	890	36.00	1.44e-04
	FSAOR	97	3.58	1.84e-01	197	10.93	5.58e-02	457	16.66	1.44e-04
512	**FSGS**	221	25.31	1.86e-01	732	153.67	5.65e-02	1635	427	1.47e-04
	FSSOR	128	20.69	5.39e-01	553	86.22	1.28e-02	1490	374.84	1.54e-04
	FSAOR	106	18.71	5.39e-01	525	83.02	1.28e-02	1357	193.83	1.53e-04
1024	**FSGS**	480	115.89	1.89e-01	1923	714.51	5.69e-02	5937	948.83	1.49e-04
	FSSOR	271	172.33	5.45e-01	1463	218	1.32e-02	4619	2210.72	1.25e-04
	FSAOR	213	168	5.45e-01	1298	198	1.32e-02	4329	2103	1.25e-04
2048	**FSGS**	1186	557.00	1.88e-01	6241	1259.31	5.85e-02	8482	5345.02	1.20e-04
	FSSOR	880	424.00	1.92e-01	2530	953.23	5.73e-02	7710	4120.81	2.30e-04
	FSAOR	815	398	1.92e-01	2506	912	5.73e-02	6520	3834	2.30e-04

5. Conclusion

This paper investigates the effectiveness the FSAOR iterative method to get numerical solutions of space-fractional diffusion equations. Throught the results obtained for examples 1 and 2, it manifestly shows that the application of FSAOR iteration which uses two accelerated parameters can reduce number of iterations and computational time as shown in Tables 1 and 2. Again this accuracy of numerical solutions of this method is comparable with the FSSOR and FSGS method. Since the FSAOR method is one of Full-Sweep iteration family, our future work will be extended to investigate the effectiveness of the Half-Sweep iteration family [19,20].

References

[1] A.D. Fitt, A.R Goodwin, K.A.Ronaldson, W.A.Wakeham.2009.A Fractional Differential Equation for a MEMS Viscometer used in The Oil Industry. *Journal of Computational and Applied Mathematics.* Vol. 229:373-381.

[2] P.J.Torvik, R.L.Bagley.1987.Fractional Derivative in The Description of Damping Materials and Phenomena. *The Role of Damping in Vibration and Noise Control Ed.L.Rogers and J.Simons, ASME,* 125-136.

[3] A.Neamaty,B.Agheli,R.Darzi.2013.Solving Fractional Partial Differential Equation by using Wavelet Operational Method. *Journal Mathematics Computational Science (TJMCS),* Vol. 7: 230-240.

[4] F.Liu, V.Anh, I.Turner.2004.Numerical Solution of The Space Fokker-Plank Equations. *Journal Computational Applied Mathematics,* Vol. 16: 209-219.

[5] S.S.Ray, K.S.Chaudhuri, R.K.Bera.2008.Application of Modified Decomposition Method for The Analytical Solution of Space-Fractional Diffusion Equations. *Journal Applied Mathematics and Computation,* Vol.196:294-302.

[6] A.Saadatmandi, M.Dehghan.2011.A Tau Approach for Solution of The Space Fractional Diffusion Equation. *Journal Computers and Mathematics with Applications,* Vol.62:1135-

1142.

[7] L.B.Feng, P.Zhuang, F.Liu, I.Turner, Q.Yang.2015.Second-Order Approximation for Space Fractional Diffusion Equation with Variable Coefficient. *Progress in Fractional Differentiation and Applications An International Journal,* Vol.1:23-35.

[8] E.Sousa.2010.Numerical Approximations for Fractional Diffusion Equation via Splines. *Compute Mathematics Applied,* Vol.62:1135-1142.

[9] Y.Zhang.2009.Finite Difference Method For Fractional Partial Differential Equation. A*pplied Mathematics and Computation,* Vol.215:524-529.

[10] H.Azizi, G.B.Loghmani.2013.Numerical Approximation for Space Fractional Diffusion Equations via Chebyshev Finite Difference Method. *Journal of Fractional Calculus and Applications,* Vol.4, No.2:303-311.

[11] A.Sunarto, J.Sulaiman, A.Saudi.2014.Implicit Finite Difference Solution for Time-Fractional Diffusion Equation Using AOR Method. *Journal of Physics Conference Series,* Vol. 495:2024-2031.

[12] A.Sunarto, J.Sulaiman, A.Saudi.2013.SOR Method for Implicit Finite Difference Solution of Time-Fractional Diffusion Equation. *Proceedings of the 11th Seminar on Science and Technology,* 132-139.

[13] A.Sunarto, J.Sulaiman, A.Saudi.2014.Full-Sweep SOR Method for Solving Space Fractional Diffusion Equations. *Australian Journal of Basic and Applied Science* Vol.8:153 - 158.

[14] A.Hadjidimos.1978.Accelerated Over-Relaxation Method. *Mathematics of Computation,* Vol.32:149-157.

[15] D.M.Young.1954. Iterative Methods for Solving Partial Difference Equations of Elliptic Type," *Trans. Amer. Math. Soc,* Vol.76:92-111.

[16] W.Hackbusch.1995.Iterative Solution of Large Sparse Systems of Equations. New York: Springer-Verlag.

[17] Y.Saad.1996.Iterative Method for Sparse Linear Systems. Boston: International Thomas Publishing.

[18] H.Tian.2003.Accelerated Over-Relaxation Method for Rank deficient Linear Systems. *Applied Mathematics and Computation.,* Vol.14:485-499.

[19] A.R.Abdullah.2007.The Four Point Explicit Decoupled Group (EDG) Method: A Fast Poisson Solver.*International Journal of Computer Mathematics*, Vol.38:61-70.

[20] J.Sulaiman, M.Othman,M.K.Hasan.2007.The Red Black Half-Sweep Iterative Using Triangle Finite Element Approximation for 2D Poisson Equation. *ICCS, Part I, LNCS 4487,Springer – Verlag Berlin Heidelberg.* 326-333

Comparative Study of Stearic Acid/Iron-Oxide Binary and Stearic Acid/Iron-Oxide/Titanium-Oxide Ternary for Use as Energy Storage Material

Rizky Andiarto[1)], Muhammad Khalish Nuryadin[1)], Rosari Saleh[1, a)]
[1]Departemen Fisika, Fakultas MIPA-Universitas Indonesia, 16424 Depok, Indonesia
[2]Integrated Laboratory of Energy and Environment, Fakultas MIPA-Universitas Indonesia,1624 Depok, Indonesia

[a]Corresponding's author: rosari.saleh@gmail.com, rosari.saleh@ui.ac.id

Abstract. In this work, a series of stearic acid/Fe_3O_4, and stearic acid/Fe_3O_4/TiO_2 nanocomposites for thermal energy storage (TES) system were synthesized through a two-step process. Fe_3O_4 nanoparticles and Fe_3O_4/TiO_2 nanocomposites were first prepared using sol-gel methods and then both samples were mixed into stearic acid by dispersion technique at three different weight % ratio to stearic acid: 5%, 10% and 15% to obtain stearic acid/Fe_3O_4, and stearic acid/Fe_3O_4/TiO_2 nanocomposites. Morphologies and structural properties of the samples were characterized by X-ray diffractometer (XRD), Fourier transform infrared spectroscopy (FTIR), field emission scanning electron microscope (FESEM) and energy dispersive X-ray spectroscopy (EDX), while thermal properties of the sample were determined by differential scanning calorimetry (DSC) and thermogravimetric analysis (TGA). The XRD patterns demonstrate, that stearic acid/Fe_3O_4 contained characteristic peaks of Fe_3O_4 and stearic acid structures, while peaks corresponded to anatase TiO_2 structures appear in stearic acid/Fe_3O_4/TiO_2 nanocomposites. From the DSC measurements, it is found that the maximum latent heat was found at samples with weight ratio of 5%. Moreover, the enhancement up to 20% of latent heat in solidifying as well as melting processes was observed. TGA measurements show high degradation temperature in the range of 246 - 251°C. The TGA results also shows that the residual mass of the sample matches the composition of Fe_3O_4 and Fe_3O_4/TiO_2 which is added to the stearic acid.

1. Introduction

Phase change materials (PCMs) employing in latent thermal energy storage (LTES) system is the most popular technique due to its ability to reduce the imbalance between the demand and the supply of energy [1-3]. As organic PCM, stearic acid (Sa) has been recommended as one of the most reliable thermal energy storage material due to its distinctive features like high storage density for given volume, nontoxic, and noncorrosive behavior [4-5]. However, for heat storage application, Sa as organic phase change material have low thermal conductivity and therefore limits their utility for large-scale application [6]. To overcome the problem, various strategies and approaches have been introduced to improve the ability of PCM to store and release heat, such as introducing metal fins, metal screen, and other metal structures. Unfortunately, these efforts resulted in limited enhancement [7-9].

Recently, some researchers have focused on dispersing nanomaterials into the PCM to enhance its ability to store and release heat [10-14]. Sahan et al. [15], studied the effect of the Fe_3O_4 nanoparticles addition in paraffin and found that the latent heat of the paraffin as a phase change material increased. Another study by Jifen et al. [16], stated that, the addition of TiO_2 nanoparticle in paraffin may enhance the latent heat ability. Both Fe_3O_4 and TiO_2 nanoparticle addition by these study proved that nanoparticle could enhanced PCM ability to store and release heat. Therefore, In this research, metal oxide Fe_3O_4 nanoparticles and Fe_3O_4/TiO_2 nanocomposites were synthesized by sol gel method for used as an additive materials for stearic acid (Sa) as phase change material to form Sa/Fe_3O_4 binary and Sa Fe_3O_4/TiO_2 ternary mixture. TiO_2 nanoparticles in Fe_3O_4/TiO_2 nanocomposites were prepared with molar ratio of Fe_3O_4 : TiO_2 was 1:5. The structural, morphology and thermal properties of stearic acid Fe_3O_4 and stearic acid Fe_3O_4/TiO_2 nanocomposites were investigated.

2. Experimental Setup

In this research, Iron (II) sulfate heptahydrate ($FeSO_4.7H_2O$,99), sodium hydroxide (NaOH), ethanol, ethylene glycol (EG), titanium dioxide and stearic acid ($C_{17}H_{35}CO_2H$) as a phase change material were purchased form Merck and were used without purification. All of the chemical reagents were of analytical grade. Fe_3O_4 nanoparticles and Fe_3O_4/TiO_2 nanocomposites were synthesized using same method as explained by our previous study [17]. Binary and ternary mixtures of Sa/Fe_3O_4 and Sa/Fe_3O_4/TiO_2 were synthesized by dispersing desired amount of nanoparticle and nanocomposite in stearic acid (5, 10, 15 wt%). All of the dispersion technique were done by using sonification water bath at 69°C for 2h. The flow chart of sample synthesize process were shown in Figure 1.

The structural properties, morphologies and elemental analyses of the samples were characterized by X-ray Diffraction (XRD), Field Emission Scanning Electron Microscope (FESEM), Energy-Dispersive X-ray spectroscopy (EDX) and Fourier Transform Infrared Spectroscopy (FTIR). The thermal properties of the sample were measured by Differential Scanning Calorimetry (DSC) to observe melting and solidification behavior and Thermal Gravimetric Analyzer (TGA) to observe sample degradation temperature.

3. Results and Discussion

XRD is an effective tool to characterize the crystal structure. The XRD patterns of Sa/Fe_3O_4 and Sa/Fe_3O_4/TiO_2 nanocomposites with various amounts of Fe_3O_4 nanoparticle and Fe_3O_4/TiO_2 nanocomposite in Sa (5, 10, 15 wt%) are presented in Figure 2. This figure also shown the XRD patterns of pure Fe_3O_4 and TiO_2 nanoparticles as well as pure Sa. The XRD peaks at $2\theta = 30.2°$, $35.5°$, $53.4°$, and $57.0°$ of pure Fe_3O_4 nanoparticles are attributed to the crystalline planes of inverse

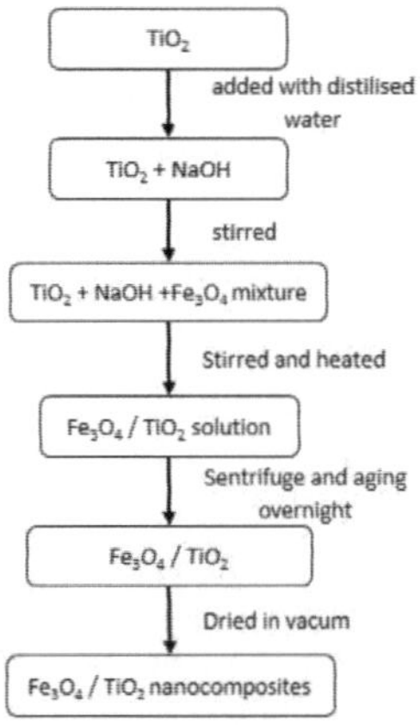

(a)

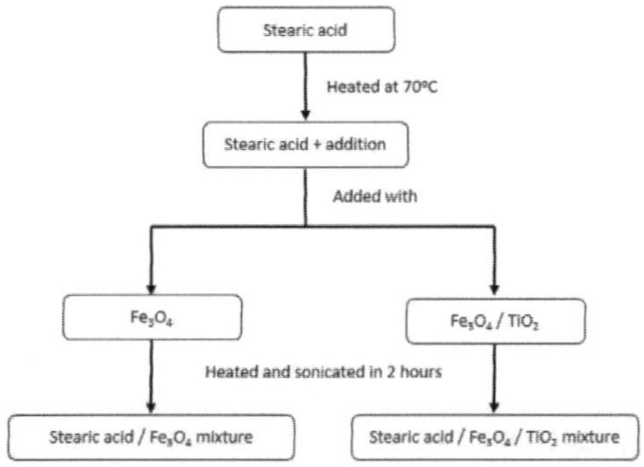

(b)

Figure.1. Synthesize process flowchart of (a) Fe_3O_4 / TiO_2 nanocomposite, (b) stearic acid / Fe_3O_4 and stearic acid / Fe_3O_4 / TiO_2 mixture

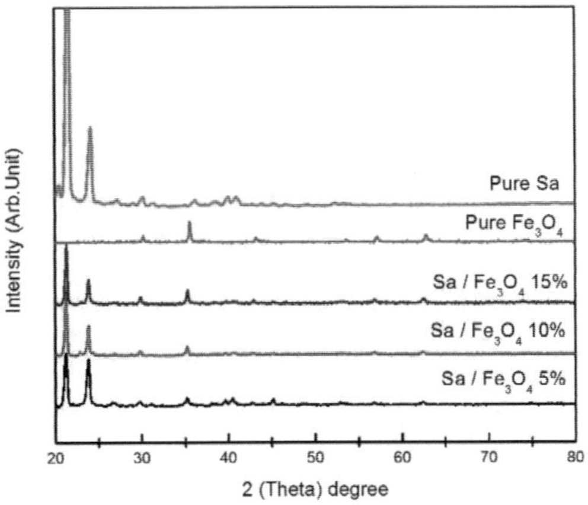

(a)

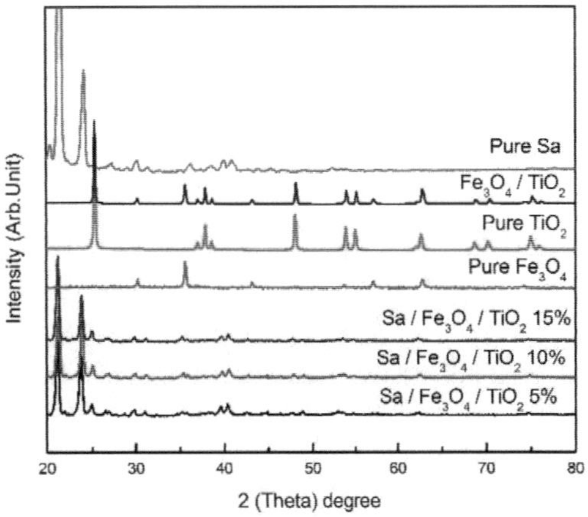

(b)

Figure. 2. (a) XRD patterns of Sa/Fe_3O_4 binary mixtures with various wt%, (b) XRD patterns of $Sa/Fe3O_4/TiO_2$ ternary mixtures with various wt%

cubic spinel structure at (220), (311), (511) and (440), while the XRD pattern of pure TiO_2 shows peaks at 2θ = 25.4°, 37.9 °, 48.1°, 54.2°, 55.1°, 62.6° and 75.1° corresponded to the crystalline planes of tetragonal anatase structure at (101), (004), (200), (105), (211), (204), and (215). The XRD peaks at 21.3° and 24.7° of the Sa are due to its regular crystallization. The XRD pattern of Fe_3O_4 /TiO_2 nanocomposites show the reflection peaks can be readily indexed to a combination of reflection peaks that can be ascribed to cubic spinel Fe_3O_4 and anatase TiO_2. The absence of other mixtures of Fe_3O_4 and TiO_2 diffraction peaks in the spectra confirms that the nanocomposites were the desired materials, Fe_3O_4/TiO_2.

As can be seen in Figure 2b, all peaks shown in Fe_3O_4, Fe_3O_4/TiO_2 and the prominent peaks of Sa are also shown in the XRD spectra of Sa/Fe_3O_4 and $Sa/Fe_3O_4/TiO_2$ nanocomposites. The peaks of Fe_3O_4 nanoparticles in Sa/Fe_3O_4 were slightly increased as the amounts of Fe_3O_4 nanoparticles dispersed in Sa were also increased. The same results were also found in $Sa/Fe_3O_4/TiO_2$ nanocomposites. The peak at 25° which corresponds to the presence of TiO_2 nanoparticles increased as the concentrations of nanocomposites in Sa were also increased.

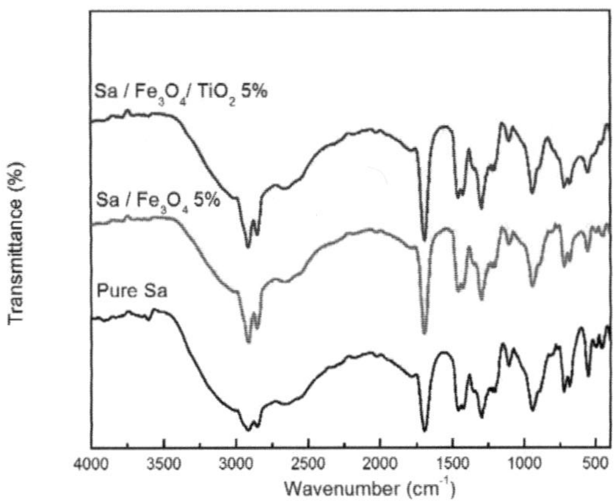

Figure. 3. FTIR spectra of Sa/Fe_3O_4 and $Sa/Fe_3O_4/TiO_2$

Figure 3 shows FTIR measurement of pure Sa, Sa/Fe_3O_4, and Sa/Fe_3O_4/TiO_2 mixtures. The transmittance at 2917 cm^{-1}, 2849 cm^{-1}, and 1703 cm^{-1} show the stretching vibration of CH_3, CH_2, and C=O molecules respectively. While 1464 cm^{-1}, 1296 cm^{-1}, 934 cm^{-1} and 720 cm^{-1} correspond to in plane, out of plane and in plane swinging of OH molecules respectively. There are neither Fe_3O_4 or TiO_2

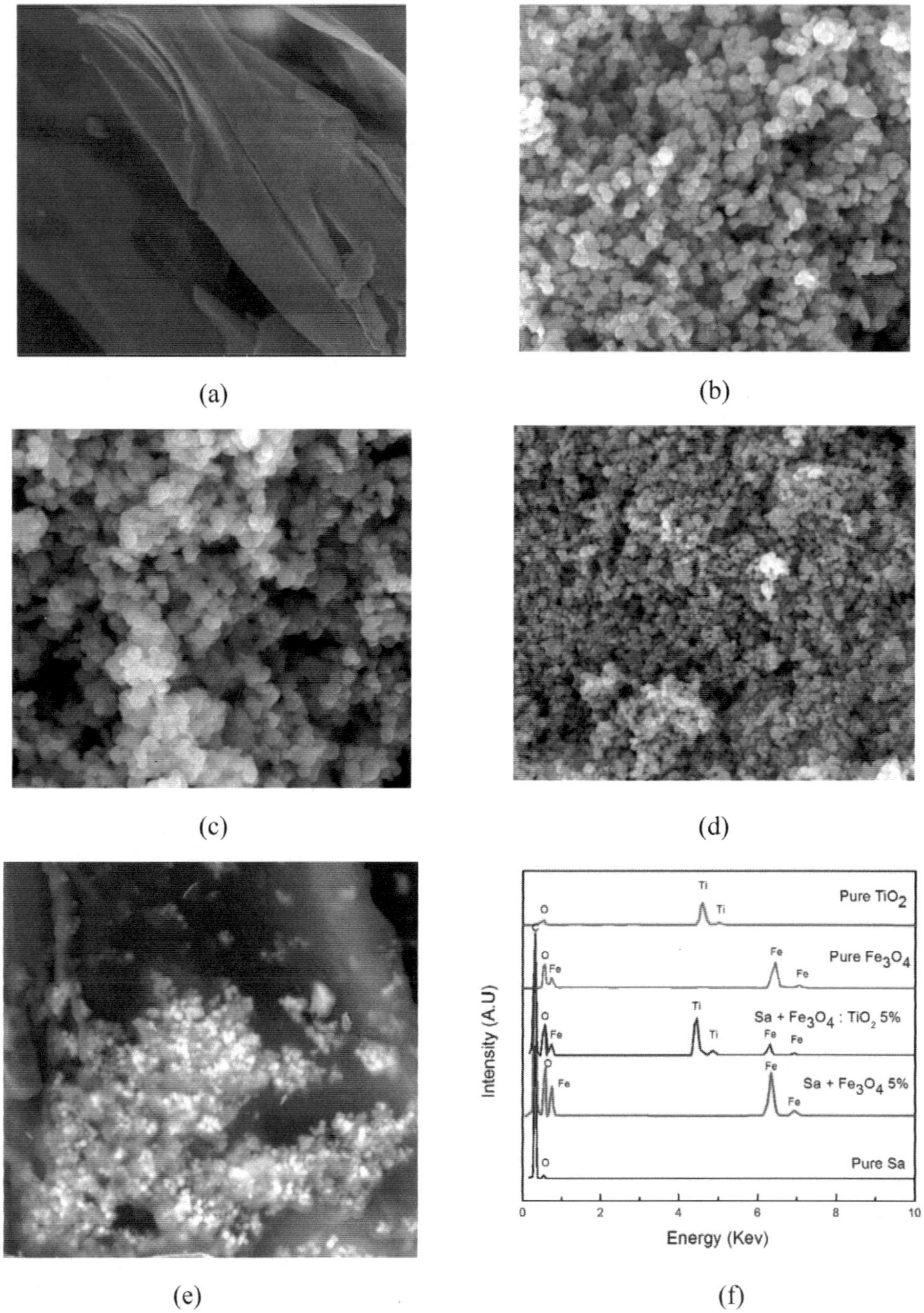

Figure. 4. FESEM images of (a) pure Sa, (b) pure Fe_3O_4 nanoparticle, (c) pure TiO_2 nanoparticle, (d) Fe_3O_4/TiO_2 nanocomposites, (e) Sa/Fe_3O_4/TiO_2 ternary mixture, (f) Corresponding EDX spectra of Sa / Fe_3O_4 and SaFe_3O_4/TiO_2

transmittance that can be found in the FTIR spectrum because the small amount of Fe_3O_4 and TiO_2 which are dispersed in Sa. However, the same value of transmittance wavenumber spectrum indicates that there are no chemical interaction between the Sa and nanoparticles/nanocomposite materials which can disrupts the phase change material itself [18].

Figure 4 shows the FESEM images of Fe_3O_4 and TiO_2 nanoparticles, Fe_3O_4/TiO_2 nanocomposites, pure Sa, and $Sa/Fe_3O_4/TiO_2$ nanocomposites. FESEM images revealed that pure Fe_3O_4 and TiO_2 as well as Fe_3O_4/TiO_2 nanocomposites have spherical-like shapes [17], while $Sa/Fe_3O_4/TiO_2$ nanocomposites showed the distribution of nanoparticles on the top of Sa.

In order to detect the atomic composition of the Sa/Fe_3O_4 and Sa/ Fe_3O_4/TiO_2 nanocomposites, the EDX spectroscopy was performed and the resulting spectra were displayed in Figure 3f. The EDX analysis confirmed that the Sa/Fe_3O_4 nanocomposites measurement consist of C, Fe, and O atoms while the $Sa/Fe_3O_4/TiO_2$ nanocomposites measurement consist of C, Fe, Ti, and O atoms. This result supports the presence Fe_3O_4 and TiO_2 by XRD measurement.

Thermal characteristics of all samples were studied with the DSC in terms of phase change temperature and latent heat. In the DSC measurements, the melting and solidifying onset temperature and peak temperature as shown in Figure 5 were observed. The onset point of melting and solidification process were determined by intersection between the slope and baseline of the curve while the peak temperature of melting and solidification point were determined by the highest and lowest peak of the DSC curve. Latent heat value of DSC measurement was calculated by integrating area under the peak of melting and solidifying process. The phase-change-characteristic data derived from DSC experiments are summarized in Table.1. As can be seen in Table.1, with the addition of Fe_3O_4 nanoparticles and Fe_3O_4/TiO_2 nanocomposites, there is no significant change in the onset temperature which the sample start to melt as well as peak temperature which shows complete process of melting. Similar things were also found in solidifying process. These results indicated that the phase change temperature process is not disrupted by addition of nanoparticle and nanocomposite.

It is known that latent-heat as a thermal property of phase change material is the most important parameters for a latent heat thermal storage system. This parameter shows the ability on how much phase change material absorbs and releases heat [19]. Based on the latent heat calculation of DSC curve in Figure 5, the presence of Fe_3O_4 nanoparticles and Fe_3O_4/TiO_2 nanocomposites were found to enhance the value of latent heat. As can be seen in Figure.6, The enhancement was found to be maximum at the addition of 5wt% of Fe_3O_4 nanoparticles and Fe_3O_4/TiO_2 nanocomposites in Sa. Moreover, the enhancement up to 20% of latent heat in solidifying as well as melting processes was observed in the

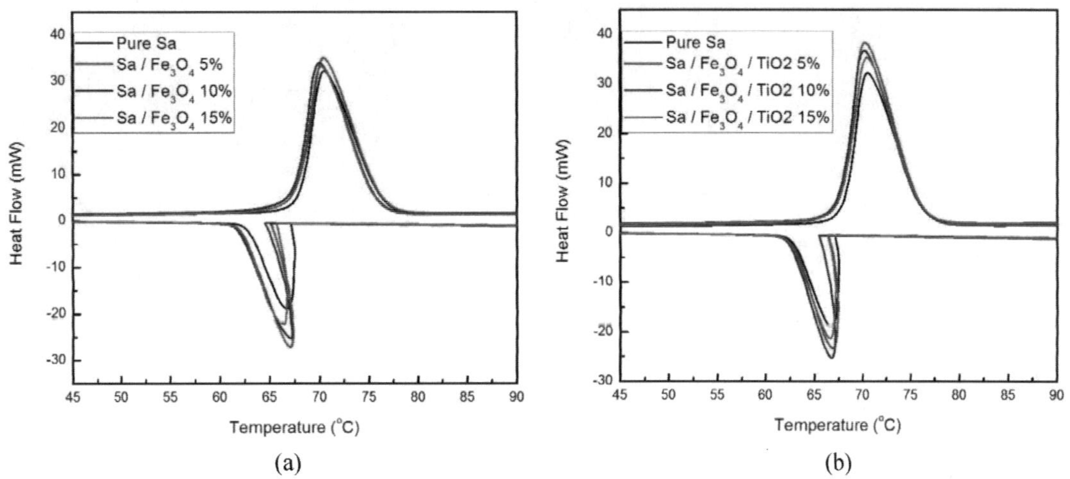

Figure. 5. Corresponding DSC curves with all various wt% of (a) Sa/Fe_3O_4, (b) $Sa/Fe_3O_4/TiO_2$

Table 1. solid-liquid, liquid-solid phase change temperature, melting, and solidifying latent heat data based on DSC measurement in Sa/Fe_3O_4 and Sa/Fe_3O_4/TiO_2

Sample Name	Solid- Liquid (°C)		Liquid-Solid (°C)		Melting Latent Heat	Solidifying Latent Heat
	Onset	peak	Onset	peak	(J/g)	(J/g)
Sa	67.7	70.5	67.2	66.6	177.635	174.135
Sa/$Fe_3O_4$5%	67.4	70.4	64.6	66.6	196.803	193.303
Sa/$Fe_3O_4$10%	67.2	70.2	63.5	66.1	188.169	184.669
Sa/$Fe_3O_4$15%	67.3	70.1	65.1	66	178.344	174.844
Sa/Fe_3O_4/5$TiO_2$5%	67.3	70.3	65.5	66.9	214.269	209.407
Sa/Fe_3O_4/5$TiO_2$10%	67.2	70.2	66.5	66.8	200.41	196.24
Sa/Fe_3O_4/5$TiO_2$15%	67.4	70.4	66.4	66.6	196.884	192.992

addition of Fe_3O_4/TiO_2 nanocomposites. The possible reason for the increment in latent heat is due to the molecular attraction between nanoparticles and Sa [20]. With this enhanced latent heat value, Sa/Fe_3O_4 and Sa/Fe_3O_4/TiO_2 nanocomposites developed a good candidate for latent heat thermal storage system.

Thermal stability of the sample was also measured by thermogravimetric analysis. As can be seen in Figure 7, all of the samples exhibit strong thermal stability up to 200°C. In table.2, the pure Sa degradation starts at 251°C with 100% weight loss. The addition of Fe_3O_4 nanoparticles and Fe_3O_4/TiO_2 nanocomposites in Sa didn't cause any favorable effect on the degradation process. However, the value of percentage loss in the Table.2 provides an information about the residual mass of the sample. Based on the results, it can be concluded that the residual mass of all the samples match the composition of Fe_3O_4 nanoparticles and Fe_3O_4/TiO_2 nanocomposites which is added into the Sa.

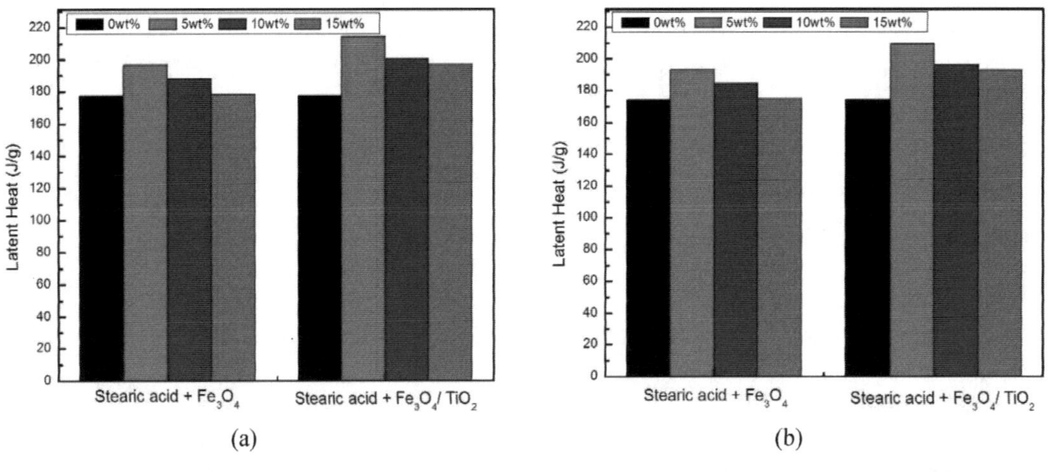

Figure. 6. Latent heat of Sa/Fe_3O_4 and Sa/Fe_3O_4/TiO_2 with all various wt% in (a) melting process, (b) solidifying process

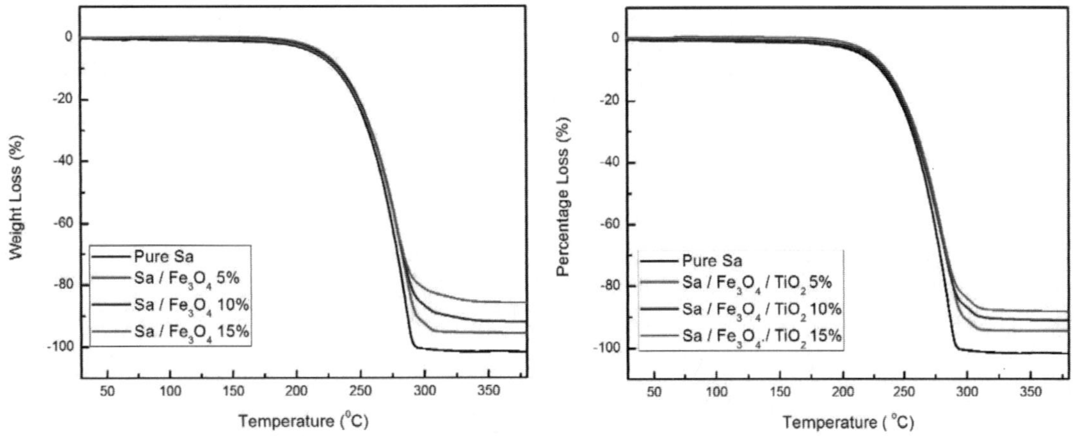

(a) (b)

Figure. 7. Corresponding TGA curves with all various wt% of (a) Sa/Fe_3O_4, (b)$Sa/Fe_3O_4/TiO_2$

Table 2. Onset temperature and percentage loss of Sa/Fe_3O_4 and $Sa/Fe_3O_4/TiO_2$ samples

Sample Name	Onset Temperature (°C)	Percentage Loss (%)
Sa	251.6	100%
$Sa/Fe_3O_4$5%	251.5	94.49%
$Sa/Fe_3O_4$10%	250.8	90.26%
Sa/ $Fe_3O_4$15%	250.1	86.75%
Sa/ Fe_3O_4 /5$TiO_2$5%	250.3	93.37%
Sa/ Fe_3O_4 /5TiO210%	249.7	90.86%
Sa/ Fe_3O_4 /5$TiO_2$15%	249.6	85.15%

4. Conclusion

Fe_3O_4 nanoparticles and Fe_3O_4/TiO_2 nanocomposites were synthesized using sol gel method and were dispersed into stearic acid (Sa) using dispersion technique. Based on the XRD data, the structure of Fe_3O_4 nanoparticle was inverse cubic spine while the structure of TiO_2 was anatase. Both XRD and EDX measurement confirmed the presence of Fe_3O_4 and Fe_3O_4/TiO_2. FESEM results show that Fe_3O_4 and Fe_3O_4/TiO_2 was attached on top of stearic acid (Sa). Based on the DSC results, it is found that the maximum latent heat was found at 5 wt.% of nanoparticles and nanocomposites in Sa. Moreover, the enhancement up to 20% of latent heat in solidifying as well as melting processes was observed in the addition of Fe_3O_4/TiO_2 nanocomposites with melting and solidification point in the range of 65-69°C. by these results, the ability of Sa to absorb and release heat were found to be enhanced by the presence of the Fe_3O_4 and Fe_3O_4 / TiO_2 nanocomposites. TGA measurement shows right composition of Fe_3O_4 and Fe_3O_4 in Sa with high thermal decomposition temperature.

References

[1] A. Sharma, V.V Tyagi, C.R. Chen, et al., Renewable and Sustainable Energy Reviews 13 (2009) 318-345
[2] M.M. Farid, A.M. Khudhair, S.A.K. Razack, et.al., Energy Conervation and management 45 (2004) 1597-1615
[3] B.Zalba, J.M marin, L.F. Cabeza, et al., Applied Thermal Engineering 23 (2003) 251-283
[4] B.Li, T.Liu, L.Hu, Y.Wang, S.Nie. Chemical Engineering 215 (2013) 819-826
[5] L.Cao, Y.Tang, G.Fang. Energy 80 (2014) 98-103
[6] Z.Chen, L.Cao, G.Fang. Energy and Buildings 62 (2013) 469-474
[7] A.Castell, C.Sole, M.Medrano, et-al., Applied Thermal Engineering 28 (2008) 1676-1686
[8] C.Y. Zhao, W.Lu, Y.Tian, Solar Energy 84 (2010) 1402-1412
[9] A.Castel, C.Sole, M.Medrano, et-al., Journal of Building Performance Simulation 3 (2010) 245-254
[10] L.Chai, X. Wang, D. Wu, Applied Energy 138 (2015) 661-674
[11] H. Zhang, X. Wang, D. Wu, Journal of Colloid and Inerface Science 343 (2010) 246-255
[12] N. Sahan, M. Fois, H. Paskoy, Solar Energy Materials & Solar Cells 137 (2015) 61-67
[13] L.Cao, F. Tang, G.Fang, Energy & Buildings 72 (2014) 31-37
[14] F.Jiang, X. Wang, D. Wu, Applied Energy 134 (2014) 456-468
[15] N. Sahan, H.O Paskoy, Solar Energy Materials & Solar Cells 126 (2014) 56-61
[16] J.Wang, H.Xie, Z.Guo, L.Guan, Y.Li, Applied Thermal Engineering 73 (2014) 1541-1547
[17] S.A. Arifin, S. Jalaludin, R. Saleh, Advanced Materials research 1123 (2015) 264-269
[18] S. Harikrishnan, S.Magesh, S.Kalaiselvam, Thermochimica Acta 565 (2013) 137-145
[19] A.F. Regin, S.C. Solanki, J.S. Saini, Renewable and Sustainable Energy Reviews 12 (2008) 2438-2458
[20] S.Shaikh, K.Lafdi, K.Hallinan, Journal of Applied Physics 103 094302 (2008)

Effects of supplementary lighting by natural light for growth of Brassica chinensis

Shih-Chuan Yeh[1]*, Hui-Ping Lee[2] , Shih-Tse Kao[3], and Ju-Lin Lu[2]
[1] De Lin Institute of Technology, Taiwan
[2] Taipei Municipal Wenhua Elementary School, Taiwan
[3] Peichi irrigation association, Taiwan
E-mail: yehshch@gmail.com

Abstract. This paper present a model of cultivated chamber with supplementary natural colour light. We investigate the effects of supplementary natural red light and natural blue light on growth of Brassica chinensis under natural white light illumination. After 4 weeks of supplementary colour light treatment, the experiment results shown that the weight of fresh leaf were not affected by supplementary natural blue light. However, those Brassica chinensis were cultivated in the chambers with supplementary natural red light obtained a significant increasing of fresh weight of leaf under both white light illuminate models. The combination of natural white light with supplementary natural red light illumination will be benefits in growth for cultivation and energy saving.

1. Introduction

Light quality is one of the most important variables affecting the efficiency for the growth of plants [1-5]. For example, blue light increased levels of anthocyanins in tomato [6], carotenoids in coffea [7] and ascorbic acid in lettuce and Komatsuna [8], and red light seems to be most effective for anthocyanin production in cranberry fruits [9]. Li and Kubota investigated the effects of different supplemental light qualities on phytochemicals and growth of baby leaf lettuce grown at a high planting density inside a growth chamber, the results demonstrated that supplemental light quality could be strategically used to enhance nutritional value and growth of baby leaf lettuce grown under white light [2]. Those results have shown the viability of optimizing light quality in increasing phytochemical concentration.

In recent decades, light emitting diodes (LEDs) have been received attention as a light source with low power consumption for indoor illumination, plant cultivation and other many applications. Lin et al., demonstrate that supplemental light quality can be strategically used to enhance the nutritional value and growth of lettuce plants grown under RBW LED lights. Precise management of the irradiance and wavelength may hold promise in maximizing the economic efficiency of plant production, quality, and nutrition potential of vegetables grown in controlled environments. Lee et al., presents a high-performance LED agricultural luminaire that uses a beam-shaping diffuser to achieve high optical efficiency and energy saving. There are many studies demonstrated that the combination of red (R) and blue (B) LED light was an effective light source for plant growth and development, and the light spectra, intensities, and durations can easily be controlled by growers in artificial growing environments [8, 10, 11-14].

[1] Corresponding author: Shih-Chuan Yeh, E-mail:yehshch@gmail.com

In addition to agricultural production, energy saving is also an important issue for human life and reducing environment degradation. Artificial lighting systems can be one of the major electrical loads in a building. In recent decades, there are many research for the utilizing daylight for interior illumination can contribute to energy savings, daylight use of solar radiation is particularly beneficial as the natural light brought by the natural light illumination system can decrease energy consumption by reducing artificial lighting use [15-18].

The utilization of supplementary lighting by natural light for growth of plants with the benefits for improving the efficient of production of plant and energy saving. However, the research about the effect of supplementary light quality on growth of plant by natural light source is still rare. This paper present a model of cultivated chamber which supplementary the transmitted natural colour light by Sunshine illuminating on the area, where is filled with colour filter, of the roof for growth of plants. The experiment results of supplementary natural red light and blue light for growth of Brassica chinensis are also shown in this paper. We describe the plant material, experiment and measurement in section 2; the results of this experiment are shown in section 3, and the final section is the conclusions of this research.

2. Material and Methods

The assessment of the efficiency of supplementary natural red light and blue light for improving growth of plants is one of aims of this research. As a result of the spectrum of red and blue are most effectively use in visible light energy for photosynthesis, the growth chambers were illuminated either by combination of natural white light and red light or combination of natural white light and red light, respectively. Brassica chinensis was the experiment crop. .

2.1. Plant material and growth conditions

Brassica chinensis (bok choy) is one of important vegetables, which is a good source of dietary fiber, vitamins A, B6, C, and K, calcium, folate and manganese, furthermore, it is applicable in hydroponics and in soil cultivation. In this research, 24 Brassica chinensis were cultivated into 6 growth chambers, which are illuminated by different supplementary natural light scheme, after the plants developed first true leaves. The plants were cultivated in the growth chambers for 4 weeks. The temperature and hours of Sunshine for the duration of experiment.is shown in Fig. 1. In this period, the average temperature is 23℃, the maximum sunshine hours is 10 hours in a day. The climate is appropriate for Brassica chinensis cultivation.

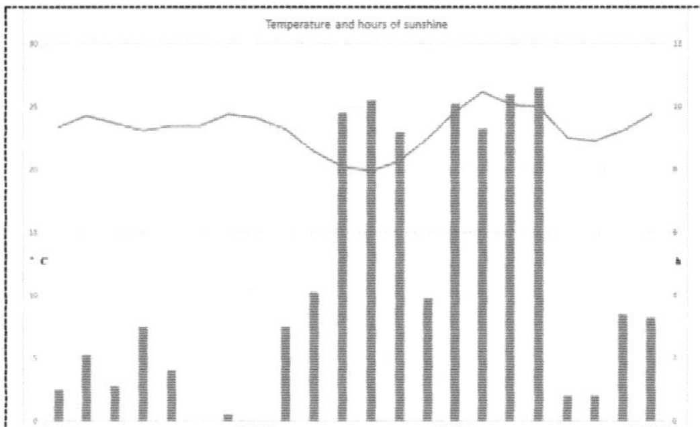

Figure 1. Temperature and hours of Sunshine for the duration of experiment.

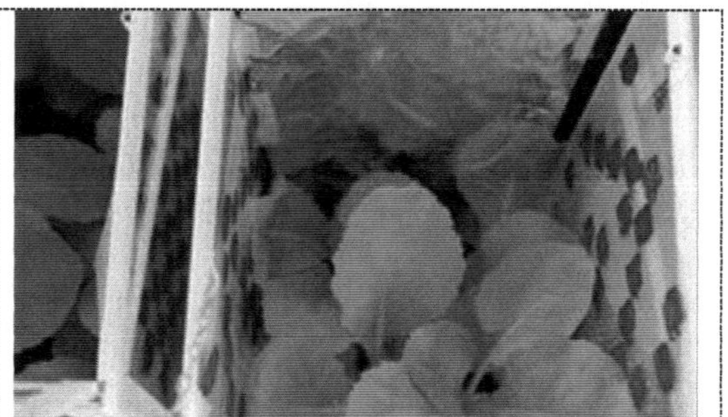

Figure 2. An illustrated for the growth chamber.

2.2. Growth chamber

In this research, natural white light as the control treatment. Two natural white light illuminating models are used in this experiment, that the ratio of white light areas are 70% (W1) and 50%(W2) of total area of the roof of the chamber respectively. The growth chambers were illuminated either by combination of natural white light and red light or combination of natural white light and red light, respectively. Both of the natural white light source and the supplementary colour light are obtained by Sunshine illuminating on the areas of transparent and colour filter of the roof respectively. Fig. 2 is an illustration for the growth chamber. The ratio of areas of colour filter on the chambers are described in table 1, where the ratio (W: B: R: K) means the relative areas of the regions where without colour filter, where is filled with blue filter, red filter and black (T=0) respectively. The areas distribution of colour filter are designed for uniformly illumination on the chambers during a day, illuminating model 1 and model 2 are shown in Fig. 3a and Fig. 3b respectively. The transmitted spectrum of blue and red colour filters are shown in Fig. 4a and Fig. 4b respectively.

Table 1. Ratio of areas of colour filter region .

chamber	**Ratio(W:B:R:K)**[a]
W1	7:0:0:3
WB1	7:3:0:0
WR1	7:0:3:0
W2	5:0:0:5
WB2	5:5:0:0
WR2	5:0:5:0

[a] Where W,B,R and K refers to white, blue, red and black respectively.

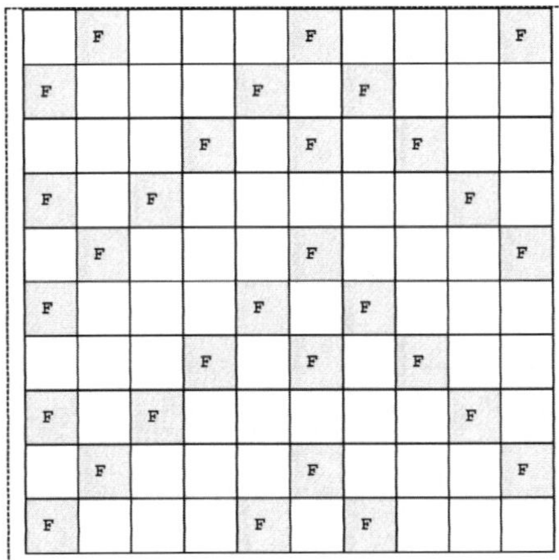

Figure 3a. The areas distribution of colour filter on white light illuminating model 1.

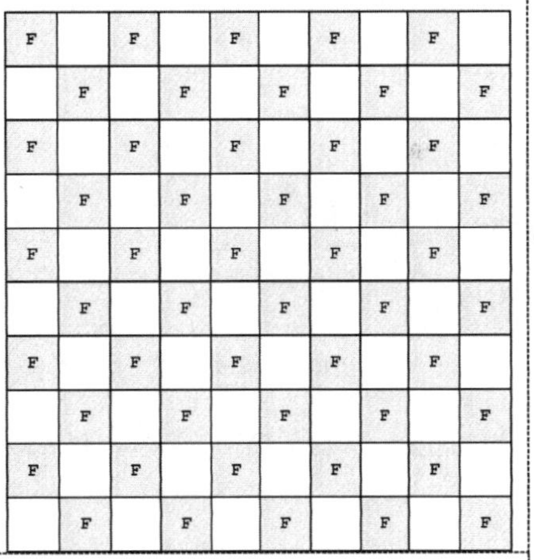

Figure 3b. The areas distribution of colour filter on white light illuminating model 2.

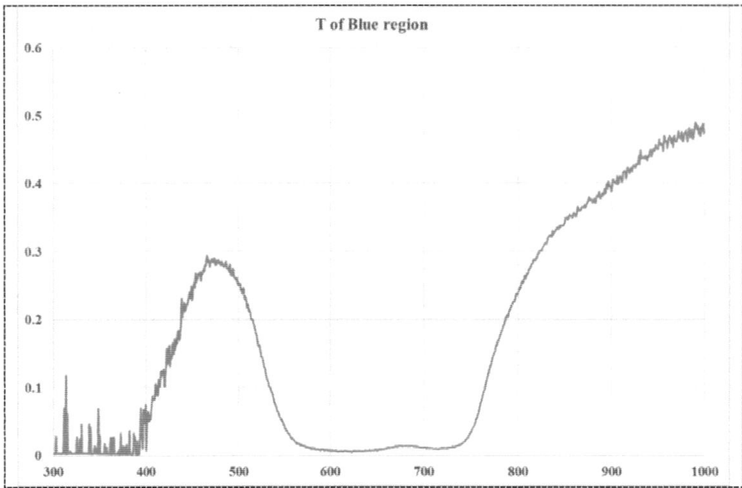

Figure 4a. The transmitted spectrum of the colour filter for blue.

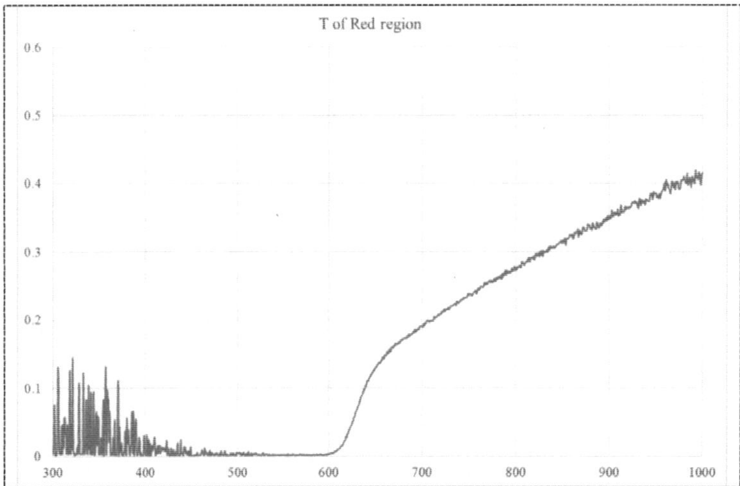

Figure 4b. The transmitted spectrum of the colour filter for red.

The ratio of supplementary light compare to natural white light can be determined by Eq. (1)~(3), where I_F, I_S, R_F, T_F and A_F are refer to the relative transmitted intensity of natural colour light, the relative intensity of natural white light, ratio of supplementary natural colour light, the average transmittance of colour filter which corresponding to active spectrum and the ratio of colour filter on the growth chamber. The ratio of supplementary colour light is described in Table 2, where A_F is obtained by the total area where the region with colour filter relative to the total area where is sunshine illuminated, T_F is the average transmittance which corresponding to the action spectrum.

$$I_F = \int I_S(\lambda)d\lambda \cdot T_F(\lambda) \tag{1}$$

$$I_W = \int I_S(\lambda)d\lambda \tag{2}$$

$$R_F = \frac{I_F A_F}{(1 - A_F) I_W} \tag{3}$$

Table 2. The ratio of supplementary colour light

Chamber	A_F	T_F	I_F
W1	30%	0	0%
WB1	30%	0.1	4%
WR1	30%	0.2	9%
W2	50%	0	0%
WB2	50%	0.1	10%
WR2	50%	0.2	20%

2.3. Measurement
We investigate the effects of supplementary natural red light and natural blue light on growth of Brassica chinensis which were grown under natural white light illumination. The stem length, length of leaf and width of leaf of each plant were measurement per week after the plants were cultivated in the growth chamber by supplementary natural colour light. After 4 weeks of supplementary colour light treatment, the leaf fresh weight was also measured in the finally measurement.

3. Results
The stem length, length of leaf and width of leaf of each plant were measurement per week after the plants were cultivated in the growth chamber, which were cultivated in the chambers with different natural light treatment. The average length of leaf and the average of leaf during experiment are described in Table 3 and Table 4 respectively. Fig. 5 and Fig. 6 shows the average values of product of leaf length and width of leaf, which is related to the area of leaf. The results shown that the Brassica chinensis which were cultivated in chambers WR1 and WR2 can be Maximum increased to about 25% and 26% relative to those were cultivated in chambers W1 and W2 during this period respectively. The values of products shown that the Brassica chinensis which were cultivated in chamber WB1 and WB2 can be increased to about 26% and 6% relative to W1 and W2 respectively.
Fig. 7 and Fig. 8 are the latest measurements in this experiment for supplementary natural colour lighting illuminated on Brassica chinensis growth. Fig.7 shows that the harvest total fresh weight of leaf was 70g, which is obtained by those Brassica chinensis were cultivated in chamber WR1. The total fresh weight of leaf which were cultivated in Chamber WR1 got an increase of about 37% and 32% relative to those were cultivated in chamber WB1 and W1 respectively. The average length of leaf of Brassica chinensis which were cultivated in Chamber WR1 got an increase of about 13% relative to those were cultivated in chamber W1. The average length of leaf of Brassica chinensis which were cultivated in Chamber WB1 got an increase of about 8% relative to those were cultivated in chamber W1. The average width of leaf of Brassica chinensis which were cultivated in Chamber WR1 got an increase of about 11% relative to those were cultivated in chamber W1. The average length of leaf of Brassica chinensis which were cultivated in Chamber WB1 got an increase of about 17% relative to those were cultivated in chamber W1.
Fig. 8 shown that the harvest total fresh weight of leaf is 58.8g, which Brassica chinensis were cultivated in chamber WR2. The fresh weight of Brassica chinensis which were cultivated in Chamber WR2 got an increase of about 14% and 21% relative to those were cultivated in chamber WB2 and W2 respectively. The average length of leaf of Brassica chinensis which were cultivated in Chamber WR2 got an reduction of about 5% relative to those were cultivated in chamber W2. The average length of leaf of Brassica chinensis which were cultivated in Chamber WB2 got an increase of about 8% relative to those were cultivated in chamber W2. The average width of leaf of Brassica chinensis which were

cultivated in Chamber WR2 got an reduction of about 9% relative to those Brassica chinensis were cultivated in chamber W2. The average length of leaf of Brassica chinensis which were cultivated in Chamber WB2 got an reduction of about 1% relative to those were cultivated in chamber W2.

Table 3. The average length of leaf during experiment (cm)

Chamber	10/4	10/11	10/18	10/25
W1	4.0	7.6	11.5	13.5
WB1	4.0	8.2	13.1	14.0
WR1	4.0	8.6	12.6	14.8
W2	4.0	7.6	12.4	13.8
WB2	4.0	8.2	11.3	12.5
WR2	4.0	7.8	12.6	14.6

Table 4. The average width of leaf during experiment (cm)

Chamber	10/4	10/11	10/18	10/25
W1	4.0	7.6	11.5	13.5
WB1	4.0	8.2	13.1	14.0
WR1	4.0	8.6	12.6	14.8
W2	4.0	7.6	12.4	13.8
WB2	4.0	8.2	11.3	12.5
WR2	4.0	7.8	12.6	14.6

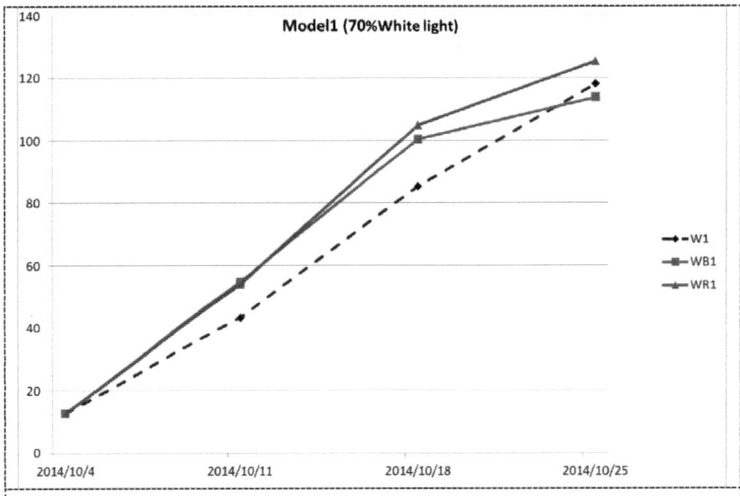

Figure 5. The average values of products of length and width of leaf which were cultivated in illuminated model 1.

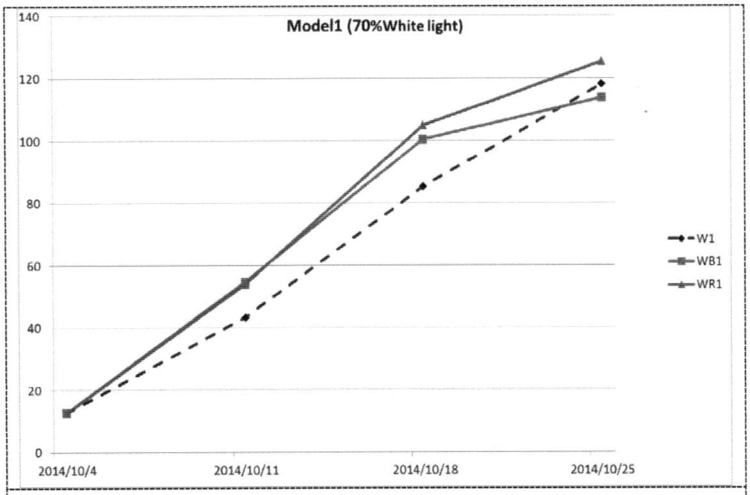

Figure 6. The average values of product of length and width of leaf which were cultivated in illuminated model 2.

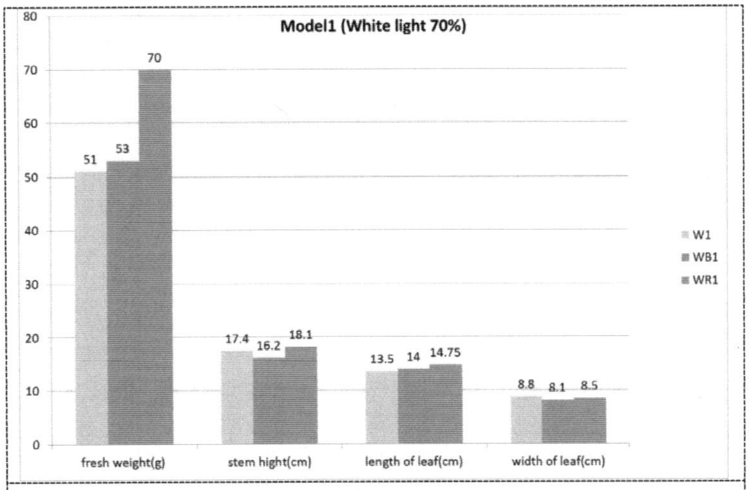

Figure 7. The fresh weight, height, length, and width of leaf after 4 weeks cultivation in illuminated model 1.

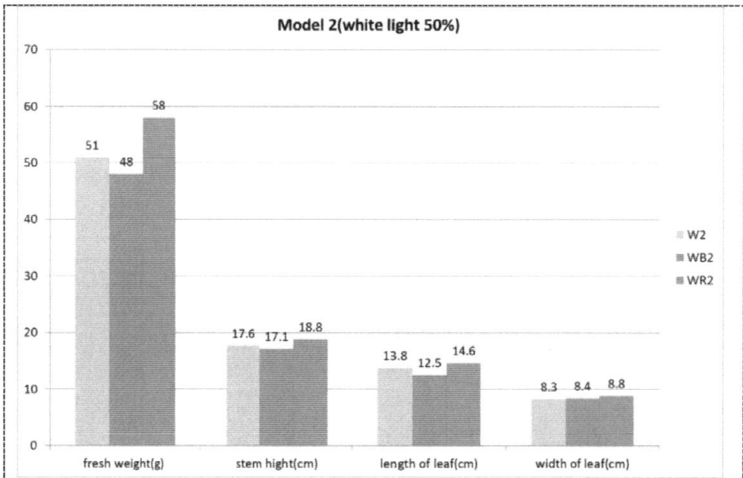

Figure 8. The fresh weight, height, length, and width of leaf after 4 weeks cultivation in illuminated model 2.

4. Conclusion

In this research, we investigate the effects of supplementary natural red light and natural blue light on growth of Brassica chinensis grown under natural white light illumination. After 4 weeks of combination of natural whit light and supplementary natural red light treatment and supplementary natural blue light treatment, the experiment results shown that the weight of fresh weight were not affected by supplementary natural blue light. However, those Brassica chinensis were cultivated in the chamber with supplementary natural red light obtained a significant increasing of fresh weight of leaf under both white light illuminate models. The combination of natural white light with supplementary natural red light illumination will be benefits in growth for cultivation and energy saving.

5. References

[1] Kopsell, D.A., Kopsell, D.E., 2008. Genetic and environmental factors affecting plant lutein/zeaxanthin. Agro Food Ind. Hi-Tech. 19, 44–46.

[2] LI, Qian; KUBOTA, Chieri. Effects of supplemental light quality on growth and phytochemicals of baby leaf lettuce. Environmental and Experimental Botany, 2009, 67.1: 59-64.

[3] Perez-Balibrea, S.,Moreno, D.A., Garcia-Viguera, C., 2008. Influence of light on healthVergeer, L.H.T.,

[4] Aarts, T.L., Degroot, J.D., 1995. The wasting disease and the effect of abiotic factors (light-intensity, temperature, salinity) and infection with labyrinthula-zosterae on the phenolics content of zostera-marina shoots.

[5] Ohashi-Kaneko, K., Takase, M., Kon, N., Fujiwara, K., Kurata, K., 2007. Effect of light quality on growth and vegetable quality in leaf lettuce, spinach and komatsuna. Environ. Control Biol. 45, 189–198

[6] Keller, Markus, and Geza Hrazdina. "Interaction of nitrogen availability during bloom and light intensity during veraison. II. Effects on anthocyanin and phenolic development during grape ripening." American Journal of Enology and Viticulture 49.3 (1998): 341-349.

[7] Ramalho, J. C., et al. "Cold acclimation ability and photosynthesis among species of the tropical Coffea genus." Plant Biology 5.6 (2003): 631-641.

[8] SHIMADA, Aoi; TANIGUCHI, Yoshio. Red and blue pulse timing control for pulse width modulation light dimming of light emitting diodes for plant cultivation. Journal of Photochemistry and Photobiology B: Biology, 2011, 104.3: 399-404.

[9] Zhou, Yu, and Bal Ram Singh. "Red light stimulates flowering and anthocyanin biosynthesis in American cranberry." Plant growth regulation 38.2 (2002): 165-171.
[10] LEE, Xuan-Hao; CHANG, Yu-Yu; SUN, Ching-Cherng. Highly energy-efficient agricultural lighting by B+ R LEDs with beam shaping using micro-lens diffuser. Optics Communications, 2013, 291: 7-14.
[11] LIN, Kuan-Hung, et al. The effects of red, blue, and white light-emitting diodes on the growth, development, and edible quality of hydroponically grown lettuce (Lactuca sativa L. var. capitata). Scientia Horticulturae, 2013, 150: 86-91.
[12] MACEDO, Andrea Furtado, et al. The effect of light quality on leaf production and development of in vitrocultured plants of Alternanthera brasiliana Kuntze. Environmental and experimental botany, 2011, 70.1: 43-50.
[13] CHANG, C. L.; HONG, G. F.; LI, Ying-Li. A supplementary lighting and regulatory scheme using a multi-wavelength light emitting diode module for greenhouse application. Lighting Research and Technology, 2014, 46.5: 548-566.
[14] OLLE, Margit; VIRŠILE, Akvile. The effects of light-emitting diode lighting on greenhouse plant growth and quality. Agricultural and food science, 2013, 22.2: 223-234.
[15] Chirarattananon S, Chaiwiwatworakul P, Pattanasethanon S. Daylight availability and models for global and diffuse horizontal illuminance and irradiance for Bangkok. Renew Energy 2002;26(1):69–89
[16] Krarti M, Erickson PM, Hillman TC. A simplified method to estimate energy savings of artificial lighting use from daylighting. Build Environ 2005;40(6):747–54.
[17] Chel A, Tiwari GN, Chandra A. A model for estimation of daylight factor for skylight: an experimental validation using pyramid shape skylight over vault roof mud-house in New Delhi (India). Appl Energy 2009;86(11):2507–19.
[18] Yeh, S. C. "A natural lighting system using a prismatic daylight collector." Lighting Research and Technology 46.5 (2014): 534-547.

Acknowledgments
The authors gratefully acknowledge the financial support of the Ministry of Science and Technology of Taiwan (MOST 104 - 2221 - E - 237 - 006 -).

Implementation of Elliptic Curve Cryptography in Binary Field

D R Susantio, I Muchtadi-Alamsyah

Algebra Research Group, Faculty of Mathematics and Natural Sciences, Institut Teknologi Bandung, Jalan Ganesha no. 10, Bandung, 40132, Indonesia.

E-mail: ntan@math.itb.ac.id

Abstract. Currently, there is a steadily increasing demand of information security, caused by a surge in information flow. There are many ways to create a secure information channel, one of which is to use cryptography. In this paper, we discuss the implementation of elliptic curves over the binary field for cryptography. We use the simplified version of the ECIES (Elliptic Curve Integrated Encryption Scheme). The ECIES encrypts a plaintext by masking the original message using specified points on the curve. The encryption process is done by separating the plaintext into blocks. Each block is then separately encrypted using the encryption scheme.

1. Introduction

Information security has become one of the oft-discussed problem in the wake of the 21st century. Most information channels that we use are insecure: they are vulnerable to leakage. Therefore, it is essential that we look for ways to secure the insecure information channel. Cryptography is one of the many ways to secure the channels.

Neal Koblitz and Victor Miller independently introduced elliptic curve cryptography (ECC) [5]. ECC uses the set of points on an elliptic curve along with an addition rule. The unique mathematical structure of the points with the addition rule enables us to perform encryption and decryption of plaintexts. Another reason that supports the feasibility of ECC is the fact that it uses significantly smaller key sizes than the RSA Cryptosystem. For a comparison between RSA and ECC key sizes, see [6].

In the previous research [3], we presented the implementation of binary field arithmetic operation algorithms. In this paper, we discuss the implementation of elliptic curve cryptography using elliptic curves over binary field. We will address several issues, among them will be case-handling for the point operations, the key generation process, and the encryption.

2. Elliptic Curves and the ECIES

An elliptic curve over $GF(2^n)$ is defined by the simplified Weierstrass equation $y^2 + xy = x^3 + ax^2 + b$, where $a \neq 0$ and $b \neq 0$ [2]. It is possible to define several operations on the points of the elliptic curve, namely point negation, addition, and doubling.

We define the point operations as follows. First, let E be an elliptic curve over $GF(2^n)$ and $P(x, y)$ a point on E. The negative of P is defined to be $-P(x, x + y)$. Now let $P(x_1, y_1)$ and $Q(x_2, y_2)$ be distinct points on E. The result of adding P and Q is the point $P + Q$ (x_3, y_3) where $x_3 = \lambda^2 + \lambda + x_1 + x_2 + a$, $y_3 = \lambda(x_1 + x_3) + x_3 + y_1$, and $\lambda = (y_2 + y_1)/(x_2 + x_1)$. In the case that P and Q are not distinct, the operation is called a point doubling. The double of point P is the point $2P$ (x_3, y_3) where $x_3 = \lambda^2 + \lambda + a$, $y_3 = x_1^2 + \lambda x_3 + x_3$, and $\lambda = x_1 + y_1/x_1$. It is also possible to define a scalar multiplication on elliptic curves. Given a positive integer n the scalar multiple of a point is defined as

$$nP = \underbrace{P + P + P + \cdots + P}_{n \text{ times}}.$$

The set of all points of an elliptic curve E over $GF(2^n)$ forms a commutative group with respect to point addition. Furthermore, according to the Mordell-Weil Theorem, this group is finitely generated [7] [8] . The group structure of the elliptic curves over $GF(2^n)$ allows its use for cryptography. In this paper, we discuss the implementation of the simplified version of Elliptic Curve Integrated Encryption Scheme (ECIES).

The simplified ECIES is a modification of the ElGamal scheme, which is based on the elliptic curve discrete logarithm problem described as follows. [9]

Definition 2.1. *Let E be an elliptic curve over $GF(2^n)$, P be a point on E with order n, and $Q \in \langle P \rangle$, the finite group generated by P. The discrete logarithm problem is the problem of determining an integer m, $0 \leq m \leq n-1$ which satisfies $Q = mP$. The integer m is then called the discrete logarithm of Q.*

Let $\mathcal{P}$ the set of all possible plaintexts, $\mathcal{C}$ the set of all possible ciphertexts, and $\mathcal{K}$ the set of all keys in the encryption scheme. We now define the simplified ECIES as follows. [9]

Definition 2.2. *Let E be an elliptic curve over $GF(2^N)$ such that E contains a cyclic subgroup $H = \langle P \rangle$ of order n in which the elliptic curve discrete logarithm problem is infeasible.*

(i) The plaintext space is $\mathcal{P} = GF(2^N)$ and the ciphertext space is $\mathcal{C} = E(GF(2^N)) \times GF(2^N)$.

(ii) The key space is $\mathcal{K} = \{(E, P, Q, m, n) : Q = mP\}$, where E, P, Q, and n are public keys and m a private key.

(iii) For every $K \in \mathcal{K}$, plaintext $x \in \mathcal{P}$, and ciphertext $(U, c) \in \mathcal{C}$, and a (secret) random number $k \in [0, n-1]$, define the encryption and decryption as follows.

(a) The encryption function is $e_K(x, k) = (kP, x \cdot x_0)$, where $kQ = (x_0, y_0)$, and $x_0 \neq 0$.

(b) The decryption function is $d_K(U, c) = c(x_0)^{-1}$, where $(x_0, y_0) = mU$.

3. Implementation

In this section we will describe several algorithms to perform elliptic curve operations, as well as some modifications made to the simplified ECIES.

Elliptic Curve Operations

We first describe the algorithm for point negation. Given a point $P(x, y)$ on an elliptic curve E, the algorithm simply takes the coordinates of P as input, then returns $-P(x, x + y)$ (see Algorithm 1).

Algorithm 1 Point negation

INPUT: point $P(x_1, y_1)$.
OUTPUT: point $-P$.
1: $X \leftarrow x_1$
2: $Y \leftarrow x_1 \oplus y_1$
3: **return** (X, Y)

Next, we describe the algorithm for point doubling. We first consider the case when $P = -P$ and P is the identity. In this case, the algorithm returns the identity. Otherwise, we compute $2P$ using the doubling formula (see Algorithm 2).

Algorithm 2 Point doubling

INPUT: point $P(x_1, y_1)$.
OUTPUT: point $2P$.

if $P = -P$ or $P = \mathcal{O}$ **then**
    return $\mathcal{O}$
else
    $\lambda \leftarrow x_1 \oplus y_1/x_1$
    $X \leftarrow \lambda^2 \oplus \lambda \oplus a$
    $Y \leftarrow x_1^2 \oplus \lambda \cdot X \oplus X$
    return (X, Y)
end if

We now describe the point addition algorithm. We first consider whether $P = Q$. If this is the case, we utilize the point doubling algorithm. Otherwise, we have several sub-cases to consider. The first sub-case is the case when $P = -Q$. In this case, return $\mathcal{O}$. The second sub-case is the case when $P = \mathcal{O}$ or $Q = \mathcal{O}$. If $P = \mathcal{O}$, return Q. If $Q = \mathcal{O}$, return P. Should all of these sub-cases fail, we conclude that P and Q are distinct and are not the identity. Therefore, we compute $P + Q$ using the addition formula.

Algorithm 3 Point addition

INPUT: points $P(x_1, y_1)$, $Q(x_2, y_2)$.
OUTPUT: point $P + Q$.

if $P \neq Q$ **then**
    if $P = -Q$ **then**
        return $\mathcal{O}$
    else if $P = \mathcal{O}$ **then**
        return Q
    else if $Q = \mathcal{O}$ **then**
        return P
    else
        $\lambda \leftarrow (y_2 \oplus y_1)/(x_2 \oplus x_1)$
        $X \leftarrow \lambda^2 \oplus \lambda \oplus x_1 \oplus x_2 \oplus a$
        $Y \leftarrow \lambda \cdot (x_1 \oplus X) \oplus X \oplus y_1$
        return (X, Y)
    end if
else
    return $2P$
end if

Finally, to calculate the scalar multiple of a point, we use the double-and-add method. This method is analogous to the square-and-multiply algorithm for calculating powers. Let P be a point on an elliptic curve and let s be a positive integer. First, let $b_n b_{n-1} \cdots b_0$ be the binary representation of s. We observe the value of b_0. If $b_0 = 1$, we begin with an initial value of P. Otherwise, we begin with $\mathcal{O}$. Then, for each integer i, where $1 \leq i \leq n$, observe b_i. If $b_i = 1$, then add $(2^i)P$. For example, suppose we want to compute $5P$. The binary representation of 5 is 101. Therefore, using this method, instead of calculating $P + P + P + P + P$, we calculate $4P + P$. The double-and-add method is more efficient than the naive repeated addition method to compute scalar multiples [9].

ECIES Parameter Generation
To begin the encryption and decryption process using the simplified ECIES, we first generate the curve. This is simply done by choosing a random number between 1 and $2^N - 1$, where N is the degree of the chosen binary field. We then take the binary representation of the random number as the curve parameters a and b.

Next, we generate the point P. We first randomly choose the x-coordinate for the point using a similar procedure to the curve parameter generation. Substituting for x in the curve equation yields an equation in the form $y^2 + by + d = 0$. We then determine whether the equation has a solution using Theorem 3.1 [4].

Algorithm 4 Scalar multiple

INPUT: point P, integer n.
OUTPUT: point nP.
$A \leftarrow P$
$R \leftarrow \mathcal{O}$
while $n > 0$ **do**
  if $n \equiv 1 \bmod 2$ **then**
    $R \leftarrow R + A$
  end if
  $n \leftarrow n \gg 1$
  $A \leftarrow 2A$
end while
return R

Theorem 3.1. *The polynomial* $t^2 + t + d \in GF\left(2^N\right)[t]$ *has a root* u *in* $GF\left(2^N\right)$, *if and only if* $\mathrm{Tr}(d) = 0$. $\mathrm{Tr}(d)$ *is called the trace of an element* d, *which is defined to be*

$$\mathrm{Tr}(d) = d + d^2 + d^{2^2} + \cdots + d^{2^{N-1}}.$$

If the equation has a root, we are then able to calculate the roots using Theorem 3.2 [1]. Otherwise, select another x-coordinate and repeat the equation checking process.

Theorem 3.2. *(Kugurakov) Let* $\delta \in GF\left(2^N\right)$ *and* $\mathrm{Tr}(\delta) = 0$. *Then the equation* $y^2 + y = \delta$ *has the explicit root* $y = 1 + \sum_{j=1}^{m-1}\left(\delta^{2^j}\left(\sum_{k=0}^{j-1} u^{2^k}\right)\right)$ *where* $u \in GF\left(2^N\right)$ *and* $\mathrm{Tr}(u) = 1$. *The other root is* $y + 1$.

To choose the element u such that $\mathrm{Tr}(u) = 1$, we select a random element of $GF\left(2^N\right)$, then calculate its trace. This process is repeated until an element with trace 1 is found.

The next step is to calculate the order of P. We use Hasse's theorem [8].

Theorem 3.3. *(Hasse) Let* $E/\mathbb{F}_q$ *be an elliptic curve over* $GF(q)$. *Then,*

$$\#E(\mathbb{F}_q) = q + 1 - t$$

where $|t| \leq 2\sqrt{q}$.

As a consequence of this theorem, we have $(\sqrt{q} - 1)^2 \leq \#E(\mathbb{F}_q) \leq (\sqrt{q} + 1)^2$. Therefore, there exists an integer $M \in [(\sqrt{q} - 1)^2, (\sqrt{q} + 1)^2]$ such that $MP = \mathcal{O}$. We can then compute n, the order of P, as the factor of M.

Finally, we can generate the private key m and point $Q = mP$. The private key m is randomly chosen from integers between 1 and $n - 1$, inclusive. Since, by Definition 2.2, the point P generates a cyclic group of order n, it suffices to choose m in this manner. After the private key is determined, we compute $Q = mP$ using the double-and-add algorithm.

Encryption and Decryption

Encryption is accomplished by first converting the plaintext into an element, or elements, of $GF(2^N)$. We take the binary representation of the ASCII code of each character in the plaintext. In this paper, it is assumed that every character has an ASCII value of 0 – 127.

The next step is to group the characters, which are already in their binary representation, into blocks. To do this, first take note that due to the assumption of the ASCII value, the size of the ASCII binary representation is at most 7 bits. We then do a padding process so that every character has a 7-bit binary representation. Once padded, the binary representations are concatenated, then split into blocks where each block has a maximum of $\lfloor N/7 \rfloor$ characters of the original plaintext.

The last step is to encrypt each block of the plaintext using the encryption function defined before.

We provide the proof of correctness for the encryption algorithm (Algorithm 5) as follows.

Algorithm 5 Encryption with the simplified ECIES

INPUT: plaintext x.
OUTPUT: ciphertext $(U(x_1, y_1), y)$.

1: **for** $char \in x$ **do**
2: $\quad char \leftarrow$ BINARYASCII($char$)
3: $\quad char \leftarrow$ PADDING($char$)
4: $\quad$ APPEND(x',$char$)
5: **end for**
6: $blocklength \leftarrow \lfloor N/7 \rfloor$
7: $x' \leftarrow$ BLOCK(x', $blocklength$)
8: $k \leftarrow$ RANDOM($[1, n-1]$)
9: $U(x_1, y_1) \leftarrow kP$
10: $V(x_2, y_2) \leftarrow kQ$
11: **for** $char \in x'$ **do**
12: $\quad cipher \leftarrow char \cdot x_2$
13: $\quad$ APPEND(y, $cipher$)
14: **end for**
15: **return** (U, y)

Theorem 3.4. *The encryption algorithm with the simplified ECIES is correct.*

Proof. We first split the algorithm as described into three blocks: lines 1 – 5, 6 – 10, and 11 – 15. We will establish the correctness of each block.

Lines 1 – 5. Let x be a string and x' be an array of binary strings of length 7. For the purposes of the algorithm, we allow x' to be empty. We also note that a string can be viewed as an array of characters. We assert that x' is a loop invariant. Note that at the beginning of the loop, x' is an empty array, which we allow. At each iteration, the algorithm will convert the nth character of x into its ASCII value in binary form by the BINARYASCII function. Its binary form is then padded by the PADDING function. Therefore, the character is now a binary string of length 7. This binary string is then appended to x', which establishes the fact that by the end of the execution of the loop, x' is an array of binary strings of length 7, verifying our assertion that x' is indeed a loop invariant. Finally, since we have a 'for' loop, and the length of x is finite, the loop is guaranteed to terminate. We have established the correctness of this block.

Lines 6 – 10. For this block, let x', the array of binary strings of length 7 and the points P and Q, the public keys, be our initial assertions. The final assertion will be x', the array of binary strings of length at most N, and points kP and kQ. The algorithm on this block first calculates the block length, which is the maximum of characters of the original plaintext that can be concatenated by the BLOCK function. Next, the BLOCK function manipulates x' by replacing each of the elements with blocks of concatenated binary representations of at most $\lfloor N/7 \rfloor$ characters from the original plaintext. Since $\lfloor N/7 \rfloor \leq N/7$, each block can only have a binary string of length at most N. The algorithm then randomly computes an integer within the $[1, n-1]$ interval. Finally, by the point scalar multiple algorithm, we obtain kP and kQ. We have completed the proof of correctness for this block.

Lines 11 – 15. Let x' be an array of blocks as previously described, U be kP, the first part of the ECIES ciphertext, and y be an array of ciphertexts which make up the second part of the ECIES ciphertext, which we allow to be empty. We first consider the 'for' loop. As with the first block, we assert that y is a loop invariant. We note the fact that y is empty at the beginning of the loop. At the execution of the loop, each element of the array is multiplied with x_2, the x-coordinate of kQ. It is then appended to y. Therefore, at the end of the execution of the loop, y is indeed an array of ciphertexts. Again, since we have a 'for' loop, the loop is guaranteed to terminate. Finally, the algorithm returns (U, y), which is the complete form of an ECIES ciphertext, establishing the correctness of this block.

Since we have established the correctness of the blocks, we have also established the correctness of the encryption algorithm. □

The decryption process is essentially the reverse of the encryption process. We first decrypt each block of the ciphertext using the decryption function. Then, we make sure that the length of each block of the decrypted ciphertext is divisible by 7, padding in if necessary. Each block is then split into sub-blocks of length 7. We can then convert each sub-block into the original character of the plaintext.

Algorithm 6 Decryption with the simplified ECIES

```
INPUT: ciphertext (U(x_1, y_1), y).
OUTPUT: plaintext x.
 1: V(x_2, y_2) ← mU
 2: for cipher ∈ y do
 3:     char ← cipher/x_2
 4:     char ← PADDING(char)
 5:     APPEND(x, char)
 6: end for
 7: x ← SPLIT(x)
 8: for char ∈ x do
 9:     char ← DEASCII(char)
10:     APPEND(x',char)
11: end for
12: return x'
```

We now provide the proof of correctness for the decryption algorithm (Algorithm 6).

Theorem 3.5. *The decryption algorithm with the simplified ECIES is correct.*

Proof. We now split the algorithm into two blocks: lines 1 – 6 and 7 – 12.

Lines 1 – 6. Let $(U(x_1, y_1))$ be the ECIES ciphertext and m be the private key. The first line of this block calculates $V = mU$, which is used to decrypt the ciphertext. Next, we observe the 'for' loop. Let x be an array of binary representation blocks of the plaintext which length is divisible by 7 and has a maximum length of N. We allow x to be empty. We assert that x is a loop invariant. At the beginning of the loop, x is indeed empty. At each iteration, we divide an element of y with x_2, the x-coordinate of V. By Definition 2.2, this decrypts that element to a block of binary representation of the plaintext. The block is then padded so that each block has a length that is divisible by 7 and appended to x. Therefore, at the end of the execution of this loop, we have an array of binary representation blocks of the plaintext which length is divisible by 7 and has a maximum length of N. Again, since we are dealing with a 'for' loop, this loop is guaranteed to terminate, completing this block's proof of correctness.

Lines 7 – 12. The first line of this block splits the array of plaintext blocks, x, into binary strings of length 7. Next, again we face a 'for' loop. Let x' be the original plaintext. Since x' is a string, we can treat x' as an array of characters, which, again, we allow to be empty. We assert that x' is a loop invariant. On loop entry, x' is empty. At each iteration, each element of the array is translated back from a binary string ASCII code into the character represented by the ASCII code. This character is appended into x'. Therefore, x' is an array of characters of the original plaintext, proving that x' is indeed a loop invariant. Since we are dealing with a 'for' loop, the loop is guaranteed to terminate, and by the termination of the loop, we have recovered all characters of the original plaintext. We have established the correctness of this block.

Since we have established the correctness of each block, we have also established the correctness of the decryption algorithm. □

4. Results

We implemented the aforementioned algorithms in an encryption-decryption simulation program written in Python. A sample output of the program is given in Figure 1. Basically, the program first randomizes the curve parameters, then picks a point and calculates its order. Afterwards, the user is asked to provide a plaintext. Once the plaintext is set, the program encrypts the plaintext, then displays the resulting ciphertext. Next, the program prompts the user to enter the ciphertext. Once the ciphertext is set, the program decrypts the ciphertext, then displays the recovered plaintext.

Several encryption and decryption simulations were conducted using the program. During the simulation, we attempt to encrypt the following plaintext:

> Lorem ipsum dolor sit amet, consectetur adipiscing elit. Aliquam in nulla quis mi maximus suscipit at sed ipsum. Maecenas tincidunt turpis diam, sed ultricies justo vulputate hendrerit. Donec eget metus nec tortor finibus varius a mattis turpis. In facilisis convallis hendrerit. Praesent pellentesque eros sed metus feugiat dictum. Donec vel nibh mollis, dignissim magna ac, tempus nisl. Vestibulum eleifend pretium lorem at consectetur. Proin lacinia pharetra ipsum, vel ornare urna hendrerit eu. Maecenas sed libero pretium, condimentum nulla tincidunt, egestas est. Sed placerat tortor a nisi lobortis vulputate. Vestibulum dignissim justo ac lectus varius, vulputate elementum sapien rhoncus. Suspendisse et condimentum nibh. Morbi eleifend nisl vel magna elementum euismod. Nunc turpis dolor, cursus vel eros at, mollis vestibulum urna.

which is a *lorem ipsum* text created by a *lorem ipsum* generator. It consists of 842 bytes of information. For the purposes of the simulation, we restrict the program to work in $GF(2^{20})$.

The results of the simulation is shown in Table 1. It is worth noticing that for all cases, the point generation time (PtGenTime), encryption time (EncTime), and decryption time (DecTime) is relatively stable and fast, given the amount of text encrypted (and decrypted). On the other hand, the point order calculation times (PtOrdTime) are extremely varied and relatively slow. This is due to the way Hasse's theorem (Theorem 3.3) is implemented in the program. The program simply tries every possible integer in the Hasse interval until it finds an integer M which satisfies $MP = \mathcal{O}$. The factorization is performed by checking from a predetermined list of primes, which is also rather inefficient.

5. Conclusion

We have established several algorithms for implementing the simplified ECIES over binary field. Several cases must be handled when adding points, which are the case where the point one (or both) points are points at infinity, and the case where the points are the same. The point generation process can be done effectively using Theorems 3.1 and 3.2. The encryption and decryption process can also be done efficiently using the provided algorithm. On the other hand, there is room for improvement to the point order calculation.

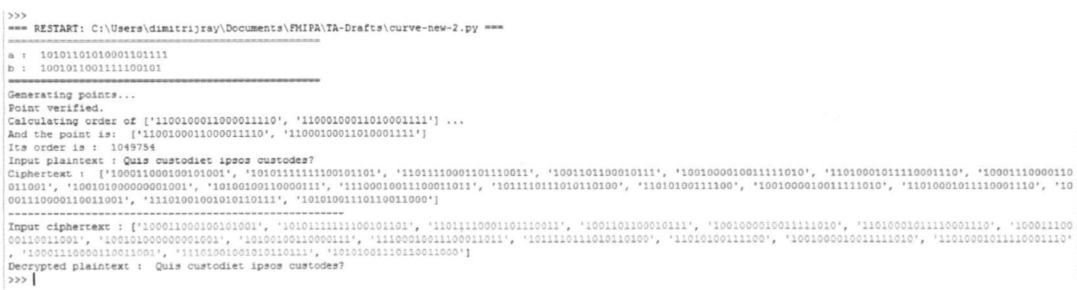

```
>>>
=== RESTART: C:\Users\dimitrijray\Documents\FMIPA\TA-Drafts\curve-new-2.py ===
==============================================================
a :  10101101010001101111
b :  10010110011111100101
==============================================================
Generating points...
Point verified.
Calculating order of ['11001000110000111110', '11000100011010001111'] ...
And the point is:  ['11001000110000111110', '11000100011010001111']
Its order is :  1049754
Input plaintext : Quis custodiet ipsos custodes?
Ciphertext :  ['100011000100101001', '10101111111100101101', '11011110001101110011', '1001101100010111', '1001000010011111010', '11010001011110001110', '10001110000110
011001', '1001010000000001001', '10100100110000111', '11100010011100011011', '1011110111010110100', '11010100111100', '1001000010011111010', '11010001011110001110', '10
001110000110011001', '11101001001010110111', '10101001110110011000']
---------------------------------------------------------------
Input ciphertext : ['100011000100101001', '10101111111100101101', '11011110001101110011', '1001101100010111', '1001000010011111010', '11010001011110001110', '100011100
00110011001', '1001010000000001001', '10100100110000111', '11100010011100011011', '1011110111010110100', '11010100111100', '1001000010011111010', '11010001011110001110'
, '10001110000110011001', '11101001001010110111', '10101001110110011000']
Decrypted plaintext :  Quis custodiet ipsos custodes?
>>> |
```

Figure 1. Encryption-decryption program, executed via the Python IDLE GUI

No.	Curve Parameters		Point chosen		Order	PtGenTime	PtOrdTime	EncTime	DecTime
	a	b	x	y					
1	11010100010100001	1111101111001110000	11010100010100001	1111101111001110000	1048870	0.0102	22.1138	0.3178	0.2799
2	10001100001010001 01	1100001011101110011 0	1110000011001111011 1	1011001010000011111 0	1048392	0.0102	17.9539	0.3177	0.2693
3	101111100100001001 1	1101010001001000010 0	1011010100010101 1	1011101001011100110 0	1049464	0.0099	28.3902	0.3238	0.2721
4	1101001110100110100 0	1101100100001011011 0	1010010011111101010 0	1000011001110000101 0	1050416	0.0926	38.0370	0.2978	0.3185
5	1110100010110101110 1	1000011111000010011 1	1001010001101001 00	1110011011000100011	1047226	0.1215	9.1655	0.3091	0.4407
6	111101100111101000	1110000101111111001 1	1010011011110000000 0	1100001110010101000 0	1048222	0.1661	28.4010	0.3173	0.3306
7	1110001110101000110 1	1010100000011100101	1000001101110000010 1	10010000111110111	1048202	0.0127	30.2198	0.2818	0.3142
8	111101011110111 10	1101100100101100100 1	1001111101101001 00	110101001110100000	210008	0.0193	57.3799	0.2630	0.5272
9	1101100010110001111 0	1101110111000011101 0	1010001011111010011 0	1001100000111111100	1047440	0.1303	15.7834	0.3129	0.4970
10	111001101011011110 1	1101100101011001000 1	110001010100100 10	1010101101011011111 0	1049842	0.0118	33.2842	0.2506	0.3056
11	111011100111010001	1000110111011010001 1	1010011011010101010	1111110111110100011 1	1047130	0.0968	7.0309	0.2690	0.3076
12	1010001101010100111 0	1111010011010111111 1	1101010001010100111 0	1001111101011101010	1050166	0.0121	37.1911	0.2341	0.2961
13	1000111010000011101 0	111000001110010000	1011001101000100011 0	1111101000000101101 0	1049396	0.0218	29.2950	0.2611	0.3748
14	111110100111010011	1011100010000000011 0	101111010100010001	1011000001101111101 1	1046846	0.0208	3.7891	0.2569	0.3157
15	110111100110100	1011000100000001010 1	1111101001011111011 1	1001101100000101100	116292	0.1288	1.2743	0.2677	0.2636
16	1100010100011100	100000101010011101 0	1110111110001001	1111001010101111110 1	116328	0.0309	4.6491	0.2794	0.3814
17	100000111001011011	100000000100011000	110101101111110100 1	100000101000110111	349762	0.0954	29.8303	0.2478	0.2640
18	111001110010000100 1	1101100011100000101 1	11000010101001100	1101000010011110101 0	209482	0.1013	8.5569	0.2518	0.2582
19	1001010001101110101 1	111100111110001011 0	100000110001101100	110101000100111011	95332	0.0238	26.3968	0.2518	0.2466
20	1010110110111001110 1	1110110011010101101 1	111001101000100111 1	1001110111100111100	13978	0.0963	19.3744	0.2600	0.2874
21	101101001111010001 1	110010001010000010 1	1101000011101010011 0	1101011101101001100 0	349548	0.1195	21.5548	0.2499	0.2529
22	1001101101100011001 1	1101010001101010001 0	1111111110001010111 1	1010100101011010010	1047328	0.0992	7.2352	0.2393	0.2555
23	100000000101111000	110100001100011111 1	1111011001111000011 1	1010011101111101011	149818	0.1355	25.2752	0.2933	0.2811
24	1011011100000001110 0	100000010000001010	110100110011011001	100001011000000101	1048438	0.1017	21.1759	0.2504	0.2529
25	101010001010111100 0	1110001110000010111 1	10011110000110001	1101000111011000011	209364	0.1075	2.6863	0.2926	0.2808

Table 1. Experiment results with the encryption-decryption program. The PtGenTime column shows the time required to generate a valid point in the curve. The PtOrdTime column shows the time required to calculate the chosen point's order. The EncTime and DecTime column shows the time required to process the encryption and decryption, respectively.

Journal of Physics: Conference Series **710** (2016) 012022 doi:10.1088/1742-6596/710/1/012022

References

[1] Cherly J, Gallardo L, Leonid V and Wheland E 1998 Solving quadratic equations over polynomial rings of characteristic two *Publicacions Matemàtiques* **42**

[2] Hankerson D, Menezes A and Vanstone S 2004 *Guide to Elliptic Curve Cryptography* (New York: Springer-Verlag)

[3] Maulana M, Senjaya W F, Rahardjo B, Muchtadi-Alamsyah I and Paryasto M W 2014 Implementation of finite field arithmetic operations for polynomial and normal basis representations *3rd Int. Conf. on Computation for Science and Technology (ICCST-3)*

[4] Pommerening K 2012 Quadratic equations in finite fields of characteristic 2 retrieved from `www.staff.uni-mainz.de/pommeren/MathMisc/QuGlChar2.pdf`

[5] Paar C and Pelzl J 2010 *Understanding Cryptography* (New York: Springer)

[6] Shanmugalakshmi R and Prabu M 2009 Research issues on Elliptic Curve Cryptography and its applications *International Journal of Computer Science and Network Security* **9**

[7] Silverman J and Tate J 1994 *Rational Points on Elliptic Curves* (New York: Springer)

[8] Silverman J 2009 *The Arithmetic of Elliptic Curves* (New York: Springer)

[9] Stinson D 2006 *Cryptography: Theory and Practice 3rd Edition*, (Boca Raton: Chapman & Hall/CRC)

Quarter-sweep Nonlocal Discretization Scheme with QSSOR Iteration for Nonlinear Two-point Boundary Value Problems

M. U. Alibubin[1], A. Sunarto[2], and J. Sulaiman[3]
[1,2,3] Faculty of Science and Natural Resources,
Universiti Malaysia Sabah, Malaysia.

E-mail: [1]mrusran@gmail.com, [2]andang99@gmail.com, [3]jumat@ums.edu.my

Abstract. The aim of this paper is to consider the Quarter-sweep Successive Over Relaxation (QSSOR) iteration for solving nonlinear two-point boundary value problems. The second order finite difference (FD) method is applied to derive the quarter-sweep nonlocal discretization scheme for the sake of transforming the system of nonlinear approximation equations into the corresponding system of linear equations. The formulation and the implementation of the methods are discussed. In addition, the numerical results by solving the proposed problems using QSSOR method are included and compared with the Full-sweep Successive Over Relaxation (FSSOR) and Half-sweep Successive Over Relaxation (HSSOR) methods.

1. Introduction

For the numerical solution of nonlinear two-point boundary value problems (TPBVP), the paper deals with the nonlocal finite difference methods which are used to determine the numerical value at the whole grid points with difference grid sizes. In fact again, the solution to a nonlinear TPBVP must satisfy the boundary conditions. Actually, the nonlinear TPBVP plays the important role to describe various physical problems in sciences, economics and engineering which are designed mathematically by using nonlinear equation model [1-6]. However in many cases, the nonlinear two-point boundary value problem does not have exact analytic solution. Therefore numerical techniques must be used to get the approximate solution of the problems. Nonlinear boundary value problems have attracted much attention from many researchers. For instance, a finite difference (FD) method has been proposed in recent works [1], finite element method [2], shooting method [3], spline approximation method [4] and Sinc-Galerkin method [5]. Literally, there are various iterative methods also have been studied to yield the fast numerical solution of linear systems [7, 8, 9]. Recently, Akhir et al [10] suggested a Half-Sweep Modified SOR (HSMSOR) method by bringing together the concept of the half-sweep iterations and the modified SOR method to solve two-dimensional Helmholtz equation. Actually, the concept of the half-sweep (HS) iteration method has been introduced by Abdullah [11] via the Explicit Decoupled Group (EDG) method to solve two-dimensional Poisson equations. Similarly, further investigations have been extensively conducted in [12, 13, 14, 15, 16]. In 2000, Othman and Abdullah continued this concept by introducing quarter-sweep (QS) iterative method via the Modified Explicit Group (MEG) method to solve two-dimensional Poisson equations [17]. Further studies to prove the effectiveness of the QS iterative methods have been carried out by [18, 19, 20, 21]. The fundamental concept of the HS and QS iterative methods is to reduce the computational complexities during iteration process in which HS and QS iterations will only consider nearly half and quarter of all

interior node points in a solution domain respectively. Thus, in this paper we examined the applications of the QS iteration concept with SOR iterative method by using approximation equation based on quarter-sweep nonlocal discretization scheme for solving proposed problems. The standard SOR iterative method is also called as the Full-Sweep Successive Over Relaxation (FSSOR) method. Meanwhile, combinations of the SOR method with HS and QS iterations are called as Half-Sweep Successive Over-Relaxation (HSSOR) and Quarter-Sweep Successive Over-Relaxation (QSSOR) methods respectively.
To investigate the performance of these three proposed iterative methods, let us consider the general form of a nonlinear two-point boundary value problem being defined as

$$-\frac{d^2U}{dx^2} = g(x,U,U'),\ a \le x \le b \tag{1}$$

subject to the boundary conditions

$$U(a) = \beta_0,\quad U(b) = \beta_1$$

and β_0, β_1, and $g(x,U,U')$ are constants and a nonlinear continuous function, respectively. As mentioned in the first paragraph, the importance of problem (1) deals with many physical problems. Therefore, the study investigates the effectiveness of QS nonlocal discretization scheme in solving the problem (1) via QSSOR iteration.

In formulating various iterative schemes such as full-, half- and quarter-sweep iterations, finite grid network can be used as a guide as well as to facilitate us in terms of development and implementation of the corresponding proposed algorithms for the proposed methods. According to the point iterations, the implementation of these three iterative methods will be applied onto the node points of the same type • until the iterative convergence fixed is achieved [22]. Based on Fig. 1, the FS , HS and QS iterative methods will compute approximate values onto node points of type • only until the convergence criterion is reached. Then, other approximate solutions at remaining points (points of the different type) can be computed using the direct method [11, 19].

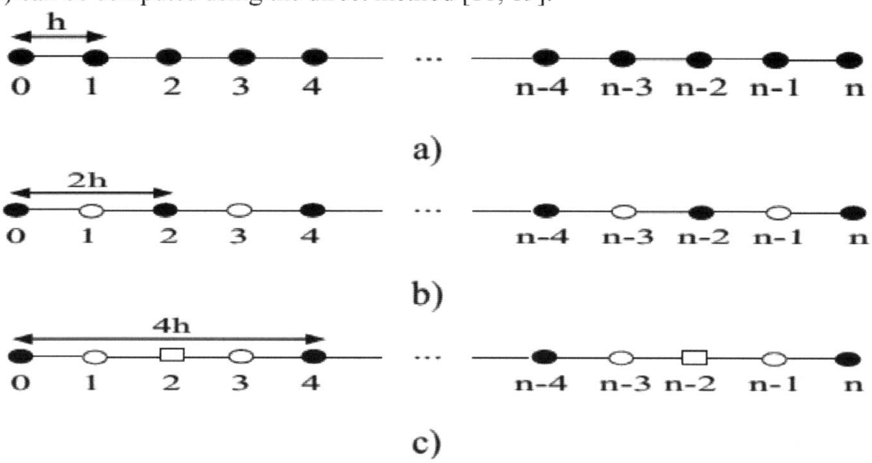

Figure 1 a) , b) and c) show distribution of uniform solid node points for the FS, HS and QS cases respectively.

Based on Figure 1, the following discussion will be restricted to divide the solution domain [a, b] of problem (1) into subintervals in which the distance of the subinterval, Δx is defined in Eq. (2).

$$\Delta x = \frac{(b-a)}{m} = h,\ n = m-1 \tag{2}$$

Based on Figure (1), clearly the distance of solid node points for quarter-sweep is $4h$ meanwhile the full-sweep and half-sweep h and $2h$ respectively as shown in Eq. (2).

2. Formulations of Quarter-sweep Nonlocal Discretization Scheme

Before constructing the Quarter-sweep FD approximation equation of problem (1), let us consider several nonlocal quarter-sweep discretization schemes being given as follows [23]

$$U_i^2 = U_i U_{i+4} \tag{3}$$

$$U_i^2 = \left(\frac{U_{i-4}+U_{i+4}}{2}\right)U_i \tag{4}$$

$$U_i^2 = \left(\frac{U_{i-4}+U_{i+4}}{2}\right)U_i^2 \tag{5}$$

By using the approach of second-order quarter-sweep FD discretization scheme, the corresponding approximation equations for problem (1) can be easily shown as

$$-U_{i-4}+2U_i-U_{i+4}-(h)^2 f_i(U_4,U_8,\cdots,U_{n-3})=0, \quad i=4,8,12,\cdots,n-3 \tag{6}$$

Where

$$f_i(U_4,U_8,\cdots,U_{n-3})=g\left(x_i,U_i,\frac{U_{i+4}-U_{i-4}}{8(h)}\right) \tag{7}$$

Actually, Eq. (7) is called as the nonlinear term of problem (1). To solve the nonlinear system in Eq. (6), the nonlocal discretization scheme is used to transform the nonlinear system into the form of a system of linear equations. In this paper, however, we consider the nonlocal discretization scheme in Eq. (4) being imposed over the nonlinear approximation equation (6). Therefore, Eq. (7) can be rewritten as follows

$$f_i(U_4,U_8,\cdots,U_{n-3})=g\left(x_i,\frac{U_{i-4}+U_{i+4}}{2},\frac{U_{i+4}-U_{i-4}}{8(h)}\right) \tag{8}$$

3. Formulations of Quarter-sweep Successive Over Relaxation method

In this section, we present on how to derive the formulation of the QSSOR methods. Based on the approximation equation in Eqs. (6) and (8), the general scheme of the QSSOR method can be stated as

$$U_i^{(k+1)}=(1-\omega)U_i^{(k)}+\frac{\omega}{2}\left(\begin{array}{l}U_{i-4}^{(k+1)}+U_{i+4}^{(k)}\\ +h^2 f_i\left(U_4^{(k+1)},U_8^{(k+1)},\cdots,U_{i-4}^{(k+1)},U_{i+4}^{(k)},\cdots,U_n^{(k)}\right)\end{array}\right), i=4,8,12,\cdots,n-3 \tag{9}$$

where ω and $U_i^{(k)}, 4,8,12,\cdots n-3$ represent as a relaxation factor and the kth estimation for corresponding approximate solutions respectively. Actually the FSSOR iterative method in form of discussed Eq. (9) was by Young [7]. Practically, the optimal value of ω in range $1 \le \omega < 2$ will be obtained by implementing several computer programs and then the best approximate value of ω is chosen in which its number of iterations is the smallest. The general algorithm for the QSSOR iterative methods to solve the linear equations (9) would be generally described in Algorithm 1 [18]

Algorithm 1 : QSSOR scheme

i. Initialize $\underset{\sim}{U}_i^{(0)} \leftarrow 0, \varepsilon \leftarrow 10^{-10}$

ii. Assign the value of ω

iii. For $i = 4, 8, 12, \ldots, n-3$, calculate

$$U_i^{(k+1)} = (1-\omega)U_i^{(k)} \frac{\omega}{2}\left(\begin{array}{l} U_{i-4}^{(k+1)} + U_{i+4}^{(k)} \\ + h^2 f_i \left(U_1^{(k+1)}, U_2^{(k+1)}, \ldots, U_{i-4}^{(k+1)}, U_{i+4}^{(k)}, \ldots, U_n^{(k)} \right) \end{array} \right)$$

iv. Check the convergence test, $\left| U_i^{(k+1)} - U_i^{(k)} \right| \le \varepsilon = 10^{-10}$. If yes, go to step (iv). Otherwise go back to step (iii).

v. Display approximate solutions.

4. Numerical Experiments

In order to validate the performance of the FSSOR, HSSOR and QSSOR iterative methods together with the nonlocal approach, three nonlinear example problems were tested. For the sake of comparison, three criteria will be considered for these three proposed iterative methods which are number of iterations, computation time (in seconds) and maximum absolute error.

Example 1 [24]

$$\varepsilon U'' + 2U' - e^{U} = 0, \qquad \text{for} \qquad 0 \le x \le 1 \tag{10}$$

subject to the boundary conditions

$$U(0) = 0 \qquad U(1) = 0$$

with exact solution were defined by

$$U(x) = \log\left(\frac{2}{1+x}\right) - \exp\left(\frac{-2x}{\varepsilon}\right)\log 2. \tag{11}$$

Example 2 [25]

$$U''(x) = \frac{3}{2}U^2, \quad \text{for} \quad 0 < x < 1 \tag{12}$$

subject to the boundary conditions

$$U(0) = -4, \quad U(1) = 1$$

with exact solution were defined by

$$U(x) = \frac{4}{(1+x)^2}. \tag{13}$$

Example 3 [26]

$$U''(x) + \frac{0.5}{x}U'(x) = e^{U(x)}\left(0.5 - e^{U(x)}\right), \quad \text{for} \quad 0 < x < 1 \tag{14}$$

subject to the boundary conditions

$$U(0) = \log[2], \quad U(1) = 0$$

From above three examples, results of numerical experiments obtained have been summarized in Table 1. In the implementation approach, the convergence test considered the tolerance error $\varepsilon = 10^{-10}$.

5. Numerical Experiments

In this paper, the performance of Quarter-sweep nonlocal discretization scheme with QSSOR method for the solution of nonlinear two-point boundary value problem associated with the second order finite difference approximation scheme has been investigated. Based on Table 1, numerical results showed that QSSOR method solved the proposed problems with least number of iterations as compared to the FSSOR and HSSOR methods. Meanwhile, in terms of computation time, QSSOR method computes with the fastest time for all considered mesh sizes. In the aspect of accuracy, numerical solutions obtained for test problems 1 to 3 are comparable for all the tested iterative methods. Finally, it can be concluded that the QSSOR method is superior to FSSOR and HSSOR methods. This is mainly because of the reduction of computational complexity in which the QSSOR method will only consider approximately quarter of all interior node points in a solution domain during iteration process.

References

[1] E. Doedel, *SIAM Journal. Numerical Analysis*, **16**,173-185 (1979).
[2] A. G. Deacon and S. Osher, *SIAM Journal. Numerical Analysis*. **16**, 756-778 (1979).
[3] S. M. Roberts and J. S. Shipman, *American Elsevier, New York*, (1972).
[4] T.Y. Na, *Academic, New York*, (1979).
[5] M. El-Gamel, S.H. Behiry, H. Hashih, *Appl. Math. Comput*, **145**, 717-734 (2003).
[6] A.A. Salama and A.A. Mansou, *Numer. Heat Tr,* **47**, 383–401 (2005).
[7] D.M. Young, *Trans. Amer. Math. Soc.,* **76**, 92-111(1954).
[8] W. Hackbuusch, *Springer-Verlag, New York,* (1995).
[9] Y. Saad, *United States America:PWS Publishing Company*, (1996).
[10] M. K. Akhir, M. Othman, J. Sulaiman, Z.A Majid, *Proceedings of Fundamental Science (FSC)*, 88-89 (2009).
[11] A.R. Abdullah, *Int. J. Comp. Math*, **38,** 61-70 (1991).
[12] M.K. Hassan, J. Sulaiman, S.A.A. Karim, M. Othman, *Journal of Applied Sciences,* **11**(7), 1255-1260 (2011).
[13] E. Aruchunan and J. Sulaiman, *Australian Journal of Basic and Applied Sciences*, **5** (3), 38-43 (2011).
[14] W. S. Yousif and D. J. Evans, *Implementation. Paral. Algo. App*, **7,** 63-71 (1995).
[15] A. R. Abdullah and A. Ibrahim, *Int. J. Comp. Math*, **58**, 253-256 (1995).
[16] M. Othman and A.R Abdullah. *Int. J. Comp. Math,* **69**, 219-229 (1998).
[17] J. Sulaiman, M.K. Hasan and M. Othman, *In Computational and Information Science, Zhang, J., He, J.H. and Fu. Yu., (Eds). Berlin: Springer-Verlag,* 57-63. (2004).
[18] N.I.M. Fauzi and J, Sulaiman, *Journal of Applied Sciences*, **12**(17), 1817-1824 (2012).
[19] M.K. Akhir, M. Othman, J. Sulaiman and Z.A. Majid, *Proceedings of International Conference on Intelligent Network and Computing, IEEE Press. ISBN*: 978-1-4244-8270-2, 416-419 (2010).
[20] M. Othman and A.R. Abdullah, *International Journal of Computer Mathematics*, **76**, 203-217(2000).
[21] M. Othman and A.R. Abdullah, *In V. Malyshkin (Ed.), Lectures Notes in Computer Science, Berlin: Springer-Verlae,* 480-489, (2001).
[22] M.K. Hasan, M. Othman, R. Johari, Z. Abbas and J. Sulaiman, *In B.M. Ali et al. (Eds.), 13th IEEE International Conference on Network,* 1002-1007 (2005).
[23] J. Sulaiman, M.K. Hasan, M.Othman, *In J. Zhang et al. (Eds.), Lectures Notes in Computer Science,* 57-63 (2004).
[24] M.K. Hasan, J. Sulaiman, and M. Othman, *In H.B. Zaman et al. (Eds.), International Symposium on Information Technology,* 1864-1869 (2008).
[25] M.K. Ahir, M. Othman, J. Sulaiman, Z. A. Majid and M. Suleiman, **5**(12), 3033-3039 (2011).
[26] K. Moaddy, I. Hashim, A. K. Alomari and S. Momani, *Sains Malaysiana,* **40**(5), 515–519 (2011).
[27] M.K. Kadalbajoo and A.S.Yadaw, *Applied Mathematics and Computation*, **201**, 504–513 (2008).
[28] S.N. Ha, *Comput. Math. App.* **42**, 1411-1420 (2001).
[29] H. Jafari, M. Ahmadi and S. Sadeghi, *Applications and Applied Mathematics. An International Journal,* **7**(1), 357-364 (2012).

TABLE 1. Comparison between number of iterations (K), the execution time (seconds) and maximum errors for the iterative methods using example at grid sizes 1024, 2048, 4096, 8192, 16384

	EXAMPLE 1			EXAMPLE 2			EXAMPLE 3		
	Number of iteration								
M	**FSSOR**	**HSSOR**	**QSSOR**	**FSSOR**	**HSSOR**	**QSSOR**	**FSSOR**	**HSSOR**	**QSSOR**
1024	3842	1949	982	3011	1527	769	5751	2880	1418
2048	7526	3842	1949	6029	2914	1527	11819	5748	2880
4096	14625	7526	3842	11463	5413	2914	22356	11297	5748
8192	27951	14625	7526	22096	10331	5413	49154	21793	11297
16384	53169	27951	14625	42248	20031	11255	98306	2880	21793
	Execution time (seconds)								
M	**FSSOR**	**HSSOR**	**QSSOR**	**FSSOR**	**HSSOR**	**QSSOR**	**FSSOR**	**HSSOR**	**QSSOR**
1024	1.07	0.44	0.29	0.57	0.39	0.26	1.78	1.15	0.45
2048	2.47	0.95	0.55	1.41	0.72	0.44	6.47	3.06	0.83
4096	7.50	2.35	1.00	5.32	2.15	0.83	24.35	8.63	2.56
8192	24.86	7.85	2.51	17.85	6.10	2.10	95.68	30.36	8.59
16384	91.06	26.63	7.81	65.21	21.84	6.03	387.77	128.83	27.66
	Maximum errors								
M	**FSSOR**	**HSSOR**	**QSSOR**	**FSSOR**	**HSSOR**	**QSSOR**	**FSSOR**	**HSSOR**	**QSSOR**
1024	1.787947e-02	1.787945e-02	9.319738e-02	3.669275e-06	1.470019e-05	5.878965e-05	8.344832e-07	3.293855e-06	1.295298e-05
2048	1.787948e-02	1.787947e-02	9.350227e-02	9.338850e-07	3.690722e-06	1.470019e-05	2.540690e-07	8.354497e-07	3.293860e-06
4096	1.787951e-02	1.787949e-02	9.365475e-02	2.701202e-07	9.428743e-07	3.690722e-06	1.740979e-07	2.226486e-07	8.354497e-07
8192	1.787961e-02	1.787951e-02	9.373101e-02	1.420423e-07	2.774780e-07	9.428743e-07	1.293783e-08	1.368791e-07	2.226486e-07
16384	1.787976e-02	1.787961e-02	9.376914e-02	2.192884e-07	2.111116e-07	2.762543e-07	3.207085e-09	1.231990e-08	1.368792e-07

Directly executable formal models of middleware for MANET and Cloud Networking and Computing

D. V. Pashchenko[1], Mustafa Sadeq Jaafar[2], S. A. Zinkin[3], D. A. Trokoz[4], T. U. Pashchenko[5] and M. P. Sinev[6]

[1] Professor, Penza State University, Penza, Russia

[2] Postgraduate, Penza State University, Penza, Russia

[3] Professor, Penza State University, Penza, Russia

[4] Assistant professor, Penza State University, Penza, Russia

[5] Assistant professor, Penza State University, Penza, Russia

[6] Assistant professor, Penza State University, Penza, Russia

E-mail: dmitry.pashchenko@gmail.com

Abstract. The article considers some "directly executable" formal models that are suitable for the specification of computing and networking in the cloud environment and other networks which are similar to wireless networks MANET. These models can be easily programmed and implemented on computer networks.

1. Introduction

Several new trends are opening up the era of Cloud Computing, which is an Internet-based development and use of computer technology. They are cheaper and more powerful processors, together with the "software as a service" (SaaS) and computing architecture, are transforming data centers into pools of computing service on a huge scale. Meanwhile, the increasing of networks bandwidth and their reliability yet flexible network connections make it even possible that clients can now subscribe some high quality services from data and software that reside solely on remote data centers or data pools. Although envisioned as a promising service platform for the Internet, the new data storage paradigm in the "Cloud" brings many challenging design issues which have profound influence on the security and performance of the overall system. One of the biggest concerns related to cloud data storage is that of data integrity verification at non trusted servers. What is more serious is that for saving money and storage space the service provider might neglect to keep whatever is necessary or delete rarely accessed data files which belong to an ordinary client. Consider the huge size of the outsourced electronic data and the client's constrained resource capability, the core of the problem can be generalized in the theme of how the client can find an efficient way to perform periodical integrity verifications without needing the local copy of data files. Considering the role of the verifier in the model where all the schemes presented before are falling into two categories: private auditability and public auditability. Although schemes with private auditability can achieve remarkably higher scheme efficiency, public auditability allows anyone, not just the client (data owner) to challenge the cloud servers for the correctness of data storage while keeping no private information.

Then, clients are capable of handing over or to be more specific, to delegate the evaluation of the service performance to some independent third party auditors (TPAs), without the devotion of their computation resources. In the cloud, the clients themselves mostly are unreliable or may not be capable of affording the overhead of performing frequent integrity checks. Router free feature connecting to files on other computers and/or the Internet without the need of a wireless router is the main advantage of using an ad hoc network. So because of this, running an ad hoc network can be more affordable and easier to construct than a traditional network because you don't have the added cost of a router. However, if you only have one node for example only one computer then it won't be possible to construct an ad hoc network. Mobility - an Ad hoc network can be created on the fly in nearly any situation with multiple nodes (wireless devices) which they are existing for the purpose of connection establishment. For example: emergency situations in some rare and remote locations make a traditional network nearly impossible, but "For example the medical team in the army can easily utilize 802.11 radio NICs in their laptops and PDAs and other wireless devices and enable broadband wireless data communication as soon as they reach their destination. "The other feature is the Speed - creating an Ad hoc network from scratch is remarkably fast and only requires a few setting tunings and some changes and no additional hardware or software.

2. The results of previous works

Mobile ad hoc networks (MANET) consist of nodes that are able to communicate through the use of wireless mediums and form dynamic network topologies. The basic characteristic of these networks is the complete lack of any kind of infrastructure and the absence of any dedicated nodes that provide network management operations as the role of the traditional routers in fixed kinds of networks. In order to maintain the connectivity in a mobile ad hoc network all the nodes have to perform routing of network traffic. The cooperation of the nodes in an ad hoc network cannot be enforced by a centralized administration authority since one does not actually exist. Therefore, a network-layer protocol that is designed for such self-organized networks must enforce the connectivity and the security requirements in order to guarantee the undisrupted other operations of the higher layer protocols. Unfortunately most of the widely used ad hoc routing protocols have no high security considerations and trust all the participants (the nodes) to correctly forward routing and data traffic.

Proposed by one of the authors of this article in [1, 2, 3] modified Ack-Based scheme (Acknowledgment-Based) for node authentication with AODV (Ad hoc On-Demand Distance Vector) existing protocol in MANET. Ack-Based scheme also provides the possibility of detecting wormhole attacks and nodes misbehavior in ad hoc networks. Different researcher proposed different scheme for Ack-Based for providing security in mobile ad hoc network. But all the other proposed techniques suffered common problem that problem is the generation of huge amount of pack overhead and node ambiguity. Actually due to this problem the given scheme is not used in a general form. So we proposed the Ack-Based scheme with the help of finite state machine against the malicious nodes. The authors propose the 2Ack based scheme to detect the malicious links and to mitigate their malicious effects. This scheme is based on a 2Ack packet that is actually assigned a fixed route of two hops in the reversed/opposite direction of the received data between the sender and the receiver. In this scheme, each sender's packet maintains the following parameters; (i) list of identifiers of data packets that have been sent out but haven't been yet acknowledged, (ii) a counter of the forwarded data packets, (iii) a counter of the missed data packets. According to the value of the acknowledgements ratio (Rack), only a certain fraction of the sent data packets will be acknowledged in order to reduce the resulted overhead. This technique overcomes some of the weaknesses of the Watchdog/path rater such as: ambiguous collisions, receiver collisions and power control transmissions. Both of the previous presented works remain vulnerable to the attacks that are launched by groups of other malicious nodes. To encounter these attacks, must provide a framework to mitigate the damage caused by the colluding black hole attacks in AODV. The proposed technique actually has a moderate

overhead that is induced by the ACK which has been sent back by the destination node during selected intervals of data transfer periods. Throughout the data packets transmissions, a flow of special data packets is transmitted at random intervals of time along with the data. The reception of these special data packets invokes the destination to actually send out an ACK through different multiple routing paths. The ACK data packets take multiple routes to reduce the actual probability that all of the ACKs are getting dropped by the malicious nodes, and also to encounter for possible loss due to several broken routes or congestions in some certain nodes. If the source node does not get back any ACK packet, then it becomes aware of the presence of some potential attackers in the forwarding path. As a reaction to that, it broadcasts a list of suspected malicious nodes in order to isolate them from the network's routing table.
Without a fixed infrastructure and with the node mobility feature in ad hoc networks it is a great challenge as concerning the security issue. For security concerns various methods are used to enhance node authentication in mobile ad hoc networks. The authentication scheme of leader agent node and members of surveillance greatly reduces the relative calculating overheads and the communication costs. Generally, when leader agent node and surveillance nodes are not getting destroyed, the united nodes can effectively ensure the reliability and the authentication results will be more reliable. Suggestion for future work is to apply the modified Ack-Based FSA (Finite State Machine) scheme to other protocols to overcome the requirement of the memory, during the path discovery and path establishment it takes a little bit much time in comparison with normal Ack-Based schemes and also future minimized route calculation with finite state machine. Here in [1, 2, 3] proposed Modified ACK-Based scheme for node authentication with AODV existing protocol in MANET. ACK-Based scheme also provides the capability of detecting the wormhole attacks and node misbehaviors in ad hoc networks. Different researcher proposed different scheme for ACK-Based for providing security in mobile ad hoc network. But all these techniques suffered from a common problem which is the generation of huge amount of pack overhead and node ambiguity. Actually due to this problem the given scheme is not used in a general form. So we proposed the ACK-Based scheme with the help of the finite state machine concept for controlling the generation of data packets and also improve the performance of AODV protocol. Here first we will discuss basic ACK-Based scheme and then we will discuss the finite state machine and finally we will discuss the combined approach for ACK-Based scheme.

When a malicious node is receiving an application packet from a node and that packet is actually destined for some other node then instead of forwarding that packet, it simply just drops that application packet. This data loss may become severe when the number of malicious nodes present in network is becoming high. In the proposed work, we overcome this problem by identifying this kind of malicious behavior of nodes and then a route via such a node is never being chosen by its neighboring node to forward an application packet in the network.

When a node wants to send an application packet to another node which is not its immediate neighboring node then it sends a RREQ (request) packet to all of its neighboring nodes. If a neighbor knows a route to the destination of this packet then it sends a RREP (reply) packet that contains the next hop address to which the mentioned neighboring node will forward the packet to. Let us call this next hop address as the next to next hop address instead. The algorithm is described as follow:

1. The sender node forwards the application packet to one of its immediate neighbors delegating the responsibility of further forwarding it to that neighboring node which is the destined one. The sender also sets a timer (which is twice the network's diameter) to receive the acknowledgement from the destination node.
2. If an acknowledgement is received before the timer expires then the route is considered to be trusted/ valid and no further action is needed to be taken.
3. If the timer expires and an acknowledgement is not yet received then the route is not considered to be trusted/ valid. Now the sender sends an application packet to the next to next hop address node and again sets a timer to receive an acknowledgement.

- If the acknowledgement is received before the timer expires then the neighboring node is considered to be trusted and no further action will be needed, otherwise the neighboring node is considered as a half-trusted. In this situation, this half-trusted node will be under observation until it shows the same malicious behavior all over again when an application packet is forwarded to it the next time.

4. If the half-trusted node does not show any other malicious behavior when the application packet is forwarded to it the next time then it will be considered as a trusted-node, otherwise this half-trusted node will be considered as a malicious one and therefore the following actions will be performed:
   - No more RREP messages from this node will be considered.
   - The application packets will no longer be forwarded to this node.
   - New routes will be found and discovered in order to forward the application packets to those destinations that are having this un-trusted (malicious) node as the next hop address in the routing table.

3. Some "reconfigurable" directly executable formal models, suitable for the specification of computing and networking in the cloud environment and other networks similar to wireless networks MANET

The distributed programming computational infrastructures of Cloud and MANET networking and computing available globally for offering uniform services has become an important topic in Computer Sciences. The challenges are coming from the necessity of dealing at once with issues like the security, privacy, failures, communications, co-operation, mobility, resources usage, etc., in a setting where the demands and the guarantees can be very different for the many different components. This has stimulated research on abstractions and models that could provide the basis for the design and the analysis of network aware programs, where dynamically changing the logical and physical architecture of systems plays a crucial role. The range of applications of reconfigurable, or changeable, models proposed in this paper concerns the problem of global computing and communications, which is especially important for MANET applications and Cloud Computing. We will give some information on the used formalism. Logical and algebraic models describe a set of properties of network programs. However, they do not give an idea of why these properties are changing. Executable or operating models help us to monitor the behavior of the object. Thus, every executable formal model is an abstract program for some virtual machines. Some formal models can be executed immediately before the phase of the network programming. Such models we conventionally call as "the directly executable". Those models are classified as follows: the models of intermediating type, possessing properties of the logic-algebraic and the executable models.

The concept of "directly executable" formal models is a metaphor, which is used in the sense of Glushkov and his followers [10-17]. It means that the design of formal specifications is the last step for the programmer before the writing of the network programs for MANET and Cloud Computing.

First, for a unified view of a number of important directly executable models let us consider (dynamically changing, many-sorted) algebraic system:

$AS = (\boldsymbol{A}, P, F, I_{F0}, I_{P0}, Rules, \boldsymbol{M}, Z)$,

where $\boldsymbol{A} = \{A_1, A_2, \ldots, A_n\}$ – a finite set of some basic finite sets, or in another word "sorts" $A_1, A_2, \ldots, A_n$;

P – a finite set of predicate symbols;.

F – a finite set of function symbols;.

I_{F0} – initial interpretation of function symbols;

I_{P0} – initial interpretation of predicate symbols;

$Rules$ – a finite set of modifications (or updates) to the interpretation of the predicate and function symbols;

$\boldsymbol{M}$ – a finite set of abstract modules;

Z: $\boldsymbol{M} \to \mathbf{P}(Rules)$ – a mapping from the set *M* to the set of subsets of *Rules*, where **P** is a Boolean symbol. Thus, as a result of the work of abstract modules of the set ***M*** dynamic algebraic system *AS* evolves, passing from one interpretations of predicate and function symbols to another.

We shall also use the known in algorithmic algebras operators-constants *E* (the identity operator, or "empty" operator) and *N* (nowhere defined operator), as well as additional new operators *H* (halt) and *Ret* (return to check conditions).

Reconfigurable dynamically changing, many-sorted algebraic system is defined as follows:

$RAS = (\boldsymbol{A}, P, F, I_{F0}, I_{P0}, Rules, \boldsymbol{M}, Z, Rrules)$,

where *Rrules* – a set of "reconfigurable" rules that change the structure or the configuration of a given formal model. The rules from the given set *Rrules* are actually very similar to the ones from the original set *Rules* but relate to other properties of the model.

Algebraic systems identified in many studies. For example, algebraic systems reviewed in the monograph [5]. Many-sorted algebraic systems are described in the monograph [6]. A similar concept of evolving algebras, or, in modern interpretation, of abstract state machines (ASMs), was introduced by the American mathematician Y. Gurevich, as well as the concept of updates of predicates and functions [7]. It is well-known that ASMs can simulate "step-by-step" any type of machines (Turing machines, RAMs, etc.). Various uses of ASMs are also described in the works [8, 9]. A systematic study of ASMs is also done in [8]. One of the modifications of ASM has been described in [9]. The authors modified Gurevich's given notion of ASM to that of EMA which is the (Evolving Multi-Algebra) simply by replacing the originally given program (which is a syntactic object) by a semantic one: a function which has to be very simply definable over the static part of the ASM.

The main difference is that the reconfiguration of the structure of our model is carried out in the course of its execution. Another major difference between the interpretation of the ASM and interpretation of a tuple *N* as the network of abstract modules, or abstract machines (NAM) is structuring by using the algebra of algorithms by Glushkov V. M. [10, 11] which is also applicable to various problems of his colleagues and followers [12-17]. In particular, in this article we are using the superposition of "alpha-disjunction": $[\alpha](A \vee B)$ – which is an analogue of the operator "if", and the "composition: $A \bullet B$ or can be simply just *A*,*B*" operations for the structured description of the abstract machines networks. Also an additional operation "←" is used for modificating the value of a variable, function or a predicate. "Reconfigurable" version of NAM we will call abbreviated as RNAM.

3.1. Formal NAM and RNAM description of the reconfigurable (deterministic or non-deterministic) Turing machines (RDTM and RNDTM)

The description of a Turing machine can be found, for example, in [18]. The work of the deterministic or non-deterministic Turing machine (NDTM) can be described by the following expression for the abstract module of NAM:

$M = Place(Head) \leftarrow 0;\ State(q_1) \leftarrow \mathbf{true};$
$[(\exists_{Any}\ q)\ State(q)]([a \leftarrow Tape(Place(Head))]$
$([\exists_{Any}(q, a, q', a', w')Program(q, a, q', a', w')]$
$(\{[State(q')\&S_{Q0}(q')](H \vee E), Tape(Place(Head)) \leftarrow a',$
$State(q) \leftarrow \mathbf{false}, State(q') \leftarrow \mathbf{true}, \{[w' = R]$
$(Place(Head) \leftarrow Inc(Place(Head))\vee$
$[w' = L](Place(Head) \leftarrow Dec(Place(Head))\vee E))\}, Ret\}\vee N)\vee N)\vee N),$

where *q* – the current state, $q \in Q$, $Q = \{q_1, q_2, \ldots, q_n\}$;

a – the symbol in the current cell of the tape, $a \in A$, $A = \{a_1, a_2, \ldots, a_m\}$;

q_1 – the initial state;

q' – the next state, $q' \in Q$;

a' – the next symbol in the current cell of the tape, $a' \in A$;

$w' \in (R, L, S)$ – the variable, that is determining the movement's direction of the head – right (*R*), left (*L*), stand on the observed cell (*S*);

$\exists_{Any}$ – the operator of non-deterministic selection of one of the tuples that are in the truth domain of the predicate *Program*;

$Program(q, a, q', a', w')$ – the predicate, the truth domain of which defines the work program of a Turing machine, *Program*: $(Q\times A)\times(Q\times A\times W) \rightarrow \{\textbf{true}, \textbf{false}\}$;

Tape – the unary function, that defines a sequence of symbols on the tape,

Tape: $\mathbf{Z} \rightarrow A$, where $\mathbf{Z}$ – the set of integers;

Place – the unary function, that defines the head position,

Place: $Hd \rightarrow \mathbf{Z}$, where Hd – a set of head's names (in this case $Hd = \{Head\}$, where *Head* is the individual constant, or the name of the head);

$State(q)$ – this predicate determines the state of the Turing machine (so that at any time the statement $(\exists!q)State(q)$ is true, *State*: $Q \rightarrow \{\textbf{true}, \textbf{false}\}$);

Inc (*Dec*) – the increment (or decrement) function;

$S_{Q0}(q)$ – a characteristic function of a subset Q_0 of final states.

A certain symbol of the external alphabet $A = \{a_1, a_2, \ldots, a_m\}$ is called a dummy symbol. This symbol is represented by # or λ.

According to the [18] at any instant of time the controlling device is in a certain state q_i, which belongs to an internal alphabet or a set $Q = \{q_1, q_2, \ldots, q_n\}$. Disjoint subsets Q_1 and Q_0 of initial and final states respectively define the start and stop operations of a Turing machine. A Turing machine stops when $q \in Q_0$ after the execution of some command. We also introduce $S_{Q0}(q)$ – a characteristic function of a subset Q_0 of final states.

The list of all quintuples or commands (q, a, q', a', w') in the truth domain of the predicate *Program* which determines the operations of the Turing machine is a program of this machine.

In our example the quintuple (q_i, a_j, q', a', w') begins with the pair (q_i, a_j); $i = 1, 2, \ldots, n$; $j = 1, 2, \ldots, m$; n – the number of states; m – the number of symbols of the external alphabet A. For a program of the deterministic Turing machine (DTM) there should be only one such the quintuple. In the case of the non-deterministic Turing machine (NDTM) may be more than one tuple, starting with a pair of the form (q_i, a_j). The values of q_i and a_j are defined as follows:

$q_j = (\exists_{Any}\, q)\ State(q)$; $a_j = Tape(Place(Head))$.

The values of q_i and a_j determined in this way immediately during the execution of the expression M.

Reconfigurable (deterministic or non-deterministic) Turing machine (RDTM and RNDTM) may further contain reconfigurable rules that change the structure or the configuration of a formal model. For example, it is possible to change the program of work of the Turing machine with the following rules from the set *Rrules*:

$Program(q_i, a_j, q_1, a_1, L) \leftarrow \textbf{false}$;

$Program(q_i, a_i, q_2, a_2, R) \leftarrow \textbf{true}$.

By incorporating these rules to the expression M we obtain the following expression RM for the abstract module of RNAM:

$RM = Place(Head) \leftarrow 0;\ State(q_1) \leftarrow \textbf{true};$
$[(\exists_{Any}\, q)\ State(q)]([a \leftarrow Tape(Place(Head))]$
$([\exists_{Any}(q, a, q', a', w')Program(q, a, q', a', w')]$
$(\{[State(q')\&SQ_0(q')](H \vee E), Tape(Place(Head)) \leftarrow a',$
$State(q) \leftarrow \textbf{false}, State(q') \leftarrow \textbf{true}, \{[w' = R]$
$(Place(Head) \leftarrow Inc(Place(Head)) \vee$
$[w' = L](Place(Head) \leftarrow Dec(Place(Head)) \vee E))\},$
$\{Program(q_i, a_j, q_1, a_1, L) \leftarrow \textbf{false};$
$Program(q_i, a_i, q_2, a_2, R) \leftarrow \textbf{true}\}, Ret\} \vee N) \vee N) \vee N).$

Replacing of the currently executed command may be performed as follows:

$RM' = Place(Head) \leftarrow 0;\ State(q_1) \leftarrow \textbf{true};$
$[(\exists_{Any}\, q)\ State(q)]([a \leftarrow Tape(Place(Head))]$

$([\exists_{Any}(q, a, q', a', w')Program(q, a, q', a', w')]$
$(\{[State(q')\&SQ_0(q')](H \vee E), Tape(Place(Head)) \leftarrow a',$
$State(q) \leftarrow \mathbf{false}, State(q') \leftarrow \mathbf{true}, \{[w' = R]$
$(Place(Head) \leftarrow Inc(Place(Head)) \vee$
$[w' = L](Place(Head) \leftarrow Dec(Place(Head)) \vee E))\},$
$\{Program(q, a, q', a', w') \leftarrow \mathbf{false};$
$Program(q, a, q_2, a_2, R) \leftarrow \mathbf{true}\}, Ret\} \vee N) \vee N) \vee N).$

Now it is easy to describe the multitape non-deterministic Turing machine described in the books [21, 22]. The multitape non-deterministic Turing machine operates in accordance with the following expression for the abstract machine module:

$MK = Place(Head^{(1)}) \leftarrow 0; Place(Head^{(2)}) \leftarrow 0; \ldots; Place(Head^{(k)}) \leftarrow 0;$
$State(q_1) \leftarrow \mathbf{true}; [(\exists_{Any}\, q)\, State(q)]([a^{(1)} \leftarrow Tape^{(1)}(Place(Head^{(1)}))]$
$([a^{(2)} \leftarrow Tape^{(2)}(Place(Head^{(2)}))] \ldots$
$\ldots ([a^{(k)} \leftarrow Tape^{(k)}(Place(Head^{(k)}))]$
$([\exists_{Any}(q, a^{(1)}, a^{(2)}, \ldots, a^{(k)}, q', a'^{(1)}, a'^{(2)}, \ldots, a'^{(k)}, w'^{(1)}, w'^{(2)}, \ldots, w'^{(k)})$
$Program(q, a^{(1)}, a^{(2)}, \ldots, a^{(k)}, q', a'^{(1)}, a'^{(2)}, \ldots, a'^{(k)}, w'^{(1)}, w'^{(2)}, \ldots, w'^{(k)})]$
$(\{[State(q')\&S_{Q0}(q')](H \vee E),$
$Tape^{(1)}(Place(Head^{(1)})) \leftarrow a'^{(1)},$
$Tape^{(2)}(Place(Head^{(2)})) \leftarrow a'^{(2)}, \ldots, Tape^{(k)}(Place(Head^{(k)})) \leftarrow a'^{(k)},$
$State(q) \leftarrow \mathbf{false}, State(q') \leftarrow \mathbf{true},$
$\{[w'^{(1)} = R](Place(Head^{(1)}) \leftarrow Inc(Place(Head^{(1)})) \vee$
$[w'^{(1)} = L](Place(Head^{(1)}) \leftarrow Dec(Place(Head^{(1)})) \vee E))\},$
$\{[w'^{(2)} = R](Place(Head^{(2)}) \leftarrow Inc(Place(Head^{(2)})) \vee$
$[w'^{(2)} = L](Place(Head^{(2)}) \leftarrow Dec(Place(Head^{(2)})) \vee E))\}, \ldots$
$\ldots, \{[w'^{(k)} = R](Place(Head^{(k)}) \leftarrow Inc(Place(Head^{(k)})) \vee$
$[w'^{(k)} = L](Place(Head^{(k)}) \leftarrow Dec(Place(Head^{(k)})) \vee E))\}, Ret\} \vee N) \vee N) \ldots \vee N)$
$\vee N) \vee N),$

where superscripts correspond to the numbers of heads and tapes in multitape non-deterministic Turing machine. Description of such a reconfigurable (deterministic or non-deterministic) Turing machine is obvious.

3.2. Formal NAM and RNAM description of the reconfigurable (deterministic or non-deterministic) finite state automata (RFSA and RNDFSA)

The present study is based on the works [19-22]. The work of the deterministic or non-deterministic finite state automata (FSA or NDFSA) can be described by the following expression for the abstract module of NAM:

1) for Mealy NDFSA automata:

$A_1 = Place(XHead) \leftarrow 0; Place(YHead) \leftarrow 0; State(a_1) \leftarrow \mathbf{true};$
$[(\exists_{Any}\, a)\, State(a)]([x \leftarrow Xtape(Place(Xhead))]$
$([\exists_{Any}(a, x, a', y)Table(a, x, a', y)]$
$(\{[State(a')\&S_{AF}(a')](H \vee E), State(a) \leftarrow \mathbf{false}, State(a') \leftarrow \mathbf{true},$
$Ytape(Place(Yhead)) \leftarrow y,$
$Place(Xhead) \leftarrow Inc(Place(Xhead)),$
$Place(Yhead) \leftarrow Inc(Place(Yhead)), Ret\} \vee N) \vee N) \vee N),$

where a – the current state ($a \in A$, $A = \{a_1, a_2, \ldots, a_n\}$ – a finite set of internal states, a_1 – an initial state);

x – the symbol in the current cell of the input tape ($x \in X$, $X = \{x_1, x_2, \ldots, x_m\}$ – a finite set of input symbols;);

a'– the next state $a' \in A$;

y – the output symbol ($y \in Y$, $Y = \{y_1, y_2, \ldots, y_k\}$ – a finite set of output symbols);

$\exists_{Any}$ – the operator of non-deterministic selection of one of the tuples that are in the truth domain of the predicate *Table*;

$Table(a, x, a', y)$ – the predicate, the truth domain of which defines the work table of NDFSA automata, $Table$: $A \times X \times A \times Y \rightarrow \{\mathbf{true}, \mathbf{false}\}$;

Xtape – the unary function, that defines a sequence of input symbols on the input tape, *Xtape*: $\mathbf{Z} \rightarrow X$, where $\mathbf{Z}$ – the set of integers;

Ytape – the unary function, that defines a sequence of output symbols on the output tape, *Ytape*: $\mathbf{Z} \rightarrow Y$;

Place – the unary function, that defines the heads positions, *Place*: $Hd \rightarrow \mathbf{Z}$, where Hd – a set of head's names (in this case $Hd = \{XHead, YHead\}$, where *XHead*, *YHead* are the individual constants, or the names of the input and output heads);

$State(a)$ – this predicate determines the state of automata (so that at any time the statement $(\exists! a)State(a)$ is true, $State$: $A \rightarrow \{\mathbf{true}, \mathbf{false}\}$);

Inc – the increment function;

A_F – a subset of the final states, $A_F \subset A$;

$S_{AF}(a)$ – a characteristic function of a subset A_F of final states, S_{AF}: $A \rightarrow \{\mathbf{true}, \mathbf{false}\}$;

2) for Moore NDFSA automata:

$A_2 = Place(XHead) \leftarrow 0;\ Place(YHead) \leftarrow 0;\ State(a_1) \leftarrow \mathbf{true};$
$[(\exists_{Any}\ a)\ State(a)]([x \leftarrow Xtape(Place(Xhead))]$
$([\exists_{Any}(a, x, a', y)Table'(a, x, a', y)]$
$(\{[State(a')\&S_{AF}(a')](H \vee E),\ State(a) \leftarrow \mathbf{false},\ State(a') \leftarrow \mathbf{true},$
$Place(Xhead) \leftarrow Inc(Place(Xhead)),$
$Place(Yhead) \leftarrow Inc(Place(Yhead)),$
$Ytape(Place(Yhead)) \leftarrow y,\ Ret\} \vee N) \vee N) \vee N);$

the work of the deterministic or non-deterministic reconfigurable finite state automata (RFSA or RNDFSA) can be described by the following expression for the abstract module of RNAM:

3) for Mealy RNDFSA automata:

$A_3 = Place(XHead) \leftarrow 0;\ Place(YHead) \leftarrow 0;\ State(a_1) \leftarrow \mathbf{true};$
$[(\exists_{Any}\ a)\ State(a)]([x \leftarrow Xtape(Place(Xhead))]$
$([\exists_{Any}(a, x, a', y)Table(a, x, a', y)]$
$(\{[State(a')\&S_{AF}(a')](H \vee E),\ State(a) \leftarrow \mathbf{false},\ State(a') \leftarrow \mathbf{true},$
$Ytape(Place(Yhead)) \leftarrow y,$
$Place(Xhead) \leftarrow Inc(Place(Xhead)),$
$Place(Yhead) \leftarrow Inc(Place(Yhead)),$
$\{Table(a, x, a', y) \leftarrow \mathbf{false},$
$Table(a, x, a_1, y_1) \leftarrow \mathbf{true}\},\ Ret\} \vee N) \vee N) \vee N),$

4) for Moore RNDFSA automata:

$A_4 = Place(XHead) \leftarrow 0;\ Place(YHead) \leftarrow 0;\ State(a_1) \leftarrow \mathbf{true};$
$[(\exists_{Any}\ a)\ State(a)]([x \leftarrow Xtape(Place(Xhead))]$
$([\exists_{Any}(a, x, a', y)Table'(a, x, a', y)]$
$(\{[State(a')\&S_{AF}(a')](H \vee E),\ State(a) \leftarrow \mathbf{false},\ State(a') \leftarrow \mathbf{true},$
$Place(Xhead) \leftarrow Inc(Place(Xhead)),$
$Place(Yhead) \leftarrow Inc(Place(Yhead)),$
$Ytape(Place(Yhead)) \leftarrow y$
$\{Table(a, x, a', y) \leftarrow \mathbf{false},$
$Table(a, x, a_1, y_1) \leftarrow \mathbf{true}\},\ Ret\} \vee N) \vee N) \vee N).$

3.3. Formal NAM and RNAM description of the reconfigurable information-inhibitor Petri net (RPNII)

A Petri net [23-26] is one of several mathematical modeling languages for the description of distributed systems. A Petri net is nothing but a directed bipartite graph, in which the nodes represent transitions (i.e. events that may occur and signified by bars) and places (i.e. conditions that are signified by circles). The directed arcs describe which places are pre- or postconditions for which transitions (signified by arrows).

Petri net usually is defined by a tuple:

$N = (P, T, F, H, M_0)$,

where $P = \{p_1, p_2, \ldots, p_n\}$– a non-empty finite set of positions, $T = \{t_1, t_2, \ldots, t_m\}$– a non-empty finite set of transitions, F, H – the incidence functions, M_0 – a function of the initial marking of the positions. We consider the first-order logical interpretation of 1-safe (or 1-bounded) Petri nets, for which three predicates are determined:

$F: P \times T \rightarrow \{\textbf{true}, \textbf{false}\}$; $H: T \times P \rightarrow \{\textbf{true}, \textbf{false}\}$; $M_0: P \rightarrow \{\textbf{true}, \textbf{false}\}$.

We shall also use well known in the literature on Petri nets two optional binary predicates

$F_{Inh}: P \times T \rightarrow \{\textbf{true}, \textbf{false}\}$ and $F_{Inf}: P \times T \rightarrow \{\textbf{true}, \textbf{false}\}$ in order to set the inhibitory and information arcs in a Petri net. Thus, information-inhibitor Petri net is defined as a dynamic algebraic system by a tuple: $NII = (P, T, F, F_{Inh}, F_{Inf}, H, M_0, R)$, where R is a finite set of rules, that can update the unary predicate $M(p)$ (current marking). In the terms of Petri nets theory such rules are called the rules of transitions firing. For this reason $|R| = |T|$.

The reconfigurable information-inhibitor Petri net now is defined as a reconfigurable dynamic algebraic system by the next tuple: $RNII = (P, T, F, F_{Inh}, F_{Inf}, H, M_0, \boldsymbol{R}, Rrules, S_{Upd})$, where $Rrules$ is a set of the "reconfigurable" rules that can update the binary predicates F, F_{Inh}, F_{Inf}, H; S_{Upd} is a set of the rules that can update the sets P and T, if it is needed.

Further consideration is expedient to carry out with an example. Let's define a fragment of the network NII as follows: $F(p_1, t_1) = \textbf{true}$, $F_{Inh}(p_2, t_1) = \textbf{true}$, $F_{Inf}(p_3, t_1) = \textbf{true}$, $H(t_1, p_4) = \textbf{true}$, $M(p_1) = \textbf{true}$, $M(p_2) = \textbf{false}$, $M(p_3) = \textbf{true}$, $M(p_4) = \textbf{false}$. Rule r_1 of firing the transition t_1 is described by the following expression for the network NAM: $r_1 = [M(p_1)\&\neg M(p_2)\&M(p_3)](\{M(p_1) \leftarrow \textbf{false},\ M(p_4) \leftarrow \textbf{true}\} \vee E)$. After firing the transition t_1 marking becomes the following: $M(p_1) = \textbf{false}$, $M(p_2) = \textbf{false}$, $M(p_3) = \textbf{true}$, $M(p_4) = \textbf{true}$.

For reconfiguration of the fragment we introduce two positions p_5 and p_6 and define reconfiguration rules: $Rrules = \{F(p_1, t_1) \leftarrow \textbf{false}, F_{Inh}(p_2, t_1) \leftarrow \textbf{false}, F_{Inf}(p_3, t_1) \leftarrow \textbf{false},$

$F_{Inh}(p_3, t_1) \leftarrow \textbf{true}, F(p_5, t_1) \leftarrow \textbf{true}, F_{Inh}(p_6, t_1) \leftarrow \textbf{true}\}$.

The rule r' firing the transition t_1 is described now by the following expression for the network RNAM:

$r' = [M(p_1)\&\neg M(p_2)\&M(p_3)](\{M(p_1) \leftarrow \textbf{false},\ M(p_4) \leftarrow \textbf{true},$

$\{F(p_1, t_1) \leftarrow \textbf{false}, F_{Inh}(p_2, t_1) \leftarrow \textbf{false}, F_{Inf}(p_3, t_1) \leftarrow \textbf{false},$

$F_{Inh}(p_3, t_1) \leftarrow \textbf{true}, F(p_5, t_1) \leftarrow \textbf{true}, F_{Inh}(p_6, t_1) \leftarrow \textbf{true}\}\} \vee E)$.

The new rule r'' firing the transition t_1 is described now by the following expression for the modified network RNAM: $r'' = [\neg M(p_3)\&\ M(p_5)\&\ \neg M(p_6)](\{M(p_5) \leftarrow \textbf{false},\ M(p_4) \leftarrow \textbf{true}\} \vee E)$.

Another principle of Petri nets modification was proposed in [27] and was associated only with multiple arcs.

4. Conclusion

In this paper we proposed some "reconfigurable directly executable formal models" that are suitable for the specification of computing and networking in the cloud environment and other networks which are similar to wireless networks MANET. These models can be easily programmed and implemented in computer networks. In this article, it was demonstrated that the directly executable formal models can be classified as the models of intermediate type, possessing properties of logic-algebraic and executable models.

The major difference between the interpretation of the ASM and the network of abstract modules, or abstract machines (NAM) is the structuring by using the algebra of algorithms. Reconfigurable version of NAM we have called abbreviated as RNAM. Reconfigurable, dynamically changing, many-sorted algebraic system is the basis for the following models of discrete systems: the formal "directly executable" NAM and RNAM; the formal "directly executable" NAM and RNAM models for the reconfigurable (the deterministic and the non-deterministic) Turing machines (RDTM and RNDTM); the formal "directly executable" NAM and RNAM models of the reconfigurable (the deterministic and the non-deterministic) finite state automata (RFSA and RNDFSA); the formal "directly executable" NAM and RNAM models of the reconfigurable information-inhibitor in Petri nets (RPNII).

The modifications of some well-known formalisms will allow us in the future to construct adequate models of systems with variable structures. Preliminary considerations of this issue as well as the concept of the reconfigurable formal models were carried out in the article [28]. We propose to focus further attention on the application of the proposed models in the field of cloud computing, which is characterized by the following problem.

Moving traditional applications and their infrastructure to the cloud has shifted the in-house control issue to a third party auditor. It posts many challenges including the security and the privacy issues which come on top along with the performance and the availability out of the security which is the number one concern. Clearly using the concept of cloud computing does not make the security issues go away. It becomes an even challenging topic. So In that case, it is not a quite usual utility concept that we are talking about in here. Still there is a problem concerning the security which is the data leakages due to the poor authentications and information's assurance. In this research study our investigation presents a framework and an efficient construction for a significant integration of these two components in our protocol design along with the present of the new approach to keep an efficient and secure cloud computing framework. Juels et al. in [29] they have described the concept of the "proof of retrievability" (POR) model, where the concepts of spot-checking and error-checking and correcting codes are used to ensure both of the possession and the retrievability of data files on the archive service systems. Specifically, some special kind of blocks called the "sentinels" which they are randomly embedded into the data file F for the detection purposes, and F is further encrypted in order to protect the positions of these special blocks mentioned.

However, like [30], mentioned that there are some kind of defects in the system like the number of queries/ tasks a client can perform is also a fixed priori along with the introduction of pre-computed "sentinels" which prevent the development of the dynamic data updates. In addition, the concept of the public auditability is not supported in their proposed scheme. Shacham et al. in [31] designed an improved the POR scheme with full proofs of security and deployed in the security model defined in [29]. They are using some kind of publicly verified homomorphic authenticators [32], on which the proofs of security can be aggregated into a small authenticator value, and the required public retrievability is achieved. Still, the authors are only considering the static type of data files and not the dynamic type and their dynamic operations and this issue is still of course affecting the security of the scheme which is the main concern now of course along with the data availability and integrity.

Our proposals for further research (part of the problem was formulated by one of the authors of this paper in [4]) can be summarized as follows:

(1) We will propose a general formal proof of retrievability (POR) model with public verifiability for the cloud's data storage system, in which block less verifications are achieved and encountered.

(2) We will also equip our proposed proof of retrievability (POR) construction with the functions of supporting for the fully kinds of the dynamic data operations especially to support the block insertion mechanism, which is missing in the most currently existing schemes. In addition to this, multiple TPA's (Third Party Auditors) in working mode in order to handle all the requests in between the clients and the CSS (Cloud's Storage System).

(3) We will propose the new approach of checklist generation for each kind of the cloud like public cloud, community cloud and private cloud with an appropriate procedure. This approach is more affordable and efficient rather than using each time specific algorithms for providing the specific required security.

The objectives of our future scope are:

1) To motivate the public auditing system of data storage security in Cloud Computing, and to propose a protocol supporting for the fully kinds of the dynamic data operations, especially to support the block insertion mechanism, which is missing in most of the currently existing schemes.

2) To extend the scheme to support the scalable and the efficient public auditing in Cloud Computing using the IT auditing and the checklist generation for each kind of the cloud. In particular, the scheme achieves the concept of the batch auditing where multiple delegated auditing tasks from different users will be performed simultaneously and significantly fast by the TPAs.

3) To prove the security of our proposed construction and to justify the performance of the proposed scheme through some significant and concrete kinds of implementations and comparisons with the state-of-the-art.

Acknowledgments
The work is performed as part of the Federal Special Purpose Program - UIN: RFMEFI57414X0045).

References

[1] Mustafa Sadeq Jaafar, Sawant H. K. ACK Based Scheme for Performance Improvement of Ad-hoc Network // International Journal of Advances in Engineering & Technology, (IJAET). Vol. 3, Issue 2, May 2012.

[2] Mustafa Sadeq Jaafar, Sawant H. K. Design and Development of ACK-Based Scheme Using FSA for Ad-hoc Networks // International Journal of Modern Engineering Research, (IJMER). Vol. 2, Issue. 2, Mar-Apr 2012, pp. 102-106.

[3] Mustafa Sadeq Jaafar. On Modification ACK-Based Scheme using FSA for Ad-Hoc Networks // Proceedings of the Eleventh International Conference of Science and Technology “New Information Technologies and Systems”, Penza, Penza State University, Russia, November 25–27, 2014, pp. 334-340.

[4] Mustafa Sadeq Jaafar. IT Auditing Based on Public Verifiability and the Cloud’s Dynamics and Data Storage Security using Multiple TPA’s // Proceedings of the XIX-th International Conference “University Education”, Penza, Penza State University, Russia, April 9–10, 2015, pp. 198-202.

[5] Malcev, A. I. Algebraic Systems. Springer-Verlag, 1973, 329 p.

[6] Plotkin, B. I. Universal algebra, algebraic logic, and databases. Dordrecht – Boston: Kluwer Academic Publishers, 1994. 453 p.

[7] Gurevich, Y. Evolving Algebras 1993: Lipari Guide // Specification and Validation Methods. E. Boerger (ed.), Oxford University Press, 1995, pp. 9-36.

[8] Boerger, E. Unifying View of Models of Computation and System Design Frameworks // Annals of Pure and Applied Logic. Vol. 133, 2005, pp. 149-171.

[9] Grigorieff, S., Valarcher, P. Evolving Multialgebras Unify all Usual Sequential Models // Symposium on Theoretical Aspects of Computer Science. Nancy, France, 2010, pp. 417-428.

[10] Glushkov, V. M. Theory of Automata and Formal Transformations of Microprograms // Kibernetika, No. 5, 1975, pp.1-10.

[11] Gluschkow, W. M., Zeitlin, G. E., Justchenko, J. L. Algebra. Sprachen. Programmierung. Akademie-Verlag, Berlin, 1980. – 340 p.

[12] Tseytlin, G. E. Glushkov Algebras and Clone Theory // Cybernetics and Systems Analysis. Vol. 39, No. 4. Springer-Verlag, New York, 2003, pp. 509-516.

[13] Cejtlin, G., Jushchenko, E. Mathematical theory of multiprocessor control systems and its applications // Computer Science Journal. Vol. 2, No. 3(6), 1994, pp. 247-261.

[14] Cejtlin, G.E. Schematics of Structural Parallel Programming and its Applications // Mathematical Institute, Czechoclovak Academy of Olomous MFCS'79. Lecture Notes in Computer Science. Vol. 74, 1979, pp. 474-481.

[15] Ivanov, P. M. Algebraic modelling of complex systems. – Moscow, 1996. – 274 p.

[16] Nepejvoda, N. N. Abstract algebras of different classes of programs // Proceedings of the 3rd international conference on applicative computation systems (ACS'2012), 2012, pp. 103-128.

[17] Nepejvoda, N. N. Algebraic approach to control. Control Sciences, 2013, No. 6, pp. 2-14.

[18] Gavrilov, G. P., Sapozenko, A. A. Selected Problems in Discrete Mathematics. – M.: Mir Publishers, 1989. – 414 p.

[19] Brauer, W. Automaten Theorie. B. G. Teubner Stuttgart, 1984. – 392 p.

[20] Cooke, D. J., Bez H. E. Computer Mathematics. Cambridge University Press, Cambridge, 1984. – 384 p.

[21] Aho, A. V., Hopcroft, J. E. Ullman, J. D. The Design and Analysis of Computer Algorithms. Addison-Wesley Publishing Company, 1976. – 470 p.

[22] Aho, A. V., Ullman, J. D. The Theory of Parsing, Translation and Compiling. Vol. 1. Prentice-Hall, Englewood Cliffs, 1973. – 309 p.

[23] Petri, C. A. Introduction of General Net Theory // Lecture Notes in Computer Science. Berlin: Springer-Verlag. Vol. 84, 1980, pp. 1-26.

[24] Peterson, J. L. Petri Net Theory and the Modeling of Systems. Prentice-Hall, Englewood Cliffs, NJ, 1981. – 264 p.

[25] Petri Nets. Fundamental Models, Verification and Applications / Ed. by M. Diaz. John Wiley and Sons, 2009. – 613 p.

[26] Petri Nets. Theory and Applications / Ed. by V. Kordic. I-TECH Education and Publishing, Vienna, Austria, 2008. – 544 p.

[27] Valk, R. Self-modifying Nets, a Natural Extension of Petri Nets // Lecture Notes in Computer Science. Berlin: Springer-Verlag. Vol. 62, 1978, pp. 464-476.

[28] Mustafa Sadeq Jafar, Zinkin, S. A. Reconfigurable formal models for MANET and Cloud networking and computing // Proceedings of the XII-th International Conference of Science and Technology "New Information Technologies and Systems". Penza State University, Russia, 2015, pp. 141-149.

[29] Juels, A., Kaliski, B. S., Jr., "Pors: proofs of retrievability for large files," in Proc. of CCS'07. New York, NY, USA: ACM, 2007, pp. 584–597.

[30] Bowers, K. D., Juels, A., Oprea, A. Proofs of retrievability: Theory and implementation // Cryptology ePrint Archive, Report 2008/175, 2008.

[31] Shacham, H., Waters, B. Compact proofs of retrievability // In Proc. of ASIACRYPT'08. Springer-Verlag, 2008, pp. 90–107.

[32] Boneh, D., Lynn, B., Shacham, H. Short signatures from the weil pairing // In Proc. of ASIACRYPT'01. London, UK: Springer-Verlag, 2001, pp. 514–532.

Journal of Physics: Conference Series **710** (2016) 012025 doi:10.1088/1742-6596/710/1/012025

A study: Effect of Students Peer Assisted Learning on Magnetic Field Achievement

Wannapa Mueanploy
Department of Science and Mathematic
Faculty of Science and Technology
Pathumwan Institute of Technology, Bangkok, Thailand 10330

Email: bpuifai06@hotmail.com

Abstract: This study is the case study of Physic II Course for students of Pathumwan Institute of Technology. The purpose of this study is: 1) to develop cooperative learning method of peer assisted learning (PAL), 2) to compare the learning achievement before and after studied magnetic field lesson by cooperative learning method of peer assisted learning. The population was engineering students of Pathumwan Institute of Technology (PIT's students) who registered Physic II Course during year 2014. The sample used in this study was selected from the 72 students who passed in Physic I Course. The control groups learning magnetic fields by Traditional Method (TM) and experimental groups learning magnetic field by method of peers assisted learning. The students do pretest before the lesson and do post-test after the lesson by 20 items achievement tests of magnetic field. The post-test higher than pretest achievement significantly at 0.01 level.

1. Introduction

The magnetic field is a vector having both direction and magnitude. The magnetic force that carry on a charge q moving with a velocity v in a magnetic field B has a magnitude given by $F = qvB\ sin\theta$. Where θ is the angle between v and B. The right-hand rule applies in order to find the direction of force. The SI unit of magnetic field is the **tesla** (T), also used the weber per square meter (Wb/m^2). An additional commonly used another unit for magnetic field is the **gauss** (G), where 1 $T=10^4$ G. If a straight conductor of length L current I, the magnetic force on that conductor when it is placed in an external magnetic field B, is $F = BIL\ sin\theta$. The right-hand rule also gives the direction of the magnetic force on the conductor. The torque τ on a current-carrying loop of wire in the magnetic field B has magnitude given by $\tau = BIA\ sin\theta$ .Where I is the current in the loop and A is its cross-sectional area. The angle θ between B line drawn perpendicular to the plane of the loop.

If a charged particle moves in a uniform magnetic field such that its initial velocity is perpendicular to the field, the particle will move in a circular path whose plane is perpendicular to the magnetic field. The radius r of the circular path is $r = \frac{mv}{qB}$ where m the mass of the particle and q is its charge.

The magnetic field at a distance r from a long, straight wire carrying a current I has a magnitude $B = \frac{\mu_0 I}{2\pi r}$ Where $\mu_0 = 4\pi \times 10^{-7}\ T.m/s$ is the permeability of free space. The magnetic field lines created by a long, straight wire are in the form of concentric circles.

Ampere's law can be used to find the magnetic field around certain simple current-carrying conductors. It is stated as $\sum B\| \, \Delta l = \mu_0 I$. Where $B\|$ is the component of B tangent to a small current element of length Δl that is part of a closed path and I is the total current that penetrates the closed path. The force per unit length between two parallel wires separated by a distance d and carrying currents I and I has a magnitude given by $\frac{F}{L} = \frac{\mu_0 I_1 I_2}{2\pi d}$. The force is attractive if the current are in

the same direction and repulsive if they are in opposite directions. The magnetic field inside a solenoid has magnitude $B = \mu_0 nI$. Where n is the number of turns of wire per unit length, $= N/L$.
The magnetic field can apply in advance further and related issues, Two different temperature dependencies of the thermal conductivity are observe for $\vec{H} \parallel \vec{b}$: linear at low magnetic field and quadratic for magnetic field larger than 1 Tesla[1]. The magnetic fields in the mT range are applied to study the effect on single particles or few-particle systems that are trap inside the sheath region. The behavior of the paramagnetic dust particles is considerably different form that of dielectric plastic particles [2]. The magnetic force microscope is a powerful tool for recording media studies. However, identifying magnetic contribution is still an issue in high-resolution studies [3].
The major objectives of teachers are to effectively use instructional strategies to improve students' cognitive and effective outcomes. Many studies have been conducted in different settings of education, using different kinds of the cooperative learning techniques. Cooperative learning methods of peers assisted learning indicates that the appropriate method of learning at the high level is Student Teams Achievement Divisions [4]. One kind of cooperative learning to approach, in order to encourage students to improve their achievement and promote more positive attitude an alternative to lecture-based teaching could be Student Teams Achievement Divisions[5].

2. Methodology

2.1 The achievement test
The achievement tests were 20 items in 5 multiple-choice about unit of Magnetic Field and Its Effects, Field Sources and Magnetic Flux and Induction Law. The tests and two parallel sets created by the author and analyzed by B-index Program.

2.2 The experiment setting
The sample in this study was selected from the 72 students. 12 students who received grade A,B^+ or B, 36 students who received grade C^+ and C and 24 students who received grade D^+ an D in Physic I Course. After that rearrange in 12 groups in each group composed 6 students: one from A, B^+ or B, three from C^+ and C and two from D^+ and D. Random sampling 6 groups (36 students) are control groups and another 6 groups (36 students) are experimental groups. Groups of student are small (six students) for this experiment.

2.3 Study in Class
Planning, teaching each cooperatives learning should take between 50-60 minutes to complete content.
2.3.1 The students in control groups (N = 36) went to learning TM in lesson of magnetic field.
2.3.2 The experimental groups (N = 36) went to learning PAL in lesson of magnetic field.

2.4 The process of learning PAL is follows:
2.4.1 The teacher introduces a lesson. The time spent in this step is about 8-10 minutes.
2.4.2 The experimental groups learning PAL are asked to discuss through group work and brainstorming. The times expected for this learning activity is about 25-30 minutes
2.4.3 All of students in experimental groups take individual post-test (first time). The times set for this step about 15-20 minutes. In addition each student can check correct answers by themselves after taking the mentioned test. Importantly, the results obtained from the post- test of each experimental group must be more than 80 percent. However, if the result of each group is lesser than 80 percent, it means that the set hypothesis is rejected. Each student is expected to obtain the result more than 80 percent that means the student can answer the correct answer more than 16 items. Under such circumstances, if the particular student gets the correct answer less than 16 items, the students are expected to go back to the second step and the third step respectively so as to take the post test that is designed in parallel to the first one. In this study, all students in the experimental groups are expected to get the scores 80 percent after going back to learn the lesson in three times.

The results of the experimental groups and control groups analyze by means, standard deviation; independent t-test significant at 0.01 level.

3. Discussion and Result

3.1 The results of this study control groups learning TM and experimental groups learning PAL in lesson of magnetic field get the score pretest are not different in figure 1 and post-test score of control groups and experimental groups are significantly at 0.01 level in figure 2. The post-test scores of experimental groups use the chanced to get back to step 2 and 3 in three times to take everyone 80 percent in figure 3

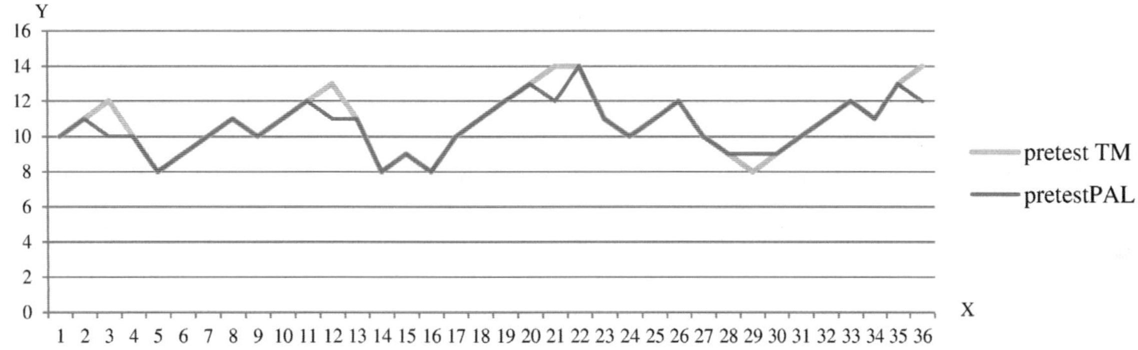

Figure1 the pretest score control groups (TM) and experimental groups (PAL)

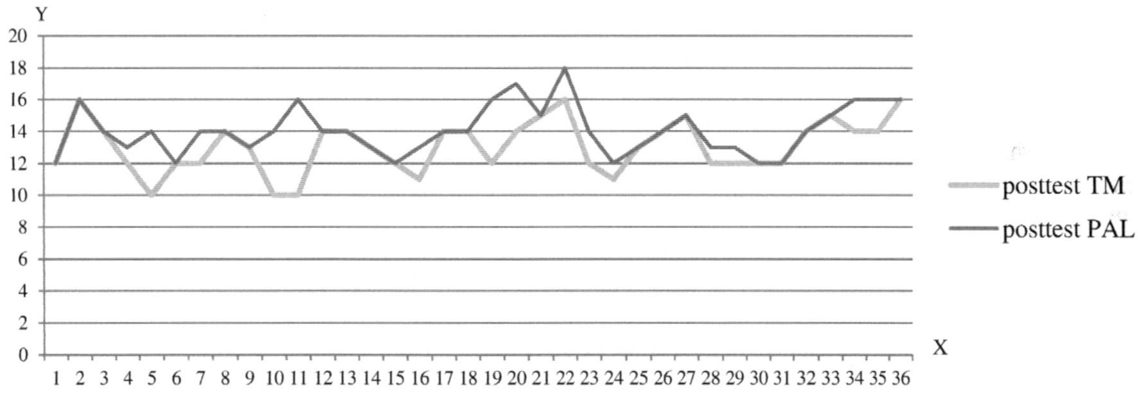

Figure2 the post-test score of control groups(TM) and experimental groups (PAL)

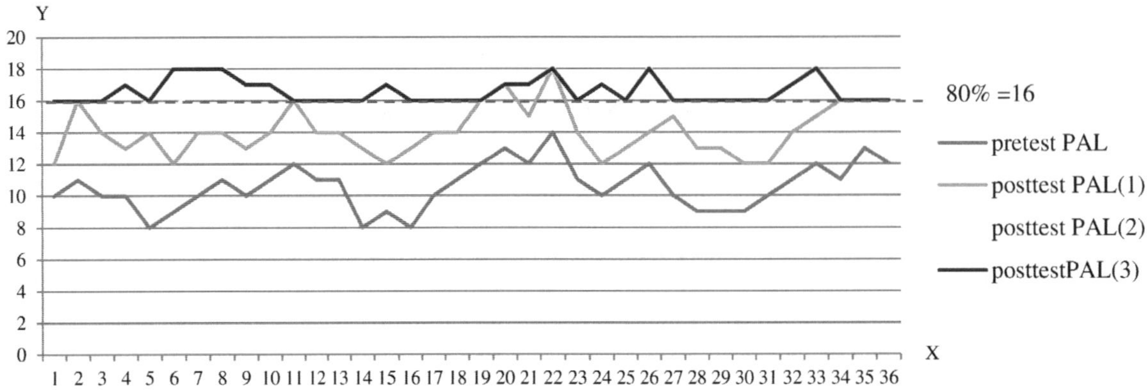

Figure3 the experimental groups pretest and post-test scores in three times

3.2 The results of analyzing pretest and post-test physics achievement magnetic field of PIT's students are as the following.

Table 1 The results of analyzing pretest physics achievement magnetic field of PIT's students

Statistics Sample	**N**	$\bar{X}$	SD	t-test
Controlled groups	36	9.70	1.70	0.50
Experiment groups	36	9.52	1.44	

From table1 finding an average of the pretest in physics achievement, magnetic field the experiment groups learning PAL is 9.52 the average of the pretest the controlled groups learning TM is 9.70. Using t-test independent sample to compare achievement finds that the experiment groups and controlled group's achievement is different insignificantly.

Table 2 The results of analyzing post-test physics achievement magnetic field of PIT's students

Statistics Sample	**N**	$\bar{X}$	SD	t-test
Controlled groups	36	11.73	1.65	
Experiment groups(1)	36	12.67	1.54	2.5^{**}
Experiment groups(2)	36	14.65	0.88	10^{**}
Experiment groups(3)	36	14.90	0.77	3.6^{**}

$^{**}p < .01$

From table (2) Finding the average of the post-test in physics achievement, magnetic field learning method of PAL is 12.67,14.65 and 14.90 the average of the post-test the controlled groups studied in TM is 11.73. Using t-test independent sample to compare achievement finds that is significantly higher at 0.01 level.

3.3 The results chart of analyzing pretest and post-test physics achievement magnetic field of PIT's students were show percentage the development in groups. In this study the p-value is used as an alternative to rejection points to provide the smallest level of significance at which the null hypothesis would be rejected.The probability of getting the results to given that the null hypothesis is true. It means the achievement of PIT's students learning PAL is higher than the achievement of PIT's students learning TM.

This study helps engineering students to understand in magnetic field to apply in advance further and related issues. The magnetic field measurement was the most widely applied technology of magnetic measurement it was not only significant to study in all areas of solid physics but also make and important contribution to study in related subject [6]. It was found that magnetic field has power on surface tension and electrical conductivity of mine water [7]. The moving body in this magnetic field exerts little influence on the magnetic intensity and the influence can be neglected. The force of the moving body about its position can be calculated in static analysis and then be imported into the impact calculation to obtain the velocity and displacement [8].

4. Conclusion

In this study, 20 items used in achievement test of magnetic field were analyzed by B-index Program, The result is Difficulty (P) = 0.43-0.63, Discrimination = 0.13-0.68 and Reliability(r) = 0.28. The achievement of PIT's students learning PAL is higher than that of PIT's students learning TM at level of 0.01 which is related to the specified hypothesis that is consistent with study of this research has shown that it is better for students' experience cognitive from directly observed experiments than by reflecting on reported experience from popularization papers or writings found on the internet [9]. The present study shows that PAL strategy is more effective to the students' learning and retention than lecture. These differences were clearer in the comprehension and application levels. PAL learning scores are significantly are higher than lecture by overall and knowledge level [10]. PAL has benefits of collaborative and cooperative learning experience which was manifest in the strong social congruence among PAL groups in the present study [11]. In addition PAL creates a learning environment which allows students to ask questions and PAL as part of the solution and not as part of the problem.
Future research will develop PAL provides a comprehensive in Physics I and Physics II to ensure continuity and coherence in learning physics.

Acknowledgment
The author would like to thank Pathumwan Institute of Technology for the financial support and permission to experiment.

Reference

[1] Ludovic Howald,Mathieu Taupinand Dai Multigap 2014 Superconductivity in the Ferromagnetic Superconductor UCoGe Revealed by Thermal Conductivity Measurements *Hindawi Publishing Corporation Physics Research International* Volume 2014, 7 pages.
[2] Marian Puttscher and André Melzer, 2014 Paramagnetic dust particles in rf-plasmas with weak external magnetic fields, *New Journal of Physics* 16 , 13 pages.
[3] Hai Zhong et al, 2008 Non-contact magnetic force microscopy of recording media by *ex situ* tip magnetization reversal method, *Journal of Physics D: Applied Physics*, Volume 41, Number 8
[4] Thida Mosikarat et.al, 2005 Development of Cooperative Learning Method of Peer Assisted Learning for Students of Sukhothai Thammathirat Open University: Case Study of Thai Studies Course Samutsakorn Province, Sukhothai Thammathirat Open University.
[5] Van Dat Tran 2013 Effects of Student Teams Achievement Division (STAD) on Academic Achievement, and Attitudes of Grade 9th Secondary School Students towards Mathematics *International Journal of Sciences* Volume 2, Issue Apr (2013) 5-15.
[6] Shi Kun Zhu Shi Kun Zhu, Study on the Methods of Measuring Magnetic Field, *Applied Mechanics and Materials* Volume 214 ,(2012) 124-129.
[7] Zhe Guo, Jian Guo Cui, Gang Wei, 2014 The Effect of Applied Magnetic Field on the Mine Water Coagulation *Applied Mechanics and Materials* (Volumes 522 - 524) 1021-1026.
[8] Ya Ping Ding, Xiao Yan Tong, Yong Tao Liu, Yong Liao 2013 A Method of Impact Simulation in Permanent Magnetic Field *Applied Mechanics and Materials* (Volume 443) 129-133.

[9] Mirko Marušić and Josip Sliško 2012 Influence of Three Different Methods of Teaching Physics on the Gain in Students' Development of Reasoning International *Journal of Science Education* Volume 34, Issue 2.
[10] Mohammadreza Abediniet.al, 2013 A New Teaching Approach in Basic Sciences: Peer Assisted Learning, *Procedia - Social and Behavioral Sciences*, Volume 83, 4 July Pages 39–43.
[11] Seema Daud and Syeda Kauser Ali,2014 Perceptions of Learners about Peer Assisted Learning and Lectures, International *Journal of Science and Study* (IJSR) Volume 3 Issue 11, November.
[12] Ngoc-Loan Nguyen and David E. Meltzer 2005 Visualization Tool for 3-D Relationships and the Right-Hand Rule, *The Physics Teacher* , vol.43,March.
[13] Antonio Svedružić, 2008 Demonstration in teaching Physics, *Metodika* 17 ,442-450.
[14] Shafqat Hussain,2011 The Effectiveness of Teaching Physics through Project Method on Academic Achievement of Students at Secondary Level -A Case Study, *Journal of Education and Practice*, Vol. 2, No 8.

Journal of Physics: Conference Series **710** (2016) 012026 doi:10.1088/1742-6596/710/1/012026

Thermodynamics properties study of diatomic molecules with q-deformed modified Poschl-Teller plus Manning Rosen non-central potential in D dimensions using SUSYQM approach

A Suparmi,C Cari and B N Pratiwi
Physics Department, Faculty of Mathematics and Science, Sebelas Maret University
Jl. Ir. Sutami 36A Kentingan Surakarta 57126, Indonesia

Email: soeparmi@staff.uns.ac.id

Abstract. D-dimensional Dirac equation of q-deformed modified Poschl-Teller plus Manning Rosen non-central potential was solved using supersymmetric quantum mechanics (SUSY QM). The relativistic energy spectra were analyzed by using SUSY QM and shape invariant properties from radial part of D dimensional Dirac equation and the angular quantum numbers were obtained from angular part of D dimensional Dirac equation. The SUSY operators was used to generate the D dimensional relativistic wave functions both for radial and angular parts. In the non-relativistic limit, the relativistic energy equation was reduced to the non-relativistic energy. In the classical limit, the partition function of vibrational, the specific heat of vibrational, and the mean energy of vibrational of some diatomic molecules were calculated from the equation of non-relativistic energy with the help of error function and Mat-lab 2011.

1. Introduction

In some area of physics, relativistic quantum mechanics play important roles. Finding an accurate exact solution of Dirac equation for a certain potential is one of its important roles. Various methods have been applied to solve the Dirac equation for some potentials, central and non-central potentials, with or without tensor coupling potentials, such as NU method, [1-6] SUSY QM method, [7-11] and Romanovski polynomial method. [12-16]

For very limited potentials, three dimensional radial Dirac equations are exactly solved only for s-wave ($l = 0$). However, the three dimensional radial Dirac equations for the spherically symmetric potentials can only be solved approximately for $l \neq 0$ states due tothe approximation scheme of the centrifugal term $\sim r^{-2}$. [17-23]

Furthermore, the extension in higher dimensional spaces for some physical problems is very important in some physics area. The D-dimensional non-relativistic and relativistic physical systems have been investigated by many authors, such as ring-shaped pseudoharmonic potential, [21] the isotropic harmonic oscillator and inverse quadratic potential, [22] Pseudoharmonic potential,[23] Kratzer-Fues potential, [24-25]hydrogen atom, [26]modified Poschl-Teller potential, [27] linierly energy dependent quadratic potential, [28] trigonometric scarf potential, [29] ring-shaped Kratzer potential. [30]

The Dirac equation for a charged particle that moves in a field governed by q-deformed Poschl-Teller potential[31] in D dimension is investigated using supersymmetric quantum mechanic (SUSY QM) with idea of shape invariance. SUSY QM method is developed based on Witten proposal [32] and the idea of shape invariant potential is proposed by Gendenshtein [33]. SUSY QM is a powerful tool to determine energy spectrum and wave function of shape invariant potentials for one dimensional Schrodinger equation. Thus the relativistic energy spectrum is obtainable by using the idea of shape invariance and the wave functions are achieved by using lowering and raising SUSY operators. Some of hyperbolic and trigonometric potentials are exactly solvable within the approximation of centrifugal term and their solutions have been reported in the previous papers [10-11].The q-deformed modified Poschl-Teller plus Manning Rosen non-central potential that govern the diatomic molecules vibration is expressed as

$$V(r,\theta) = t^2\left(\left(a(a-1)/\sinh_q^2 tr\right) - \left(b(b+1)/\cosh_q^2 tr\right)\right) - \left(1/r^2\right)\left[\left(p(p+1)/\sin^2\theta\right) - 2s\cot\theta\right] \tag{1}$$

with $a > 1, b > 0$; $\alpha > 0$; $p > 0$, $s > 0$, a, b, p and s are positive constants that control the depth of the potential, t is positive constant which controls the width of the potential, $0 < r < \infty$, q causes the deformation of the potential shape, $0 < q < 1$. In the non-relativistic limit, the relativistic energy equation reduces the non-relativistic energy equation. In the non relativistic limit is the condition when the energy is subtracted by the mass equals to the non-relativistic energy which is usually obtained from Schrodinger equation solution of the system, and the condition where the sum of the energy and the mass equals to twice of masses. The non-relativistic energy is used in calculation of thermal properties in classical limit. In classical regime, the thermal properties including vibrational partition function Z, mean energy U, and specific heat C [34,35] are determined by applying the non-relativistic energy equation.

This paper is organized as follows. Brief review of SUSY quantum mechanics is presented in section 2, solution of Dirac equations and its application to study thermodynamical properties are presented in section 3 and conclusion is presented in section 4.

2. Review of Supersymmetric Quantum Mechanics Approach Using Operator

2.1. Supersymmetry Quantum Mechanics (SUSY QM)

According to the definition proposed by Witten, in a SUSY QM there are super charge operators Q_i which commute with the Hamiltonian H_{ss} and obey to anti commutation algebra [32]

$$[Q_i, H_{ss}] = 0; \ \{Q_i, Q_j\} = \delta_{ij} H_{ss} \text{ with, } i = 1, 2, 3, \ldots \tag{2}$$

with H_{ss} is called supersymmetric Hamiltonian. Witten proposed that the SUSY QM is the one dimensional model of SUSY field theory and he stated that the simplest SUSY QM system has $N=2$[32] where the two charge operators are given as

$$Q_1 = \left(1/\sqrt{2}\right)\left(\sigma_1\left(p/\sqrt{2m}\right) + \sigma_2\varphi(x)\right); Q_2 = \left(1/\sqrt{2}\right)\left(\sigma_2\left(p/\sqrt{2m}\right) + \sigma_1\varphi(x)\right) \tag{3}$$

where σ_i are the usual Pauli spin matrices, $p = -i\hbar(\partial/\partial x)$ is the usual one dimensional momentum operator, and $\phi(x)$ is superpotential. By inserting equation (3) into the second equation of equation (2) we get,

$$H_{ss} = \begin{pmatrix} -\frac{\hbar^2}{2m}\frac{d^2}{dx^2} + \frac{\hbar}{\sqrt{2m}}\frac{d\varphi(x)}{dx} + \varphi^2(x) & 0 \\ 0 & -\frac{\hbar^2}{2m}\frac{d^2}{dx^2} - \frac{\hbar}{\sqrt{2m}}\frac{d\varphi(x)}{dx} + \varphi^2(x) \end{pmatrix}; H_{ss} = \begin{pmatrix} H_+ & 0 \\ 0 & H_- \end{pmatrix} \tag{4}$$

From equation (4) we have

$$V_-(x) = \varphi^2(x) - \frac{\hbar}{\sqrt{2m}}\varphi'(x); \qquad V_+(x) = \varphi^2(x) + \frac{\hbar}{\sqrt{2m}}\varphi'(x) \tag{5}$$

Here H_- and H_+, are supersymmetry partner of the Hamiltonian, $V_-(x)$ and $V_+(x)$ are the supersymmetry partner potential. To simplify the determination of the energy spectrum and the wave functions, the new operators, raising and lowering operators, are introduced as

$$A^+ = -\frac{\hbar}{\sqrt{2m}}\frac{d}{dx} + \phi(x) \text{ and } A = \frac{\hbar}{\sqrt{2m}}\frac{d}{dx} + \phi(x) \tag{6}$$

By inserting equation (6) into equation (5) we get the SUSY Hamiltonian as

$$H_-(x) = A^+A \text{ and } H_+(x) = AA^+ \tag{7}$$

to factorize the usual Hamiltonian as

$$H = H_- + E_0 = -\frac{\hbar^2}{2m}\frac{d^2}{dx^2} + V_-(x; a_0) + E_0 \tag{8}$$

By using equation (5) and equation (8) it is obtained that

$$V(x) = V_-(x; a_0) + E_0 = \phi^2(x; a_0) - \frac{\hbar}{\sqrt{2m}}\phi'(x; a_0) + E_0 \tag{9}$$

where $V(x)$ is the effective potential, while $\phi(x)$ is determined hypothetically from equation (9) based on the shape of effective potential from the associated system.

2.2. Shape Invariance

It is observed that the supersymmetry only gives the relationship between the eigenvalues and eigenfunctionsbetween the two Hamiltonian partners but does not provide the actual spectrum. [36].The energy spectrum is only obtainable by implementing SUSY charged operators properties and the condition of shape invariant proposed by Gendenshtein.[33]. If a pair of potentials $V_{\pm}(x)$ defined in equation (5) are similar in shape but different in the parameters, then they are called to be shape invariant. More specifically, if $V_{\pm}(x,a_0)$ satisfy the condition that

$$V_+(x;a_j) = V_-(x;a_{j+1}) + R(a_{j+1}) \tag{10}$$

with
$$V_+(x;a_j) = \varphi^2(x;a_j) + \frac{\hbar}{\sqrt{2m}}\varphi'(x;a_j); V_-(x;a_{j+1}) = \varphi^2(x;a_{j+1}) - \frac{\hbar}{\sqrt{2m}}\varphi'(x;a_{j+1}) \tag{11}$$

where $j = 0,1,2,..$, and a is a parameter in our original potential, V-, whose ground state energy is zero, $a_j = f_j(a_0)$ where f $j$is a function applied j times, the remainder $R(a_j)$ is a's dependence but is independent of x, then $V_{\pm}(x,a_0)$ is said to be shaped invariant. The energy eigenvalue of the Hamiltonian H_- is given by [33]

$$E_n^{(-)} = \sum_{k=1}^{n} R(a_k) \tag{12}$$

and by using equation (8) and equation (12) we get the energy spectra of the system given as,

$$E_n = E_n^{(-)} + E_0 \tag{13}$$

Based on the characteristics of lowering operator, the ground state wave function of H-, whose ground state energy is zero, is obtained from condition that,

$$H_-\psi_0^{(-)} = 0 \rightarrow A\psi_0^{(-)} = 0 \tag{14}$$

Subsequently, the excited wave function, $\psi_1^{(-)}(x,a_0),...\psi_n^{(-)}(x,a_0)$ of H_- are obtained by using raising operator operated to the lower wave function, [37] given as

$$\psi_n^{(-)}(x;a_0) \approx A^+(x;a_0)A^+(x.a_1)....A^+(x,a_{n-1})\psi_0^{(-)}(x;a_n) \tag{15}$$

In the Dirac equation we have applied the SUSY potential partner and the SUSY operator equations (5-6) as

$$V_-(x) = \phi^2(x) - \phi'(x) \;;\; V_+(x) = \phi^2(x) + \phi'(x) \tag{16}$$

$$A = \frac{d}{dx} + \varphi(x); A^+ = -\frac{d}{dx} + \varphi(x) \tag{17}$$

The SUSY QM and the idea of shape invariance potential are suitable to be used in solving one dimensional Schrodinger equation problems. One dimensional Dirac equation reduces to one dimensional Schrodinger type equation by suitable change in parameters which are the coefficient of the potential function and the energy term.

By obtaining the super-potential, the potential partners, $V_+(x,a_0)$ and $V_-(x,a_0)$, and the SUSY operators, A^+ and A are obtained and so the energy spectrum and the wave function.

3. Solution of Dirac Equation for Non-central Potential in D dimension

3.1. Solution of radial part of D dimensional Dirac Equation

The Dirac equation with the scalar potential $S(\vec{r})$ and magnitude of vector potential $V(\vec{r})$ is given as in Hu et al. [38]

$$\{\vec{\alpha}\vec{p} + \beta(M + S(\vec{r}))\}\psi(\vec{r}) = \{E - V(\vec{r})\}\psi(\vec{r}) \tag{18}$$

where M is the relativistic mass of the particle, E is the total relativistic energy, and $\vec{p}$ is the three-dimensional momentum operator, $-i\nabla$

$$\vec{\alpha}=\begin{pmatrix}0 & \vec{\sigma}\\ \vec{\sigma} & 0\end{pmatrix};\beta=\begin{pmatrix}I & 0\\ 0 & -I\end{pmatrix} \tag{19}$$

with $\vec{\sigma}$ are the three-dimensional Pauli matrices and I is the2 × 2 identity matrix. The potential in equation (18) is spherically symmetric potential,it does not only depend on the radial coordinate $r=|\vec{r}|$, and we have taken $\hbar=1$ and $c=1$. The Dirac equation expressed in equation (18) is invariant under spatial inversion and therefore its eigenstates have definite parity. By writing the spinor in D dimension as

$$\psi(\vec{r})=\begin{pmatrix}\zeta(\vec{r})\\ \Omega(\vec{r})\end{pmatrix} \tag{20}$$

and if we insert equations (19) and (20) into equation (18) and use matrices multiplication, we achieve

$$\vec{\sigma}.\vec{p}\Omega(\vec{r})=\{-M-S(\vec{r})+E-V(\vec{r})\}\zeta(\vec{r}) \tag{21}$$

$$\vec{\sigma}.\vec{p}\zeta(\vec{r})=\{M+S(\vec{r})+E-V(\vec{r})\}\Omega(\vec{r}) \tag{22}$$

In the exact spin symmetric case, when the scalar potential is equal to the magnitude of vector potential $S(\vec{r})=V(\vec{r})$, then the upper Dirac spinor obtained from equations (21) and (22) are

$$\vec{\sigma}.\vec{p}\frac{\vec{\sigma}.\vec{p}}{M+E}\zeta(\vec{r})=\{-M+E-2V(\vec{r})\}\zeta(\vec{r}) \tag{23}$$

By applying the Pauli matrices, it is simply shown that if $(\vec{\sigma}.\vec{p})(\vec{\sigma}.\vec{p})=p^2$, then equation (23) becomes

$$p^2\zeta(\vec{r})+2V(\vec{r})(M+E)\zeta(\vec{r})=(E^2-M^2)\zeta(\vec{r}) \tag{24}$$

Since $p^2=-\Delta_D=-\nabla_D^2$ with the hypersphericalLaplacian ∇_D^2 is given by [39-40]

$$\nabla_D^2=r^{1-D}\frac{\partial}{\partial r}\left(r^{D-1}\frac{\partial}{\partial r}\right)+\frac{1}{r^2}\left[\frac{1}{\sin^{D-2}\theta_{D-1}}\frac{d}{d\theta_{D-1}}\left(\sin^{D-2}\theta_{D-1}\frac{d}{d\theta_{D-1}}\right)-\frac{L_{D-2}^2}{\sin^2\theta_{D-1}}\right] \tag{25}$$

with $L_{D-2}^2=l_{D-2}(l_{D-2}+D-3)$.

To reduce equation (24) into Schrodinger like equation we set $V\to(1/2)V$ and if V is modified Poschl-Teller plus trigonometric Manning-Rosennon-central potential then equation (24) becomes

$$r^{1-D}\frac{\partial}{\partial r}\left\{r^{D-1}\frac{\partial}{\partial r}(\zeta(\vec{r}))\right\}+\frac{1}{r^2}\left[\frac{1}{\sin^{D-2}\theta_{D-1}}\frac{d}{d\theta_{D-1}}\left(\sin^{D-2}\theta_{D-1}\frac{d(\zeta(\vec{r}))}{d\theta_{D-1}}\right)-\frac{L_{D-2}^2}{\sin^2\theta_{D-1}}(\zeta(\vec{r}))\right]$$
$$-\left(t^2\left(\frac{a(a-1)}{\sinh^2 tr}-\frac{b(b+1)}{\cosh^2 tr}\right)-\frac{1}{r^2}\left[\frac{p(p+1)}{\sin^2\theta}-2s\cot\theta\right](M+E)\right)((\zeta(\vec{r}))=-(E^2-M^2)((\zeta(\vec{r})) \tag{26}$$

The upper component of Dirac spinor in equation (20) is given as

$$\zeta(\vec{r})=\frac{1}{r^{\frac{D-1}{2}}}F_l(r)Y_{\ell_1\ldots\ell_{D-2}}^{\ell}(\hat{x}=\theta_1,\theta_2,\ldots,\theta_{D-1})\text{ and }Y_{\ell_1\ldots\ell_{D-2}}^{\ell}(\hat{x}=\theta_1,\ldots,\theta_{D-1})=\Phi(\theta_1=\phi)H(\theta_2,\ldots,\theta_{D-1}) \tag{27}$$

where $\vec{x}$ is a D dimensional position vector in hyper-spherical Cartesian coordinate[40,41], the unit vector along $\vec{x}$ vector is denoted as $\hat{x}=\vec{x}/r$.The hyper-spherical Cartesian coordinate components x_1, x_2, x_3, ... are given as

$$x_1=r\cos\theta_1\sin\theta_2\sin\theta_3\ldots\sin\theta_{D-1};x_2=r\sin\theta_1\sin\theta_2\sin\theta_3\ldots\sin\theta_{D-1};$$
$$x_j=r\cos\theta_{j-1}\sin\theta_j\sin\theta_{j+1}\ldots\sin\theta_{D-1}\quad 3\le j\le D-1 \tag{28}$$

$$x_{D-1}=r\cos\theta_{D-2}\sin\theta_{D-1};x_D=r\cos\theta_{D-1} \tag{29}$$

The simultaneous eigenfunctions are given as

$$L_1^2 Y^{(\ell)}_{\ell_1\ldots\ell_{D-2}} = m^2 Y^{(\ell)}_{\ell_1\ldots\ell_{D-2}}; m=\ell_1; L^2_{D-1} Y^{(\ell)}_{\ell_1\ldots\ell_{D-2}} = \ell_{D-1}\left(\ell_{D-1}+D-2\right) Y^{(\ell)}_{\ell_1\ldots\ell_{D-2}} \quad (30)$$

$$L_j^2 Y^{(\ell)}_{\ell_1\ldots\ell_{D-2}} = \ell_j\left(\ell_j + j - 1\right) Y^{(\ell)}_{\ell_1\ldots\ell_{D-2}}; 1<j<D-1 \quad (31)$$

$$L_k^2 = \sum_{a<b=2}^{k+1} L_{ab}^2 = -\left[\frac{1}{\sin^{k-1}\theta_k}\frac{d}{d\theta_k}\left(\sin^{k-1}\theta_k\frac{d}{d\theta_k}\right) - \frac{L^2_{k-1}}{\sin^2\theta_k}\right]; 2\le k\le D-2 \quad (32)$$

By applying the variable separation method we have radial and angular parts of D dimensional Dirac equation which are given as

$$\frac{d^2F(r)}{dr^2} - t^2\left(\frac{a(a-1)(E+M)+\lambda_l+(\frac{D-1}{2})(\frac{D-3}{2})}{\sinh_q^2 tr} - \frac{(E+M)b(b+1)}{\cosh_q^2 tr}\right)F(r) + (E^2-M^2-t^2\left(\lambda_l+(\frac{D-1}{2})(\frac{D-3}{2})\right)d_0)F(r)=0 \quad (33)$$

$$\left[\frac{1}{\sin^{D-2}\theta_{D-1}}\frac{d}{d\theta_{D-1}}\left(\sin^{D-2}\theta_{D-1}\frac{d}{d\theta_{D-1}}\right)+\lambda_\ell - \frac{1}{\sin^2\theta_{D-2}}\left(L^2_{D-2}\right) - \left[\frac{p(p+1)}{\sin^2\theta_{D-1}} - 2s\cot\theta_{D-1}\right](M+E)\right]H(\theta_{D-1})=0 \quad (34)$$

In order to solve the radial Dirac equation in Eq.(33), we use the approximation value for the centrifugal term as in Greene and Aldirch, and in Ikdhair [20-21], $1/r^2 \cong t^2\left(d_0 + 1/\sinh_q^2 tr\right)$, for $tr << 1$ and $d_0 = 1/12$. By setting

$$\left(a(a-1)\right)\left(E+M\right)+\lambda_l+(\frac{D-1}{2})(\frac{D-3}{2}) = c(c-1) \quad ; \quad b(b+1)(E+M) = b'(b'+1) \quad (35)$$

$$E^2 - M^2 - t^2\left(\lambda_l + (\frac{D-1}{2})(\frac{D-3}{2})\right)d_0 = E' \quad (36)$$

equation (33) becomes

$$-\frac{d^2F(r)}{dr^2} + t^2\left(\frac{c(c-1)}{\sinh_q^2 tr} - \frac{b'(b'+1)}{\cosh_q^2 tr}\right)F(r) = E'F(r) \quad (37)$$

Equation (37) is solved using SUSY QM and by introducing the hypothetical super-potential as in [26-27]

$$\phi(r) = tP\coth_q tr + tB\tanh_q tr \quad (38)$$

using equation (9), equation (37) and equation (38) we get

$$t^2\left(\frac{c(c-1)}{\sinh_q^2 tr} - \frac{b'(b'+1)}{\cosh_q^2 tr}\right) = t^2(P^2+P)(q\,\mathrm{csc}h_q^2 tr) - t^2(B^2+B)(q\,\mathrm{sec}h_q^2 tr) + E'_0 + \left(2BP+P^2+B^2\right)t^2 \quad (39)$$

that gives

$$P = -(1/2) - \sqrt{(c(c-1)/q)+1/4}\,;\; B = -(1/2) + \sqrt{(b'(b'-1)/q)+1/4}\,; E'_0 = -(P+B)^2 t^2 \quad (40)$$

The superpotential,super-partner potentials, ground state energy and raising-lowering operators obtained from equations (5), (6),(12), (13), and (38) are

$$\varphi(r) = \left(-\frac{1}{2}t - t\sqrt{\frac{c(c-1)}{q}+\frac{1}{4}}\right)\coth_q tr + \left(-\frac{1}{2}t + t\sqrt{\frac{b'(b'-1)}{q}+\frac{1}{4}}\right)\tanh_q tr \quad (41)$$

$$V_-(r,a_0) = q\left\{t^2P(P+1)\right\}\mathrm{csc}h_q^2 tr - qt^2B(B+1)\mathrm{sech}_q^2 tr + t^2(P+B)^2 \quad (42)$$

$$V_+(r,a_0) = q\left\{t^2P(P-1)\right\}\mathrm{csc}h_q^2 tr - qt^2B(B-1)\mathrm{sech}_q^2 tr + t^2(P+B)^2 \quad (43)$$

$$A^+ = -\frac{d}{dr} + tP\coth_q tr + tB\tanh_q tr;\; A = \frac{d}{dr} + tP\coth_q tr + tB\tanh_q tr \quad (44)$$

By comparing the super-partner potential $V_-(r,a_0)$ and $V_+(r,a_0)$ we get the translation parameters as

$$a_0 = P,\ a_1 = P-1, \ldots\ a_n = P-n, \text{ and } b_0 = B,\ b_1 = B-1, \ldots\ b_n = B-n, \tag{45}$$

By implementing equations (10-13) and (42-43) and (45) we get the equation of relativistic energy spectra given as

$$E_n^{'} = -t^2\left(\left(-\sqrt{c(c-1)/q+1/4}\right)+\left(\sqrt{b'(b'+1)/q+1/4}\right)-1-2n\right)^2 \tag{46}$$

The relativistic energy equation obtained from equations (35-36) and (46) is given as

$$\left(E^2-M^2\right)=\left\{\left[\lambda_l+(\frac{D-1}{2})(\frac{D-3}{2})\right]t^2 d_0\right\}-t^2\left(\left(-\sqrt{\frac{c(c-1)}{q}+\frac{1}{4}}\right)+\left(\sqrt{\frac{b'(b'+1)}{q}+\frac{1}{4}}\right)-1-2n\right)^2 \tag{47}$$

with $c(c-1), b'(b'+1)$ is expressed in equation (35) and

$$\lambda_l = -\left[(D-2)/2\right]^2+(p'-n)^2-\frac{s'^2}{(p'-n)^2};\quad p' = -1/2+\sqrt{\left(\ell_{D-2}+(D-3)/2\right)^2+(E+M)p(p+1)} \tag{48}$$

which are obtained from angular solution in the next section.From the relativistic energy equation in equation (47), we can obtain the numerical value of relativistic energy by using Mat-Lab in Table 1.

Table 1. Relativistic energy for various dimensions level

$\left(t=0.8, s=0.3, d_0=1./12, a=6, b=2, p=4, s=2, q=0.75, M=5(1/fm), l_{D-2}=2\right)$

D	E_0	E_1	E_2	E_3	E_4
3	-5.9455	-7.1363	-7.9716	-9.2079	-9.2079
4	-5.8748	-7.0824	-7.9306	-8.6091	-9.1800
5	-5.8075	-7.0212	-7.8823	-8.5695	-9.1467

By using equations (14-15) and (44) we obtain the un-normalized relativistic ground state and first exited state wave functions for upper Dirac spinnor

$$F_0 = C(\sinh_q tr)^{-P}(\cosh_q tr)^{-B};\quad F_1 = 2\left(tP\coth_q tr + tB\tanh_q tr\right)(\sinh_q tr)^{-P}(\cosh_q tr)^{-B} \tag{49}$$

with the values of P and B are expressed in equation (40). By using raising operator in equation (44) we obtain all exited states of wave functions in exact spin symmetric case.

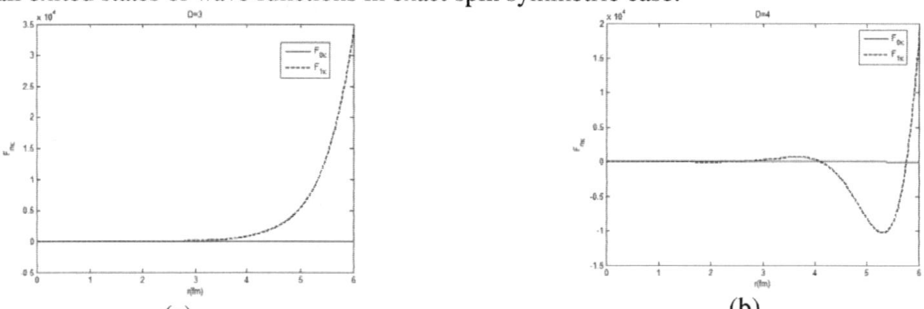

(a) (b)

Figure 1.Ground state (solid line) and first exited (dash line) radial wave functions for (a)$D = 3$,(b)$D = 4$

3.2. Solution of Anggular part of D dimensional Dirac Equation

The angular part of D dimensional Dirac equation for Manning Rosen angular potential are

$$\left[\frac{1}{\sin^{j-1}\theta_j}\frac{d}{d\theta_j}\left(\sin^{j-1}\theta_j\frac{d}{d\theta_j}\right)+\ell_j(\ell_j+j-1)-\frac{\ell_{j-1}(\ell_{j-1}+j-2)}{\sin^2\theta_j}\right]H(\theta_j)=0;\quad j\in\left[2, D-2\right] \tag{50}$$

$$\left[\frac{1}{\sin^{D-2}\theta_{D-1}}\frac{d}{d\theta_{D-1}}\left(\sin^{D-2}\theta_{D-1}\frac{d}{d\theta_{D-1}}\right)+\lambda_{\ell}-\frac{\left(L_{D-2}^2\right)}{\sin^2\theta_{D-1}}-(E+M)\left(\frac{p(p+1)}{\sin^2\theta_j}-2s\cot\theta_j\right)\right]H(\theta_{D-1})=0 \quad (51)$$

These two equations are solvable by using SUSY QM approach by reducing it into one dimensional Schrodinger like equation by suitable substitution of wave functions in equations (50) and (51) as

$$H(\theta_j)=Q(\theta_j)/\sin^{(j-1)/2}\theta_j \quad \text{and} \quad H(\theta_{D-1})=Q(\theta_{D-1})/\sin^{(D-2)/2}\theta_{D-1} \quad (52)$$

By inserting equation (52) into equations (50) and (51) we obtain

$$\left[d^2Q(\theta_j)/d\theta_j^2\right]+\left[\ell_j(\ell_j+j-1)+((j-1)/2)^2-\ell_{j-1}(\ell_{j-1}+j-2)+[(j-1)/2][(j-3)/2]/\sin^2\theta_j\right]Q(\theta_j)=0 \quad (53)$$

and

$$d^2Q(\theta_{D-1})/d\theta_{D-1}^2+\left[\lambda_\ell+[(D-2)/2]^2-\frac{\left(L_{D-2}^2\right)+[(D-2)/2][(D-4)/2]}{\sin^2\theta_{D-1}}-(E+M)\left(\frac{p(p+1)}{\sin^2\theta_{D-1}}-2s\cot\theta_{D-1}\right)\right]Q(\theta_{D-1})=0 \quad (54)$$

By setting the parameters in equation (53) as

$$\ell_j(\ell_j+j-1)+((j-1)/2)^2=\varepsilon\,;\ \ell_{j-1}(\ell_{j-1}+j-2)+[(j-1)/2][(j-3)/2]=o(o+1) \quad (55)$$

then we have

$$\left[d^2Q(\theta_j)/d\theta_j^2\right]+\left[\varepsilon-o(o+1)/\sin^2\theta_j\right]Q(\theta_j)=0; \quad (56)$$

where ε is assumed to stand for the energy.The hypotheticalsuperpotential for equation (56) is set as

$$\phi(\theta_j)=I\cot\theta_j \quad (57)$$

By using equations (9), (56), and (57) we get the values

$$I=-(o+1) \text{ and } \varepsilon_0=(o+1)^2=\left(-1/2+\left(\ell_{j-1}+(j-2)/2\right)\right)^2 \quad (58)$$

The superpotential,super-partner potentials, and raising-lowering operators obtained from equations (5), (6),(12), (13), and (57) are

$$V_-(a_0,\theta)=o(o+1)/\sin^2\theta-(o+1)^2\,;V_+(a_0,\theta)=(o+2)(o+1)/\sin^2\theta-(o+1)^2 \quad (59)$$

$$A=d/d\theta_j-(o+1)\cot\theta_j;\ A^+=-d/d\theta_j-(o+1)\cot\theta_j \quad (60)$$

By shifting $o\to o+1$ in equation (59) and by applying equations (10-13) with equations (59-60) we obtain

$$\varepsilon=\ell_j(\ell_j+j-1)+((j-1)/2)^2=(o+n+1)^2\to\ell_j=(-1+\ell_{j-1}+n) \quad (61)$$

The ground state and first excited state wave functions are obtained by using equations (14-15), and (60) as

$$Q_0(\theta_j)=C\left(\sin\theta_j\right)^{o+1};Q_1(\theta_j)=-(o+1)\cos\theta_j)\left(\sin\theta_j\right)^{o} \quad (62)$$

The angular wave function for the highest level of angular component is obtained from the solution of equation (54) by setting

$$\left(L_{D-2}^2\right)+[(D-2)/2][(D-4)/2]+(E+M)p(p+1)=p'(p'+1) \quad (63)$$

$$(E+M)s=s'\,;E'=\left\{\lambda_\ell+[(D-2)/2]^2\right\} \quad (64)$$

such that equation (54) become

$$d^2Q(\theta_{D-1})/d\theta_{D-1}^2-\left[p'(p'+1)/\sin^2\theta_{D-1}-2s'\cot\theta_{D-1}\right]Q(\theta_{D-1})=-E'Q(\theta_{D-1}) \quad (65)$$

By setting the hypothetical superpotential for equation (65) as

$$\phi(\theta_{D-1}) = N\cot\theta_{D-1} + B/N \tag{66}$$

and by using equations (9) and (65-66) we have

$$N = -1/2 + (p'+1/2) = p'\,;\; -s' = B\,;\, p'^2 - s'^2/p'^2 = E'_0 \tag{67}$$

By combining equations (5-7) and (65-67) we have the super-potential, super partner potential and SUSY operator as given as

$$\varphi(\theta_{D-1}) = p'\cot\theta_{D-1} - s'/p' \tag{68}$$

$$V_-(\theta_{D-1},a_0) = \frac{p'(p'+1)}{\sin^2\theta_{D-1}} - 2s'\cot\theta_{D-1} - p'^2 + \frac{s'^2}{p'^2}\,;\; V_+(\theta,a_0) = \frac{p'(p'-1)}{\sin^2\theta_{D-1}} - 2s'\cot\theta_{D-1} - p'^2 + \frac{s'^2}{p'^2} \tag{69}$$

$$A = d/d\theta_{D-1} + p'\cot\theta_{D-1} - s'/p'\;;\; A^+ = -d/d\theta_{D-1} + p'\cot\theta_{D-1} - s'/p' \tag{70}$$

By shifting the parameter $p' \to p'-1$ in equation (69) and by applying equation (10-13) with equations (68-69) we obtain relativistic energy equation

$$E'_n = (p'-n)^2 - s'^2/(p'-n)^2 \tag{71}$$

with p' in equation (63).

By using equations (63-64), and (71) we get orbital quantum number equation given as

$$\lambda_l = -[(D-2)/2]^2 + (p'-n)^2 - s'^2/(p'-n)^2 \tag{72}$$

By using equations (8), (17), (18) and (89) we get the relativistic angular wavefunctions for ground state and first exited state which are obtained using equations (14-15) and (70) are ,

$$Q_0^{(-)} = (\sin\theta_{D-1})^{-p'} e^{\frac{s'}{p'}\theta_{D-1}}\;;\quad Q_1 = 2(p'\cot\theta_{D-1} - s'/p')(\sin\theta_{D-1})^{-p'} e^{\frac{s'}{p'}\theta_{D-1}} \tag{73}$$

The total relativistic ground state and first excited state wave function for any D dimension is obtained by combining equations (49), (62), and (73).

3.3. Thermodynamical Properties

In non-relativistic condition, the relativistic energy equation expressed in equation (47) reduces into non-relativstic energy by taking $(M+E) \to 2\mu$ where μ is the non-relativistic mass, $(E-M) \to E_{NR}$, E_{NR} is the non-relativistic energy, and if d_0 is small then equation (47)

$$E_{nr} = -(t^2/2\mu)\left(\left(-\sqrt{c_{nr}(c_{nr}-1)/q+1/4}\right) + \left(\sqrt{b_{nr}'(b_{nr}'+1)/q+1/4}\right) - 1 - 2n\right)^2 \tag{74}$$

with $(a(a-1))2\mu + \lambda_l + ((D-1)/2)((D-3)/2) = c_{nr}(c_{nr}-1)$; $b(b+1)2\mu) = b_{nr}'(b_{nr}'+1)$ (75)

In classical regimes [43], the vibrational partition function, vibrational mean energy, and specific heat are obtained from the non-relativistic energy equation in equation (74).The vibrational partition function is defined as

$$Z(\varsigma,\beta) = \sum_{n=0}^{\varsigma} e^{-\beta E_{nl}}, \beta = \frac{1}{kT} \tag{76}$$

k is Boltzman constant, *Enr* is non-relativistic energy spectrum of the system. When the temperature, *T*, is high enough, then the value of ς is high , β is small, and if $1/\delta^2 = 2t^2\beta/\mu$ then equation (76) becomes

$$Z(\xi,\beta) = \sum_{n=0}^{\frac{\xi}{\delta}} e^{y^2} = \delta\int^{\frac{\xi}{\delta}} e^{y^2}\,dy = \delta\frac{\sqrt{\pi}}{2} erfi\left(\frac{\xi}{\delta}\right) \tag{77}$$

where $E_{NR} = -(2t^2/\mu)(\xi-n)^2$; $\xi = 1/2\left\{\left(-\sqrt{c_{nr}(c_{nr}-1)/q+1/4}\right) + \left(\sqrt{b_{nr}'(b_{nr}'+1)/q+1/4}\right) - 1\right\}$ (78)

with $y=(n-\xi)/\delta$ and *erfi* is the imaginary error function [44].

The vibrational mean energy and the vibrational specific heat are defined as

$$U(\beta,\xi)=-(\partial/\partial\beta)\ln Z(\xi,\beta);C=-(\partial/\partial T)U=-k\beta^2(\partial/\partial\beta)U \tag{79}$$

By using equations (77-79) we obtain the vibrational mean energy and specific heat equation given as

$$U(\beta,\xi)=(-1/\beta)\left((-1/2)+\left((1/\sqrt{\pi})\left(\xi t\sqrt{2\beta}/\sqrt{\mu}\right)\exp\left(-\xi^2t^2 2\beta/\mu\right)\right)/erf\left(\xi t\sqrt{2\beta}/\sqrt{\mu}\right)\right) \tag{80}$$

and

$$C(\xi,\beta)=k\left(\frac{1}{2}-\frac{\left(\frac{1}{\sqrt{\pi}}\frac{\xi t\sqrt{2\beta}}{\sqrt{\mu}}\exp\left(-\frac{\xi^2t^2 2\beta}{\mu}\right)\right)}{erf\left(\frac{\xi t\sqrt{2\beta}}{\sqrt{\mu}}\right)}\left(\frac{1}{2}+\frac{\xi^2t^2 2\beta}{\mu}\right)-\frac{\left(\frac{\xi^2t^2 2\beta}{2\sqrt{\pi}\mu}\exp\left(-2\frac{\xi^2t^2 2\beta}{\mu}\right)\right)}{\left(erf\left(\frac{\xi t\sqrt{2\beta}}{\sqrt{\mu}}\right)\right)^2}\right) \tag{82}$$

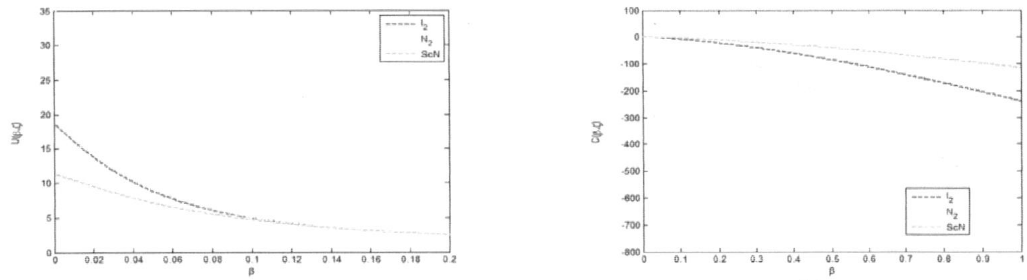

(a) (b)

Figure 3.Graph of (a) mean energy U as a function of β ,(b) specific heat as a function of β (for $D=4, n=5, l=4, a=6, b=3, p=4, s=2, q=0.75$)

The graphs of mean energy and specific heat as a function of β are shown in figure 3 (a and b). From figure 3 (a) and (b) we see that for larger values of β the values of vibrational mean energy and specific heat are constant. The specific heat for system whose is governed by q-deformed modified Poschl-Teller plus trigonometric Manning-Rosen non-central potential are negatives. The negative specific heat may occur for the astronomical objects [45].

4. Conclusion

The Dirac equation in D dimensions of q-deformed modified Poschl-Teller potential plus trigonometric Manning-Rosen non-central potential is solved using SUSY QM. The radial part of D-dimensions of the Dirac equation reduces to one dimensional Schrodinger type equation in centrifugal approximation scheme. There are two solutions of angular Dirac equations, the first one is for hyper-spherical harmonic and the other one is for angular potential functions.

In the exact spin symmetric case, the relativistic energy equation reduces to the non-relativistic energy in the non-relativistic condition. In the classical regime , some thermodynamics properties are derived from the non-relativistic energy equation. The mean energy and specific heat are numerically calculated from non-relativistic energy equation by using Mat-Lab.

Acknowledgement

This research was partly supported by Higher Education Project Grant with contract no. 698/UN27/PN/2015.

References

[1] Hamzavi M and Rajabi A A 2013 *Adv. High En. Phys.* **2013**1-12 ID 196986.
[2] Kurniawan A, Suparmi A and Cari C 2015*Chin. Phys. B* **24** (3) 030302
[3] Hassanabadi H, Maghsoodi E and Zarrin kamar S 2012 *Commun. Theor. Phys.* **58** 807–814

[4] Ikhdair S M and Sever R 2008 *Int. J. Mod. Phys. C***19**(9)1425-1442
[5] Zhou F, Wu Y and Guo J Y 2009*Commun. Theor. Phys.* (Beijing, China)**52**813-816
[6] Deta U A, Suparmi, Cari, Husein A S, Yuliani H, Khaled IK A, Luqman H,and Supriyanto 2014 *AIP Conf. Proc.***1615**, 121 doi: 10.1063/1.4895872
[7] Sukumar C V 1985 *J. Phys. A: Math. Gen.* **18** p. L697
[8] Chen G, Chen Z D and Lou Z M 2004 *Chin. Phys.* **13**(2) 279-284
[9] Cari C and Suparmi A 2014*AIP Conference Proceedings***1615** 101 doi: 10.1063/1.4895870
[10] Suparmi and Cari 2014*J. Math. Fund. Sci.***46** 3205
[11] Onate C A, Oyewumi K J and Falaye B J 2013 *Af. Rev.Phys.* **8**:0020129-137
[12] Suparmi A and Cari C 2013 *At. Ind. J.***39**(3)112-123
[13] Cari C, Suparmi A, Deta U A and Werdiningsih I S 2013 *Makara. J. of Sci.* **17**(3) 93-102
[14] Suparmi A, Cari C and Deta U A 2014 *Chin. Phys. B*, **23**(9)090304-1-10
[15] Suparmi A, Cari C, Deta U A , Husein A S and Yuliani H 2014 *J.Phys.Conf. Ser.***539**012004
[16] Suparmi A, Cari C and Angraini L M 2014 *AIP Conf. Proc.* 1615**111** doi: 10.1063/1.4895871
[17] Ikhdair S M and Server R 2008 Int. *J. Mod. Phys.* C**19**(9)1425-1442
[18] Ikot A N and A k pabio L E 2010 *App. Phys. Res***2**(2) 202-208
[19] Agboola D 2011 *Pramana***76**(6)875–885
[20] Greene R I and Aldrich C 1976 *Phys. Rev. A*, **14** 2363
[21] Ikhdair S M and Sever R 2008 arXiv:0801.4857v3 [quant-ph] 14 Feb 2008
[22] Oyewumi K J 2003 *The Arab. J. for Sci. and Eng.***28** No. 2A 173
[23] Oyewumi K J, Akin pelu F O and AgboolaA D 2008 *Int. J. Theor. Phys.* **47** 1039
[24] Oyewumi K J 2005 *Found. Phys. Lett.* **18** No. 1 75
[25] Agboola D 2011 *Acta Phys. Pol. A***120** No. 3, 371
[26] Al-Jaber S 1998*Int. J. Theor. Phys.* **37**(4) 1289
[27] Agboola D 2009 arXiv:0811.3613v3 [math-ph] 5 Nov 2009
[28] Hassanabadi H, Zarrin kamar S and Rajabi A A 2011 *Commun. Theor. Phys.* **55**(4) 541–544
[29] Falaye BJ and Oyewumi K J 2011 *Af. Rev. Phys.* **6**:0025 211
[30] Ikhdair S M and Sever R 2007 †arXiv:0704.0489v1 [quant-ph] 4 Apr 2007
[31] Derezinski J and Wrochna M 2010 arXiv:1009.0541v2 [*math-ph*]
[32] Witten E 1981 *Nucl. Phys. B*, **188** 513-554
[33] Gendenshtein L E 1983 *JETP Lett.* **38** 356-359
[34] Ikhdair S M, Falaye B J and Ade poju A G2013 arXiv:1308.0155v1 [quant-ph] 1 Aug 2013
[35] Dong S H, Lozada-Casson M, YuJ, Jimenez-Angles Fand RiveraAL 2007*Int. J. Quant. Chem.* **107** 366.
[36] Khare A and Bhaduri R K 1993 arXiv:hep-th/9310104v1 1-16, 1993.
[37] Dutt R, Khare A and Sukhatme U P 1998 *Am. J. Phys.* **56** 163-168
[38] Hu X Q, Luo G, Wu Z M, Niu L Band MaY 2010*Commun. Theor. Phys.* **53** 242-249
[39] Falaye B J and Oyewumi K J 2011 *Af. Rev. Phys.* 6:0025 211
[40] Ikhdair S Mand SeverR 2008 arXiv:0801.4857v3 [quant-ph] 14 Feb 2008
[41] Dong, S-H 2011 *Wave equations in Higher Dimensions*Springer Dordrecht Heidelberg London New York
[42] Cari 2013*Quantum Mechanics-Solutions of Non-Central Potentials Using SUSY Quantum Mechanics, Hypergeometry Method, Nikiforov-Uvarov Method, and Romanovski Polynomial*, p. 240, Surakarta UNS Press
[43] Ikhdair S M 2013 arXiv:1308.0155v1 [quant-ph] 1 Aug 2013
[44] Abramowit z M and Astegun I A 1972 (Eds.) *Handbook of Mathematical Functions with Formulas,Graphs, and Mathematical Tables*, 9th edition. New York: Dover, pp. 295 and 297
[45] Lynden-Bell D 1998 arXiv:cond-mat/9812172v1[cond-mat.stat-mech]10 Dec 1998

doi:10.1088/1742-6596/710/1/012027

Fabrication and Characterization of *Sansevieria trifasciata*, *Pandanus amaryllifolius* and *Cassia angustifolia* as Photosensitizer for Dye Sensitized Solar Cells

Cari[a)], Agus Supriyanto[b)], Ulfa Mahfudli Fadli[c)], Ashari Bayu Prasada[d)]

[1]Physics Department Faculty of Mathematics and Science, Sebelas Maret University, St. Ir. Sutami 36A Kentingan Surakarta 57126, Indonesia

[a]carinln@yahoo.com, [b]*agusf22@yahoo.com, [c]oasis.190712@gmail.com, [d]asharidulhasyim@yahoo.co.id

Abstract: Dye sensitized Solar Cells (DSSC) is one of the electric cells photochemical consisting of photoelectrode, dye, counter electrode, and electrolyte. The aims of the research to determine of the optical and electrical characteristic of the extract *Sansevieria trifasciata*, *Pandanus amaryllifolius*, and *Cassia angustifolia*. The study is also aimed to determine the effect of natural dyes extract to increase the efficiency of solar cells based DSSC. Sandwich structures formed in the sample consisted of working electrode pair Titanium dioxide (TiO_2) and the counter electrode platinum (Pt). Dye extraction process is performed by stirring for 1 hour and then allowed to stand for 24 hours. Absorbance test is measure by using UV-Vis spectrophotometer Lambda 25, conductivity test by using a two-point probes Elkahfi 100, and characterization of current and voltage (I-V) by using a Keithley 2602A. The results showed that the greatest efficiency of 0.160% at Dye *Pandanus amaryllifolius*.

Keywords: **Dye Sensitized Solar Cells, TiO_2, Photosensitizer, Natural Dyes.**

1. Introduction

Solar cells based on current technological developments and manufacturing materials can be divided into three: first, solar cells made from silicon single and multi-crystalline silicon, second, the type of thin-layer solar cells, and third, organic solar cells (DSSC). Conventional solar cells in the form of a p-n junction made of a semiconductor material such as silicon, is still expensive to develop because it uses sophisticated technology. Until invented by Gratzel that organic solar cells, DSSC as a solar cell with a dye sensitizer of organic materials can be developed at low cost and low fabrication [1].

DSSC is different from a commercial silicon-based solar cells, in which the solar cells DSSC is a separation between the functions of light absorption to transport cargo carrier [2]. DSSC advantages compared to commercial silicon-based solar cells including a cheap, easy to manufacture, and has a high efficiency even at low light intensity [3]. DSSC is a photoelectrochemical solar cells that use an electrolyte as charge transport medium. DSSC is divided into sections consisting of nanopori TiO_2, dye molecules are adsorbed on the surface of TiO_2 and catalysts which are all deposited between two conductive glass. DSSC structure looks like Figure 1.

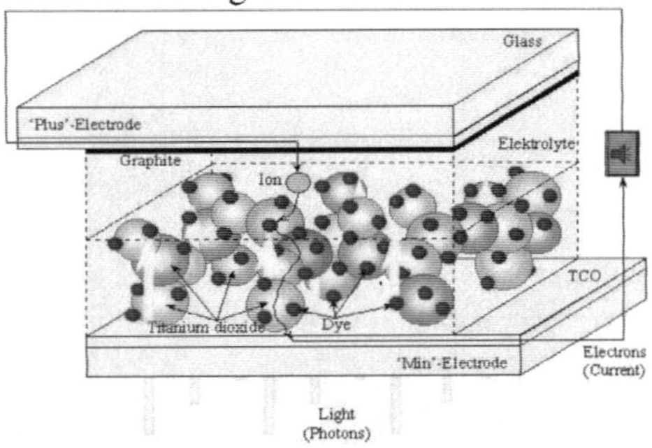

Figure 1. Structure of Dye-Sensitized Solar Cells (DSSC) [3]

Basically the working principle of DSSC convert light into electrical energy in a molecular scale in the form of electron transfer reactions. The first process starts with the excitation of electrons in the dye due to the absorption of energy photons. Where it is one of the roles of the nature of TiO_2. When photons of sunlight impinge on the working electrode DSSC, the photon energy is absorbed by the electron dye attached to the surface of TiO_2. Thus get energy for the excited electron. Photoexcitation of the dye results in the injection of electron into the conduction band of TiO_2. TiO_2 acts as an electron acceptor and dye acts as an electron donor. Furthermore, the electrons are transferred past the outer circuit to the reference electrode of platinum layer. Electrolytes containing I^-/I_3^- which acts as an electron mediator so that it can produce in the cell cycle process. Triodide ions capture electrons originating from outside the circuit with the help of platinum molecules as catalyst. Excited electrons back into the cell and aided by the platinum that can react with the electrolyte which causes the addition of iodide ions on the electrons. Then one of iodide ions in the electrolyte to deliver electrons that carry the energy to the oxidized dye. Electrolytes provide replacement electrons to oxidized dye molecules. So that the dye back to its initial state. As shown in Figure 2.

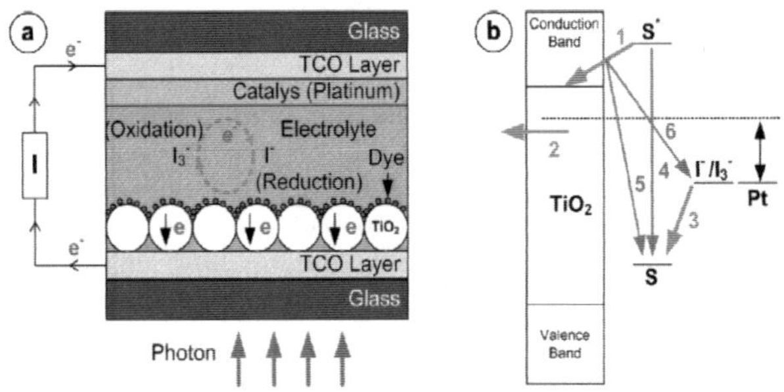

Figure 2. (a) The working principle and (b) diagrams of solar cell energy [5]

DSSC is inseparable from Dye, in this case the color has an important role as an absorber of sunlight and convert it into electrical energy. In the studies that have been conducted, the ruthenium complex dye of the compound can achieve efficiencies of 11-12% [6]. However, the amount of ruthenium complex dye is limited and the price is quite expensive. For these reasons, developing research towards finding natural dyes extracted from flowers, leaves, and fruits [7]. Natural dye that is used as a sensitizer in DSSC classified as environmentally friendly, making it too easy and cheap despite its low lifetime [8]. Several studies using material from platinum (Pt) as the counter electrode in the DSSC.

In this study, used platinum (Pt) as the counter electrode in DSSC made from organic extracts of *Sansevieria trifasciata*, *Pandanus amaryllifolius*, and *Cassia angustifolia*.

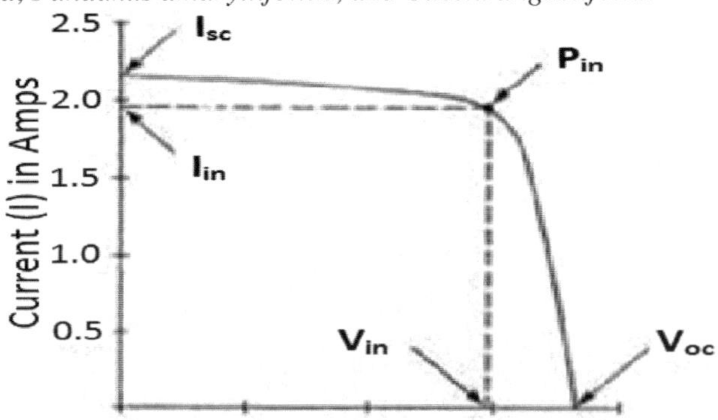

Figure 3. Graph Characteristics I-V DSSC [4]

Performance solar cells is the ability to convert light into electrical energy. Performance solar cells is measured using Keithley 2602A to find the value of short circuit current (I_{sc}) and open circuit voltage (V_{oc}). Figure 3. shows I_{sc} vs V_{oc}. The value of I_{sc} is determined based on the data in which the voltage value starts from a negative to a positive. Simillary for the value of V_{oc} is determined by the current value starts from a negative to a positive value. Then the value maximum voltage (V_{max}) and maximum current (I_{max}) are obtained by multiplication of voltage and current to find the maximum value of power. Performance solar cell is determined by comparing maximum power and multiplication result V_{oc} and I_{sc}.

Fill factor (FF) is a measure of the quality of the performance of the solar cells [9], as well as the size of the square area of $I_{sc} - V_{oc}$ curve. The value of FF can be obtained using equation (1):

$$FF = \frac{V_{maks} \times I_{maks}}{V_{oc} \times I_{sc}} \qquad (1)$$

The maximum power (P_{max}) produced by the solar cell can be obtained through the equation (2):

$$P_{maks} = V_{oc} \times I_{sc} \times FF \qquad (2)$$

Efficiency (η) produced by solar cells can be obtained through the equation (3):

$$\eta = \frac{P_{maks}}{P_{in}} = \frac{V_{oc} \times I_{sc} \times FF}{I(t) \times A_c} \qquad (3)$$

From equation (3), the solar cell efficiency is a quantity of the ratio between the maximum power generated cells (P_{maks}) and the power of the incoming light (P_{in}).

2. Experiment

2.1. Preparation of TiO_2 Solution.

TiO_2 nano powders as much as 0.5 grams dissolved in 2 ml of ethanol, then stirred using a vortex stirrer with a speed of 200-300 rpm for 30 minutes. Pasta TiO_2 that has been formed is inserted into the bottle covered with aluminum foil and stored in an area free from direct sunlight to reduce evaporation process.

2.2. Preparation of Natural Dye Sensitizers.

Sansevieria trifasciata, *Pandanus amaryllifolius*, and *Cassia angustifolia* using digital scales weight 10 grams. Furthermore, these materials crushed and pulverized using a mortar. The materials that have been refined dissolved in 40 ml of ethanol, then stirred for 60 minutes using a vortex stirrer with a rotational speed of 300 rpm at 60°C. Once the material is dissolved, allowed to stand for 24 hours and filtered with a filter paper quality no. 42. Extraction result is stored in closed containers and protected from the sun.

2.3. Preparation of Electrolyte

Potassium iodide (KI) 0.5 M of 0.8 grams in the form of solids are mixed into 10 ml polyethylene glycol then stirred. Further to the solution is added iodine (I_2) 0.05 M of 0.127 grams was stirred with a vortex stirrer at 300 rpm for 30 minutes. Electrolytes solution that is so, is stored in closed containers lined with aluminum foil.

2.4. Preparation of Counter Electrode

The counter electrode in the form of FTO conductive glass that has been coated with a thin layer of platinum (Hexachloroplatinic (IV) acid 10%). The step of making the counter electrode is 1 ml Hexachloroplatinic (IV) acid 10% was mixed with 207 ml of isopropanol was stirred using a vortex stirrer at 300 rpm for 30 minutes. FTO glass is heated using a hotplate at a temperature at 250^0C for 15 minutes and then a few drops of a solution of 3 ml of platinum onto the surface of the glass substrate FTO with drip method. Glass that has been spilled platinum is then cooled at room temperature.

2.5. Preparation of TiO_2 Electrode.

The working electrode is made of glass on which the FTO conductive paste deposited TiO_2 nano with spin coating techniques. At the FTO glass with size 2 x 2 cm formed for area deposition of TiO_2 at 1 x 1 cm on the conductive surface. FTO side masking tape affixed as a barrier. Pasta TiO_2 dripped on FTO glass that has been glued in the spinner, then stirring with a speed of 200-300 rpm. FTO glass that has been coated with TiO_2 is heated using a hotplate at temperature of 500^0C for 60 minutes, then cooled at room temperature.

2.6. Assembly of Sandwich DSSC

The composition of the DSSC in the form of glass FTO that has been coated with TiO_2 and has been soaked in a dye solution is called the working electrode. The working electrode is etched with an electrolyte solution and then covered with glass that has been coated with platinum, called the counter electrode. Then the composition of DSSC is clamped with a clamp on both sides of the right and left before a measuring to get a data.

3. Result and Discussion

3.1. Absorption Spectra of extract natural dye

Before being used as a sensitizer, extract *Sansevieria trifasciata*, *Pandanus amaryllifolius* and *Cassia angustifolia*, first characterized using UV-Vis Spectrophotometer. Absorbance spectra measured in the wavelength range of 200-800 nm.

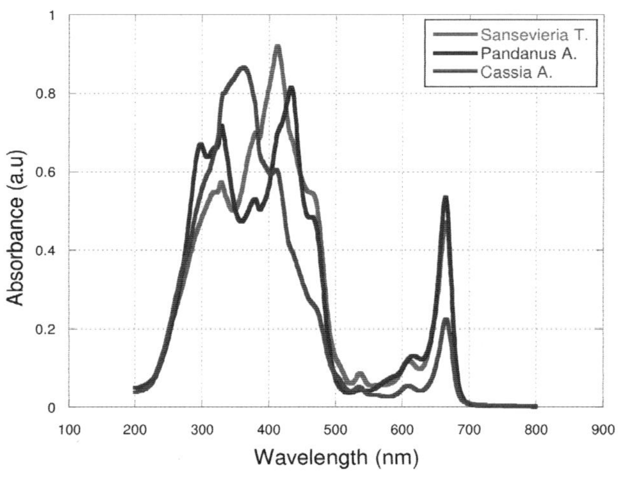

Figure 4. Absorption Spectrum of extract *Sansevieria trifasciata*, *Pandanus amaryllifolius*, and *Cassia angustifolia*

Figure 4 shows a maximum absorption around 300-670 nm, absorbance maximum extract *Sansevieria trifasciata* at 0.919 and 0.471 a.u with wavelength between 412 nm and 665 nm. Extract of *Pandanus amaryllifolius* at 0.815 and 0.535 a.u with wavelength between 433 nm and 664 nm, and extract *Cassia angustifolia* is at 0.865 and 0.224 a.u with wavelength between 362 nm and 666 nm. Absorbance dye of the three materials showed that the value of wavelength of these material are able to work on the visible light. From Figure 4, it can be concluded that the third material of organic compounds can be used for DSSC.

3.2. Conductivity of extract natural dye

Further characterization is conductivity testing of dye using Elkahfi 100/IV-Meter under irradiation halogen lamp 100 mW/cm^2 and the energy intensity of 680.3 W/m^2.

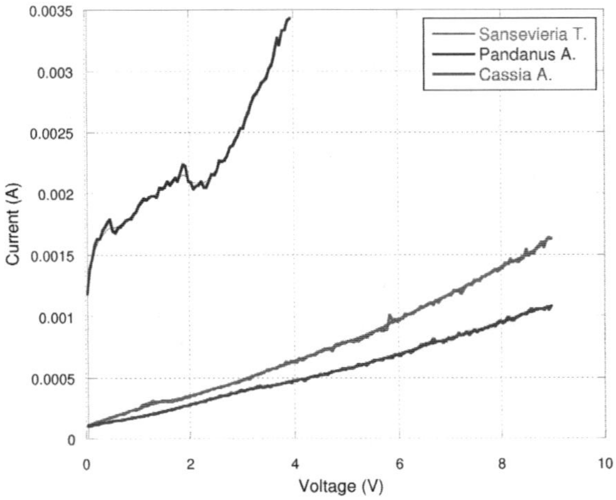

Figure 5. Conductivity Curve for *Sansevieria trifasciata*, *Pandanus amaryllifolius* and *Cassia angustifolia* extract sensitized solar cell

Figure 5 shows the conductivity value of the *Sansevieria trifasciata* is 8.34 x 10^{-3} $\Omega^{-1}m^{-1}$, *Pandanus amaryllifolius* is 9.73 x 10^{-2} $\Omega^{-1}m^{-1}$, and *Cassia angustifolia* is 6.39 x 10^{-3} $\Omega^{-1}m^{-1}$.

3.3. Photovoltaic Properties

Characterization of the current-voltage (I-V) is a method to determine how much ability DSSC can convert light into electrical energy. Measurements using Keithley 2602A conducted in the dark and the light that is under irradiation with a halogen light intensity of 1000 W/m^2. The test results I-V with a variety of immersion during 12 hours and 24 hours.

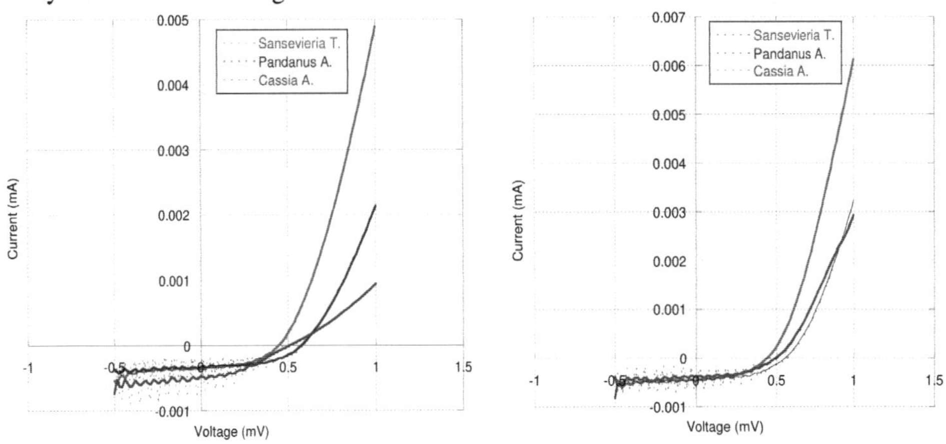

(a) (b)

Figure 6. Current-Voltage curve for; (a) DSSC natural dye within 12-hours immersion, (b) DSSC natural dye within 24-hours immersion

Based on the graph in Figure 6, it can be shown the area of *Pandanus amaryllifolius* dye the largest, because the value of V_{oc} intercept at 5,50 x 10^{-1} mV and the value I_{sc} intercept at 4,79 x 10^{-4} mA. Therefore, it has the largest value of efficiency. So based on the calculations of equation (1-3) have been carried out, The results of the efficiency of the extract *Sansevieria trifasciata*, *Pandanus amaryllifolius* and *Cassia angustifolia* by 12 hours and 24 hours immersion, are shown in Table 1.

Table 1. Photovoltaic performance with natural dyes from different sources

Dye Source	Isc (mA)	Voc (mV)	FF	immersion	η (%)
Sansevieria trifasciata	4,25 x 10^{-4}	4,60 x 10^{-1}	1,07 x 10^{-7}	12 hours	0,116
	4,25 x 10^{-4}	4,75 x 10^{-1}	8,83 x 10^{-8}	24 hours	0,098
Pandanus amaryllifolius	4,79 x 10^{-4}	5,50 x 10^{-1}	1,39 x 10^{-7}	12 hours	0,160
	1,74 x 10^{-4}	5,95 x 10^{-1}	3,80 x 10^{-8}	24 hours	0,130
Cassia angustifolia	3,74 x 10^{-4}	5,20 x 10^{-1}	8,80 x 10^{-8}	12 hours	0,122
	3,26 x 10^{-4}	5,05 x 10^{-1}	7,70 x 10^{-8}	24 hours	0,119

The value of efficiency is calculated by using equation 3, see Table 1. From Table 1 and figure 6, it can be concluded that the DSSC that produces the best performance is dye of *Pandanus amaryllifolius* with efficiency value is 0.160% in 12-hour immersion. In general, the value efficiency of the third material of organic compounds is still very small (under 1%). There are two kind of efforts to increase the performance DSSC. The first, using different counter electrode with polyaniline (PANI), because PANI can increase efficiency from 6,90% using platinum to 7,15% using PANI [10]. The second, using the kind of electrolyte which has viscosity is smaller than iodine electrolyte such as PEO polymer gel [11].

4. Summary

From the results of this study concluded that the solar cell DSSC has been successfully created by using dye *Sansevieria trifasciata*, *Pandanus amaryllifolius* and *Cassia angustifolia*. The results of the absorbance dye of the three materials showed that the value of wavelength these material are able to work on the visible light. The largest conductivity value is *Pandanus amaryllifolius* dye, whereas the greatest efficiency in achieved by using the extract samples *Pandanus amaryllifolius* on 12 hour immersion in the amount of 0.160%.

Acknowledgement

This research was partly supported by LPPM Postgraduate UNS Grant with contract no. 698/UN27/PN/2015 May 11th, 2015.

References

[1] V. Ekasari, G. Yudoyono, Fabrikasi DSSC dengan Dye Ekstrak Jahe Merah (*Zingiber Officinale Linn Var. Rubrum*) Variasi Larutan TiO_2 Nanopartikel Berfase Anatase dengan Teknik Pelapisan *Spin Coating*, vol. 2, no. 1, 2013, pp. B-15-B-20.
[2] A. Hagfeldt, B. Didriksson, T. Palmqwist, H. Lindstrom and S. Lindsquist, Solar Energy Matter and Solar Cells Verification of high efficiencies for the Gratzel-cell a 7% efficient solar cell based on dye-sensitized colloidal TiO2 films, vol. 5, 1994, pp. 481-488.
[3] M. Akiyama, E. Abe and I. Imai, High efficiency dye-sensitized solar cell based on a nitrogen doped nanostructured titania electrode, vol. 5, 2005, pp. 2543-2547.
[4] Alfa, B. *et al.*, *Fabrication and Characterisation of Titanium Dioxide Based Dye Sensitized Solar Cell using Flame of the Forest Dye*, Applied Physics Research, vol.4, no.1, 2012, pp 48-56.
[5] Xu, F., and Sun, L. 2011. *Solution-Derived ZnO Nanostructures for Photoanodes of Dye-Sensitized Solar Cells*, Energy Environ. Sci. 4, 818.
[6] S. Agustini, D.D. Risanti dan D. Sawitri, Fabrikasi *Dye Sensitized Solar Cell* (DSSC) Bersadarkan Fraksi Volume TiO_2 Anatase-Rutile dengan *Garnicia mangostana* dan *Rhoeo Spathacea* sebagai *Dye Fotosensitizer*, vol. 2, no. 2, 2013, pp. B-131-B-136.
[7] H. Zhou, W. Liqiong, G. Yurong and M. Tingli, Dye-Sensitized Solar Cells Using 20 Natural Dyes As Sensitizers, *Journal of Photochemistry and Photobiology A Chemistry,* 2011, pp. 188-194.
[8] J.K. Lee and Y. Mengjin, Progress In Light Harvesting and Charge Injection of Dye-Sensitized Solar Cells, *Material Science and Engineering* B 176, 2011, pp. 1142-1160.
[9] Cari, Nurussaniah, Biosandi, Anita, A. Supriyanto dan R. Suryana, Studi Penggunaan Poly(3-Hexylthiophene) P3HT dan Grafit Terhadap Kinerja Solar Cells, vol. 3, no. 1, 2013.
[10] Qinghua Li. Jihua Wu. Qunwei Tang. Zhang Lan. Pinjiang Li. Jianming Lin. Leqing Fan, *Application of Microporous Polyaniline Counter Electrode for Dye Sensitized Sollar Cells*, Electrochemistry Communications, Elseiver, 2008.
[11] Duangkaew, P., Chindaduang, A., Tumcharem, Efficiency and Stability Enhanchement of Dye Sensitized Solar Cell using PEO Polymer Gel and Imidazolium-Based Ionic Liquid Electrolyte, Science Jurnal Ubonratchathani University. Vol 1. No 1 (January-June), 2010, pp. 9-14.

IMPACT OF FOREIGN DIRECT INVESTMENT ON THE UNEMPLOYMENT RATE IN MALAYSIA

Hamidah Muhd Irpan[1], Rosfadzimi Mat Saad[1], Abu Hassan Shaari Md Nor[2], Abd Halim Md Noor[1], and Noorazilah Ibrahim[1]

[1]Universiti Teknologi MARA, Km 26, Jalan Lendu, 78000 Alor Gajah, Melaka,
[2]Universiti Kebangsaan Malaysia, 43600 UKM, Bangi, Selangor

Email: hamidahirpan@melaka.uitm.edu.my

Abstract. Malaysia as a developing country needs support from other countries for economic growth. This is done by receiving massive foreign direct investment (FDI) which contributes to a higher employment rate. Higher employment leads to a better living among Malaysians while increasing its gross domestic product (GDP). During 2009, Malaysia faced a downward trend on the FDI. In many studies, decreasing FDI affects employment rate significantly. This study focuses on the impact of FDI on employment rate in Malaysia. Other factors such as the number of foreign workers, gross domestic product (GDP) and exchange rate (EXCR) are also included in the study. Data used in the study is annual data spanning from 1980 to 2012. Autoregressive distributed lag (ARDL) model is used to determine the long run relationship between the variables. The study finds that FDI, number of foreign workers, and GDP significantly influence the unemployment rate in Malaysia.

1. Introduction

According to the global report published by International Labour Organization (ILO), Malaysia was ranked as the 20th country with the lowest unemployment rate of 3.1% in 2013 in the world. The highest unemployment rate in 2013 is in Mauritania (30.8%), followed by Macedonia (29.8%), Bosnia (29.2%), Greece (28.1%) and Spain (27.2%). A large increase on unemployment rate occurred in Greece and Spain compared to the lowest rate of 7.7% and 8.4% since 2007 respectively. In 2012, about 396,300 Malaysian were unemployed which counted 3.0% of unemployment rate.

Based on the world statistics of unemployment rate, it is possible for a country to achieve lower unemployment rate of less than 1% such as Cambodia (0.3%), Qatar (0.5%), Rwanda (0.6%), and Thailand (0.7%). Researchers in Malaysia has studied the factors leading to high unemployment rate and one of the strategies was to attract foreign investors to invest in Malaysia so that they can bring in new technology and expertise. Based on past studies, inward foreign direct investment (IFDI) is found to be important in helping to decrease the unemployment rate in Malaysia. It can be seen in Figure 1.1 where the fluctuation of employment rate in Malaysia after 1986 reduced significantly with the rapid increment of IFDI and maintain below 4% ever since showing the successfulness of the initiatives done to overcome the unemployment problem.

Other factors which affected the unemployment significantly were exchange rate [13] gross domestic product [8] and foreign labour. Unemployment rate increased with the increasing of exchange rate while unemployment rate decreased with the increasing of gross domestic product. The penetration of foreign labour into the labour market system avoiding the shortage of labour. Somehow, the eccessive of illegal foreign labour will affect the labour system and causing unemployment problem among the local labours for a lower wages compared to minimum wages of RM900 in peninsula and RM800 in Sabah, Sarawak and the Federal Territory of Labuan.

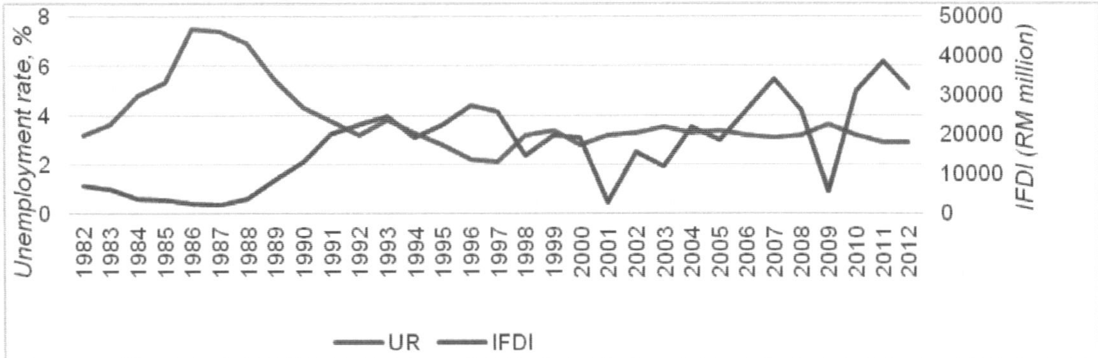

Figure 1.1: Fluctuation of unemployment and inward FDI in Malaysia
Source: Unemployment rate by Department of Statistics Malaysia (DoSM) and inward FDI by United Nations Conference on Trade and Development (UNCTAD)

The rising unemployment rate in 1982 to 1986 is mainly caused by the low level of labour utilization. At this time, unemployment was very high among the young, particularly those in the age of 15 to19 and 20 to 24 age groups which were about three times of the overall unemployment rate. Most of the unemployed were first-time job seekers, hence improvement in education and skill level of the labour force were implemented. The wider usage of new technology and equipment resulted in higher labour productivity and value added as well as employability among them.

Major economic sector in Malaysia was agriculture and shortfall in job creation occurred due to the slower performance, rural labour supply constraints, and low investment in this sector. Shortage of labour during late 1940s, particularly in plantation sector, continued into early 1980s. In addition to that, the world tin market crisis reduced the employment in mining from 80,100 in 1980 to 60,500 in 1985. The tin mining was major activity in mining sector. In response to this, the government acted by converting the mining land for agricultural purposes, constructing feeder roads to support agriculture and constructing low-cost housing units to increase utilization of local material and employment opportunities. In doing so, foreign labours were needed to avoid labour shortage and to maximize the agricultural productivity. These foreign workers provided the needed skills and new technologies.

IFDI helps decreasing the unemployment rate until 1997 when the global financial crisis affected the Malaysian economy resulted in increasing unemployment rate of 2.1 per cent in 1997 to 3.2 per cent in 1998. Malaysia faced negative growth of real GDP (-7.4 per cent) in 1998 and retrenchment soared to 83,865 workers in 1998 as compared to approximately 19,000 in 1997 [5].

Then in 2001 IFDI reached the lowest IFDI (RM 2,762 million) since 1987 (RM2,173 million) compared to the average of RM17,206 million. This is affected by the economic recession and 911 incident of the US World Trade Centre which affect the Malaysian unemployment rate with the lower level of exports to the US and lower IFDI resulted in slow growth in the economy as well as increasing in unemployment rate. In 2009, IFDI once again dropped to RM 5,573.6 million due to global financial crisis.

In general, large amount of IFDI created wider job market for local labour as well as technological assistance and expertise to Malaysia. This financial assistance helps Malaysia to strengthen the economy with the optimal level of employment and local production. Somehow, high IFDI allowing rapid development on the manufacturing industry which caused the shortage of labour that encouraged foreign investor to bring in the foreign labour for the production and also attracting the illegal foreign labour to work in Malaysia, creating the disequilibrium of labour market system due to unreliable data of labour force. Furthermore, the dependency on foreign investment has caused Malaysia to face great impact on the unemployment especially during the economy recession.

Based on previous studies, there were many factors which influenced the unemployment rate. Factors which influenced unemployment are also different between places due to social behaviour and economic development. Many African and Latin American countries face high unemployment rates due to under development of infrastructures and economies [6]. [15] reveals that high unemployment contribute to high crime rate in Malaysia. While according to Okun's Law, unemployment has a negative effect to the output or economic growth. Hence the study of unemployment is very important and need to be revised from time to time in alignment to the development and economic condition of the country.

In this study, relationship and impacts of both inward and outward foreign direct investment (IFDI and OFDI), Gross Domestic Product (GDP), exchange rate (EXCR), and foreign labour (FL) are to be determined. Other factors not included in the study were oil price, capital stock, status of development, rural or urban area, wages, labour utilization, money supply and reduction of demand. Thus, this study aims to firstly identify the long run relationship between IFDI, OFDI, GDP, EXCR and FL on unemployment and then secondly to investigate the short and long run impact of the factors on unemployment in Malaysia. This study might help the policymakers to make a suitable revision and stimulate the employability strategy so that the unemployment rate as well as crime rate can be decreased.

2. Literature review

Both inward foreign direct investment (IFDI) and outward foreign direct investment (OFDI) are important in developing country. Projects funded by foreign investors produced more job market for both skilled and unskilled labours. This is supposed to reduce the unemployment rate [1, 7, 12, 14, 16]. High OFDI is also important as it utilized many labours, both local and foreign, for the production to be exported out.

However, some foreign investors prefer to bring in unskilled foreign labours as they were cheaper in terms of wages. Starting from 2012 Malaysia has implemented the minimum wage gazette of RM900 for workers in Peninsula at the private sector and RM800 for workers in Sabah, Sarawak and Federal Territory of Labuan which attract foreign labours to penetrate Malaysia's job market [4].

As a developing country, foreign direct investment is important to boost our economy. Aligned with that, demands for cheaper, unskilled workers are getting higher from Vietnam, Indonesia, Cambodia and China compared to local labours. This reducing the chances of unskilled local labour to find jobs as the wages offered to foreign labour is cheaper. Even worse foreign investors tend to invest to the neighbouring countries instead of Malaysia and decelerating the growth of the Malaysian economy. This situation may shrink the job market and increase the unemployment rate as the development injected by foreign investors fund provided huge skilled and unskilled job market [4].

Most of the foreign labours in Malaysia are from Indonesia. Study on the push and pull factors on it has been done by [3]. Result showed that low income (GDP) and high unemployment in the home country lead them to move in order to find jobs in Malaysia with higher GDP and low unemployment rate.

Larger FDI produced more job market and thus improving the gross domestic product (GDP). GDP represents the income of a country. Higher income leads to low unemployment rate [7] and stable economy while low income leads to high unemployment rate. During the economic recession between 2008 and 2009, [8] stated that Malaysia GDP growth was -4.23%. Majority of developing countries faced growth deceleration with positive growth. However, some of the developing countries faced positive impact on unemployment rate while the rest faced negative impact. Malaysia faced an increasing on unemployment rate at 11.11%. This situation is not good for social development hence larger domestic economies and having healthier reserves were seen as the strategy to withstand the shock of the recession.

Exchange rate is influenced by economic stability of a country. Bad reputation on economics by decelerating growth or negative growth may cause exchange rate to increase, resulted in the high unemployment. This is supported by [7, 13] in their study which stated that exchange rate positively influences the unemployment rate.

3. Methodology

Data used in the study is annual data from 1982 until 2012 collected from Department of Statistics Malaysia, International Monetory Fund (IMF) and United Nations Conference on Trade and Development (UNCTAD) . Eviews software was used to analyse the data.

Autoregressive Distributed Lags (ARDL) was used for the analysis. It is originally introduced by [10] and further extended by [11]. This model was used as it is applicable if variables are integrated at level I(0) and first difference I(1). Furthermore, it can be applied for small sample size data and for different variables to have different optimal number of lags.

Linear model specification can be presented as:

$$UR_t = \alpha_0 + \alpha_1 IFDI_t + \alpha_2 OFDI_t + \alpha_3 GDP_t + \alpha_4 EXCR_t + \alpha_5 FL_t + u_t \quad (1)$$

Where

UR is the unemployment rate reported by Department of Statistics Malaysia (DoSM),

IFDI is inward foreign direct investment reported by UNCTAD,

OFDI is outward foreign direct investment reported by UNCTAD,

GDP is gross domestic product reported by IMF,

EXCR is Malaysia exchange rate reported by IMF,,

FL is foreign labour registered reported by DOSM, and

u is the error term.

An ARDL representation of equation (1) is formulated as follows:

$$\Delta UR_t = \alpha_0 + \sum_{i=1}^{n} \alpha_{1i}\, \Delta UR_{t-i} + \sum_{i=1}^{n} \alpha_{2i}\, \Delta IFDI_{t-i} + \sum_{i=1}^{n} \alpha_{3i}\, \Delta OFDI_{t-i} + \sum_{i=1}^{n} \alpha_{4i}\, \Delta GDP_{t-i} + \sum_{i=1}^{n} \alpha_{5i}\, \Delta EXCR_{t-i} + \sum_{i=1}^{n} \alpha_{6i}\, \Delta FL_{t-i} + \beta_1 UR_{t-i} + \beta_2 IFDI_{t-i} + \beta_3 OFDI_{t-i} + \beta_4 GDP_{t-i} + \beta_5 EXCR_{t-i} + \beta_6 FL_{t-i} + e_t \quad (2)$$

Where

Δ denotes the first difference operator,

α_0 is the drift component,

e_t is the usual white noise residuals.

The left-hand side is the unemployment rate. The first until fifth expressions (β_1 – β_6) on the right-hand side correspond to the long-run relationship. The remaining expressions with the summation sign (α_1 – α_6) represent the short-run dynamics of the model.

Co-integration test was done to investigate the long run relationship among the variables. A series of time series variables are co-integrated when some linear combination of them is stationary. For co-integration test, bound testing was used due to several reasons that includes all variables are assumed to be endogenous and simultaneous estimation of short and long run coefficient can be determined. The presence of co-integration denotes that there is a long run relationship among the variables. Co-integrated variables exhibit mean reverting spread that force the variables to move around the common stochastic trends. The test is based on F-Test which rely on the hypothesis of:

$H_0: \beta_1 = \beta_2 = \beta_3 = \beta_4 = \beta_5 = \beta_6 = 0$

There is no co-integration among the variables

$H_0: \beta_1 \neq \beta_2 \neq \beta_3 \neq \beta_4 \neq \beta_5 \neq \beta_6 \neq 0$

There is co-integration among the variables

The ARDL bound test is based on the Wald-test (F-statistic). The asymptotic distribution of the Wald-test is non-standard under the null hypothesis of no co-integration among the variables. Two critical values are given by [11] for the co-integration test. The lower critical bound assumes all the variables are I(0) meaning that there is no co-integration relationship between the examined variables. The upper bound assumes that all the variables are I(1) meaning that there is co-integration among the variables. When the computed F-statistic is greater than the upper bound critical value, then the H0 is rejected (the variables are co-integrated). If the F-statistic is below the lower bound critical value, then the H0 cannot be rejected (there is no co-integration among the variables). When the computed F-statistics falls between the lower and upper bound, then the results are inconclusive.

In the meantime, the unrestricted error correction model (UECM) was developed based on the assumption made by [11]. From the unrestricted error correction model, the long-run elasticities are the coefficient of the one lagged explanatory variable (multiplied with a negative sign) divided by the coefficient of the one lagged dependent variable.

Thus, equation (2) in the ARDL version of the error correction model can be expressed as equation (3). If the co-integration exists, the error correction part will be included in the model to identify the duration for the system to be in equilibrium condition. Error correction (EC_{t-1}) is the error of linear regression between variables in the study. The error correction version of ARDL model pertaining to the variables in equation (2) is as follows:

$$\Delta UR_t = \alpha_0 + \sum_{i=1}^{n} \alpha_{1i} \Delta UR_{t-i} + \sum_{i=1}^{n} \alpha_{2i} IFDI_{t-i} + \sum_{i=1}^{n} \alpha_{3i} \Delta OFDI_{t-i} + \sum_{i=1}^{n} \alpha_{4i} \Delta GDP_{t-i} + \sum_{i=1}^{n} \alpha_{5i} \Delta EXCR_{t-i} + \sum_{i=1}^{n} \alpha_{6i} \Delta FL_{t-i} + \lambda EC_{t-i} + e_t \quad (3)$$

where λ is the speed of adjustment parameter and EC is the residuals that are obtained from the estimated co-integration model of equation (1). The error correction model shows the change in unemployment is related to the change in other variables, as well as the gap between the variables in the previous period.

4. Findings

The analysis was started with co-integration checking among the variables involved in the study. The first practice in applying any co-integration technique is to determine the degree of integration of each variable. For this reason, the ADF test was employed. The test results are presented in table 4.1.

Table 4.1: Results of ADF tests

Variable	Level		First differences	
	Constant	Constant and trend	Constant	Constant and trend
UR	-1.4217(0)	-2.1923(0)	-4.0284(0)***	-3.5171(1)*
IFDI	-2.3433(0)	-3.2017(0)	-7.3803(0)***	-7.2441(0)***
OFDI	-1.1315(0)	-3.3578(0)*	-7.4372(0)*	-7.3323(0)***
GDP	-1.0183(0)	-1.1154(0)	-4.3866(0)***	-4.4055(0)***
EXCR	-1.5852(0)	-1.2571(0)	-4.6341(0)***	-4.7245(0)***
FL	-1.4220(0)	-1.7408(0)	-5.0587(0)***	-5.0077(0)***

Notes:
1. *, **, *** imply significance at the 1%, 5%, 10% level, respectively.
2. The numbers within parentheses for the ADF (Dickey-Fuller 1979) statistics represents the lag length of the dependent variable used to obtain white noise residuals.
3. The lag length for the ADF was selected using Akaike Information Criterion (AIC).

The results of table 4.1 indicate that the variables are integrated at order I(0) and I(1). For this reason ARDL model approach is used for the co-integration of the model.

The main advantage of this approach lies on the fact that it obviates the need to classify variables into I(1) or I(0). Moreover, compared to the standard co-integration test, there is no need for unit root pre-testing.

The ARDL co-integration test, assumed that only one long run relationship exists between the dependent variable and the exogenous variables [11].

Table 4.2 presents the co-integration results of the variables used in equation 2, using F-test with the [9] which are quoted in small samples. The results of table 4.2 show that F-statistic is greater than critical values, indicating that there is co-integration between the variables, hence 2 lags used in the analysis.

Table 4.2: The results of F-Test for Co-integration

Order of lag	*1*	*2*	*3*
UR	2.4560	4.8109**	8.2985

Notes:
1. *, **, *** denotes that the F-statistic falls above the 90% upper bound, above the 95% upper bound, and above the 99% upper bound respectively.
2. According to [9] the existing critical values reported in [11] cannot be used for small sample sizes because they are based on large sample sizes. [9] provides a set of critical values for sample sizes ranging from 30 to 80 observations. They are 2.496 - 3.346 at 90%, 2.962 – 3.910 at 95%, and 4.068 – 5.250 at 99%.

In the second stage, Akaike's information criterion (AIC) was employed in selecting the lag length on each first differenced variable. Results shows that ARDL (1,2,1,1,2,0) is the most suitable model to be used in the study. Based on the ARDL model, long run estimation was estimated. As can be seen in Table 4.3, all the long run coefficients are listed.

In the long run, LIFDI and LFL significantly affect the unemployment rate where 1% increase of these items may reduce the unemployment rate in Malaysia by 0.255% and 0.625% respectively.

Table 4.3: Estimates of the long run coefficients based on ARDL model

Long run coefficients			
AIC – ARDL (1,2,1,1,2,0)			
Dependent variable: LUR	Coefficient	Standard error	t-stat
LIFDI	-0.2553	0.0731	-3.4954***
LOFDI	-0.4877	0.0657	-0.7422
LEXCR	-0.4435	0.3815	-1.1626
LGDP	0.5525	0.3261	1.6946
LFL	-0.6254	0.1938	-3.2263***
C	-0.9384	3.1087	-0.3019

Notes:
1. *, **, *** indicate significance at 10%, 5%, and 1% level respectively

The results of the short-run dynamic and the various diagnostic tests are presented in Tables 4.4. Panel A reports the coefficient estimates of all lagged first differenced variables in the ARDL model (short-run coefficient estimates). A negative and significant coefficient of EC_{t-1} will be an indication of co-integration. Panel B reports some diagnostic statistics. The diagnostic tests include the test of serial autocorrelation (χ^2 Auto), normality (χ^2 Norm), and heteroscedasticity (χ^2 White).

Table 4.4: Error correction representation of ARDL(1,2,1,1,2,0) model

Panel A: AIC – ARDL (1,2,1,1,2,0)			
Dependent variable: LUR			
	Coefficient	***Standard error***	***t-stat***
ΔLIFDI	-0.1223	0.0368	-3.3249***
ΔLIFDI(-1)	0.0725	0.0298	2.4311**
ΔLOFDI	0.0586	0.0302	1.9408*
ΔLEXCR	0.3358	0.03404	0.9863
ΔLGDP	-1.3010	0.8152	-1.5959
ΔLGDP(-1)	-2.3883	-0.6382	-3.7420***
ΔLFL	-0.3416	0.1398	-2.4439**
Constant	-0.5125	1.6885	-0.3036
ECM(-1)	-0.5462	0.1348	-4.0527***
ECM = LUR + 0.2554*LIFDI + 0.4877*LOFDI + 0.4435*LEXCR – 0.5525*LGDP + 0.6254*LFL + 0.9384*C			
R-squared	0.8912	DW statistic	2.3087
F (8,20)	16.3887***	RSS	0.0868
Panel B: Diagnostic test			
χ^2_{Auto}	1.1459[0.301]	χ^2_{White}	0.1442[0.707]
χ^2_{Norm}	0.0313[0.862]		

Notes:
1. *, **, *** indicate significance at 10%, 5%, and 1% level respectively
2. χ^2 Auto is the Breusch–Godfrey LM test for autocorrelation.
3. χ^2 Norm is the Jarque–Bera normality test.
4. χ^2 White is the White test for heteroscedasticity

Not much interpretation could be attached to the short-run coefficients. The result shows the dynamic adjustment of all variables. Surprisingly, LIFDI, LGDP and LFL also have impact in the short run, showing that all these factors are the strong causal effect of the unemployment rate in Malaysia.

As can be seen from table 4.4 at panel A, the EC (-1) carries an expected negative sign, which is highly significant, indicating that, variables are co-integrated. The absolute value of the coefficient of the error-correction term indicates that about 55% percent of the disequilibrium in the unemployment rate is offset by short-run adjustment in each year. This means that low unemployment rate is followed in the next period by a changes in the other variables. Thus, it is important to reduce the existing disequilibrium over time in order to maintain long-run equilibrium.

The diagnostic tests presented in the lower panel B of Table 4.4 show that there is no evidence of diagnostic problem with the model. Measuring the explanatory power of the equations by their adjusted R^2 show that roughly 89.12% of the variation in unemployment rate can be explained by the independent variables. The Lagrange Multiplier (LM) test of autocorrelation suggests that the residuals are not serially correlated. According to the Jarque-Bera (JB) test, the null hypothesis of normally distributed residuals cannot be rejected. The White heteroscedasticity test suggest that the disturbance term in the equation is homoscedastic.

The existence of a stable and predictable relationship between the unemployment rate and its determinants is considered a necessary condition to reduce poverty and crime rates [15]. The stability of the long-run coefficients are used to form the error-correction term in conjunction with the short run dynamics. Some of the problems of instability could stem from inadequate modelling of the short-run dynamics characterizing departures from the long run relationship. Hence, it is important to incorporate the short run dynamics for constancy of long run parameters. In view of this we apply the CUSUM and CUSUMSQ tests.

The CUSUM test is based on the cumulative sum of recursive residuals based on the first set of n observations. It is updated recursively and is plotted against the 18 break points. If the plot of CUSUM statistic stays within 5% significance level, then estimated coefficients are said to be stable. Similar procedure is used to carry out the CUSUMSQ that is based on the squared recursive residuals. A graphical presentation of these two tests is provided in Figures 4.1 and 4.2.

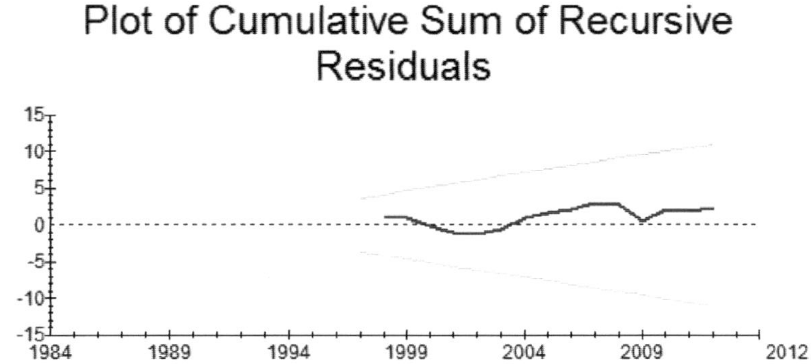

Figure 4.1: Cumulative sum of recursive residuals

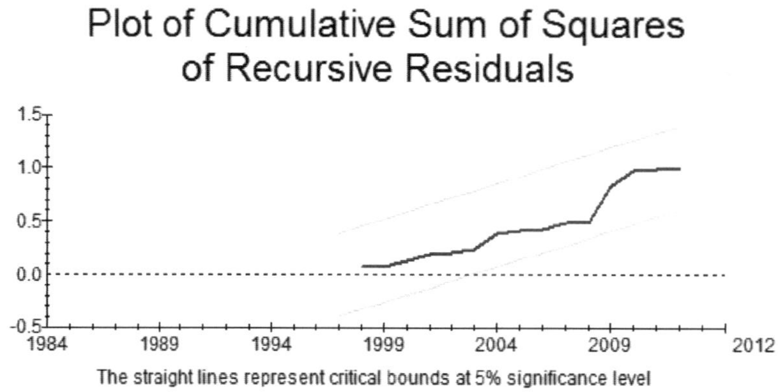

Figure 4.2: Cumulative sum of squares of recursive residuals

Since the plots of CUSUM and CUSUMSQ statistic marginally cross the critical value lines, we are safe to conclude that unemployment rate is stable.

5. Conclusion and Recommendation

Unemployment rate should be reduced to avoid shortage of national production and waste of human resources. Okun's Law stated that there is a negative relationship between unemployment and output (GDP). The lower the unemployment the higher GDP should be produced. [17] said unemployment is significantly affecting GDP in Malaysia where one percent decrease in unemployment contribute to 1.75% increase in the GDP. On the other hand, the decreasing of unemployment may reduce the crime rate in Malaysia [15].

From the result of the study, among all potential factors contribute to unemployment, FDI support and foreign workers are significantly reducing the unemployment rate. This is clearly seen from 1982 to 1991 where the unemployment reached 7.5% and rapid inward FDI reduced it significantly. It also shows that domestic plan to facilitate construction work would not provide enough job market for the Malaysian. Outward FDI also important as the opportunities to develop network between countries. International trade will allow the domestic market to supply goods for international demand, create more job opportunities and reduce unemployment rate. The increasing number of foreign labours in Malaysia are aligned with the increasing of FDI inflows as many foreign companies hire cheaper labours from other countries instead of local labours.

Finally, Malaysia as a developing country, should focus on attracting foreign direct investment inward and outward to reduce the unemployment rate. Somehow, with the advancement of technology and higher financial aids, foreign investors may monopoly the market and affect the local producer as well as bringing more foreign workers to Malaysia. The increasing number of foreign labours contributes both positive and negative effect to the country. As they contribute to the local production and new technology learning, the government is also facing the problem of the increasing estimated government subsidy on fuel, transportation, and education. Hence, government need to control the penetration of foreign investment on the local market by imposing the joint venture policy so that the local producers have the opportunities to learn new technologies and ideas as well as sharing the profit earned.

Acknowledgement
The authors would like to express the acknowledgment to the Research Management Centre (RMC), UiTM for providing the Research Acculturation Grant Scheme (RAGS) to complete the study. This paper is written based on a part of research under the RAGS, Ministry of Higher Education (MOHE) Malaysia.

References

[1] Balcerzak, A. P., & Żurek, M. (2011). Foreign direct investment and unemployment: VAR analysis for Poland in the years 1995-2009. *European Research Studies*, 4-14.

[2] Department of Statistics. (2015). Metadata of labour force survey. Putrajaya, Malaysia. Retrieved from Department of Statistics Malaysia.

[3] Djafar, F., & Hassan, M. K. (2012). Dynamic of push and pull factors of migrant workers in developing countries: The case of Indonesian workers in Malaysia. *Journal of Economics and Behavioural Studies*, 703-711.

[4] Elangkovan, K. (2012). Minimum wage laws in Malaysia: An answer to the increasing rate of unemployment. *International Journal ofAcademic Research in Economics and Management Sciences*, 135-146.

[5] Essays, UK. (November 2013). Cause And Affect Of Unemployment In Malaysia Economics Essay. Retrieved from http://www.ukessays.com/essays/economics/cause-and-affect-of-unemployment-in-malaysia-economics-essay.php?cref=1

[6] Experimental Economics Center. (2006). Comparing Unemployment Among Countries.

[7] Habib, M. D., & Sarwar, S. (2013). Impact of foreign direct investment on employment level in Pakistan: A Time Series Analysis. *Journal of Law, Policy and Globalization* , 46-55.

[8] Majid, N. (2009). *The Global Recession and Developing Countries.* Geneva: International Labour Office.

[9] Narayan, P. K. (2005). The saving and investment nexus for China: evidence from cointegration tests. *Applied Economics*, 1979-1990.

[10] Pesaran, M. H., & Shin, Y. (1999). An Autoregressive Distributed-Lag Modelling Approach to Cointegration Analysis. In S. Strøm, *Econometrics and Economic Theory in the 20th Century* (pp. 371-413). Cambridge : Cambridge University Press.

[11] Pesaran, M. H., Shin, Y., & Smith, R. J. (2001). Bounds testing approaches to the analysis of level relationships. *Journal of Applied Econometrics*, 289–326.

[12] Pinn, S. L., Ching, K. S., Kogid, M., Mulok, D., Mansur, K., & Loganathan, N. (2011). Empirical analysis of employment and foreign direct investment in Malaysia: An ARDL bounds testing approach to cointegration . *Advances in Management & Applied Economics*, 77-91.

[13] Shaari, M. S., Hussain, N. E., & Rahim, H. A. (2013). The effect of oil price changes and exchange rate volatility on unemployment: evidence from Malaysia. *International Journal of Research in Business ans Social Science*, 72-83.

[14] Subramaniam, T., & Baharumshah, A. Z. (2011). Unemployment and speed of adjustment in Asean-3 Economies: A cointegration analysis. *The Singapore Economic Review*, 327-347.

[15] Tang, C. F. (2010). An exploration of dynamic relatinship between tourist arrival, inflation, unemployment and crime rates in Malaysia. *International Journal of Social Economics*, 50-69.

[16] Yabuuchi, S. (1999). Foreign direct investment, urban unemployment and welfare. *The Journal of International Trade & Economic Development: An International and Comparative Review*, 359-371.

[17] Zaleha M. N., Norashidah M. N., & Judhiana Abdul Ghani (2007). The relationship between output and unemployment in Malaysia: Does Okun's Law exist?. *International Journal of Economics and Management,* 337-344.

The Research on Computerized Adaptive Testing

Peng Lu[1,a], **Xiao Cong**[2,b]
[1]Department of Media Technology and Communication, Northeast Dianli University, Jilin, Jilin,132012, P.R.China;
[2]College of Science, Northeast Dianli University, Jilin, Jilin, 132012, P.R.China

E-mail: [a]lup595@nenu.edu.cn; [b]peng.lu2008@gmail.com

Abstract. In recent years, computerized adaptive testing is becoming the focus of the field of modern educational evaluation. In this form of the test, the response relationship between the examinee with the item by IRT modelling, then use the computer to estimates the ability level of the examinees and real-time select item. Computerized adaptive test development process were reviewed in the paper, and discuss the latest research results and pointed out that the current problems and future trends.

1. Introduction

The computerized Adaptive Testing(CAT) is a test form which according to the personalization features of the students. As can be seen from the test development process, as old as the ideas of adaptive testing to students and the test itself, and its prototype is oral in the field of education and psychology in the diagnosis [1], the teachers follow the principle of comparability to students for testing or diagnostic.

The earliest examples of adaptive testing can be traced back to the early 20th century by psychologists Binet. Alfred who developed an intelligence test, the test was conducted precise description to the relationship between the response of student and items. In his view, as long as the same item selection rules which can make a reasonable assessment of all students for a standardized test, and do not need to provide the same items for all students. Therefore, the most important innovation of Benet's test is the intuitive response model[2]. And later, in the study of quantification method, the Psychologist Louis L. Thurstone[3]was used the data sets of the Binet's test. But, at this time in the field of education and psychological tests have been widely used groups test which based on classical test theory, and using observation scores to maintain comparability of scores. This test form to facilitate the implementation, but does not allow any adaptations. For students, the items included in the test are too difficult or too easy and resulting in lower test efficiency.

The development of adaptive testing dependent on two aspects, the most important is the development of the basic theory of test, and it appears theory which called Item Response Theory(IRT). IRT use project characteristic curve (ICC) to explain the response of the probability distribution of the student on the item. The first item response model is two-parameter Normal Ogive Model proposed by Lord [4], and this model as shown in the following equation 1:

$$P_i(\theta) = \int_{-\infty}^{a_i(\theta - b_i)} \frac{1}{\sqrt{2\pi}} e^{-\frac{y^2}{2}} dy, \qquad (1)$$

Where, parameter θ represents the ability of student; Discrimination parameter a_i represents the value which is proportional to the slope K at the inflection point of the ICC; difficulty parameter b_i indicates the ability value corresponding to the inflection point at ICC.

Another well-known model is the Rasch model IRT[5], the model includes the ability parameters θ and item difficulty parameter b, and the probability of correct responses of student on an item i is expressed as equation 2:

$$P_i(\theta) = \frac{e^{(\theta - b_i)}}{1 + e^{(\theta - b_i)}}, \tag{2}$$

Where, parameter θ represents the ability of student; parameter b_i is difficulty of item i, and has to be estimated sufficiently accurate by check.

In addition to the above item response model, and Birnbaum also proposed Logistic model, Lazarsfeld and Henry proposed potential linear model and potential distance model, Samejima proposed continuous response model, Masters proposed part of the scoring model, etc.. IRT's evolving provides a solid theoretical foundation for adaptive testing.

In addition to the basic theory, on the other hand is the development of computer technology. In the 1970s, Scholars had meaningful attempt on computerized adaptive testing. For example, David Weiss's research on CAT at the University of Minnesota 's done pioneering work [2], and Lord were studied in the computerized estimate the parameters of the items and item selection methods. With the development of computer technology and the continuous improvement of computing power, it has become possible as a testing tool. The large-scale application of adaptive testing appeared in the mid-1990s, it's not only used in the psychological tests, but also widely used in university admission exams, professional certification, military and other fields.

With depth application in practice, the researchers conducted a more in-depth thinking for the basic theory and the actual situation faced in implementation process, it promotes two aspects of continuous development of basic theory of test and computer technology application. We will summarize recent progress and the future development trend of the prospects.

2. The development of basic theory

In recent years, with the application of IRT in the adaptive testing, the researchers have constantly improving for the item response model by adding important parameter, modify the basic assumptions of IRT etc., and in order to be a more scientific evaluation of students, the main contents in detail.

2.1. Item Response Time

Typically, in the test items and students have different response time (RT), and harder items requires more time. And the adaptive algorithm is often providing more time-consuming items for student who has a high level of competency, as a result, the test is not scientific and reasonable. Therefore, automatically recorded the RT of the students in CAT, and the differences of RT into the test results is a new research focus. Representative results in this respect is four-parameter logistic timing model proposed by Tianyou Wang and Bradley A. Hanson [6], as shown in the following equation:

$$P = c_i + (1 - c_i)\frac{1}{1 + e^{-1.7a_i\left[\theta_m - (\rho_m d_i / t_{mi}) - b_i\right]}}, \tag{3}$$

Where, ρ_m is speed parameter of student m, the larger the value, the more time student m needed in resolving item; d_i is speed parameter of item i, the higher the value, the more time required for correct solution item i; t_{mi} is the response time of student m on item i.

The model shows that, when the speed parameters ρ_m and d_i unchanged, increases the probability P of correct responses with the increase of the response time t_{mi} in the item; when then response time t_{mi} in the item unchanged, the greater ρ_m d_i the smaller probability of student m to answer the item i; When t_{mi} tends to infinity, index converges to $a(\theta - b)$. This means that even when time is unlimited it does not guarantee that students will answer item in correct. Therefore, this model is more suitable for proficiency test. With this model can be reasonably estimate of item parameter and the ability for limited testing, and ability to estimate quickly converge to the true ability level. In addition, it can also study whether students can solve items and the accuracy and the best strategy of test.

It should be noted, this model is only a partial description of the test procedure, and a more complete description should include RT distributed on all items in the pool and form a more appropriate statistical model. Therefore, some researchers proposed using a hierarchical framework to improve a convergence rate of the ability estimate at the joint distribution of abilities parameters θ and the speed parameters ρ [7], and proposed the lognormal RT model based on the assuming that the item response time follow a normal distribution of students[8].

Another benefit is that you can use RT abnormalities that may exist in the CAT, such as functional differences, answers deception, and previous knowledge of the items. For these actions, the traditional model-based testing lose effectiveness, and the RT including more information on abnormal behaviour, therefore not affected by this [9]. The RT model with parameter structure allows us to adjust the RT based on the student's actual speed, and check test results their response to the items whether consistent with time-sensitive mode. So, even for those who have experienced cheating, it is impossible to find a regular pattern.

2.2. Multi-grade scoring

So far, most of the operational CAT are based on secondary item response model to deal with objective items[10]. In practice, more attention to students' response on subjective items of test, and many items are multi-grade scores of items, such as calculation items, essay items and so on. The use of multi-grade scoring items can get more information than use two scoring items [11]. Therefore, the use of multi-point scoring model is more reasonable. Most famous multi-point scoring model is Graded Response Mode (GRM) proposed by Samejima[12], the model is described as follows:

Set the full mark of item i is $S(S \geq 1)$, there are $S+1$ number of scoring points, $x = 0,1,2,\ldots,S$; $P_{ix}^{*}(\theta)$ represents the probability that the scores of students who has the ability is θ are not less than x , let scores for all students greater than or equal to x marked "Pass" , and let scores for all students less than x marked "Fail", then $P_{ix}^{*}(\theta)$ has become an item characteristics function to binary scores. And let $P_{i0}^{*}(\theta) = 1$, $P_{i,S+1}^{*}(\theta) = 0$. In this model, uses two stages to obtain the probability of classification score of a student on a certain item.

(1) The first stage, it is compute the cumulative probability to the item i of student who has the ability level of θ , let $P_{ix}^{*}(\theta)$ is 3PLM:

$$P_{ix}^{*}(\theta) = c_i + (1 - c_i)\frac{1}{1 + e^{[-Da_i(\theta - b_{ix})]}}, \tag{4}$$

Where, b_{ix} is a threshold parameter which associated with level x of item i, this parameter is subject to the constraint $b_{ix-1} < b_{ix} < b_{ix+1}$. For an item, all response curves share the same discrimination a_i .

(2) The second stage, the response probability obtained on a given level of student. And this probability by subtracting the cumulative probability obtained, as shown below.

$$P_{ix}(\theta) = P_{ix}^{*}(\theta) - P_{ix+1}^{*}(\theta), \tag{5}$$

Figure 1 and Figure 2 show the operator characteristic curve and graded response curve of item has three grades (a=2, b_1 =-1, b_2=0, b_3=1.5, c=0.1).

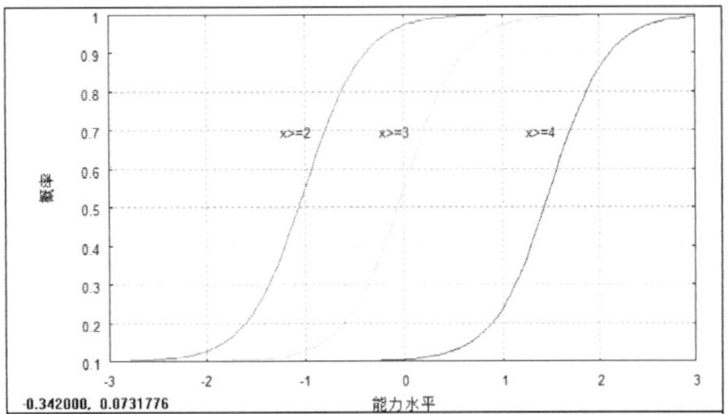

Figure 1. Item characteristic curve

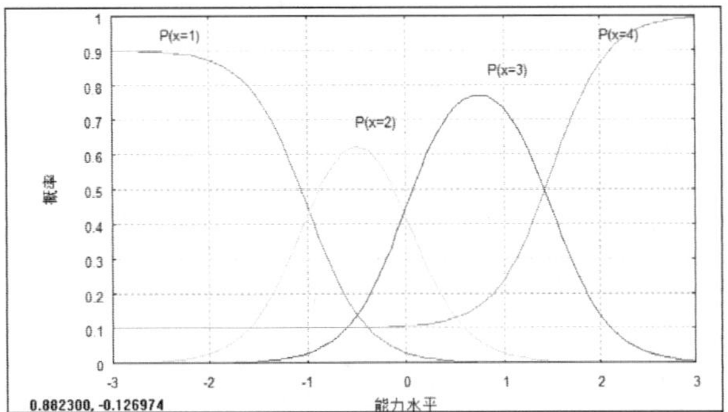

Figure 2. Item Grade response curve

IRT based on multi-level scoring model break out the limit of item response model can be used only two scoring items in the past. Therefore, researchers are increasingly inclined to replace the traditional two score method with multi-level score, also caused the research of more complex scoring models. GRM more applicable to the general items of subjective scoring form and mathematical treatment more convenient. In the coming decades, GRM must have an even broader prospect and good promotional value in applied and theoretical exploration.

2.3. Multidimensional Adaptive Testing

With IRT widely used in practical work, researchers gradually found that the traditional assumption of unidimensional IRT with many actual psychological or educational tests do not match [13]. The multidimensional of test data is consistent with students need to work with a variety of capacities at the completion of a test and few tests measure only a single ability or trait [14]. Therefore, the research on the multidimensional item response theory (MIRT) is very necessary [15], so the researchers' attention gradually shifted to the multidimensional item response models. However, this situation has greatly changed with the development of statistics and computer technology. Therefore, in recent years to study and use of multidimensional item response theory became popular.

Based on the basis of other researchers, Reckase and Mckinley proposed the most practical Logistic multidimensional item response model, and the item response function of this model shown in the following equation [16]:

$$P_{ij}\left(U_{ij}=1\left|a_i,d_i,c_i,\theta_j\right.\right)=c_i+\left(1-c_i\right)\frac{1}{1+\exp\left(a_i\theta_j+d_i\right)}, \quad (6)$$

Where, $\theta_j=\left(\theta_{j1},...,\theta_{jk},...,\theta_{jm}\right)$ is the parameter of ability vector m of student j, $a_j=\left(a_{j1},...,a_{jk},...,a_{jm}\right)$ is a parameter vector related to the ability to discrimination of items; d_i is a parameter related with difficulty of item, but the meaning is different from the one-dimensional model of the b_i; and c_i is guess parameter of item. The method of determining all parameters values is through sample tests, collect data and analysed, then determine the value of each parameter.

All items need to show a satisfactory of fit for response model, so the multidimensional items in adaptive test are more dominant role, and should be modelled by multidimensional model and adjust adaptive testing algorithms. If the purpose is for diagnosis, then each dimension should be carefully tested. The change from one-dimensional to multi-dimensional adaptive testing involves an important modification that item selection criteria. For example, the item selection method based on the information function have many parameters during the test, however, due to the information function is replaced by a matrix $p \times p$ (p is the number of item parameters), and the item selection process becomes significantly complex[17]. This not only reflects the estimation accuracy, but also reflects their relevance.

How to reduce multidimensional entity to the one-dimensional standard depends on the purpose of the actual test. For a two-dimensional test, it is should distinguish three different objectives: Firstly, the ability parameters are all the main in two dimensions and should accurately estimate; Secondly, only one main parameter and the other is the interference parameter; Thirdly, merge two parameters to compute their weighted average. For these different goals, it can use the optimize the design principles in statistics for item selection and the rules of assembled item pools to optimize experimental design or sampling procedures.

3. Depth application of computer technology
It is still face many practical problems in the implementation process of CAT, such as in order to maintain the validity of the test, it makes item selection become complex sequences optimization problems with a lot of constrained. And item pool facing high risk and high cost, and how to automatically generate high-quality items in large-scale? The solution to these problems depend on the depth application of computer technology in implementation process of adaptive testing.

3.1. Processing constraints
In the application, the adaptive testing must also meet the safety, comprehensive and many other practical requirements. This methods take test requirements as the constraints of item selection process, and take the adaptive testing as an instance of combinatorial optimization problems for constrained[18].Next, we discuss several major methods.

3.1.1. Weighted deviation modeling method
The weighted deviation modelling (WDM) is proposed by Swanson Stocking [19], it is a heuristic method of item selection. The objective function for item selection is defined the sum of weighted deviation, as shown in the following equation:

$$WDM = \sum_{k=1}^{K} w_k \left(d_{l_k} + d_{u_k} \right) + w_I d_I \quad , \tag{7}$$

Where, K is the number of constraints, $\mathbf{C}$ is the relationship matrix between the items and the constraints, w_k represents the weight of constraint k, w_I represents the weight of the amount of information of test; l_k and u_k represent the lower and upper bounds of a constraint k; d_{l_k} is the difference compare to lower bound of constraint, d_{u_k} is the excess compare to upper bound of constraint, d_I is the difference compare to information I_t ; e_{l_k} is the excess compare to lower bound of constraint, and e_{u_k} is the difference compare to upper bound of constraint.

The *WDM* method take the item selection constraint problem into a mathematical programming, to select the appropriate item is continuous calculated by the upper and lower bounds. The overall objective function is sum of the difference of all constraints and the difference of information. However, the difference between the various constraints and information are not on the same scale, as exposure constraint is expressed as a percentage, and the content constraint is expressed as the number, therefore, they should not be compared together. In addition, this method requires continuous adjustments to various boundary to achieve the desired result, it is very time consuming [20].

3.1.2. Stratification method

The stratification method is proposed by University of Illinois Professor Hua-Hua Zhang is by Champaign Urbana campus educational psychology at University of Illinois [18]. For exposure of items and content balance problems in CAT, they have proposed STR_A stratification, STR_B stratification and STR_C stratification method.

- STR_A method. The exposure of items which has high distinguish are relatively high, therefore, the ideas of STR_A stratification is stratified item pool based on item discrimination, then the test is divided into several corresponding stages[21]. The studies have shown that this method improves the utilization of low-discrimination items for in the conditions of accuracy.
- STR_B method. Researchers have also proposed STR_B method [22]. The item pool is divided into several blocks according to the item difficulty, then take the items in the various blocks merged into different stratification follow the item discrimination and the item pools generated have similar difficulty distribution in every stratification.
- STR_C method. To solve the problem of the content balance, the researchers proposed STR_C method [23]. The method considers three factors in the process of the stratification of the item pool, such as item discrimination, difficulty and test content. The pool is divided into several groups according to the contents of test; then use STR_B method to get all stratifications in each group.

The stratification approach has made great progress in terms of exposure control and content balance. However, it is need operations such as sort, group, block, stratification and assembly before testing and therefore requires a lot of compute time. It need to reorder and assembly and takes teachers to use too much time in the early preparation, so it is not easy to implement.

3.1.3. Shadow-test method

For the constraint problems faced by item selection process, van der Linden proposed the Shadow-test method [24]. Firstly, it choose a complete test from the item pool and it satisfies all the constraints; then choose the best item from this test based on the initial ability level θ_0 of student; then record student's responses to the item and re-estimate his/her ability level; and according the new ability level θ_1 reassemble test; repeat the above steps until the end of the test.

In the implementation process, the CAT is continuously reassembling Shadow-test, they are only as an intermediate step for item selection, as shown in Figure 3.

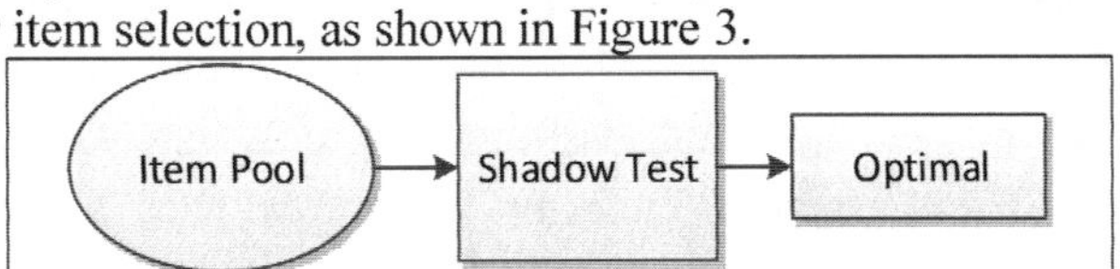

Figure 3. Shadow-test method for adaptive testing

Since each Shadow-test meets all the constraints and use the best item from Shadow-test in the testing process, so adaptive test is also optimal. The method to select the optimal item much faster than an unconstrained adaptive testing. In addition, control the rate of students [25], to ensure that the item exposure in pool and many other practical constraints were studied.

3.1.4. Rotating item pools method

For constraint problem, some researchers from the perspective of the build the item pools proposed rotating item pools (RIP) [26]. In this method, constructed item pools by two stages to handle all of the constraints, as shown in Figure 4.

- The first stage: Assign the items in item pool to a temporary collection. It is calculation and minimize the difference between any two items in item pool by objective function, and take the item pool divided into two similar temporary sets, and so on. Specifically, it can use sequential allocation or simultaneous allocation with heuristic methods.
- The second stage: Assign the items in temporary collection to the child item pools. Depending on the different purpose, it can be constructed of non-overlapping item pools or overlapping item pools. Specific methods can be used random allocation method or mathematical programming methods.

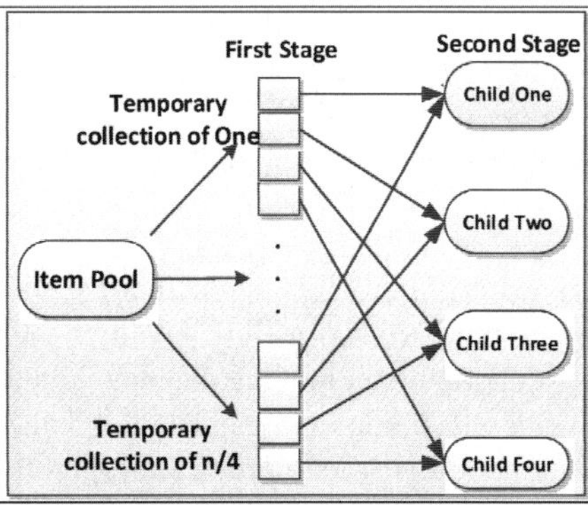

Figure 4. The allocation process of Rotating item pools method

The rotating item pools from design perspective to solve constraint problems. Before the test, take the item pool is divided into several sub-pools which various parameters are almost the same for students to use. And during the test, the system chooses the appropriate item from different sub-pools for students. However, this method also has some problems, such as takes a lot of time in division of the sub-pools before test and required design the number of sub-pools according to students.

3.1.5. The maximum priority index method

The maximum priority index (MPI)[27] take all the constraints variables are weighted by multiplying the value of the maximum amount of information of item and form product PI. Use this maximize product PI as an indicator of measure item instead MI, and the higher the product the greater the priority of the item. The compute of item priority index by the following formula:

$$PI_i = I_i \times \prod_{k=1}^{K} (w_k f_k)^{c_{ik}}, \quad (8)$$

Where, K is the total number of all constraints; **C** represents the correlation matrix of items and constraints; I_i is information of item i; The rest is the product of weight of constraints associated with the item, f_k represents the remaining quota of constraint k, w_k represents the weight of constraint k. In practice, using a two-stage item selection framework to deal with each flexible constraints which involve upper bounds and lower bounds. It is process the lower bound in the first stage, and process the upper bound in the second stage, but the methods of computing f_k of each phase is different.

Next, the above methods are compared and analysed from the constraint handling, method type, complexity and accuracy affect, and the results shown in Table 1.

Table 1. Comparison of item selection methods

	Constraint handle	Type	Complexity			Accuracy affect
			Prophase	Testing	Maintain	
WDM	many	heuristic method	low	higher	low	yes
STR	less	direct method	high	lower	high	yes
Shadow	more	direct method	low	higher	low	no

RIP	more	direct method	high	lower	high	no
MPI	many	heuristic method	low	higher	low	no

Through the above analysis it can be concluded, the heuristic method can handle multiple constraints such as item exposure, content balance and so on, and the other method is mainly for some constraint problems. However, taking into account the problem of multiple constraints, the calculating of item selection of heuristic method is larger. Overall, the heuristic method is better handling constraints and is the focus of future research.

3.2. Rule-based items generation technology

In traditional tests, it is needs for a specific test to development items and the corresponding pre-tested. If the CAT also follow this thought, so every time a new test will need to replace the entire item pool, this will involve a lot of resources; and if use the same item pool of continuous testing, it will lead to security risk. For this problem, the early solution is to use item exposure control techniques to ensure test security and use of the item pool better. But the researchers then realized, although these methods can improve the utilization rate of project pool project, but it is often difficult to use [28].

Researchers have studied the various types of item automatic generation method; the most representative method is to use an item template. In this method, some of the elements of each item (e.g.: part of stem, correct answer, etc.) have been replaced by the corresponding set. The formation of a number of item family based on the target and use the rules, thus, the entire item pool will be composed by a number item family, shown in Figure 5.

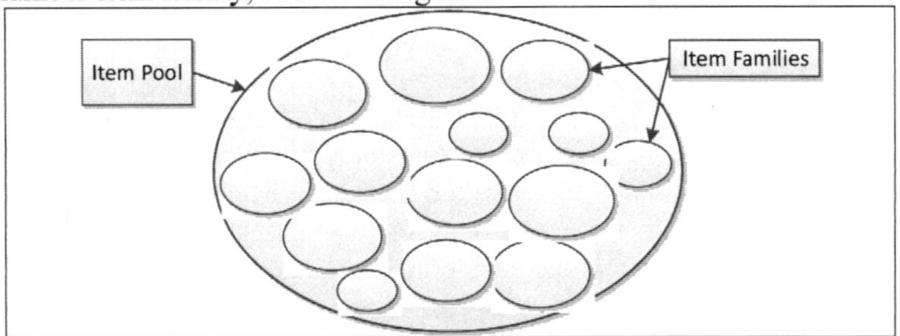

Figure 5. Item family in pool

In practice, the use of two stages of the item selection process for adaptive testing. First of all, according to the current estimate of the student's ability θ, utilizing the difference between the different item families of the item pool to select a best item family; then from this family randomly generated item to students for test, the aim is desired to produce a small difference in the interior of family. Therefore, it need to modify the item response model for a model with a two-stage structure. Set the item family is $f=1,\dots,F$, the item in family f is expressed as i_f .Then for the Rasch model, the appropriate two structures is expressed as:

$$P_{i_f}=\frac{e^{(\theta-b_{i_f})}}{1+e^{(\theta-b_{i_f})}}, \tag{9}$$

Where, the difficulty parameter b is normal in each family f, and the mean and variance are μ_f and σ_f^2.So, it can get the difference between the different item families through parameter μ_f and get the difference between the item internal families through parameter σ_f^2.

Rule-based method for generating an item unresolved problem is the cost of the pre-test. Some researchers believe that the parameters of other items automatically generated from the parameter of an item, so it can save some pre-test activities. In fact, due to the automatically generated items involves more extensive items review, items validation as well as the item parameters vary widely different among item families. In this model, items calibration is turn into item families' calibration.

Therefore, the cost savings of item calibration associated with the number of samples of items and the total number of items generating from the item family.

4. Conclusion and prospect

With the development of item response theory and computer technology also contributed to the development of the CAT and get a lot of important research results. Through computer technology, makes that it is possible to use of sophisticated statistical methods for real-time estimation of ability and to select the best item. Although adaptive testing has made considerable progress, but it has gradually exposed some problems and deficiencies in the years of practice and need to be further studied and improved, it is focused on the following aspects:

(1) More reasonable item response model. For most of the items are multi-grade scores and student will use the ability to use multiple ability to solve items, and response time also reflects the actual situation on the students' ability differences. It is necessary to item response model for a more in-depth research which combine cognitive psychology and statistics, and proposed item response model that meet the characteristics of student and can application in actually.

(2) General constraint handling methods. Like standardized linear tests and in order to maintain the validity, CAT had to meet a wide range of constraints. It should be proposed a universal constraint handling methods to process each constraints using a special algorithm, and can still guarantee a relatively high efficiency and does not affect the user experience. The heuristics algorithm in the field of artificial intelligence computing efficiency, and it can avoid two issues that excessive computing and the feasibility, and provides a viable solution for constraint process.

(3) The organizations of item pool. It can reduce the development costs of the items through using item templates and rule-based automatic generation of high-quality items and maintenance the item parameters by analyse the test record. In addition, using ontology technology to build domain ontology and complete description the details of the relationship between knowledge points and of the contents.

These are the development trend in the next stage of CAT, it also be our further research and exploration. Therefore, through in-depth research on the basic theory and related support technology of CAT, it is can promote the development of the CAT and has important significance for students' assessment of comprehensive quality and the development of education.

Acknowledgments

The work was supported by the Jilin Provincial Education Department "12th Five-Year Plan " scientific and technological research key project(2015239)

References

[1] Linda Crocker, James Algina. Introduction to Classical and Modern Test Theory. Thomson Learning Asia Pte Ltd. (2004).

[2] WimJ. van der Linden. (2008). Some New Developments in Adaptive Testing Technology. Journal of Psychology, 216(1) (2004) 3-11.

[3] Thurstone, L.L. A method of scaling educational and psychological tests. Journal of Educational Psychology, 16 (1925) 433-451.

[4] Peng Lu, Dongdai Zhou, Shaochun Zhou, Xiao Cong. Design and Implementation of Computerized Adaptive Testing System for Multi-Terminals. Modern Educational Technology, 22(6) (2012) 88-92.

[5] Rasch, G. Probabilistic models for some intelligence and attainment tests. Chicago: University of Chicago Press (1960).

[6] Tianyou Wang, Bradley Allanson . Development and calibration of an Item Response Model that Incorporates Response Time . The annual meeting of the American Educational Research Association in Seattle (2001).

[7] van der Linden, W.J. Using response times for item selection in adaptive testing. Journal of Educational and Behavioral Statistics, 33(1) (2008) 5-20.
[8] van der Linden, W.J. A lognormal model for response times on test items. Journal of Educational and Behavioral Statistics, 31(2) (2006) 181-204.
[9] van der Linden, W.J., & Guo, F.. Bayesian procedures for identifying aberrant response-time patterns in adaptive testing. Psychometrika, 73 (2008) 365-384.
[10] Fen Luo,Shuliang Ding,Xiaoqing Wang. Dynamic and Comprehensive Item Selection Strategies for Computerized Adaptive Testing Based on Graded Response Model. Acta Psychologica Sinica, 44(3) (2012) 400-412.
[11] Ping Chen, Shuliang Ding,Haijing Lin,Jie Zhou. Item Selection Strategies of Computerized Adaptive Testing based on Graded Response Model. Acta Psychologica Sinica, 38(3) (2006) 461-467.
[12] Samejima, F. Estimation of ability using a response pattern of graded scores. Psychometrika, 35(1) (1970) 139-139.
[13] Reckase, M. D. Multidimensional Item Response Theory. Springer-Verlag New York (2009).
[14] Chunhua Kang,Tao Xin. New Development in Test Theory: Multidimensional Item Response Theory. Advances in Psychological Science, 18(3) (2010) 530-536.
[15] Dongbo Tu, Yan Cai, Haiqi Dai, Shuliang Ding. Parameters Estimation of MIRT Model and Its Application in Psychological Tests. Acta Psychologica Sinica, 43(11) (2011) 1329-1340.
[16] Reckase, M. D., & McKinley, R. L. Some Latent Trait Theory in a Multidimensional Latent Space. Iowa City, IA: American College Service (1982).
[17] Mulder, J., & van der Linden, W.J. Multidimensional adaptive testing with optimal design criteria for item selection . Psychometrika, 74 (2009) 273-296.
[18] Peng Lu, Dongdai Zhou, Shaochun Zhou, Xiao Cong. Research on Item Selection Method For CAT Based On Simulated Annealing . Computer Applications and Software, 29(10) (2012) 175-179.
[19] Swanson, I.., & Stocking, M. I. A model and heuristic for solving very large item selection problems . Applied Psychological Measurement, 17 (1993) 151-166.
[20] Leung C. K., Chang, H. H. and Hau, K. T. Computerized adaptive testing: A mixture item selection approach for constrained situations. British Journal of Mathematical & Statistical Psychology, 58 (2005) 239-257.
[21] Hau, K. T., & Chang, H. Item selection in computerized adaptive testing: Should more discriminating items be used first?. Journal of Educational Measurement, 38 (2001) 249-266.
[22] Chang, H., Qian, J., & Ying, Z. a-stratified multisage CAT with b-blocking. Applied Psychological Measurement, 25 (2001) 333-341.
[23] Yi, Q., & Chang, H. a-stratified CAT design with content blocking. British Journal of Mathematical and Statistical Psychology, 56 (2003) 359-378.
[24] van der Linden, W.J. Linear models for optimal test design. New York: Springer-Verlag (2005).
[25] van der Linden, W.J., Breithaupt, K., Chuah, S.C., & Zhang, Y. Detecting differential speed in multistage testing. Journal of Educational Measurement, 44 (2007) 117-130.
[26] Ariel, A., Veldkamp, B. P., & van der Linden, W. J. Constructing rotating item pools for constrained adaptive testing. Journal of Educational Measurement, 41(2004) 345-360.
[27] Cheng, Y., & Chang, H. The maximum priority index method for severely constrained item selection in computerized adaptive testing. British Journal of Mathematical and Statistical Psychology, 62 (2009) 369-383.
[28] Chang, H., & Ying, Z. a-Stratified multistage computerized adaptive testing. Applied Psychological Measurement, 23 (1999) 211-222.

Synthesis, Characterization and Catalytic Properties of Perovskite $LaFeO_3$ Nanoparticles

Nur Afifah[1,2] **and Rosari Saleh**[1,2,a]
[1]Departemen Fisika, Fakultas MIPA-Universitas Indonesia, 16424 Depok, Indonesia
[2]Integrated Laboratory of Energy and Environment, Fakultas MIPA-Universitas Indonesia, 16424 Depok, Indonesia
[a]Corresponding's author: rosari.saleh@gmail.com, rosari.saleh@ui.ac.id

Abstract. Orthorhombic structure of rare earth perovskite-type oxide $LaFeO_3$ nanoparticles with light absorption properties in the visible range was prepared by co-precipitation method. The prepared nanoparticles was characterized by several measurements such as X-ray diffraction, energy dispersive X-ray, field emission scanning electron microscope, diffuse reflectance spectroscopy, vibrating sample magnetometer and electron spin resonance spectroscopy. Methylene blue was taken as the model pollutant to evaluate its photocatalytic, sonocatalytic, photosonocatalytic degradation in the presence of $LaFeO_3$ nanoparticles. The degradation of methylene blue followed the pseudo-first order model in all three processes. The observed rate constants indicate that the photosonocatalytic was faster than the respective individual. Radical scavenger experiments revealed that holes were the predominant oxidative species involved in all three processes.

1. Introduction

With an increase of the industrialization, environmental pollution problem are getting worse. Many contaminants are often discharged from industrial and wastewater treatment plants and contaminate groundwater and surface water. Therefore, it is urgent for the scientist to find solutions to mitigate the damage that is caused by these pollutants. Several studies using catalysts have been employed to decompose harmful organic and inorganic pollutant into harmless chemicals [1-3]. The remarkable progress of photocatalytic has been limited to ultraviolet light although visible light range is far more useful in the solar spectrum. Hence, the development of catalyst that can be activated in the visible region has become important topic research today [4-6].

Perovskite-type materials with the general formula of ABO_3 have attracted considerable attention in the field of catalysis [7-8]. It is reported that perovskite oxide exhibits high photocatalytic activity towards degradation of organic contaminant under visible light irradiation and stable in various environment [9]. Therefore, perovskite oxide with appropriate crystal structure, morphology and optical properties for improvement of photocatalytic efficiency is paid much attention by researchers [6-7].

Recently, it was also reported that ultrasonic bath is used for the degradation of organic contaminants. It is known that during the ultrasonic irradiation sonoluminescence, cavitational and hot spot were occurred, leads to the dissociation of water molecules and generated active radicals. These active radicals have a higher reaction activity could completely decompose pollutant molecule structures in wastewaters [10].

Hence, based on the above reason, $LaFeO_3$ nanoparticles as a typical ABO_3-type perovskite structure were employed as catalysts under visible light and ultrasonic irradiation separately and simultaneously to degrade methylene blue as a model of organic contaminant. The combination of the ultrasonic with the visible light assisted photocatalytic technique is expected to enhance the catalytic performance and to reduce the operating cost needed for degradation of organic contaminants [10-12]. $LaFeO_3$ nanoparticles were synthesized by co-precipitation methods. The structural and morphological properties of synthesized $LaFeO_3$ nanoparticles were characterized by X-ray diffraction, energy dispersive X-ray spectroscopy, field emission scanning electron microscope, energy band gap was determined from diffuse reflectance spectroscopy, while magnetic properties were measured using vibrating sample magnetometer and electron spin resonance spectroscopy. The influence of several experimental parameters on the catalytic efficiency was evaluated. In order to understand the main active oxidative species responsible for the degradation of methylene blue the control experiment with added different types of scavenger were conducted.

Hence, perovskite $LaFeO_3$ nanoparticles can result in a promising catalyst for the removal the organic pollutants under the combination of both irradiations.

2. Experimental details

Chemicals: All chemicals were analytical grade and were used without further purification. For co-precipitation synthesis the following reagents were used $LaCl_3.7H_2O$ (Merck), $FeCl_2.4H_2O$ (Merck), and NaOH (Merck).

Preparation of $LaFeO_3$ nanoparticles: In a typical procedure, 0.2 mol L^{-1} $LaCl_3.7H_2O$ and 0.2 mol L^{-1} $FeCl_2.4H_2O$ were dissolved in 300 ml distilled water to form solution. The solution was sonicated using an ultrasonic cleaner operating at 57 kHz for 1 h and then stirred continuously for 0.5 h at 60 °C. Then, appropriate amounts of NaOH solution with 1.0 mol L^{-1} concentration was added drop wise into the above solution under vigorous stirring at 80 °C for 2 h until the pH value of 13 was reached. The mixtures were allowed to stand at room temperature for 4 h and a large amount of precipitate was formed immediately. The precipitate was centrifuged and washed with distilled water and ethanol several times, then dried at 120°C overnight. Finally, $LaFeO_3$ nanoparticles was obtained after calcinated at 800°C for 2 h under vacuum condition.

Characterization: The crystal structure of $LaFeO_3$ nanoparticles was determined using X-ray diffraction (XRD) measurements using a monochromatic Cu-K_α (λ = 1.54060 Å) radiation operated at 30 kV and 20 mA in the range of 10° to 80°. The X-ray diffraction patterns were analyzed using the MAUD program and Rietveld whole profile fitting to determine the crystal structure and lattice parameters. Elemental analyses and the morphology of the samples were determined using energy dispersive X-ray (EDX) spectroscopy, and field emission scanning electron microscopy (FESEM). The optical characteristic was determined using UV-Vis diffuse reflectance spectrophotometer with an integrating sphere and a spectral reflectance standard over a wavelength range of 200 to 800 nm. Magnetic measurements were performed on vibrating sample magnetometer (VSM). Electron spin resonance (ESR) was performed at room temperature and an X-band spectrometer equipped with a 9.1 GHz field modulation unit.

Catalytic activity: The catalytic activity of $LaFeO_3$ nanoparticles was evaluated by measuring the degradation of methylene blue at room temperature under visible light (photocatalytic) and ultrasonic irradiation (sonocatalytic) separately and the combination between visible light and ultrasonic irradiation (photosonocatalytic) simultaneously. Photocatalytic study was conducted in a cylindrical glass vessel equipped with a magnetic stirrer under a 40 W Xe lamp light positioned horizontally 30 cm above the colloid surface. Sonocatalytic experiments were carried out using a commercially ultrasonic bath operated at fixed frequency and power of 40 kHz and 150 W, respectively. Prior to irradiation, the suspensions were sonicated in the dark for 30 min to obtain a colloidal suspension.

Under ambient conditions and with stirring, the solution in the glass vessel was exposed to light or/and ultrasonic wave. Subsequently, the solution was irradiated for 2 h at regular intervals. The solution was analyzed using UV-visible spectrophotometer. The percentage degradation of organic pollutants was recorded as C_t/C_0, where C_t is the concentration of organic pollutant in each irradiated time interval and C_0 is the initial concentration of organic pollutant after adsorption-desorption equilibrium is reached. In order to study the active species involved in catalytic degradation of organic pollutant, different substances such as ammonium oxalate, sodium sulfate, and tert-butyl alcohol, recognized as scavengers for holes, electrons and hydroxyl radical species, respectively were used.

3. Results and Discussion

Fig. 1a shows the XRD pattern of $LaFeO_3$ nanoparticles. The XRD peaks are well defined and it indicates that synthesized nanoparticles are well crystallized. The XRD reflections appeared are well indexed by an orthorhombic unit cell of $LaFeO_3$ without any impurity phases. The lattice parameters estimated from XRD pattern using the Rietveld refinement as well as crystallized size using the Scherrer's formula [13] are tabulated in Table.1. The values of the lattice parameters and the crystallite sizes are in good agreement with that of orthorhombic $LaFeO_3$ published elsewhere [13-15]. In the inset of Fig. 1a shows the field emission scanning electron images of the synthesized $LaFeO_3$ nanoparticles. It is seen that the synthesized samples consist of agglomerated particles. The individual particles appear to be spherical. In Fig. 1b, the EDX spectra were plotted. The EDX analysis was employed to determine the element of the synthesized materials. As shown in the Fig.1b, the elements present are La, Fe and O with a mole ratio of 1:1:3 corresponding to the stoichiometric composition of $LaFeO_3$.

The magnetic properties of $LaFeO_3$ nanoparticles are investigated, as shown in Fig.2a. The nanoparticles show ferromagnetic behavior, with clear hysteresis loops. The saturation magnetization, remnant magnetization and coercive field are tabulated in Table. 2. This result is of great interest since bulk $LaFeO_3$ has antiferromagnetic behavior [16-17]. Compared with other nanoparticles, magnetization obtained in this study has a higher value with that from other synthesis methods [14, 18-22]. It is reported that the origin of ferromagnetism in $LaFeO_3$ nanoparticles could be due to several reason. The first possibility is due to the uncompensated canted spins from the surface as well as from internal canted spins of Fe^{3+} as the source of magnetic moment [14]. The uncompensated surface spins can be generated due to the increase of surface area or the decrease of crystallite size [14]. This behavior has also been observed in other perovskites nanoparticles such as $BiFeO_3$ [23-24] and $YFeO_3$ [25]. Another possibility is due to clustering such as Fe, FeO and Fe_2O_3, since those have ferromagnetic materials. Finally, ferromagnetic behavior may be originated from double exchange interaction [14]. From our XRD result, it is seen that a secondary phase does not observed in our sample. Therefore, ferromagnetic origin from clustering can be excluded in our sample. In order to study the oxidation state Fe in our sample, ESR measurement was performed. The ESR spectrum of $LaFeO_3$ (Fig. 2b) showed a broad unsymmetry signal. We have carefully studied the signal and found that the spectra considered as a superposition of three overlapping signals, a line at a g value of 2.9978 attributed to La^{3+} ions and another line at g value of 2.2288 and 2.0044 assigned to Fe^{2+} and Fe^{3+} ions. From the XRD and ESR measurements, it is suggested that the origin of ferromagnetism in our $LaFeO_3$ nanoparticles might be due to uncompensated canted spins or double exchange interaction. However, further study needs to clarify this effect quantitatively.

It is known that diffuse reflectance spectroscopy can give information about the electronic interaction near the optical band gap region. The band gap energy of $LaFeO_3$ nanoparticles was calculated from diffuse reflectance spectra by performing a Kubelka-Munk analysis using the following equation: $F(R) = (1-R)^2/2R$ [26], with R as the diffuse reflectance. According to this function the band gap energy can be obtained by plotting the $F(R)^2$ vs. the energy in electron volts.

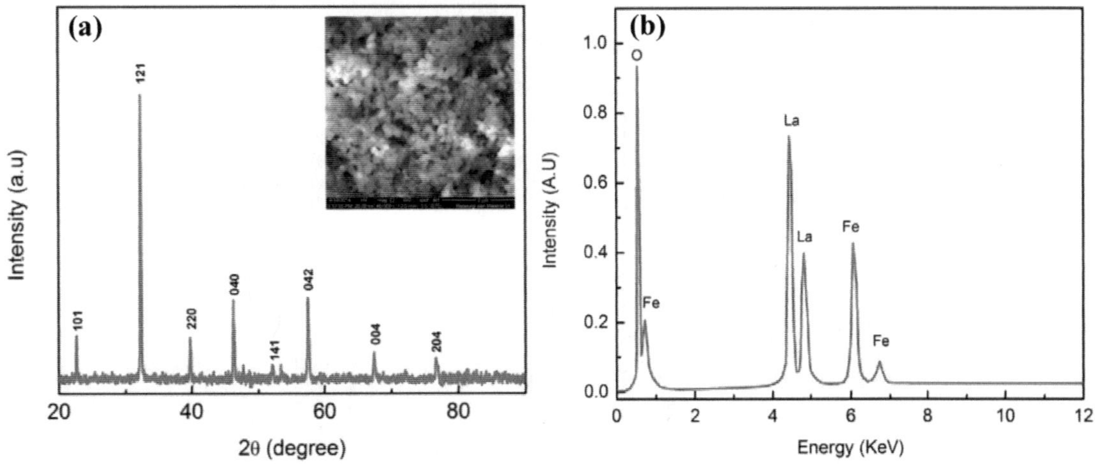

Fig. 1. (a) XRD pattern of $LaFeO_3$ nanoparticles; and (b) EDX spectra of $LaFeO_3$ nanoparticles.

Table 1. The lattice parameter, volume, and grain size of $LaFeO_3$.

Sample	Lattice Parameter			Volume ($Å^3$)	<D> (nm)
	a (Å)	b (Å)	c (Å)		
$LaFeO_3$	5.5436	7.8456	5.5528	241.5072	58

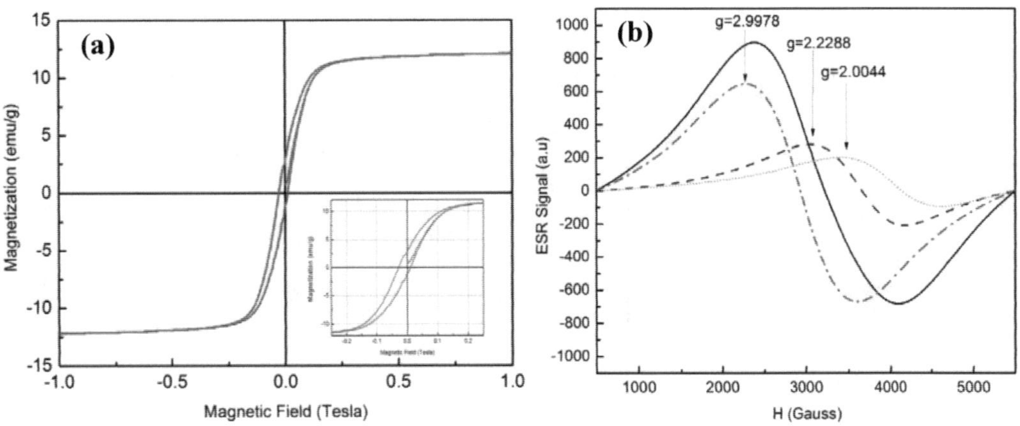

Fig. 2. (a) VSM and (b) ESR spectra of $LaFeO_3$ nanoparticles.

Table 2. The saturation magnetization, remnant magnetization, coercive field and band gap of $LaFeO_3$.

Sample	M-S (emu/g)	Magnetic Remanent (emu/g)	Coercivity (Tesla)	Band Gap (eV)
$LaFeO_3$	12	2.12	0.19	2.45

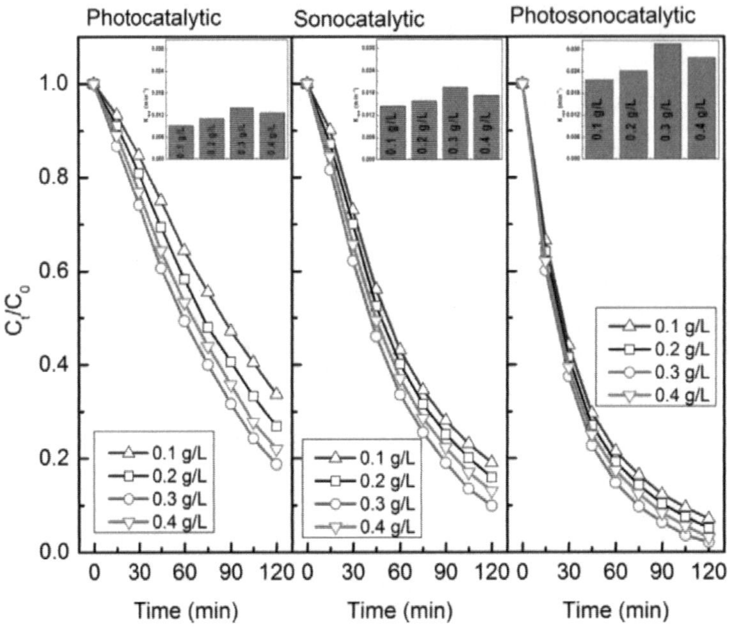

Fig. 3. The photo-, sono- and photosono- catalytic degradation of methylene blue by $LaFeO_3$ nanoparticles.

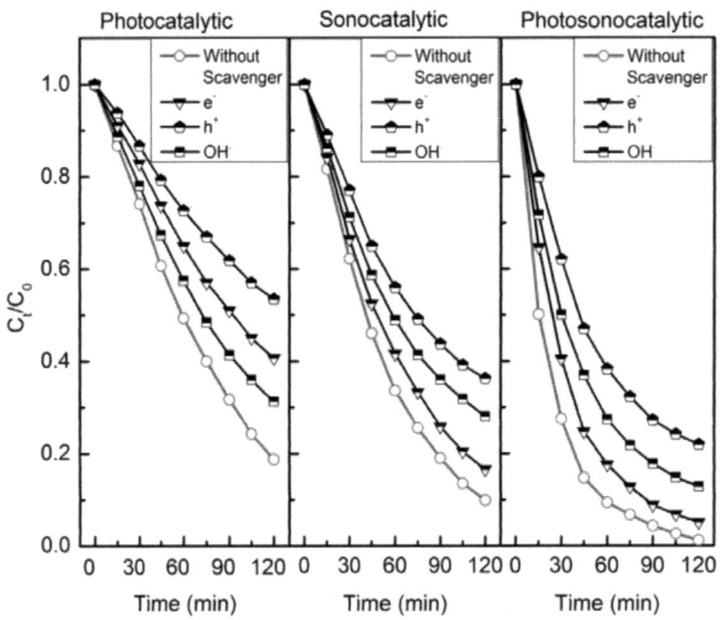

Fig. 4. The effect of various scavengers on the photo-, sono- and photosono- catalytic degradation of methylene blue by $LaFeO_3$ nanoparticles.

The linear part of the curve was extrapolated to $F(R)^2 = 0$ to calculate the direct band gap energy. The result is tabulated in Table. 2. This value is closed to the value reported in the literature [26-28].

The influence of $LaFeO_3$ nanoparticles on the degradation of methylene blue is illustrated in Fig. 3. It can be seen that methylene blue degradation in the presence of both nanoparticles increased with increasing irradiation time. As clearly seen, the degradation of methylene blue occurs appreciably faster under simultaneously irradiation of visible light and ultrasonic than under separately irradiation in a similar experiment condition. The maximum degradation of 100% is obtained for the simultaneously irradiation of visible light and ultrasonic within 2 h. The photo-, sono- and photosono-catalytic degradation of methylene blue by $LaFeO_3$ nanoparticles obeyed pseudo-first order kinetics with respect to the concentration of methylene blue: $\ln(C_t/C_0)= k_{app}\, t$, where k_{app} is the apparent reaction constant for catalysis used as the basic parameter for different catalysts. The apparent reaction constant for different catalyst was studied and the results are displayed in the inset of Fig. 3. The methylene blue degradation under combination of visible light and ultrasonic irradiation exhibits faster degradation rate compared to visible light or ultrasonic irradiation alone. In order to detect the main active oxidative species responsible in the catalytic activity, control experiments with added different type of scavengers for hydroxyl radicals, holes and electrons were conducted with light or/and ultrasonic irradiation. The degradation of methylene blue in the presence of tert-butyl alcohol, ammonium oxalate and $Na_2S_2O_8$ scavenger for hydroxyl radicals, holes and electrons, respectively, is displayed in Fig. 4. It can be seen that the presence of ammonium oxalate scavenger for holes inhibits the removal of methylene blue, which confirmed the indispensable role of holes in the degradation process of methylene blue.

The method of visible light and ultrasonic irradiation combined with catalytic activity can generate reactive oxygen species to degrade organic pollutant. It is known that sonocatalytic process is based on the ultrasonic cavitation effect that can result in sonoluminescence (wide range of wavelength) and hot spot. It is reported that hot spots can cause H_2O molecules to pyrolyze to form •OH radicals. Moreover, sonoluminescence involves intense UV light, which makes $LaFeO_3$ nanoparticles behave like photocatalyst during sonication. Therefore, in sonocatalytic activity ultrasonic irradiation not only induces sonolysis but also generate electron and hole pairs in the catalyst. The photo-generated electrons in the conduction band migrate to the surface where they are scavenged by molecular oxygen O_2 to yield $•O_2^-$ superoxide radical. Meanwhile holes can react with hydroxyl groups (OH^-), and •OH radicals are then formed. Under simultaneous excitation of ultrasonic and visible light, the catalytic activities of $LaFeO_3$ nanoparticles were enhanced, due to the high efficiency charge separation. Since the band gap of $LaFeO_3$ nanoparticles is in the visible range, electron in the conduction band and hole in the valence band can be easily generated by visible light. However, not all photo-generated electrons stay in the conduction band; some of the electrons will be transferred to electron-trapping states, which effectively separate the electrons and holes and suppress the probability of electron–hole recombination. From the ESR spectra, it is seen that there is a paramagnetic centers of La^{3+}, Fe^{2+} and Fe^{3+}. In this process, the electron collected in the paramagnetic centers can later react with the absorbed molecular oxygen to yield O_2^-. Simultaneously, the photogenerated holes in the valence band can then react with the hydroxyl group (OH^-) to form a hydroxyl radical, •OH. Thus, it is concluded that holes play the most important role in the degradation of MB, which is in agreement with the results of control experiments with added scavenger, as shown in Fig.4.

4. Conclusion

In conclusion, this study has shown that methylene blue as a model of organic pollutant can be effectively degraded in the presence of $LaFeO_3$ nanoparticles by using visible light and ultrasonic irradiation simultaneously (photosonocatalytic). The structural characterization of $LaFeO_3$ nanoparticles showed that the structure has an orthorhombic single phase structure with crystallite size of 58 nm and the band edge of 2.45 eV. Study of magnetic property using vibrating sample magnetometer showed that $LaFeO_3$ nanoparticles has ferromagnetic behavior with magnetization of 12 emu/gr at 0.19 Tesla. Based on XRD and ESR measurements we believed that the origin of ferromagnetism in our $LaFeO_3$ nanoparticles is due to the uncompensated canted spin on the surface.

References

[1] R. Andreozzi, V. Caprio, A. Insola and R. Marotta, Advanced oxidation processes (AOP) for water purification and recovery, *Catalysis Today*, vol. 53, pp. 51-59, 1999. muhammad iqbal farabi

[2] K. Pirkanniemi and M. Sillanpaa, Heterogeneous water phase catalysis as an environmental

application: a review, *Chemosphere*, vol. 48, p. 1047–1060, 2002.
[3] A. Srinivasan and T. Viraraghavan, Decolorization of dye wastewaters by biosorbe nts: A review, *Journal of Environmental Management*, vol. 91, pp. 1915-1929, 2010.
[4] P. Tang, Y. Tong, H. Chen, F. Cao and G. Pan, Microwave-assisted synthesis of nanoparticulate perovskite $LaFeO_3$ as a high active visible-light photocatalyst, *Current Applied Physics*, vol. 13, pp. 340-343, 2013.
[5] Q. Peng , B. Shan, Y. Wen and R. Chen, Enhanced charge transport of $LaFeO_3$ via transition metal (Mn, Co, Cu) doping for visible light photoelectrochemical water oxidation, *International Journal of Hydrogen Energy*, vol. 40, pp. 15423-15431, 2015.
[6] R. Abazari, S. Sanati and L. A. Saghatforoush, A unique and facile preparation of lanthanum ferrite nanoparticles in emulsion nano reactors: Morphology, structure, and efficient photocatalysis, *Materials Science in Semiconductor Processing*, vol. 25, p. 301 –306, 2014.
[7] F. Li, Y. Liu, R. Liu, Z. Sun, D. Zhao and C. Kou, Preparation of Ca-doped $LaFeO_3$ nanopowders in a reverse microemulsion and their visible light photocatalytic activity, *Materials Letters*, vol. 64, p. 223–225, 2010.
[8] M. Yang, A. Xu, H. Du, C. Sun and C. Li, Removal of salicylic acid on perovskite-type oxide $LaFeO_3$ catalyst in catalytic wet air oxidation process, *Journal of Hazardous Materials*, vol. B139, p. 86–92, 2007.
[9] S. Thirumalairajan, K. Girija and I. Ganesh, Controlled synthesis of perovskite $LaFeO_3$ microsphere composed of nanoparticles via self-assembly process and their associated photocatalytic activity, *Chemical Engineering Journal*, vol. 209, p. 420–428, 2012.
[10] P. Sathishkumar, R. V. Mangalaraja and S. Anandan, Review on the recent improvements in sonochemical and combined sonochemical oxidation processes – A powerful tool for destruction of environmental contaminants, *Renewable and Sustainable Energy Reviews*, vol. 55, p. 426–454, 2016.
[11] M. Martín de Vidales, S. Barba , C. Sáez, P. Cañizares and M. Rodrigo, Coupling ultraviolet light and ultrasound irradiation with Conductive-Diamond Electrochemical Oxidation for the removal of progesterone, *Electrochim Acta*, vol. 140, pp. 20-26, 2014.
[12] K. Jyothi, S. Yesodharan and E. Yesodharan, Ultrasound (US), ultraviolet light (UV) and combination (US + UV) assisted semiconductor catalysed degradation of organic pollutants in water: oscillation in the concentration of hydrogen peroxide formed in situ, *Ultrason Sonochem*, vol. 21, pp. 1787-1796, 2014.
[13] R. Abazari and S. Sanati, Perovskite $LaFeO_3$ nanoparticles synthesized by the reverse microemulsion nanoreactors in the presence of aerosol-OT: Morphology, crystal structure, and their optical properties," *Superlattices and Microstructures* , vol. 68, p. 148–157, 2013.
[14] S. Phokha, S. Pinitsoontorn, S. Maensiri and S. Rujirawat, Structure, optical and magnetic properties of $LaFeO_3$ nanoparticles prepared by polymerized complex method," *Journal of Sol-Gel Science and Technology,* vol. 71, pp. 333-341, 2014.
[15] M. Milanova, J. Zaharieva, R. Todorovska and D. Todorovsky, Polymetallic citric complexes as precursors for spray-pyrolysis deposition of thin $LaFeO_3$ films, *Thin Solid Films,* vol. 562, p. 43–48, 2014.
[16] W. Lee, H. J. Yun and J. Yoon, Characterization and magnetic properties of $LaFeO_3$ nanofibers synthesized by electrospinning, *Journal of Alloys and Compounds*, vol. 583, p. 320–324, 2014.
[17] D. Treves, Studies on Orthoferrites at the Weizmann Institute of Science, *Journal of Applied Physics*, vol. 36, p. 1033, 1965.
[18] Z. Kaiwen, W. Xuehang, W. Wenwei , X. Jun, T. Siqi and L. Sen, Nanocrystalline $LaFeO_3$ preparation and thermal process of precursor, *Advanced Powder Technology,* vol. 24, p. 359–363, 2013.
[19] A. Cristo´ bal , P. Botta, P. Bercoff and J. P. Lo´ pez, Mechanosynthes is and magnetic properties of nanocrystalline $LaFeO_3$ using different iron oxides, *Materials Research*

Bulletin, vol. 44, p. 1036–1040, 2009.

[20] J. Huang, P. Su, W. Wu, Y. Li, X. Wu and L. Tao, Preparation of nanocrystalline $BiFeO_3$ and kinetics of thermal process of precursor, *Journal of Thermal Analysis and Calorimetry,* vol. 111, p. 1057–1065, 2013.

[21] X. Qi, J. Zhou, Z. Yue, Z. Gui and L. Li, Auto-combustion synthesis of nanocrystalline $LaFeO_3$, *Materials Chemistry and Physics,* vol. 78, pp. 25-29, 2002.

[22] D. Jia, J. Xu, H. Ke, W. Wang and Y. Zhou, Structure and multiferroic properties of $BiFeO_3$ powders, *Journal of the European Ceramic Society,* vol. 29, p. 3099–3103, 2009.

[23] R. Mazumder, S. Ghosh, P. Mondal, D. Bhattacharya, S. Dasgupta, N. Das, A. Sen, A. Tyagi, M. Sivakumar, T. Takami and H. Ikuta, Particle size dependence of magnetization and phase transition near TN in multiferroic $BiFeO_3$, *Journal of Applied Physics,* vol. 100, p. 033908, 2006.

[24] F. Gao, Y. Yuan, K. Wang, X. Chen, F. Chen, J. Liu and Z. Ren, Preparation and photoabsobtion characterization of $BiFeO_3$ nanowires, *Applied Physics Latter,* vol. 89, p. 102506, 2006.

[25] R. Maiti, S. Basu and D. Chakravorty, Shyntesis of nanocrystalline $YFeO_3$ and its magnetic properties, *Journal of Magnetism and Magnetic Materials*, vol. 321, pp. 3274-3277, 2009.

[26] R. Köferstein, L. Jäger and S. G. Ebbinghaus, Magnetic and optical investigations on $LaFeO_3$ powders with different particle sizes and corresponding ceramics, *Solid State Ionics,* vol. 249–250, p. 1 –5, 2013.

[27] J. Li, X. Pan, Y. Xu, L. Jia, X. Yi and W. Fang, Synergetic effect of copper species as cocatalyst on $LaFeO_3$ for enhanced visible-light photocatalytic hydrogen evolution, *International Journal of Hydrogen Energy*, vol. 40, p. 13918–13925, 2015.

[28] K. Parida, K. Reddy, S. Martha, D. Das and N. Biswal, Fabrication of nanocrystalline $LaFeO_3$: An efficient sol-gel auto-combustion assisted visible light responsive photocatalyst for water decomposition, *International Journal of Hydrogen Energy*, vol. 35, pp. 12161-12168, 2010.

Journal of Physics: Conference Series **710** (2016) 012031 doi:10.1088/1742-6596/710/1/012031

Experimental Study on a Standing Wave Thermoacoustic Prime Mover with Air Working Gas at Various Pressures

Ikhsan Setiawan[1], Wahyu N. Achmadin[1], Prastowo Murti[2], and Makoto Nohtomi[3]

[1]Department of Physics, Faculty of Mathematics and Natural Sciences, Universitas Gadjah Mada, Sekip Utara BLS 21, Yogyakarta 55281, Indonesia
[2]Department of Mechanical and Industrial Engineering, Faculty of Engineering, Universitas Gadjah Mada, Jl. Teknika Utara, Yogyakarta, Indonesia
[3]Graduate School of Environment and Energy Engineering, Waseda University, Nishi-tomita 1011, Honjo, Saitama, Japan

E-mail: ikhsan_s@ugm.ac.id

Abstract. Thermoacoustic prime mover is an energy conversion device which converts thermal energy into acoustic work (sound wave). The advantages of this machine are that it can work with air as the working gas and does not produce any exhaust gases, so that it is environmentally friendly. This paper describes an experimental study on a standing wave thermoacoustic prime mover with air as the working gas at various pressures from 0.05 MPa to 0.6 MPa. We found that 0.2 MPa is the optimum pressure which gives the lowest onset temperature difference of 355 °C. This pressure value would be more preferable in harnessing low grade heat sources to power the thermoacoustic prime mover. In addition, we find that the lowest onset temperature difference is obtained when r_{h}/δ_k ratio is 2.85, where r_h is the hydraulic radius of the stack and δ_k is the thermal penetration depth of the gas. Moreover, the pressure amplitude of the sound wave is significantly getting larger from 2.0 kPa to 9.0 kPa as the charged pressure increases from 0.05 MPa up to 0.6 MPa.

1. Introduction

Thermoacoustic prime mover is an alternative energy conversion device which converts thermal energy into acoustic work (sound wave) by applying thermoacoustic effect. The thermoacoustic effect is a mutual thermodynamic interaction between sound wave and solid surface that possess a temperature gradient. One of the advantages of thermoacoustic prime movers is that they can work with ordinary air as the working fluid, because it is free, safe, and we do not need to vacuum the device before we fill it with the working gas. Another benefit, the thermoacoustic prime movers do not produce any exhaust gases, so they would serve as environmentally friendly machines, even when there is a gas leak in the system, especially when air is used as the working gas. In addition, Their structure are simple and they do not use any moving parts in generating the oscillating pressures of the acoustic wave. Hence, the combination of the simple structure and the use of air as the working gas makes the thermoacoustic prime mover can be built and maintained with a low cost.

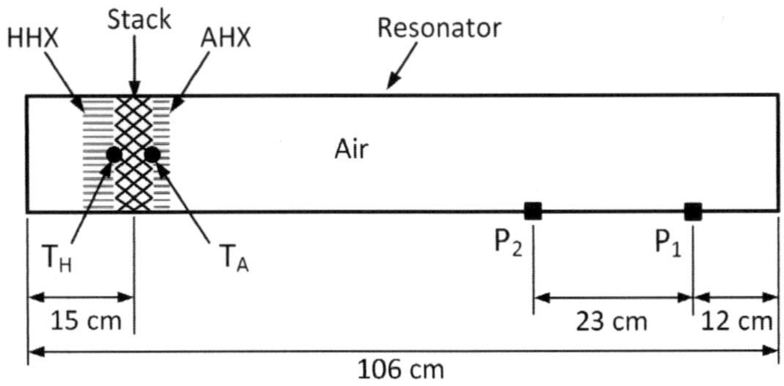

Figure 1. The schematic diagram of the standing wave thermoacoustic prime mover. HHX and AHX are the hot- and ambient- heat exchangers, respectively. T_H, T_A are temperature measurement locations of the hot-end and ambient-end of the stack, respectively. P_1 and P_2 is the mounting locations of pressure transducers.

We have constructed a standing wave thermoacoustic prime mover with air as the working gas. The schematic diagram of the prime mover is shown in Fig. 1. It consists of a straight closed resonator made of stainless-steel pipes, a stack, and a hot- and an ambient- heat exchangers which are installed at both ends of the stack to provide a large temperature gradient across the stack. When a temperature difference across the stack reaches an onset point, the stack starts to produce acoustic waves with a certain pressure amplitude. A low onset temperature difference is desired when low grade heat sources will be used as the thermal energy provider. Whereas high pressure amplitude is needed when we intend to apply it to drive a linear alternator to generate electricity, for instance.

In this paper, we describe our experimental study on the influence of various charged pressures of air inside the resonator on the onset temperature difference and pressure amplitude of the output sound wave. As will be shown later, the results recommend to apply 0.2 MPa pressure when we want the prime mover works with the lowest onset temperature difference, whereas when higher pressure amplitude is more desired, higher charged pressure should be used.

2. Air Properties at Various Pressures

Because we will use air with various pressures as the working gas, we need to know the sound speed and thermal diffusivity in air for different pressures. They are needed to calculate the Lautrec number (N_L), or $\omega\tau$ parameter ($= N_L^2$), which are non-dimensional quantities that describes the nature of heat exchange process between stack and working gas [1]. Here, ω is the angular frequency of the sound wave, τ is the thermal relaxation time in the cross section of the stack's channel, and the Lautrec number is written as [2]

$$N_L = \frac{r_{\text{h}}}{\delta_k}, \tag{1}$$

where r_{h} is hydraulic radius of the stack and δ_k is thermal penetration depth of the working gas, as will be described later.

The sound speed (a) and thermal diffusivity (α) in air at pressure P can be estimated by assuming that the air is a gas mixture which is composed of 78% nitrogen (N_2), 21% oxygen (O_2), 0.9% argon (Ar), and 0.04% carbon dioxide CO_2, namely

$$a = a_{\text{mix}} = \sqrt{\frac{\gamma_{\text{mix}} P}{\rho_{\text{mix}}}} \tag{2}$$

Table 1. The calculated properties of air at different pressures.

P (MPa)	ρ_{mix} (kg m^{-3})	$c_{p,\text{mix}}$ (kJ kg^{-1}K^{-1})	γ_{mix}	k_{mix} ($\times10^{-3}$ W m^{-1}K^{-1})	a (m s^{-1})	α ($\times10^{-7}$ m^{2}s^{-1})
0.05	0.574	1.0060	1.4006	26.400	349.20	456.94
0.10	1.149	1.0068	1.4015	26.424	349.28	228.47
0.20	2.298	1.0083	1.4031	26.474	349.44	114.25
0.30	3.448	1.0099	1.4048	26.523	349.61	76.17
0.40	4.599	1.0114	1.4066	26.573	349.78	57.13
0.50	5.750	1.0130	1.4083	26.622	349.95	45.71
0.60	6.901	1.0146	1.4100	26.672	350.12	38.09

and

$$\alpha = \alpha_{\text{mix}} = \frac{k_{\text{mix}}}{\rho_{\text{mix}} c_{p,\text{mix}}}, \tag{3}$$

where γ_{mix}, ρ_{mix}, k_{mix}, and $c_{p,\text{mix}}$ are the ratio of specific-heats, density, thermal conductivity, and isobaric specific heat of the gas mixture at pressure P, respectively.

The density of the gas mixture can be expressed as

$$\rho_{\text{mix}} = \sum_{i=1}^{4} x_i \rho_i, \tag{4}$$

where i represents the individual component of the gas mixture (i.e. N_2, O_2, Ar, and CO_2), and x_i is the corresponding mole fraction. Similarly, the isobaric specific heat of the gas mixture is

$$c_{p,\text{mix}} = \sum_{i=1}^{4} x_i c_{p,i}, \tag{5}$$

whereas the ratio of specific heats is calculated as

$$\gamma_{\text{mix}} = \frac{c_{p,\text{mix}}}{c_{v,\text{mix}}}, \tag{6}$$

as can be found elsewhere [3]. Furthermore, the thermal conductivity of a gas mixture (k_{mix}) can be approximated by [4]

$$k_{\text{mix}} = \sum_{i=1}^{4} \frac{x_i k_i}{\sum_{j=1}^{4} x_j \phi_{k,ij}} \tag{7}$$

where

$$\phi_{k,ij} = \frac{1}{2\sqrt{2}}\left(1 + \frac{M_i}{M_j}\right)^{-\frac{1}{2}} \left[1 + \left(\frac{k_i}{k_j}\right)^{\frac{1}{2}} \left(\frac{M_j}{M_i}\right)^{\frac{1}{4}}\right]^2 \tag{8}$$

with i and j represent the individual component of the gas mixture, and M_i is molar mass of the i-th component. The properties of the individual component of the gas mixture is obtained from NIST Chemistry Webbook [5]. The calculated values of the air properties within the pressure range of 0.05 MPa – 0.6 MPa are summarized in Table 1.

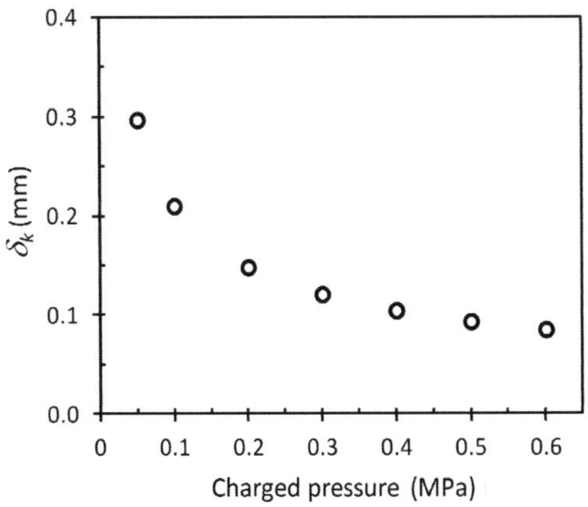

Figure 2. Variation of the thermal penetration depth (δ_k) of air within the pressure range of 0.05 MPa – 0.6 MPa.

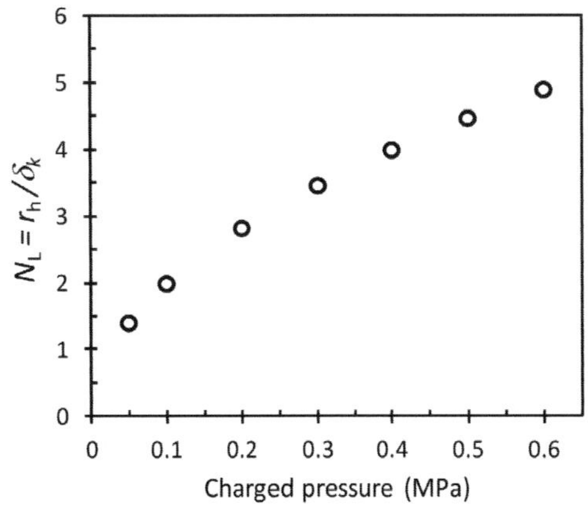

Figure 3. Variation of r_h/δ_k ratio of the standing wave thermoacoustic prime mover in the pressure range of 0.05 MPa – 0.6 MPa.

3. Experimental Condition and Method

The schematic diagram of the standing wave thermoacoustic prime mover used in this study is depicted in Fig. 1. The length and inner diameter of the resonator is L_{res} = 106 cm and D_{res}= 6.8 cm, respectively. This resonator is filled with air with various pressures in the range of 0.05 MPa – 0.6 MPa. The fundamental resonance frequency of the resonator is estimated as

$$f_1 = \frac{a}{2L_{res}}. \tag{9}$$

By using the sound speed values in Table 1, the calculated resonance frequencies of the resonator within the pressure range mentioned above are almost the same, only range from 164.7 Hz to 165.2 Hz. On the other hand, the thermal penetration depth (δ_k) of the air within the resonator is calculated as [6]

$$\delta_k = \sqrt{\frac{\alpha}{\pi f_1}}, \tag{10}$$

which is presented in Fig. 2 for various pressures from 0.05 MPa to 0.6 MPa. The figure shows that the thermal penetration depth is rapidly getting smaller when the pressure increases from 0.05 MPa to 0.3 MPa, and then decreasing more slowly as the pressure increases further.

The stack is made of a pile of stainless-steel wire-mesh screens with mesh number (n) of 16 mesh per inch and wire diameter (D_{wire}) of 0.35 mm. The stack has 5 cm length and is placed inside the resonator so that its center is located at a distance of 15 cm from the left end of the resonator in Fig.1. The porosity (ϕ) of the wire-mesh stack is estimated as $\phi = 1 - \pi n D_{wire}/4$, whereas its hydraulic radius (r_h) is calculated as [6]

$$r_h = D_{wire}\frac{\phi}{4(1-\phi)}, \tag{11}$$

and hence, our stack has $\phi = 0.83$ and $r_h = 0.42$ mm.

By knowing the δ_k and r_h, the Lautrec number N_L is then calculated by using Eq. (1); the result is depicted in Fig. 3 which shows that the r_h/δ_k ratio is getting larger as the pressure increases. In addition, it is found that $N_L \gtrsim 1$ in our experimental condition, which is appropriate for the operation of *standing wave* thermoacoustic prime mover [1, 2].

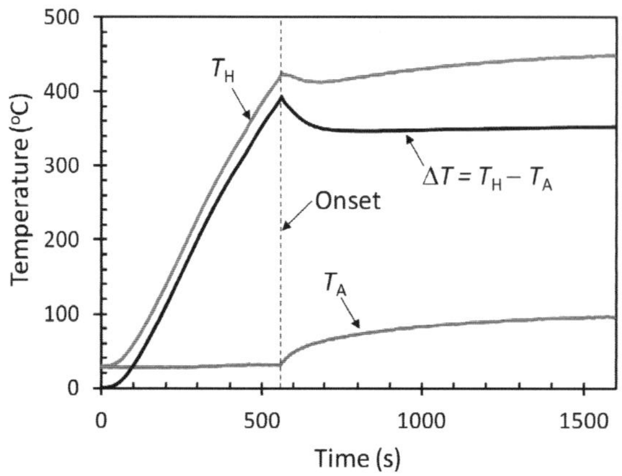

Figure 4. Measurement results of the temperatures at the hot end T_{H} and ambient end T_{A} of the stack as functions of time, with air at 0.4 MPa pressure as the working gas. The temperature difference between the both ends is $\Delta T = T_{\mathrm{H}} - T_{\mathrm{A}}$.

The hot heat exchanger is comprised of a cable electric-heater which is wound around a cylindrical copper block which has many small holes parallel to the resonator axis. The heater has 400 W maximum electric power. The length of copper block is 4 cm. The ambient heat exchanger is composed of a water jacket outside the resonator pipe and cylindrical copper block which also has many small holes and is placed inside the resonator. Both copper blocks are installed tightly at both ends of the stack (see Fig. 1).

To measure the hot temperature T_{H} and ambient temperature T_{A} of the stack's ends, a type-K thermocouple is mounted at each end of the stack. In addition, two pressure transducers (Kyowa PGM-10 KH) are mounted on the resonator wall at a distance of 12 cm and 35 cm from the right end of resonator and used to measure the dynamic pressures of the generated sound wave inside the resonator. The thermocouples and pressure transducers are connected to a data logger which is controlled by a computer.

The experiment is carried out by supplying heat through the 400 W electric heater and measuring the temperatures T_{H} and T_{A} as functions of time until a steady state is reached which is indicated by the constant temperature difference $\Delta T = T_{\mathrm{H}} - T_{\mathrm{A}}$. In the steady condition, the acoustic signals, namely the oscillating pressures, at P_1 and P_2 locations are recorded.

4. Results and Discussion

Figure 4 shows the measurement results of temperatures T_{H} and T_{A}, and the temperature difference between the stack ends ΔT, which are obtained with air at 0.4 MPa pressure as the working gas. In the figure, the onset condition is indicated by the sudden change in temperatures. Before onset, the temperature of the stack's hot-end T_{H} is increasing as the heat is supplied by the electric heater, while the temperature of the ambient end T_{A} is still remain at the room temperature because the heat from the hot end has not yet arrived there, and therefore the temperature difference ΔT is also increasing with almost the same rate as the increasing T_{H}. When the onset condition is reached, the stack starts to produce the acoustic wave. This makes the heat spreads out from the hot end quickly, and hence the temperature T_{H} decreases, whereas T_{A} increases. Consequently, the temperature different ΔT is drop soon after onset. However, because the heat is continuously given by the electric heater, the hot temperature T_{H} is increasing again but more slowly than that before onset. At the same time, the ambient temperature T_{A} is also increasing due to the heat transfer from the stack hot end by conduction through the stack

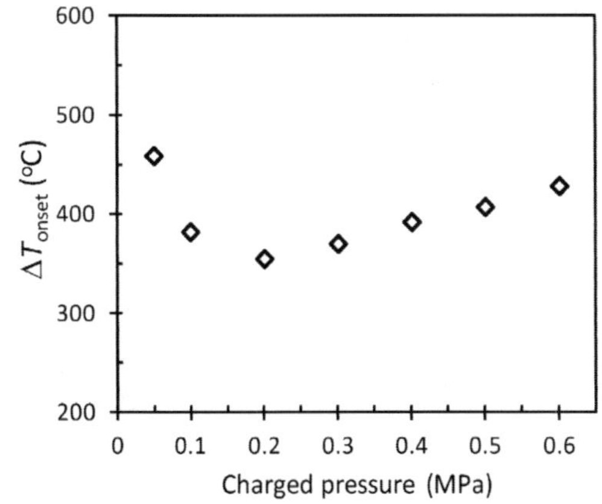

Figure 5. The onset temperature difference $(\Delta T)_{\mathrm{onset}}$ for different pressures of air inside the resonator.

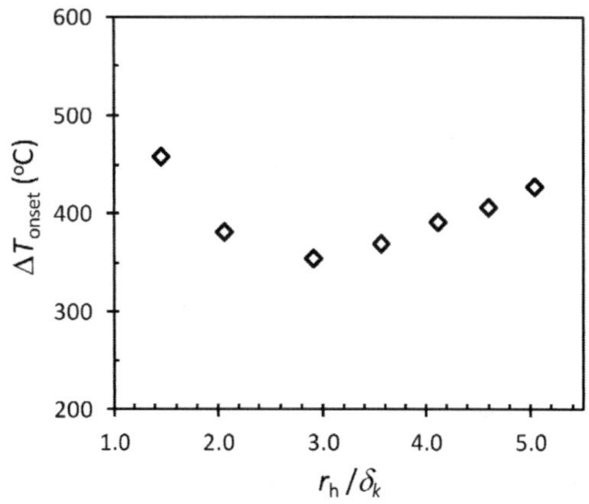

Figure 6. The onset temperature difference $(\Delta T)_{\mathrm{onset}}$ for various r_{h}/δ_k ratios in the standing wave prime mover.

material and by convection via the oscillating gas. Because the increasing rate of the T_{H} and T_{A} are almost the same, then the ΔT tends to be constant. After a long time, all temperatures and temperature difference are constant due to the thermal equilibrium among the parts of the thermoacoustic prime mover and the environment is established.

The influence of the charged pressure in the range of 0.05 MPa – 0.6 MPa on the onset temperature difference $(\Delta T)_{\mathrm{onset}}$ is shown in Fig. 5. It can be seen that the $(\Delta T)_{\mathrm{onset}}$ rapidly decreases from 460 °C to 355 °C as the air pressure increases from 0.05 MPa to 0.2 MPa, and then linearly rises to 428 °C when the pressure further increases until 0.6 MPa. In the other words, the charge pressure of 0.2 MPa is the optimum pressure which gives the minimum $(\Delta T)_{\mathrm{onset}}$ of 355 °C. At pressures lower than 0.2 MPa, the lower density of the working gas makes the $(\Delta T)_{\mathrm{onset}}$ higher, because the average distance among the gas molecules is too large, and hence it is more difficult to start the gas oscillations. Extremely, when the resonator is vacuum, the sound wave will not be generated even by a very large $(\Delta T)_{\mathrm{onset}}$. On the other hand, at pressures higher than 0.2 MPa, the gas density is higher so that the gas stiffness is larger and therefore the gas needs stronger force to oscillates. That is why the $(\Delta T)_{\mathrm{onset}}$ is larger when the charged pressure is higher.

We only focus within 0.05 MPa – 0.6 MPa pressure range because of the fact that the optimum pressure is found at 0.2 MPa giving the lowest $(\Delta T)_{\mathrm{onset}}$, and that the $(\Delta T)_{\mathrm{onset}}$ tends to rise linearly as the pressure increases, for pressures higher than its optimum value. If the pressure increases further, the $(\Delta T)_{\mathrm{onset}}$ might be constant, as obtained by Hao *et al* [7] and predicted by Ikhsan *et al* [3] via the critical temperature difference for gases other than air.

Because the increase of pressure also means the increase of r_{h}/δ_k ratio, as has been depicted in Fig. 3, then the results presented Fig. 5 can be viewed in another way, as shown in Fig. 6 which describes the influence of r_{h}/δ_k ratio on the onset temperature difference $(\Delta T)_{\mathrm{onset}}$. We can see that there is an "optimum" ratio of r_{h}/δ_k in the sense of giving the lowest onset temperature difference. In this case, the optimum r_{h}/δ_k ratio is around 2.85. It means that the thermoacoustic effect most effectively occurs in the stack to start the acoustic oscillations is when the hydraulic radius r_{h} of the stack's channels is 2.85 times the thermal penetration depth δ_k of the gas inside the stack's channels.

Note that the δ_k in Fig. 6 are calculated by using the measured frequencies, they are slightly different from those in Fig.3 where the estimated frequencies (see Eq. 9) are used to calculated

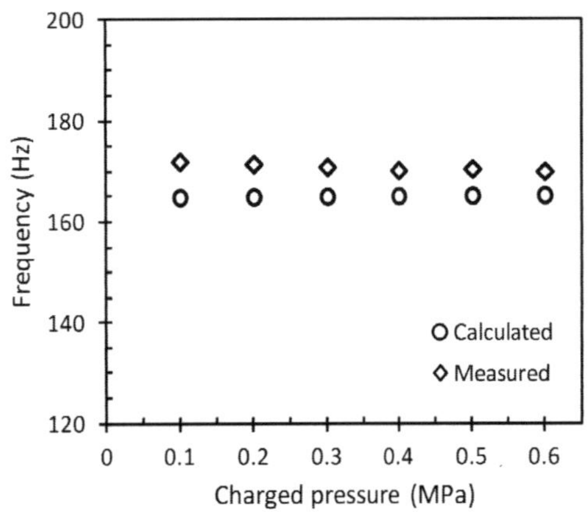

Figure 7. The calculated and measured frequencies for various charged pressures of the air as the working gas.

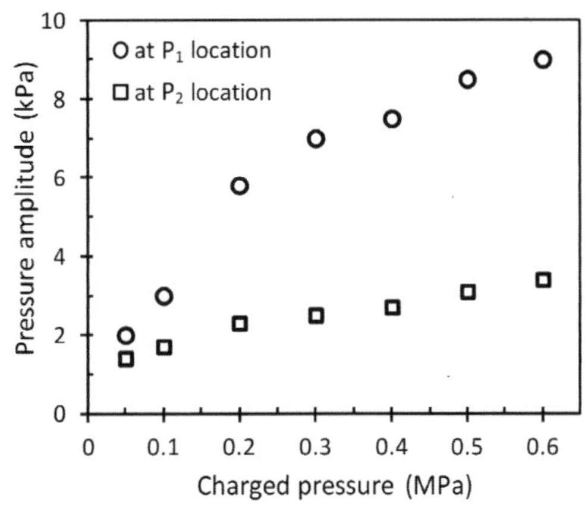

Figure 8. The influence of charged pressure on the pressure amplitudes of the sound wave at P_1 and P_2 locations (see Fig. 1).

them. The difference between the estimated and measured frequencies can be seen in Fig. 7 which shows that the measured frequencies are higher than the estimated ones approximately by 3.6% for each charged pressure. This frequency shift is mainly caused by the existence of the stack and heat exchangers inside the resonator. However, this frequency shift raised the r_h/δ_k ratio only by about 2%.

Figure 8 shows the influence of charged pressure on the pressure amplitude of the sound wave at P_1 and P_2 locations. It can be seen clearly that the pressure amplitudes increase along with the charged pressure increases. It implies that if we want to obtained the output with high pressure amplitude we should apply high mean pressure of the working gas inside the resonator. Furthermore, the pressure amplitude at P_1 is larger than that at P_2 because P_1 is located closer to the pressure antinode at the closed end of the resonator than the P_2 location. In addition, the difference in locations of P_1 and P_2 causes the pressure amplitude at P_1 is increasing faster than the increase of pressure amplitude at P_2 as the charged pressure increases, recalling that the distribution of the pressure amplitude along the resonator approximately follows a sinusoidal pattern.

From the results above, it can be said that air with 0.2 MPa pressure as the working gas is preferable in utilizing low-grade heat sources to power the thermoacoustic prime mover, by which the onset temperature difference is 355 °C. However, if we can provide more heat, which is capable to create temperature difference of 430 °C for example, it is recommended to use higher pressures up to 0.6 MPa due to a significant enhancement in pressure amplitude of the output acoustic work that would be obtained.

5. Conclusions

A standing wave thermoacoustic prime mover with air as the working gas has been experimentally studied. The onset temperature difference, pressure amplitude of the sound wave, and the thermal penetration depth of air working gas in a standing wave thermoacoustic prime mover are significantly influenced by the charged pressure of the working gas. The lowest onset temperature difference of 355 °C is obtained by using 0.2 MPa charged pressure. At this pressure, the ratio of the stack's hydraulic radius to the thermal penetration depth of air is 2.85, which is the optimum ratio to have the lowest onset temperature difference. The onset temperature difference rises linearly as the pressure increases further. In addition, the pressure amplitude of the sound wave significantly gets larger from 2.0 kPa up to 9.0 kPa when the charged pressure increases from 0.05 MPa to 0.6 MPa.

Acknowledgments

The research is supported by Laboratory of Atomic and Nuclear Physics, Department of Physics, Faculty of Mathematics and Natural Sciences, Universitas Gadjah Mada, Indonesia, and Graduate School of Environment and Energy Engineering, Waseda University, Japan. The authors wish to acknowledge the assistance provide by technical staffs of the Laboratory of Atomic and Nuclear Physics: Mr. Rizki, Mr. Jamhari, Mr. Cipto, and Mr. Farid.

References

[1] Yazaki T *et al* 1998 Traveling wave thermoacoustic engine in a looped tube, *Phys. Rev. Lett.* **81** 31283131.
[2] Garrett S L 2004 Resource letter: TA-1: Thermoacoustic engines and refrigerators, *Am. J. Phys.* **72** 11.
[3] Ikhsan Setiawan *et al* 2015 Critical temperature difference of a standing wave thermoacoustic prime mover with various helium-based binary mixture working gases, *J. Phys.: Conf. Ser.* **622** 012010.
[4] Mason E A and Saxena S C 1958 Approximate formula for the thermal conductivity of gas mixtures, *Phys. Fluids* **1** 361.
[5] Lemmon E W *et al* 2015 "Thermophysical Properties of Fluid Systems" in NIST Chemistry WebBook, NIST Standard Reference Database Number 69, Eds. P.J. Linstrom and W.G. Mallard, National Institute of Standards and Technology, Gaithersburg MD, 20899, http://webbook.nist.gov, (retrieved December 17, 2015).
[6] Swift G W 2002 *Thermoacoustics: A unifying perspective for some engines and refrigerators*, Acoust. Soc. Am., New York.
[7] Hao X H *et al* 2011 Influence of working fluid on the performance of a standing-wave thermoacoustic prime mover, *Cryogenics* **51** 559.

Journal of Physics: Conference Series **710** (2016) 012032 doi:10.1088/1742-6596/710/1/012032

Potential of Solar Energy in Kota Kinabalu, Sabah: An Estimate Using a Photovoltaic System Model

F M Markos and J Sentian

Climate Change and Research Group (CCRG), Faculty of Science and Natural Resources (FSSA), Universiti Malaysia Sabah, Jalan UMS, 88400 Kota Kinabalu, Sabah, Malaysia

E-mail: flaviour89@yahoo.com

Abstract. Solar energy is becoming popular as an alternative renewable energy to conventional energy source, particularly in the tropics, where duration and intensity of solar radiation are longer. This study is to assess the potential of solar energy generated from solar for Kota Kinabalu, a rapidly developing city in the State of Sabah, Malaysia. A year data of solar radiation was obtained using pyranometer, which was located at Universiti Malaysia Sabah (6.0367° N, 116.1186° E). It was concluded that the annual average solar radiation received in Kota Kinabalu was 182 W/m^2. In estimating the potential energy generated from solar for Kota Kinabalu city area, a photovoltaic (PV) system model was used. The results showed that, Kota Kinabalu is estimated to produce 29,794 kWh/m^2 of electricity from the solar radiation received in a year. This is equivalent to 0.014 MW of electricity produced just by using one solar panel. Considering the power demand in Sabah by 2020 is 1,331 MW, this model showed that the solar energy can contribute around 4% of energy for power demand, with 1 MW capacity of the PV system. 1 MW of PV system installation will require about 0.0018% from total area of the city. This assessment could suggest that, exploration for solar power energy as an alternative source of renewable energy in the city can be optimised and designed to attain significant higher percentage of contribution to the energy demand in the state.

1. Introduction

Energy use nowadays has become a very critical concern in the last decades because of rapid increase in energy demand. The increased concerns for environmental impacts of conventional fossil fuels most importantly those related to global warming have been the major aspect driving the transformation towards green energy. This generation of power most favorably from renewable energy resources those is abounding and free [1]. Governments and industries from all around the world are progressively looking for ways to reduce the greenhouse emissions from their actions. One major way is to focus on the application and installation of sustainable renewable energy systems [2]. Among all the renewable energy resources, solar power attracted more attentions as one of the greatest promising option since it has many benefits compare to other resources such as wind and hydro energy. Solar energy is naturally free available and clean energy source which derived from the sun that can be harnessed directly to generate electricity [3]. No emission of pollutant, low maintenance and high reliability, with life span expectation of 20 until 30 years made solar power a favorable source of energy to be utilized in the future time [4].

Malaysia introduced new fuel mixed strategy from Four to Five-Fuel Diversification Strategy in year 2002 [5]. Since then renewable energy (RE) is considered as the 5th fuel for the new alternative source Malaysia plans to attain 985 MW contribution of renewable energy in the energy mix in year 2015 [5]. At the present, renewable energy provides less than 1% to the energy mix in Malaysia. By 2020, the aim is for renewable energy to comprise 11% of overall electricity generation in the country. This aim can help to prepare to achieve much long ambitious target to 25% of total usage of renewable energy by year 2050 [6].

Malaysia as an equatorial country has an approximate of 4,000–5,000 Wh/m^2 of daily average solar radiation and a yearly average of 1,643 kWh/m^2 of the radiation received [7]. Considering that Malaysia gets on an average 4–8 hour of free sunshine every day, the potential for solar power generation is very huge and estimated at four times the world fossil fuel resources [8]. Between 1998 and 2002, six pilot grid-connected PV systems were installed with power system capacity vary from 2.8 kWp to 3.8 kWp [9]. By the end of 2005 there were almost 470 kWp of grid-connected PV systems installed in Peninsular Malaysia which one of the most significant is the 362 kWp system at Technology Park Malaysia (TPM). The PV installation at TPM showed Malaysian ability to handle and manage large scale of PV installations [9].

The involvement from the pilot grid-connected projects indicated that such PV applications in Malaysia are indeed reliable and could produce 1.3 times more electricity when compared with similar installations in Germany. The most important is the energy yields are rather stable from month to month since Malaysia is enjoying the tropical weather throughout the whole year. The utilization of solar energy through PV has a huge potential providing several advantages such as BIPV (building-integrated photovoltaic). BIPV system does not need extra land and it generates electricity at the point of use thus reducing electricity transmission losses [10]. Taking into account that only the lower PV capacity value of 1 kWp for every 10 m^2 of available building roof surfaces in these areas the technical potential is around 11,000 MWp which could supply more than 12,000 GWh of solar-generated electricity. This is equivalent to 20% of the national energy demand [10].

1.1. Advantages of solar energy
Solar energy is clearly environmentally advantageous compared to any other energy source. It does not consume natural resources, does not cause CO_2 or other gaseous emission into the air or produces liquid or solid waste products. Relating to sustainable energy development, the main direct or indirectly derived advantages of solar energy are the following [11–14]:

- no emissions of greenhouse
- reclamation of degraded land
- scaling down of transmission lines from electricity grids
- improvement of quality of water resources
- increase of regional energy independence
- diversification and security of energy supply
- acceleration of rural electrification in developing countries

1.2. Photovoltaic (PV) systems
The process of photovoltaic consist a solar cell converts energy in the photons from the sunlight into electricity by means of the photoelectric phenomenon found in certain types of semiconductor materials such as silicon and selenium [15]. The efficiency of solar cells depends on temperature, solar radiation, spectral characteristics of sunlight and so on. Currently, the efficiency of photovoltaic cells is about 12–19% at the most promising conditions. Table 1 presents PV technology goals that have been accomplished between 2000 and 2005 [15]. Solar cell research continues to improve the efficiency of solar cells with targets aimed towards the present accepted limit of 25-30% [16]. The efficiency of another renewable energy such as biomass is much, much lower – perhaps less than 1%. Table 2 shows the cost comparison of energy sources in 2015 [17]. It shows that solar energy is still considered as one of the most expensive among all the renewable energy costs. But the cost of the PV modules is decreased and expected to be keep decreasing in the future [18]. The basic PV module is interconnected and enclosed panel of PV cells has no moving parts and it can last more than 30 years.

The best way to ensure and extend the life and effectiveness of the PV system and any other energy source system is by having it installed and maintained properly [19]. Energy Return on Investment (EROI) is used to compare the quantity of energy supplied to the quantity of energy used in supply process. The higher the ratio the less energy is consumed in producing the energy. Energy balance is the amount of time needed to pay back the energy consumed. The lower the payback time, the sooner the machine starts producing energy that is renewable. Energy Return on Investment = Quantity of energy supplied / Quantity of energy used in supply process [20].

Table 1. PV technology goals been accomplished

Parameters	1995	2000	2005
PV modules efficiency (%)	7-17	8-18	10-20
PV modules cost ($/Wp)	7-15	5-12	2-8
System life (years)	10-20	>20	>25

Table 2. Cost comparison of energy sources 2015

Power Plant Type	Cost ($/kW-hr)
Coal	$0.10-0.14
Natural Gas	$0.07-0.13
Nuclear	$0.10
Wind	$0.08-0.20
Solar PV	$0.13
Solar Thermal	$0.24
Geothermal	$0.05
Biomass	$0.10
Hydro	$0.08

PV systems mainly categorized into 2 main groups which are stand-alone and grid connected systems [21, 22]. Figure 1 shows the components of the PV systems. In summary, a PV solar system consists of three parts which are PV modules or solar arrays, balance of system and electrical load [23]. Large-scale application of this technology is mainly hindered by the high cost of PV panels.

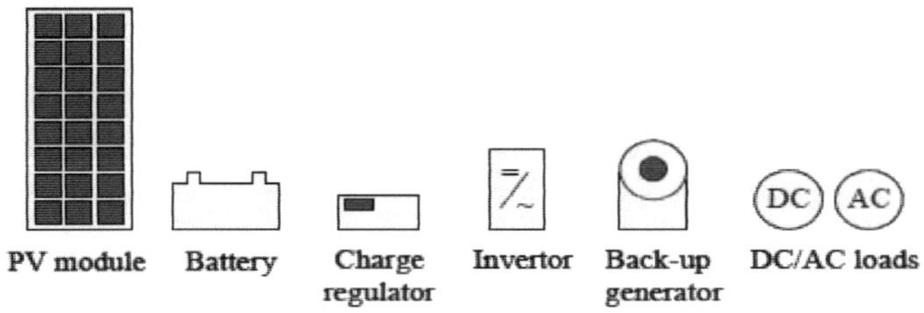

Figure 1. The components of a PV system

2. Study Area

Kota Kinabalu city is the capital of the state of Sabah in Malaysia. This state experiences a typical equatorial humid climate with high humidity, heavy rainfall usually in the afternoon, and the temperatures are uniformly high and extremely invariable throughout the year. The climate is generally hot and sunny all year round. Kota Kinabalu is also under the influence of monsoons which are the northeast monsoon (NEM) from November to March, southwest monsoon (SWM) from May to September and inter-monsoon (April and October) [24]. Kota Kinabalu is the most populous city in Sabah state with 462,963 totals of population with area size of 351 km^2 [25].

2.1. History of electricity supply performance in Sabah
Malaysia is divided into West and East Malaysia by the South China Sea as shown in Figure 2. By the geographical partition, the electric power supply is distributed by three independent grids which are West Malaysia Grid serving all the states in West Malaysia, Sarawak Grid serving Sarawak state in east Malaysia and Sabah Grid serving Sabah state in east Malaysia [26]. Since Sabah Grid is not inter-connected to the other power systems, electricity is supplied from all the power plants within Sabah state. The existing Sabah Grid consists of the East Coast grid and West Coast grid, as shown in Figure 3.

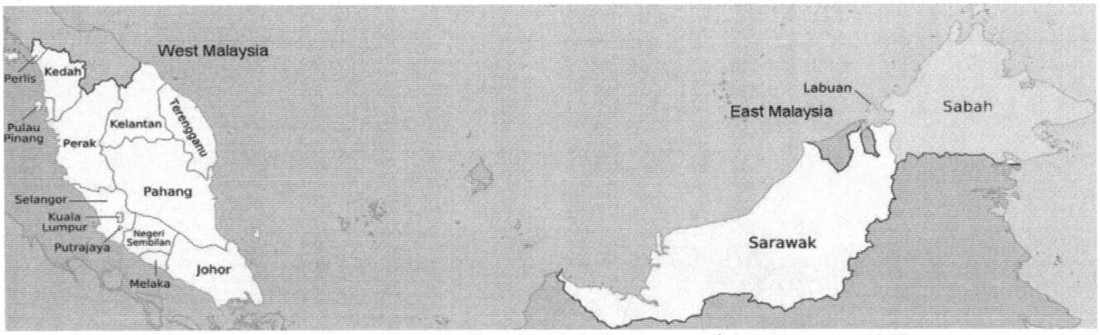

Figure 2. Map of Malaysia with Sabah as one of the 13 states [27]

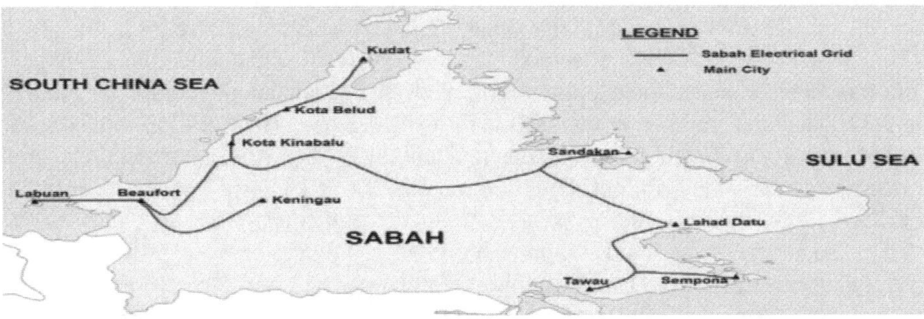

Figure 3. Sabah grid consisting West Coast grid, East Coast grid and East-West grid inter-connection [28]

2.2. Peak demand for electricity
Peak demand for electricity in Sabah from 2014-2023 is expected to grow at a rate of 5.13% per annum. Historically, peak demand growth from 2007-2013 was at an annual average of 6.78%. Table 3 shows the long term load forecast of electricity generation and peak demand for Sabah until the year of 2020. It is shown that by 2020 the electricity peak demand is forecasted to be 1,331 MW [29].

Table 3. Long term load forecast

Electricity	Year	Generation (GWh)	Peak Demand (MW)
Historical	2010	4,726	773
	2011	4,940	830
	2012	5,147	828
	2013	5,506	874
	2014	5,831	917
	2015	6,253	983
	2016	6,687	1,050
	2017	7,132	1,119

Forecast	2018	7,593	1,190
	2019	8,068	1,263
	2020	8,511	1,331

2.3. Electricity generation mix

Sabah is relying heavily on gas-based electricity generation. It is apparent in terms of generation mix as shown in Figure 4. For the generation mix recorded in 2013, gas based generation had the highest share of 67%, followed by marine fuel oil (MFO) and diesel at 21%, hydro at 8% and biomass at 4% [29]. From here it can be seen that renewable energy resources contributed at 12% which combined from both biomass and hydro energy resources. Currently, there is no utility-scale solar PV or large solar power energy that contributing in electricity generation in Sabah.

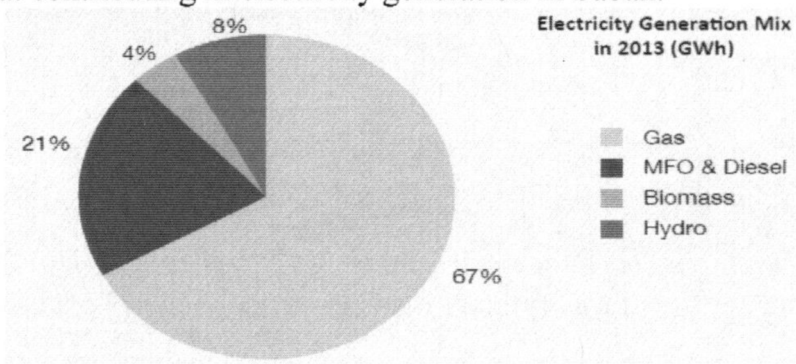

Figure 4. Electricity Generation Mix in 2013 (GWh)

With the rapid increasing of development and as well the annual population in Kota Kinabalu, Sabah, the power demand in this city certainly will also increase. The question is the current power supply would be enough to fulfill this power demand and until when. Clearly fossil fuels are finite and it's only a matter of when they run out. By 2050, one-third of the world's energy will need to come from renewable energy resources such as solar, wind, hydro and other. This paper focuses on studying the potential of solar energy as another renewable energy power source to produce and contribute providing electricity in this area. Therefore, this research was purposely to estimate the solar energy that can be produced in this city.

3. Methodology

There are two phases of methodology in this study. The first is to analyse a year climatic data of solar radiation received in Kota Kinabalu, Sabah and the second is to estimate the solar energy potential by using a PV system model. Statistical analysis of the PV system is presented next to show and compared the accuracy of the PV system model used with study that had been done before.

3.1. Solar radiation data analysis

Solar radiation data are available for many locations on an hourly basis. For that reason many simulation studies of solar energy systems have generally used these hourly values to predict long-term annual performance. However, solar radiation can displays broad difference during an hour. The analyses presented by Vijayakumar et al. [30] confirmed that the variations in solar radiation within an hour cannot be considered insignificant when conducting performance analyses of solar energy systems. Therefore, a short-term data (for every 10 minutes) is considered to carry out this study.

The most accurate measurements of solar radiation are obtained by a pyranometer or pyrometer plant at a location for a number of years, usually more than a decade in order to measure the direct radiation every few minutes. However, the volume of data generated by this process makes it impractical to supply the full data set for each location for PV system design. Instead, the data collected can be presented in some other formats. The most conceptually straight forward method of reducing and arranging the data set is to average the data over the measuring period. This form of data is called average daily, monthly or yearly radiation data. The average solar radiation data, particularly for each month of the year is also extensively used in rough prediction on the total of PV panels required [31]. In this study, annual average solar radiation is used to estimate the amount of PV panels using a PV system model.

3.2. The photovoltaic (PV) system model

This PV system model is basically a calculation to find the solar energy output of a PV system. The global formula to estimate the electricity generated in output of PV system is [32];

E = A * r * H * PR

Where;

E	is the energy output (kWh)
A	is the total solar panel area (m^2)
r	is the solar panel yield (%)
H	is the annual average irradiation on tilted panels (shadings not included) (kWh/m^2)
PR	is the performance ratio, coefficient for losses

Table 4 shows more detailed how to estimate the solar energy produced by using this formula. Below are the characteristics of the solar panel that been used in this PV system model.

One solar panel area = 1.6 m^2

Electrical power = 250 Watt peak (Wp)

The yield of the solar panel (r) is given by the ratio: electrical power (in kWp) of one solar panel divided by the area of one solar panel. Total power of the PV system can be calculated by multiply the electrical power of one solar panel to the total of solar panels used. H is the annual average of solar irradiation that received in Kota Kinabalu, Sabah, as analyzed in the first phase of this study methods. PR is the performance ratio of the whole PV system. This is the coefficient for losses (default value = 0.75). These losses are depending on the site of the system, technology used, and sizing of the system. As shown in the table, the losses are include inverter losses, temperature losses, DC (direct current) and AC (alternating current) cable losses, shadings, weak irradiation and dust.

Table 4. PV system model

a = One solar panel area (m^2)	1.6
b = Total solar panel used	
A = Total solar panel area (m^2)	
d = Electrical power of one solar panel (Wp)	250
r = Solar panel yield (%)	15.63
Total power of the system (kWp)	
H = Annual average irradiation on tilted panels (shadings not included) (kWh/m^2)	
Losses details (depend of site, technology, and sizing of the system) - Inverter losses (6% to 15 %)	8%
- Temperature losses (5% to 15%)	8%
- DC cables losses (1 to 3 %)	2%
- AC cables losses (1 to 3 %)	2%
- Shadings 0 % to 40% (depends of site)	3%
- Losses weak irradiation (3% to 7%)	3%
- Losses due to dust, snow etc. (2%)	2%
- Other losses	0%
PR = Performance ratio, coefficient for losses (range between 0.9 and 0.5, default value = 0.75)	0.75
E = Energy Output (kWh)	

3.3. Statistical analysis of the photovoltaic system model

Statistical analysis of the PV system model can referred to a study by Lim *et al.* [33]. Based on the measurement data on the solar radation at Kota Kinabalu, Sabah, the annual energy output of 1 kWp of PV panel is 1,600 kWh. The study showed that when Kota Kinabalu receiving 1,600 kWh of solar radiation, 1 kWp capacity of PV system will produce 1,600 kWh of energy. The performance of those PV panels systems is estimated with 100% efficiency of solar cell yield with no energy is considered lost during the photoelectric occurred for every area of 1 m^2 of the solar cell. There is also no energy assumed to be lost in the whole sized of PV system (0% losses in inverter, temperature, DC&AC cables, *etc.*). A study by Borhanazad *et al.* [34] not only calculated the average solar radiation but also the peak solar hours (PSH). PSH is used to indicate solar irradiation in a specific location when the sun is shining at its maximum value for a certain number of hours. Since the peak solar radiation is 1 kW/m^2, the number of peak sun hours is numerically equal to the daily solar radiation in kWh/m^2. For example, the daily output of solar array can be estimated to be 535.3 Wh, if 100 Wp solar array is installed in Sabah with an average solar radiation of 5.35 $kWh/m^2/d$. The PV system model that been used in this study is more specific and detailed where some loss is considered occurring in the system.

4. Results and Discussion

The results and discussion of this study are presented in two parts, which the first is the output of analyzed data of solar radiation received in Kota Kinabalu, and the second is the solar energy potential produced estimated by using the PV system model.

4.1. Solar radiation in Kota Kinabalu, Sabah

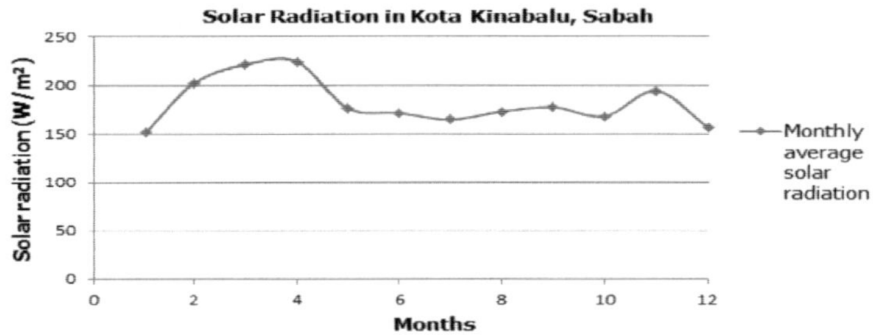

Figure 5. Monthly average solar radiation in Kota Kinabalu, Sabah

A year climatic data of solar radiation was obtained from pyranometer that located at Universiti Malaysia Sabah. Figure 5 shows the monthly average solar radiation received at Kota Kinabalu in 2014. Among the highest solar radiation received was during in February, March and the peak time during in April with 201.77, 221.26 and 224.25 W/m^2 respectively. November received slightly lower among the higher readings with 193.78 W/m^2. January and December was received the less solar radiation with 151.77 and 156.81 W/m^2 respectively. May until October received almost the same amount of solar radiation with 176.51, 171.63, 164.84, 173.10, 177.47 and 168.35 W/m^2 respectively. Overall, the annual average of solar radiation received in Kota Kinabalu was 182 W/m^2.

Despite of rainy season (NorthEast monsoon) in the South East Asia during month November until March they still have tendency increase of solar radiation compare to other months. This is caused by the difference of rainfall regime in this region. The northwest coast of Sabah (including Kota Kinabalu) experiences a rainfall regime of which two maxima and two minima that can be distinctly identified. The primary maximum occurs in October and the secondary one in June. As shown in Figure 5, month October and June are among the months with low amount of solar radiation. The primary minimum occurs in February and the secondary one in August. While the difference in the rainfall amounts received during the two months corresponding to the two maxima is small, the amount received during the month of the primary minimum is substantially less than that received during the month of the secondary minimum. Therefore, month February has tendency higher solar radiation since it received less rain compare month August as the secondary minimum. It can be deduced here that month November until March has higher average of solar radiation because in northwest coast of Sabah they received less rainfall compare with the other months. In some areas, the difference of rainfall is as much as four times.

4.2. Solar energy potential

Before using the photovoltaic (PV) system model, all the solar radiation data retrieved from the pyranometer in W/m^2 unit has to be converted into kWh/m^2 unit. Figure 6 shows the daily and monthly average solar radiation after the conversion. The annual average solar radiation received in Kota Kinabalu was 1, 590 kWh/m^2. This means that Kota Kinabalu receiving around 4.4 kWh/m^2 daily average of solar radiation. This high amount of solar radiation is obviously giving solar energy a huge potential to be harnessed to produce electricity. Higher solar radiation will produced more electric thorough the PV system practically, but technically it will still depend on its system performance and the impact of environmental on it.

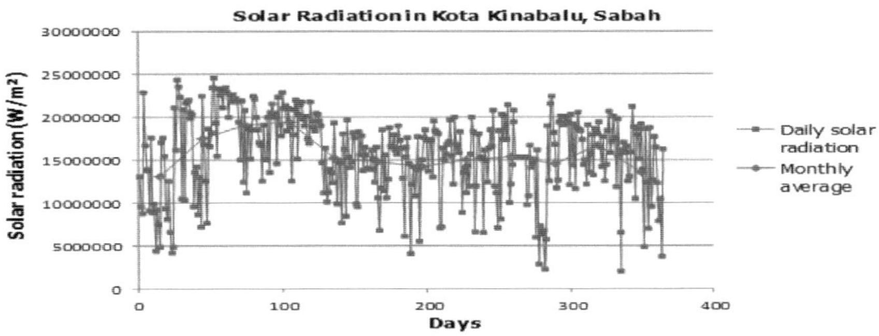

Figure 6. Solar radiation received (kWh/m^2)

In this PV system model, only one solar panel was used to estimate the energy produced. One solar panel will size a 0.25 kWp of total capacity for the PV system. The losses details are shown as in the model and these are only an estimated value of percentage of the energy that been lost in the system. These values always can change depending on the site of PV location, technology used and total sizing of the system. The result of the PV system model shows that, with an annual average solar radiation of 1, 590 kWh/m^2, 0.25 kWp capacity of PV system will produce around 29,794 kWh/m^2 of electricity in a year. This is equivalent to 0.014 MW of electricity produced just by using one solar panel.

To build a PV system with 1 MW capacity, this model will need 4,000 totals of solar panels. If one solar panel has the area of 1.6 m^2, then with 4,000 solar panels it would require 6,400 m^2 of area to install the PV system. Considering Kota Kinabalu area size is 351 km^2, 1 MW capacity of PV system installation will take only about 0.0018% from the total area of the city. The model showed that with 1 MW capacity of PV system, 55 MW of electricity can be produced in a year. Considering the power demand in Sabah by 2020 is 1,331 MW, 1 MW capacity of PV system can contribute around 4% to generate electricity for Sabah. Currently, there is still no large utility-scale solar PV plant or solar energy that contributing in electricity generation in Sabah. Other than the high cost of the PV system installation, the awareness and economics are among the two significant barriers in solar energy implementation in Sabah [35].

The challenge to become a green and resource efficient economy entirely independent of fossil fuels by 2050 has to be accepted one way or another. This model can help sizing the PV plant to make solar power as the main energy source in the future for Kota Kinabalu. Considering the power demand by 2020 is 1,331 MW, the model showed that the PV solar plant will need 100,000 totals of solar panels to generate 1,361 MW of electricity. This is equivalent to 25 MW capacity size of PV solar plant where it would take about 0.05% from Kota Kinabalu total area for the solar plant installation. This is one huge potential energy that can be harnessed within the city.

5. Conclusion

The study showed that Kota Kinabalu received an annual average of 182 W/m^2 solar radiations in a year which is equivalent to 1, 590 kWh/m^2. By using a photovoltaic (PV) system model, Kota Kinabalu is estimated to produce around 29,794 kWh/m^2 of electricity in a year. This is equivalent to 0.014 MW of electricity produced just by using one solar panel. Considering the power demand in Sabah by 2020 is 1,331 MW, 1 MW capacity of PV system can contribute about 4% to generate electricity for Sabah. 1 MW capacity of PV system installation will take only 0.0018% from the total area of the city. This assessment could suggest that, exploration for solar power energy as an alternative source of renewable energy within the city can be optimised and designed to achieve significant higher percentage of contribution to the energy demand in the state.

Recommendation

The results of this paper would recommend to a further feasibility study into implementation of solar energy in Kota Kinabalu, Sabah. This will cover the aspect of costing, project design, obtaining and implementation phases. All components selected must be assured to be mutually compatible and manufactured to provide a professionally engineered and integrated power system, thereby enhancing the photovoltaic system reliability, performance and longevity.

References

[1] Mekhilef S, Saidur R and Safari A 2011 A review on solar energy use in industries *Renewable and Sustainable Energy Reviews.* **15** 1777–1790

[2] Abdelaziz E A, Saidur R and Mekhilef S 2011 A review on energy saving strategies in industrial sector *Renewable and Sustainable Energy Reviews.* **15** 150–168

[3] Saidur R 2010 A review on electrical motors energy use and energy savings *Renewable and Sustainable Energy Reviews.* **14** 877–898

[4] Saidur R and Mekhilef S 2010 Energy use energy savings and emission analysis in the Malaysian rubber producing industries *Applied Energy.* **87** 2746-2758

[5] Energy sector embracing climate change by Pusat Tenaga Malaysia, In: National conference on climate change preparedness towards policy changes 2007 Online: http://www.undp.org.my/uploads/a Energy sector PTM.pdf. May 2011

[6] Online: http://envdevmalaysia.wordpress.com/2011/03/26/renewable-energyneeds- a-push/. May 2011

[7] Overview of Policy Instruments for the Promotion of Renewable Energy and Energy Efficiency in Malaysia Background Report, Online: http://www.serd.ait.ac.th/cogen/62/reports/countries/malaysia.pdf. May 2011

[8] Oh T H, Pang S Y and Chua S C 2010 Energy policy and alternative energy in Malaysia: issues and challenges for sustainable growth *Renewable and Sustainable Energy Reviews.* **14** 1241-1252

[9] Online:http://www.renewableenergyworld.com/rea/news/article/2007/01/equatorialsunshine-the-malaysia-bipv-programme-51560. May 2011

[10] Rosnazri A, Ismail D and Soib T 2012 A review on existing and future energy sources for electrical power generation in Malaysia *Renewable and Sustainable Energy Reviews.* **16** 4047-4055

[11] Tsoutsos T, Frantzeskaki N and Gekas V 2005 Environmental impacts from the solar energy technologies *Energy Policy.* **33** 289-296

[12] Karapanagiotis N 2000 Environmental impacts from the use of solar energy technologies *THERMIE*

[13] Wang Q and Qiu H-N 2009 Situation and outlook of solar energy utilization in Tibet, China *Renewable Sustainable Energy Review.* **13** 2181-2186

[14] Saidur R, Islam M R, Rahim N A and Solangi K H 2010 A review on global wind energy policy *Renewable Sustainable Energy Review.* **14** 1744-1762
[15] Fiorenza G, Sharma V K and Braccio G 2003 Techno-economic evaluation of a solar powered water desalination plant *Energy Conversion and Management.* **44** 2217-2240
[16] Online: http://www.pveducation.org/pvcdrom/design/efficiency-and-cost
[17] US DOE Annual Energy Outlook 2014 Online: http://www.eia.gov/forecasts/aeo/pdf/electricity_generation.pdf
[18] IRENA 2015 Renewable Power Generation Costs in 2014
[19] Online: http://www.gosolarcalifornia.ca.gov/solar_basics/faqs.php
[20] Murphy D J and Hall C A S 2010 Year in review – EROI or energy return on (energy) invested. Ann. N.Y. Acad. Sci. **1185** 102-118
[21] Libo W, Zhao Z M and Liu J Z 2007 A single-stage three-phase grid-connected photovoltaic system with modified MPPT method and reactive power compensation *IEEE Transactions on Energy Conversion.* **22** 881-886
[22] Park J H, Ahn J Y, Cho B H and Yu G J 2006 Dual-module based maximum power point tracking control of photovoltaic systems *IEEE Transactions on Industrial Electronics.* **53** 1036-1047
[23] Zeman M SOLAR CELLS Chapter 9 Photovoltaic systems *Delft University of Technology*
[24] Cooke F M 2003 Living at the Top End: Communities and Natural Resources Use in the Kudat/ Banggi Region of Northern Sabah *WWF Malaysia, Kota Kinabalu*
[25] Online: https://en.wikipedia.org/wiki/Greater_Kota_Kinabalu. September 2015
[26] Interim Report on the Performance of the Electricity Supply in Malaysia for the First Half Year of 2009 2009 *Energy Commission*
[27] Online: http://en.wikipedia.org/wiki/Malaysia. September 2015
[28] Online: http://www.sesb.com.my/corporate_profile.cfm
[29] Sabah Electricity Supply Industry Outlook 2014 *Energy Commission*
[30] Vijayakumar G, Kummert M, Klein S A and Beckman W A 2005 Analysis of short-term solar radiation data *Solar Energy.* **79** 495–504
[31] Online: http://www.pveducation.org/pvcdrom/properties-of-sunlight/analysis-of-solar-irradiance-data-sets. October 2015
[32] Online: http://photovoltaic-software.com/PV-solar-energy calculation.php. September 2015.
[33] Lim Y S, Lalchand G and Gladys M S L 2008 Economical, environmental and technical analysis of building integrated photovoltaic systems in Malaysia *Energy Policy.* **36** 2130–2142
[34] Borhanazad H, Mekhilef S, Saidur R and Boroumandjazi G 2013 Potential application of renewable energy for rural electrification in Malaysia *Renewable Energy.* **59** 210-219
[35] Sulaiman J, Azman A and Saboori B 2014 Development of solar energy in Sabah Malaysia: The case of Trudgill's Perception *International Journal of Sustainable Energy and Environmental Research.* **3** 90-99.

The Preemptive Stocker Dispatching Rule of Automatic Material Handling System in 300 mm Semiconductor Manufacturing Factories

C N Wang[1], H S Lin[1], H P Hsu[2], Yen-Hui Wang[3], Y P Chang[1]
[1] National Kaohsiung University of Applied Sciences, Kaohsiung, Taiwan
[2] National Kaohsiung Marine University, Kaohsiung, Taiwan
[3] Chihlee University of Technology, New Taipei City, Taiwan

E-mail: cn.wang@cc.kuas.edu.tw

Abstract. The integrated circuit (IC) manufacturing industry is one of the biggest output industries in this century. The 300mm wafer fabs is the major fab size of this industry. The automatic material handling system (AMHS) has become one of the most concerned issues among semiconductor manufacturers. The major lot delivery of 300mm fabs is used overhead hoist transport (OHT). The traffic jams are happened frequently due to the wide variety of products and big amount of OHTs moving in the fabs. The purpose of this study is to enhance the delivery performance of automatic material handling and reduce the delay and waiting time of product transportation for both hot lots and normal lots. Therefore, this study proposes an effective OHT dispatching rule: preemptive stocker dispatching (PSD). Simulation experiments are conducted and one of the best differentiated preemptive rule, differentiated preemptive dispatching (DPD), is used for comparison. Compared with DPD, The results indicated that PSD rule can reduce average variable delivery time of normal lots by 13.15%, decreasing average variable delivery time of hot lots by 17.67%. Thus, the PSD rule can effectively reduce the delivery time and enhance productivity in 300 mm wafer fabs.
Keywords—AMHS, 300 mm Semiconductor manufacturing, OHT, Stocker

1. Introduction

The integrated circuit (IC) industry is one of the big global industries in twenty first century. Continuous technology innovation of IC industry is the growth momentum of electronics industry and facilitates improvement in product functions. Semiconductor industry is capital-intensive and highly competitive. To sustain IC companies' competitive advantage, it is very important to maximize the output under the resource limitations and enhance the productivity [1].

The semiconductor industry is at a significant turning point. To maintain industry profitability and growth, new ideas for cost effective manufacturing must be aggressively pursued that go beyond simply enlarging the scale of production to achieve efficiency. As the size of wafer increased from 200 mm to 300 mm, weight of wafer grew from 4 kilograms to 10 kilograms. To prevent the occupational accident and increase yield rate and cleanness, automated material handling system (AMHS) has recently become the indispensable equipment in wafer fabs. The major lot delivery of 300mm fabs is used overhead hoist transport (OHT), which is one of the major vehicles of AMHS. Factory

1 Correspondence: C N Wang, cn.wang@cc.kuas.edu.tw

automation has translated into unmanned plant; therefore, machine operation is fully automated. However, frequently blocking of OHT is happened in intrabay. The moving speed of OHT is very fast around 1 meter per second. The OHT is always moving in the same direction, excluding it is doing the hoisting work. When OHT executes a hoisting work, the other OHTs behind the hoisting one will be blocked. This issue will have great delay for the delivery time of lots. AMHS will directly affect the transportation efficiency of the lots delivery [1].

In semiconductor fab, some products are very important, and they demand short cycle time and on time delivery. These products of high priority are called hot lots. Hot lots have high preemptive against other lots. Due to the increased size and weight of 300 mm wafers, the factories faces more challenges. Pilot production are normally arranged to be hot lots. They are more frequently emerging due to the increased variety of new products with small production volume. The 300 mm automatic transport system should differentiate its services to different priorities to match with frequent process changes and fine tune of the small production volumes of products. This paper objective is to provide the transport services to minimize the delivery delay of hot lots with minimum impacts for normal lots in 300 mm OHT environment [1].

The dispatching rules are very helpful in AMHS environment. proposed A routing algorithm was presented to facilitate both routing and control of the vehicles. The most commonly rules for zone blocking, loading/unloading and traffic control and communications was outlined [2]. Liao and Wang[1] developed a rule for the material handling of high priority porducts called differentiated preemptive dispatching (DPD) policy to provide differentiated priority handling services in a 300 mm fab. Its concept was a lot with higher priority should have its privilege of transportation against those with lower priority.

Some researchers have studied various approaches to improve delivery time performance. Lee et al. [3] implemented efficient scheduling and dispatching policies to gain competitiveness for modern semiconductor manufacturing systems. Fowler et al. [4] remarked that the presence of bottlenecks implies that careful attention and control will not directly have big effect on system performance. Koo et al. [5] described that a system's throughput was usually determined by bottleneck resources, which was the fundamental principle of the theory of constraints (TOC). This paper presents a vehicle dispatching policy to improve bottleneck resource.

However, none of the researchers have ever put attention on the stocker arrangement of 300mm OHT transportation. Therefore, this research proposes a heuristic OHT dispatching rule, preemptive stocker dispatching (PSD) to minimize the transport delay of lots a for lots transport services.

The remainder of this paper is organized as follows. The OHT dispatching rule is proposed in Section 2. rule. Section 3 conducts the experiment designs based on generic data from 300 mm fab of international SEMATECH. Section 4 demonstrates the experiment results and analysis. Finally, conclusions and future research directions are made in Section 5.

2. Preemptive Stocker Dispatching Rule

This study defines a transport job as a macro of transfer commands. The OHT dispatching is an assignment of a transport job to an empty OHT. The OHT variable delivery time is defined as the time to complete a transport job. The dispatched OHT is reserved to the job when it is dispatched, and becomes empty again after completing this job. The objective of OHT dispatching is to minimize OHT delivery time.

The non-reserved OHT will keep moving along the same direction with a specific distance to its front OHT, and all OHTs are equal distribution in each bay. When a set of transport jobs are ready for and waiting to be transferred by OHTs. An empty OHT (reserved OHT or non-reserved OHT) is dispatched to the highest priority job and the transportation of the OHT is preemptive. Once an empty OHT is dispatched to the highest priority job, any other ongoing transport operations, which may block this OHT, will become delay until this highest priority job is completed. If no such highest priority job exists, a non-reserved OHT follows the rule of first-meet and first-serve for normal transport jobs. But if there are several non-reserved OHTs move together, and a lot job is issued before them, the job will be dispatched to the last OHT for avoiding those OHTs blocked by the first OHT. Figure 1 shows the flow of PSD rule.

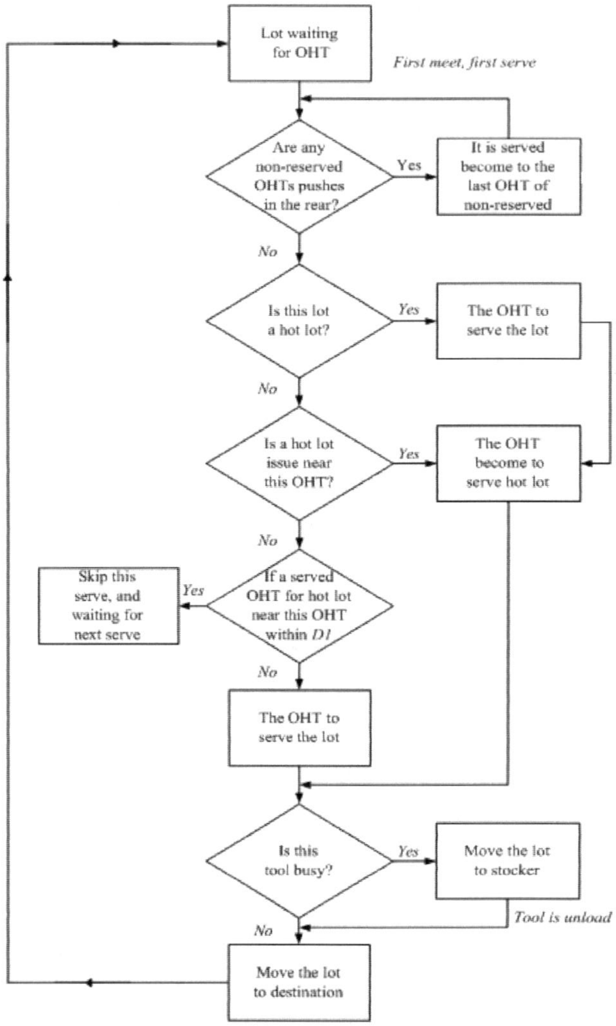

Figure 1. The PSD rule

The PSD algorithm is explained in detail as below:

(1) The overall rule

Step1. OHTs are assigned to each bay evenly, and each intrabay has the same quantity of OHTs.

Step2. When the quantity of OHTs assigned to a certain intrabay is less than the specified number, the empty OHT's of the interbay will go to the intrabay when approaching it. When the quantity of OHTs assigned to a certain intrabay is more than the specified number, the surplus of empty OHT's will go to the interbay automatically.

Step3. The non-reserved empty OHT always follows the "first-meet and first-serve" on transport jobs. Carry the lot to the intrabay of the next process, selecting an unoccupied machine. If all machines are occupied, then carry the lot to the nearest stocker, if the nearest stocker is full, then go to the second nearest, and so forth.

Step4. When any machine of an intrabay is available, if there is any waiting lot at the stocker, the stocker will issue a signal, then the reserved empty OHT will carry the lot from stocker to the available machines according to the rule of "first meet, first serve".

Step5. There are several non-reserved empty OHTs in front of a reserved empty OHT. When a lot job is issued, it is reserved to the last non-reserved OHT, not follows "first-meet and first-serve".

Step6. If a hot lot job is issued, follow the hot lot rule.

(2) The hot lot rule

Step1. When an reserved empty OHT is going to carry the normal lot, as long as a hot lot gives a signal, this reserved empty OHT will abandon the original assignment and change to carry this hot lot.

Step2. The AMHS controller, which is named as material control system (MCS), checks the location of all non-reserved empty OHTs. Reserve the nearest empty OHT to execute this job.

Step3. Even an empty OHT is reserved for a hot lot, if another empty OHT occurs and is closer to this hot lot than the original reserved OHT, the MCS changes the reservation to the nearest OHT.

Step4. If there are no any empty OHT, the MCS waits until the first empty OHT appear. Then MCS reserves this only empty OHT to execute that job.

Step5. If more than one hot lot jobs are issued, an empty OHT follows the "first hot lot meet and first hot lot serve".

Step6. If an reserved empty OHT or an OHT carrying a hot lot are approaching any OHT within a distance *D1*, loading/unloading operations are prohibited to this front OHT for any normal lot. *D1* is designed to prevent any blocking to the OHT with a hot lot by other OHTs. *D1* is equal to the average moving speed multiplied by the hoisting time of lots.

3. Simulation experiments

The AMHS is a quite complex system in the semiconductor fab. It is too complex to complete the real simulation situation. Therefore, This research simplify systems under the International SEMATECH Reports [6][7] for 300mm factory. The generic models are enough for the performance evaluation of the different handling rules. This simulation models are implemented with the discrete-event simulation package - Flexsim from Canyon park technology center. Flexsim is an object-oriented simulation tool with analysis modules that can help engineers and planners make decisions in the design and operation of a system.

Besides the PSD rule, this research also choose differentiated preemptive dispatching (DPD) rule for the performance comparison. The DPD rule utilizes the straightforward idea of first serve with the high priority. Now it has been suggested as a good rule to different preemptive product in OHT applications [1].

Observing the real environment of the OHT system, this research consider three domination control variables: bay loading ratio, hot lots ratio, and the quantity of OHTs. As the number of OHTs increase, system performance will usually be changed due to the increased resources. Therefore, this research considers for configurations of the three kinds OHT numbers, 15, 18 and 21. The increasing hot lots ratio will impose long time delays on the normal lots drastically. Three distributions of hot lots, 2%, 6% and 10% are designed for the tests. The hot lots ratio is the average number of hot lots divided by the average number of total lots in whole factory. Moreover, the heavy bay loadings are used to highlight the effect of the rules in resource limitation. The bay loading is defined as the average quantity of hourly lot input divided by the maximum quantity of hourly lot output of the bottleneck bay. Three loading ratios, 90, 95 and 100% of the design specification, are used in the simulation. The total number of simulation experiments performed is 3 (3 different OHT quantities) × 3 (3 different bay loading ratios) × 3 (3 different hot lots ratio) × 3 (replication) = 81. The simulation horizon is set to 14 days long with a one day pre-run for each experiment.

Moreover, the performance index are defined also following by Liao and Wang's literature [1]. The average lot delivery time = the theoretical moving time + loading and unloading time + product waiting time. The theoretical moving time and loading and unloading time cannot be changed. But waiting time is waste and can be reduced. A good rule can reduce the waiting time. Therefore, a good method can shrink the product delivery time. So, this research take "variable delivery time" as the main index of performance to demonstrate the efficient result.

4. Simulation results and analysis

All simulation results of average variable time for PSD and DPD are demonstrated on Table 1.

Table 1. Average lot variable delivery time in OHT of experiment results

System configuration			Average lots variable delivery time (in seconds)			
# of HOTs	Bay loading	Hot lot ratio	PSD		DPD	
			Normal lots	Hot lots	Normal lots	Hot lots
15	90%	2%	682.705	464.151	810.165	582.736
15	95%	2%	737.646	511.872	830.901	577.993
15	100%	2%	800.646	529.769	865.579	596.828
15	90%	6%	728.800	448.139	846.681	580.723
15	95%	6%	779.033	455.547	930.354	568.331
15	100%	6%	885.845	486.732	971.958	571.988
15	90%	10%	789.306	523.394	890.923	611.834
15	95%	10%	871.854	534.448	984.921	612.430
15	100%	10%	910.069	537.782	997.958	623.750
Average of 15 vehicles			798.434	499.093	903.271	591.846
18	90%	2%	647.169	458.101	782.583	544.208
18	95%	2%	723.671	464.937	841.151	567.685
18	100%	2%	771.190	487.515	844.616	580.333
18	90%	6%	715.300	438.756	827.428	514.394
18	95%	6%	753.622	463.517	900.046	560.138
18	100%	6%	834.093	479.023	958.124	573.280
18	90%	10%	751.102	473.768	870.328	583.065
18	95%	10%	828.670	500.144	965.528	605.588
18	100%	10%	889.746	516.246	971.472	607.425
Average of 18 vehicles			768.285	475.779	884.586	570.680
21	90%	2%	621.922	405.343	737.727	528.472
21	95%	2%	702.836	416.705	813.343	515.513
21	100%	2%	714.409	437.624	816.353	569.101
21	90%	6%	681.807	402.135	794.888	503.144
21	95%	6%	731.446	424.136	844.877	553.001
21	100%	6%	797.992	463.812	939.593	578.811
21	90%	10%	697.103	455.648	841.160	566.363
21	95%	10%	759.838	470.103	943.241	594.096
21	100%	10%	853.492	486.995	968.752	598.147
Average of 21 vehicles			728.983	440.278	855.548	556.294
Total average			765.234	471.716	881.135	572.940

The experimental results show that the average variable delivery time of PSD of is faster than DPD for both hot lots and normal lots in all scenarios. In the 15 OHTs portion, the time for normal lot of PSD is fast compared to that for the DPD (from average 903.271 seconds to average 798.434 seconds), and the time for hot lot of PSD is shorter than DPD (from 591.846 seconds to 499.093 seconds). Overall, the PSD reduces the average delivery time of normal lots by 13.15% (from 881.135 seconds to 765.234 seconds) and the hot lots reduces the average variable delivery time is 17.67%

(from 572.940 seconds to 471.716 seconds) in average of all scenarios. This research finds that the PSD performs significantly well as compared to the DPD. In order to compare these two rules, this study displays the tendency charts of average lots variable delivery time in different configurations. Figure 3 demonstrates the results comparison in different configurations. It demonstrates the average results of 3 kinds of system loading ratio, 3 different hot lot ratio, and 3 different OHT number. When increasing loading or hot lot ratio, all curves trend upward, and when increasing the OHT number, each curve trends downward. Obviously, the Figure 2 manifests that PSD is better than DPD.

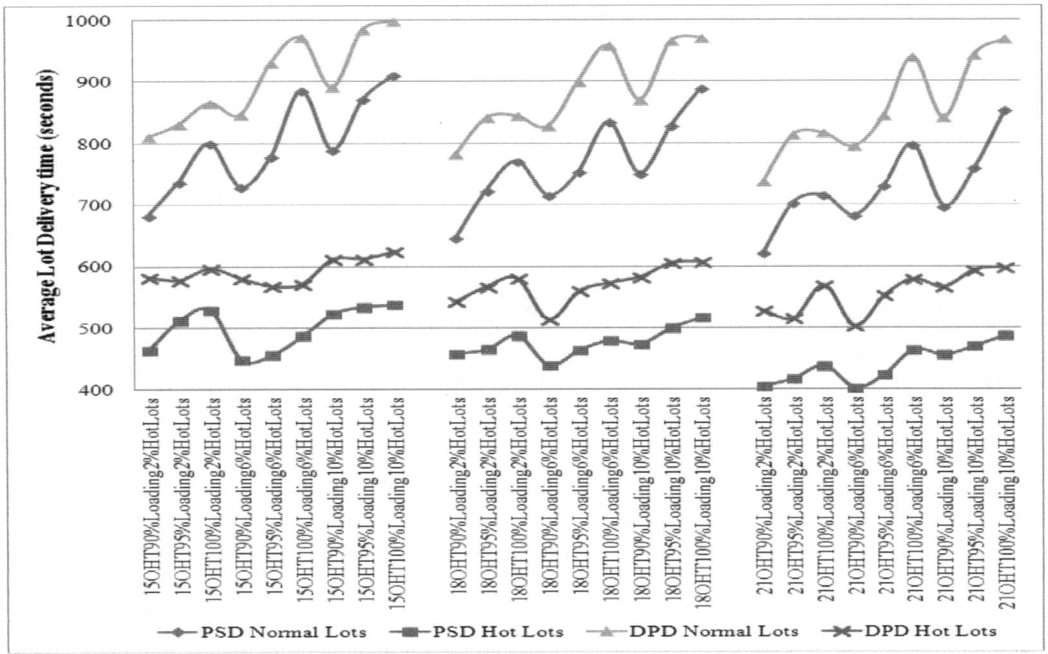

Figure 2. Simulation average results in different configurations

Statistical testing is conducted to distinguish the major difference of these two rules. **Error! Reference source not found.** and **Error! Reference source not found.** shows rule comparison between the DPD and PSD rules for normal lots and hot lots. Compare with DPD and PSD rules by statistics testing, the average transport time of normal lots and hot lots do have significant difference. It depicts the results of testing with pair data for the average variable delivery time. **Error! Reference source not found.** shows the testing with pair data of variable delivery time for normal lots. The t-value is 40.74, and the p-value is 0. There are significantly differences between DPD and PSD rules. This means the normal lots variable time can be significantly reduced by PSD. **Error! Reference source not found.** depicts the results of testing with pair data of variable delivery time for hot lots. Its t-value is 48.78, and the p-value is 0, which means that these two rules are also significantly different. This means the PSD rule still performs better than the DPD rule on hot lot delivery. According to the results above, the PSD is performs better significantly less than the DPD rule in minimizing variable delivery times for both normal and hot lots.

Above the synthesis tests of the result, proposed PSD rule to be able effectively to carry out the lots transportation. It reduces the possible detention of transportation. Simultaneously tests discovered when the OHT resource is sufficient, the PSD rule is effective. However, in real situation of a semiconductor fab, the OHT quantity mostly is sufficient. The PSD rule is a better dispatching rule for shortening the lot variable time for both lots and normal lots when sufficient OHT vehicles are present.

Table 2. Rule comparison between the DPD and PSD rules for normal lots

Normal lots								
Rule	N	Mean	STDEV	SE Mean	The Lower 95% Confidence Interval	DF	*t*-value	*p*-value
DPD	81	881.135	71.431	7.939				
PSD	81	765.234	74.819	8.313				
DPD-PSD	81	115.902	25.602	2.845	110.241	80	40.74	0.000

Table 3. Rule comparison between the DPD and PSD rules for hot lots

Hot lots								
Rule	N	Mean	STDEV	SE Mean	The Lower 95% Confidence Interval	DF	*t*-value	*p*-value
DPD	81	572.940	30.631	3.403				
PSD	81	471.716	37.652	4.184				
DPD-PSD	81	101.223	18.674	2.075	97.094	80	48.78	0.000

5. Conclusions

This paper proposes an effective OHT dispatching rule to transportation services for lots under 300 mm wafer manufacturing. The objective is to minimize the transportation delay of lots and reduce job waiting times. The results of simulation experiments demonstrate that the PSD dominates the DPD in lots delivery performances . Compared with DPD, the PSD rule can reduce the average lot variable delivery time of normal lots by 13.15% (from881.135 seconds to 765.234 seconds) and 17.67% (from 572.940 seconds to 471.716 seconds) for lot lots. From statistical testing, there are significantly differences between the PSD and DPD rules. Therefore, the PSD is better than the DPD. When increasing hot lot ratio or bay loading, the results will get upward trend for normal lots and hot lots. When increasing the OHT number, the trend of results will decline. The result shows the PSD has the shorter variable time of normal lots and hot lots in each environment. The proposed PSD rule is very useful to the shop floor control in 300mm AMHS factory.

For the future researches, it can conform to transmit demand for all levels of products and use PSD rule as the basis of 450 mm wafer fab in transport policy. Moreover, high products mix or segmented dual-track bidirectional loop (SDBL) could be also be consider. Expand simulation model to a full-scale factory environment could be also a good application.

References

[1] Liao D Y and Wang C N 2006 Differentiated preemptive dispatching for automatic materials handling services in 300mmsemiconductor foundry *International Journal of Advanced Manufacturing Technology.* **29 (9)** 890-896

[2] Vosniakos G C and Mamalis A G 1990 Automated guided vehicle system design for FMS applications *International Journal of Machine Tools and Manufacture.* **30(1)** 85-97

[3] Lee Y F, Jiang Z B and Liu H R 2009 Multiple-objective scheduling and real-time dispatching for the semiconductor manufacturing system *Computers and Operations Research.* **36(3)** 866-884

[4] Fowler J W, Hogg G L and Mason S J 2002 Workload control in the semiconductor industry *Production Planning and Control.* **13(7)** 568-578

[5] Koo P H, Jang J and Shu J 2005 Vehicle dispatching for highly loaded semiconductor production considering bottleneck machines firs *International Journal of Flexible Manufacturing Systems.* **17(1)** 23-38
[6] Quinn T and Bass E 1999 *300 mm factory layout and material handling modelling: phase I report* Austin, TX, International SEMATECH
[7] Campbell E and Ammenheuser J 1999 *300 mm factory layout and material handling modelling: phase II report* Austin, TX, International SEMATECH

Encryption On Grayscale Image For Digital Image Confidentiality Using Shamir Secret Sharing Scheme

Rodiah, Dyah Anggraini, Fitrianingsih, Farizan Kazhimi

Department of Informatics, Gunadarma University, Depok, Indonesia.

{rodiah,d_anggraini,fitrianingsih}@staff.gunadarma.ac.id,farizankazhimi@gmail.com

Abstract The use of high-frequency internet in the process of exchanging information and digital transaction is often accompanied by transmitting digital image in the form of raster images. Secret sharing schemes are multiparty protocols that related to the key establishment which provides protection against any threats of losing cryptography key. The greater the key duplication, the higher the risk of losing the key and vice versa. In this study, Secret Sharing Method was used by employing Shamir Threshold Scheme Algorithm on grayscale digital image with the size of 256x256 pixel obtaining 128x128 pixels of shared image with threshold values (4,8). The result number of shared images were 8 parts and the recovery process can be carried out by at least using 4 shares of the 8 parts. The result of encryption on grayscale image is capable of producing vague shared image (*i.e.,* no perceptible information), therefore a message in the form of digital image can be kept confidential and secure.

1. Introduction

Confidentiality is very crucial aspect in information exchange. It requires a security process to keep the privacy hidden. One of the forms of information that can be maintained to be hidden or confidential is in the form of images. As the use of internet is frequently high in the process of exchanging information and digital transaction, sending digital image in the form of raster images is often carried out. One of the methods used in image encryption is Secret Image Sharing. The Secret Image Sharing Method is one of the types of encryption that used to encrypt information in the form of images. This method divides the image into several sections or subsets. Each section (subset) does not have any perceptible information (*i.e.,* vague) and the original image can only be generated by combining the divided sections (subsets) [1]. Secret Sharing Schemes are multiparty protocols that related to the key establishment which provides protection against the threat of losing cryptography key. The greater the duplication of the key makes the risk of losing the key become higher and vice versa. Secret Sharing Scheme overcomes this problem without increasing the number of risks. It can also be used to distribute trust or shared control of critical activity with the intervention of only t of n users [2].

Previous studies related to Secret Image Sharing, among others, is the implementation of Secret Image Sharing on compressed files using Huffman Algorithm [3]. This Algorithm was analogized as the key to open the Huffman file. Shamir Secret Sharing was sufficiently adequate to be used to share general data. Unfortunately, in terms of size, Huffman Algorithm was less effective as the size of the shared data was equal to or greater than the size of the confidential data. Hence, the required size of the overall shared data was n times greater than the size of the source. Secret sharing method was used normally in encoding. The header bytes of the data were encrypted using the Shamir secret sharing, or representing Huffman tree (graph) as matrix and then secret sharing, specifically for matrix, was conducted or in other words variation of visual cryptography principle in which pixels were replaced by bits was used. In this study, to obtain secured secret sharing, secret sharing which was a modified version of visual cryptography for the header and naive secret sharing was carried out [3].

The study of secret sharing on image is extended by dividing the image into several subsets [4]. Every subset of the image was a subset of the original image. Scheme in this study generating two divider images of the original image (x), black and white image, in which the image of x_1 stands for subset 1 and the image of x_2 intended for subset 2. x_1 and x_2 were random distributions of black and white pixels and did not show any perceptible information. When x_1 and x_2 were layered or stacked, the information would be perceptible. It is the same as the original image. If there was only x_1, then the information of x would not be perceptible without x_2. The results showed that binary image generate a share image, namely shared 1 and shared 2 which did not display the information as in the original image. The white shared image had same combinations, while the black shared image had different combinations. White pixel combinations did not fully display white color. Whereas, the black pixel combinations fully displayed black color because black was the color of the image information [4].

Another study was also conducted using Dynamic Embedding Method to produce better visual quality on stego-image [5]. During the experiment, one of the authentication bits on one of pixel blocks on stego-image was changed to determine whether the authentication process was successful. The pixel block on stego-image was randomly selected and one of the authentication bits on the randomly selected pixel block was changed. At each phase of verification, the stego-image was detected as fraudulent stego-image. This proved that the Authentication-chaining Method had high probability for a stego-image, of which pixel bits had been altered, passing through verification phases by using 2 authentication bits. The value of Peak Signal-to-Noise Ratio of stego-image generated by Dynamic Embedding method was always higher than the stego-image generated by other methods. The method of Authentication-Chaining used to authenticate stego-image was able to have high probability in detecting stego-image by using 2 authentication bits [5].

Secret Sharing method was used by employing Shamir Threshold Scheme algorithm on grayscale digital image of 256x256 pixel size. The algorithm generates a shared image of 128x128 pixels with threshold values (4.8). The resulting number of shared was 8 parts and the recovery process can be carried out by at least using 4 shares of the 8 parts. The result of encryption on grayscale image is capable of producing vague share image (no perceptible information), therefore a message in the form of digital image can be kept confidential.

2. Method

In this study, an image is encrypted into several shared images, then the shared images would be decrypted into the original image. Secret sharing is a method in which the keys of cryptographic results are divided into several subsets without increasing any confidentiality risks. Secret sharing also manages any problems relating to distribution of the keys by only allowing access to t of n users in which $t \leq n$ to perform the initial key establishment. The idea of secret sharing is to divide the secret keys into several subsets, called shares, then distribute them to several parties. Only subsets of those parties can or are allowed to re-establish the initial keys [6]. The method used in this experiment is Secret Sharing by employing Shamir Threshold Scheme algorithm. The workflow process of impleenting Shamir Threshold Scheme algorithm.

2.1. Secret Sharing Algorithm

The algorithm used in this paper is Shamir's Threshold Scheme applied to perform encryption or to divide an image into multiple Share images. Shamir secret sharing scheme is a threshold scheme based on polynomial interpolation. A number of k points in two-dimensional space (x_1, y_1), (x_2, y_2), ..., (x_k, y_k) with different x_i will form exactly just one polynomial equation $q(x)$ with degree of $k - 1$ so that $q(x_i) = y_i$ is applied for all $1 \leq i \leq k$.

From the above statement and without diminishing the interpretation of it generally, it can be assumed that D data are numbers, and will be divided into several parts with a number of n, then a polynomial equation with degree of k - 1 is randomly selected as in equation (1).

$$q(X) = a_0 + a_1X + \dots + a_{k-1}X^{k-1} \quad (1)\ [6]$$

In which $a_0 = D$, and for every share which is formed as equation is calculated (2).

$$D_1 = q(1), \dots, D_i = q(i), \dots, D_n = q(n), \quad (2)\ [6]$$

In any subsets of k of Di, the coefficient of polynomial equation $q(x)$ can be searched by performing interpolation, and then calculate $D = q(0)$. However, a number of $k - 1$ of Di cannot or is not sufficient to be calculated in order to get D. Shamir secret sharing scheme method uses modulo (modular arithmetic) as a substitute for real arithmatic. As an illustration, can be seen in Figure 1. For D data, a prime number of p which is greater than D and n is selected. Coefficients of $a_1, a_2, a_3, \dots, a_{k-1}$ in the equation of $q(x)$ is randomly selected from the set of integers in $[0, p)$ and values of $D_1, D_2, \dots, D_n$ are calculated using modulo p [6]. The threshold value (t, n) that is used in our experiment is 4.8. There are several major variable components which are used:

1. M is the original image in which will be kept confidential with the size of H x W (Height x Weight).
2. n is the number of participants or the number of encrypted results (Share).
3. t is part of n or the so-called threshold value (t, n) in which $t \leq n$ and $t = 1/2\, n$ [7].
4. S_i or $S_0, S_1, S_2, \dots, S_{t-1}$ are random integer values for mod P which are the coefficient in the polynomial equation.
5. P is a prime number of which value is greater than M, S_i, n.
6. X_i atau $X_1, X_2, \dots, X_n$ are random values for the participants.

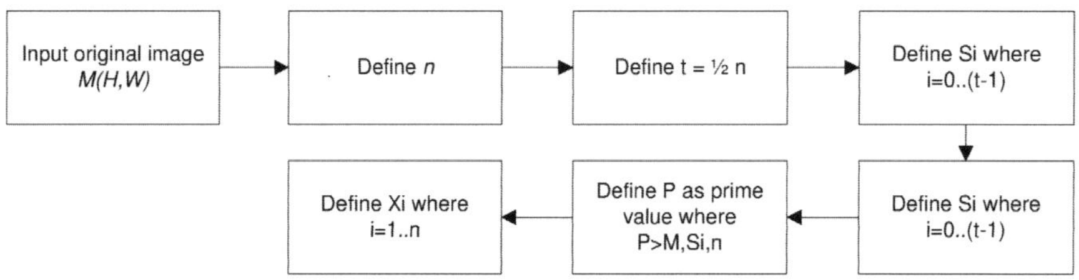

Figure 1. Flowchart shamir threshold scheme

The implementation of Shamir Threshold Scheme in this application is on grayscale images as the secret images or the confidential messages. The gray values in the grayscale images are between 0 and 255, so the variable value of the prime number P which is used is the closest prime number of 255 (the maximum value of gray), with $P = 257$ [6].

2.2. Sharing Process on Shamir Threshold

The Sharing process is aimed at dividing the original image into several share images. An original image M can be divided into n parts for participants by using modulus P basis [6].

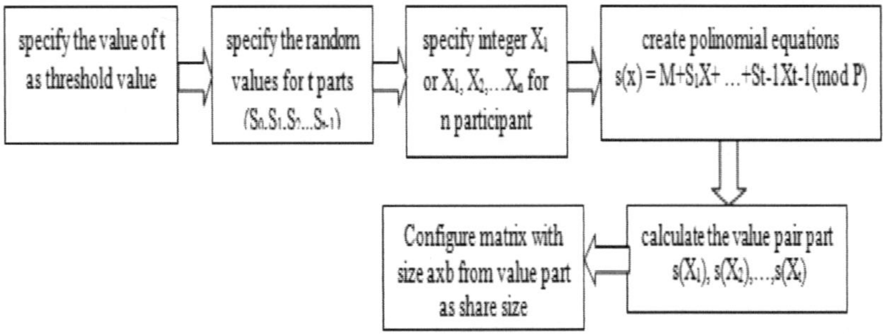

Figure 2. Scheme of Sharing Process on Shamir Threshold Scheme [6]

2.3. Recovery Process on Shamir Threshold

Recovery process is aimed at performing decryption or re-composing share images within the minimum number of *t* participants into the original image [6]. The steps in decrypting the share images can be seen in Figure 3.

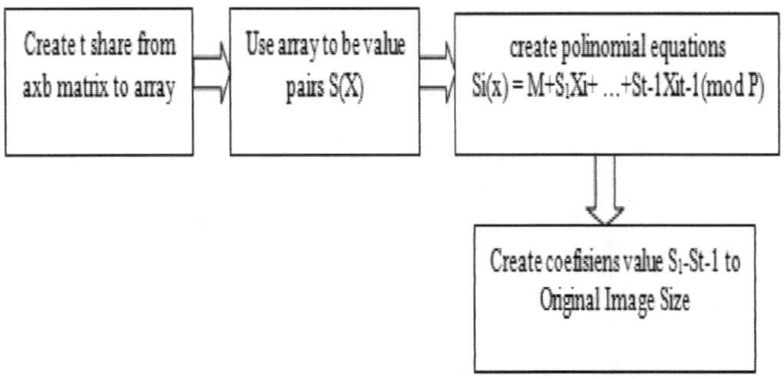

Figure 3. Scheme of recovery process on Shamir Threshold Scheme [6]

2.4. Security Analysis on Shamir Threshold

In restoring an image from *t* share, component of S_i coefficients from polynomial of $s(x) = M + S_1X + \ldots + S_{t-1}X^{t-1}$ (mod P) is required. Each polynomial must have at least *t* number of the coefficients. That is if it has coefficients which are less than *t*, then the composition cannot be known precisely. The resulting probability to estimate the values to be coefficiently correct is 1/256. Consequently, for that reason, it is quite impossible to restore the image with combination of $t - 1$ or the number of shares which is less than the *t* value [6].

3. Results and Discussion

In this study, the implementation of secret image sharing is specifically aimed at encrypting grayscale images with size of 256x256 pixels with threshold value of 4.8. The experiments were conducted on 24 types of grayscale images with 3 different sizes. The format of the images that being used were .tif or .bmp file. The reason for conducting the experiments using those data are due to indicate differences on processing times as well as the results. The followings are 2 of the 24 samples:

1. Image 1. Grayscale image size 256x256 pixels with .bmp format. File name 'clock_gray_256.bmp'
2. Image 2. Grayscale image size 512x512 pixels with .tif format. File name 'cameraman.tif'.

3.1. Image Experiments

Experiment on Image 1

The image on experiment on image 1 was a grayscale image with the size of 256x256 pixels. The format of the image is .bmp format namely 'clock_gray_256.bmp'. The experiment was conducted by using the following variables:

1. M = 256x256
2. $t = 4$
3. $n = 8$
4. Threshold values for participant $x_1 = 1$, $x_2 = 3$, $x_3 = 5$, $x_4 = 8$.
5. $P = 257$

From above variables, the following steps were taken:

a. Confidentiality on M was changed into integer value M = 256 x 256 = 65536
b. Insert components on polynomial equation as follows:

- $S(x) = M + S_1x^t + S_2x^t + \ldots + S_{t-1}x^{t-1} \pmod{257}$
- $S(1) = 65536 + S_1(1) + S_2(1)^2 + S_3(1)^3 \pmod{257}$
- $S(3) = 65536 + S_1(3) + S_2(3)^2 + S_3(3)^3 \pmod{257}$
- $S(5) = 65536 + S_1(5) + S_2(5)^2 + S_3(5)^3 \pmod{257}$
- $S(8) = 65536 + S_1(8) + S_2(8)^2 + S_3(8)^3 \pmod{257}$

Process of dividing image 1 into subsets can be seen in Figure 4.

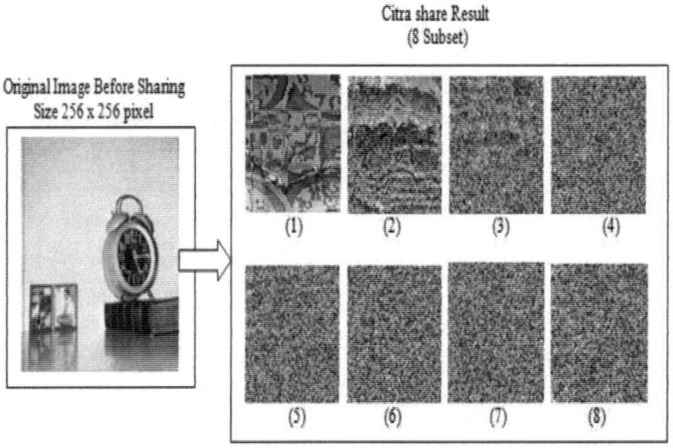

Figure 4. Original image *clock_gray_256.bmp* into 8 Shared Images

Experiment on Image 2

Experiment on image 2, the image used was grayscale image size 512x512 pixels with .bmp format under the file name of 'cameraman.tif'. The experiment was conducted by using the following variable components:

1. M = 512x512
2. $t = 4$
3. $n = 8$
4. Threshold values for participant $x_1 = 1$, $x_2 = 3$, $x_3 = 5$, $x_4 = 8$.
5. Modulus basis = 521

From above variables, the following steps were taken:

a. Confidentiality on M was changed into integer value M = 512x512 = 262144
b. Insert components on polynomial equation as follows:

- $S(x) = M + S_1x^t + S_2x^t + \ldots + S_{t-1}x^{t-1} \pmod{521}$
- $S(1) = 262144 + S_1(1) + S_2(1)^2 + S_3(1)^3 \pmod{521}$
- $S(3) = 262144 + S_1(3) + S_2(3)^2 + S_3(3)^3 \pmod{521}$
- $S(5) = 262144 + S_1(5) + S_2(5)^2 + S_3(5)^3 \pmod{521}$
- $S(8) = 262144 + S_1(8) + S_2(8)^2 + S_3(8)^3 \pmod{521}$

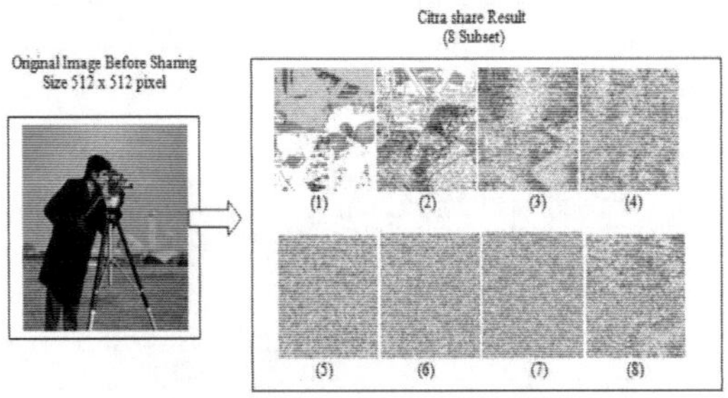

Figure 5. Original Image *'cameraman.tif'* into 8 Shared Images

3.2. Workflows of the Application

3.2.1. Encryption

Figure 6 detail the design flow secret image sharing for encryption process.

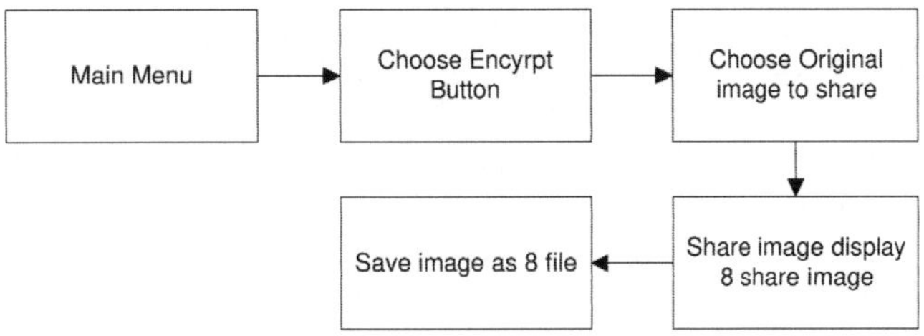

Figure 6. Encryption Process in Application

From the main menu, select the button with the scrambled rubric image (left button). If the button is clicked, the window will display the image encryption/image sharing.

Figure 7. Display of Secret Image Sharing Application Menu

From the window of image encryption/image sharing, click the "Open" button to open the image to be encrypted, as can be seen in Figure 8 (a). The next step is to click the "Share" button to encrypt the opened image into 8 share images as can be seen in Figure 8 (b). To save any of the encrypted share images, click the "Save Share 1", "Save Share 2", ..., "Save Share 8", as needed can be seen in Figure 8 (c).

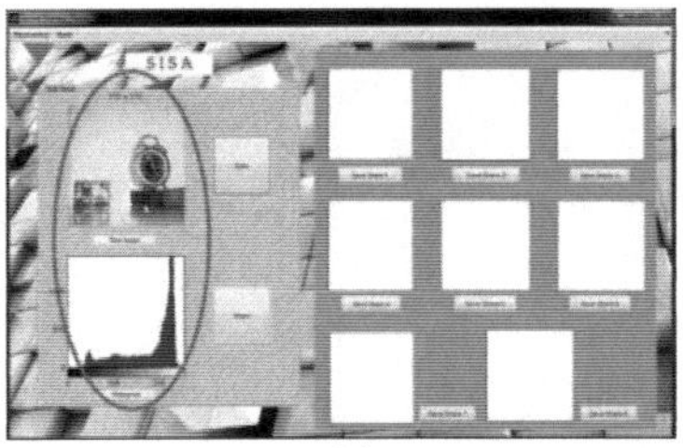

(a)

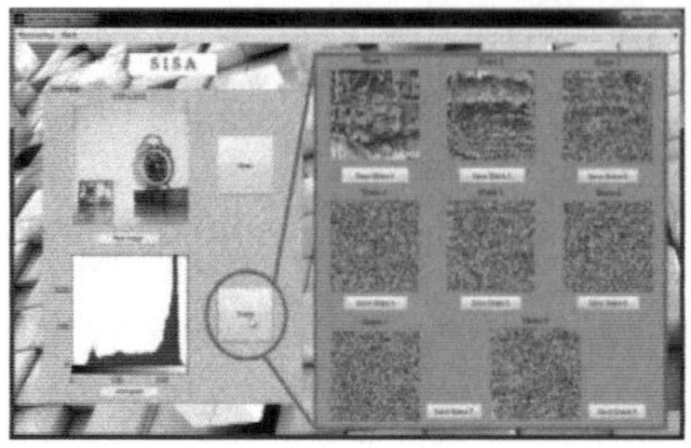

(b)

(c)

Figure 8(a). Display after successfully opening image. (b) Result of Encryption into 8 Shared Images. (c). Display of Saving the encrypted shared images

3.2.2. Process of Decryption

Figure 9 detail the design flow secret image sharing for decryption process.

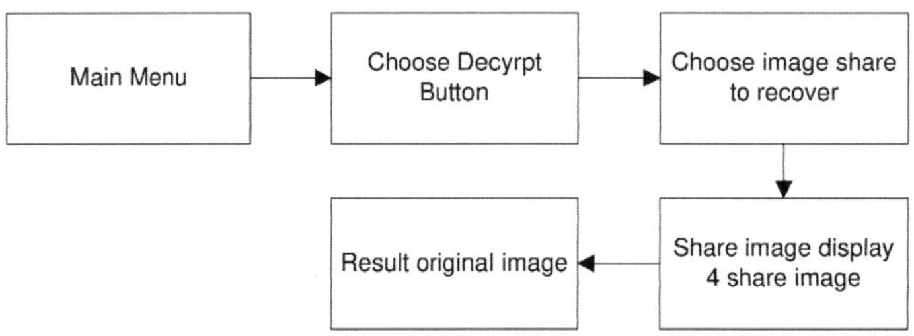

Figure 9. Decryption Process in Application

For decryption process, from the main menu, select the button with a perfectly composed rubric image (right button). If the button is clicked the window will display of image decryption/image recovery can be seen in Figure 10(a). From Figure 10(b) the button of "Open Share 1", "Open Share 2", "Open Share 3", and "Open Share 4", to open 4 of 8 share images/encrypted images are clicked from image decryption/image recovery window. Figure 10(c) shows "Recover" button to perform decryption of the opened shared images into the original image which will be displayed on the right column. Wait until the decryption process is complete. Once it is complete, then the original image will be displayed as can be seen in Figure 10(d).

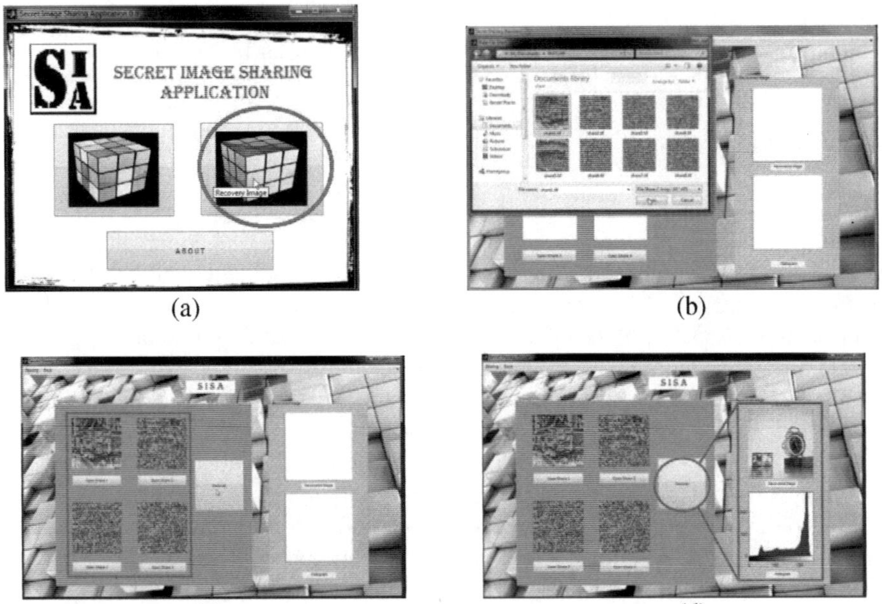

(a) (b)

(c) (d)

Figure 10. (a). Select Right Button for Decryption. (b). Display of Open Shared Figure. (c). Display of the Opened Shared Images. (d) Display of Result of Image Decryption/Image Recovery

3.2.3. Table of Experiment Results

The experiments were conducted on 24 image samples with different formats file. The results are displayed in the following table.

Table 1. Results of Experiments

No	Image Name	Format	Size (Pixel)	Share Result	Recovery Time
1	Cameraman	.bmp	256 x 256	8	2,5 minute
2	Cameraman	.tif	256 x 256	8	2,5 minute
3	Cameraman	.bmp	512 x 512	8	11 minute
4	Cameraman	.tif	512 x 512	8	11 minute
5	Cameraman	.bmp	1024 x 1024	8	43 minute
6	Cameraman	.tif	1024 x 1024	8	43 minute
7	Clock	.bmp	256 x 256	8	2,5 minute
8	Clock	.tif	256 x 256	8	2,5 minute
9	Clock	.bmp	512 x 512	8	11 minute
10	Clock	.tif	512 x 512	8	11 minute
11	Clock	.bmp	1024 x 1024	8	43 minute
12	Clock	.tif	1024 x 1024	8	43 minute
13	Lena	.bmp	256 x 256	8	2,5 minute
14	Lena	.tif	256 x 256	8	2,5 minute
15	Lena	.bmp	512 x 512	8	11 minute
16	Lena	.tif	512 x 512	8	11 minute
17	Lena	.bmp	1024 x 1024	8	43 minute
18	Lena	.tif	1024 x 1024	8	43 minute
19	Plane	.bmp	256 x 256	8	2,5 minute
20	Plane	.tif	256 x 256	8	2,5 minute
21	Plane	.bmp	512 x 512	8	11 minute
22	Plane	.tif	512 x 512	8	11 minute
23	Plane	.bmp	1024 x 1024	8	43 minute
24	Plane	.tif	1024 x 1024	8	43 minute

Recovery time was influenced by the size of the image, as can be seen in Table 1. From the results of the implementation, the average recovery time for image with 256x256 pixels was 2.5 minutes. Whereas, for image with the size of 512x512 pixels, the average recovery time required 11 minutes and for image with 1024x1024 pixels, it required 43 minutes. It took longer time to recover an image of which pixel size is bigger because the matrix size is greater to be processed.

4. Conclusion and Suggestion

Based on the results of experiments using images 1 and images 2 with the size of 256x256, 1024x1024 and 512x512 pixels, the number of shared images was 8. The threshold that used in this experiment was 4.8 by assuming that 4 is ½ of the threshold value generated by each image and 8 is the number of shared images of the original image. In this study, encryption on grayscale images by implementing Shamir Threshold Scheme algorithm was successful in producing vague share images (no perceptible information) so confidentiality of an information within digital image can be maintained. The experiments as can be seen in Table 1 specify the file size affect the recover time. Shamir Secret Sharing Method implemented in this study is able to anticipate the loss of a key when decrypted process. The trial results showed 4 images combined share of total 8 images share the results of the encryption process is able to restore the original image.

In order to develop the application, it is recommended to implement the experiment on other types of images such as stegano images which have additional confidential information inside the images or on color images that have a large range of colors.

References

[1] L. Bai. A reliable (k,n) image secret sharing scheme. In DASC. pages 31–36. 2006

[2] Alfred J. Menezes, Paul C. van Oorschot and Scott A. Vanstone. Handbooks of Applied Cryptography. CRC Press. ISBN: 0-8493-8523-7. 816 Pages. 1996

[3] C. Huang and C. Li. Secret Image Sharing Using Multiwavelet Transform. Journal of Information Science and Engineering, 733-748. 2011.

[4] Jagdeep Verma, dan Vineeta Khemchandani. A Visual Cryptographic Technique to Secure Image Shares, International Journal of Engineering Research and Applications (IJERA), ISSN : 2248-9622, Vol. 2, Issue 1, pp.1121-1125. 2012

[5] Widyadhana, Arya, dan Muchmamad Husni. Penerapan Secret Image Sharing Menggunakan Steganografi dengan Metode Dynamic Embedding dan Authentication-Chaining, Jurnal Teknik ITS, Vol.1. ISSN: 2301 – 9271. 2012

[6]. Wang, Shuang. Distributed Storage scheme Based on Secret Sharing Schemes, University of Oklahoma, Tulsa, OK, USA. 2012

[7] Gonzalez and Woods. Digital Image Procssing. Second Edition. Prentice Hall. 2002

Evaluation of Condition Assessment of Tunnel Lining using Inspection Manual of CIRIA and FHWA

Fathoni Usman[1], Nursimaa Banuar[1], Mohd Nadzari Ismail[2] and Nor Azura Othman[2]

[1]Civil Engineering Department, Universiti Tenaga Nasional, Malaysia
[2]TNB Research Sdn. Bhd, Malaysia

E-mail: fathoni@uniten.edu.my

Abstract. Maintenance need to be provided as the tunnel has serves such a long time and it is to ensure no severe failure in the tunnel will occur that would lead to the unwanted causalities. This paper presents visual condition assessment of tunnels of hydroelectric power station to determine deterioration level of shotcrete lining based on FHWA, and CIRIA report C671. From the assessment, it is found that 40% of leakage condition is in rating 2 for MAT and 63% also leakage in rating 2 for CVT. The most severe area is along Chainage 0+300 for MAT and Chainage 0+400 for CVT. Shotcrete lining is also found able to function properly to provide drainage diverting the leaking.

1. Introduction

There are three large main sections in the tunnel of a hydroelectric power station which are Main Access Tunnel (MAT), Cable and Ventilation Tunnel (CVT), and Cavern where the turbine and machine hall are located. The matter of water in this assessment is considerably concern since it may result in further instability of the tunnel lining and induce to rock fall if it is not taken seriously. As shotcrete act to protect and cover weak formation of rock face, inspections on shotcrete lining have to be conducted in order to maintain the shotcrete lining condition to remain in a good condition.

The common defect in tunnel is leakage. Leakage is characterized by water seeping through the lining and is further classified based on observed inflow. According to Federal Highway Administration (FHWA), level of leakage is divided into 3 which is minor (i.e. the concrete surface is wet although there are no drips), moderate (i.e. active flow at a volume of less than 30 drips/min), and severe (i.e. active flow at volume greater than 30 drips/min). FHWA has proposed a threshold value of leaking where the leaking still in a range of tolerance. The limit of allowance actually has developed since 1960's by tunnel's owner. A maximum value of 0.8 liters/minute per 75 linear meters is one of the allowable limits [1]. This limit is for reference purposes only, with the main emphasis for determining repair required on the location of the leak and the condition of the tunnel components that are affected. Meanwhile, CIRIA, a research and information organization dedicated to improvement in all aspects of the construction industry categorized the leakage into 5 conditions from the damp patch on the leaking surface to continuous leak when they fall at a rate of about 300 drops/minute [2].

This assessment focus on MAT and CVT where the MAT function as access way to the cavern while CVT act to dilute and remove contaminants, control temperature, smoke control and improve visibility in the tunnel. The CVT also provide fresh air to prevent dangerous or harmful accumulation of dust, mists and vapors or gases. The objective of this study is to determine condition of the shotcrete lining along the MAT and the CVT subjected to deterioration using the both inspection manual of FHWA and CIRIA.

2. Methodology

Visual inspection on the surface of shotcrete lining along MAT of 1050 m and CVT of 750 m length is conducted to record and determine its condition. Figure 1 shows the process in a form of flow chart diagram. It starts with preparation of standard form and tools to be used in the inspection processes. The details recorded on every defect observed including type of defects, deposited color, number of drippings due to leakage, dimension of deteriorated areas and location of defects as in Figure 2. Also, some samples are collected (i.e. water drippings, deposited materials, and soot). Figure 3 shows the area of observation along cross section the tunnel.

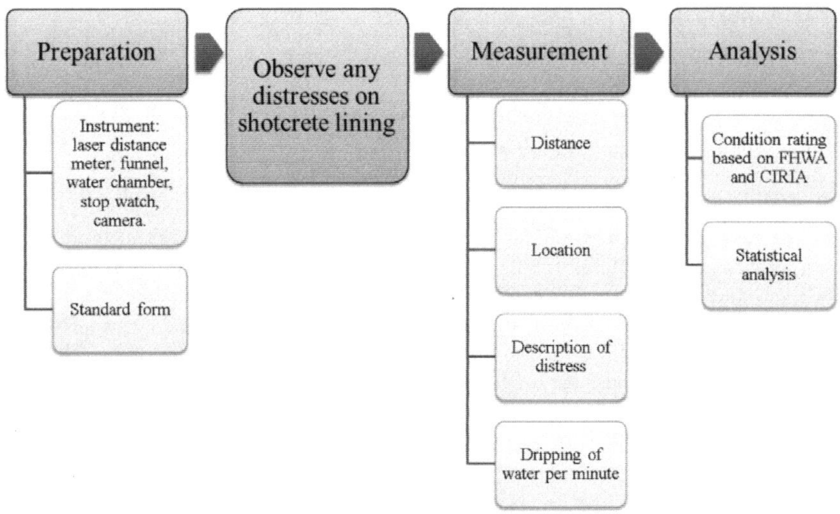

Figure 1. Research methodology

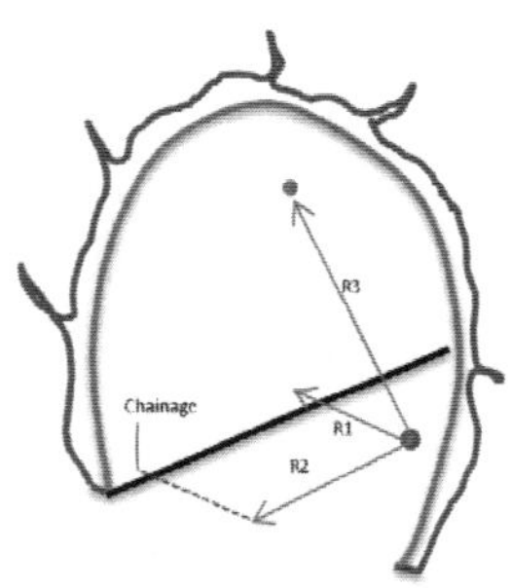

Figure 2. Method to determine location of defect on shotcrete lining

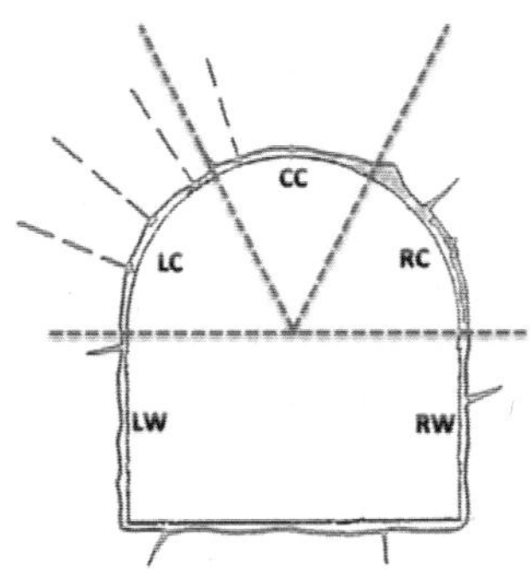

Figure 3. Area of observation in the tunnel.

After the completion of visual inspection on both MAT and CVT, analysis on the data obtained during the visual inspection take place. Descriptive statistical analysis is used. The level of deterioration shall be determined. Thus, the areas of severe defects manage to be discovered.

Journal of Physics: Conference Series **710** (2016) 012035 doi:10.1088/1742-6596/710/1/012035

3. Analysis and discussion

The obtained data consist of defects location, description of defects, images, level of leakage based on FHWA and CIRIA respectively, and deposited color. Majority the type of defects observed along the MAT and CVT are leakage and corroded anchor bolts are also observed. The summary of the tabulated data to be analyzed is as shown in Table 1. Graphs of frequency of leaking observed are used to determine the common type of leakage and the most severe area along the MAT and CVT.

Table 1. Example of tabulated data from the visual assessment.

Ref.no	Running dimension	Leakage (FHWA)	Leakage (CIRIA)
LW-45	0 + 247.26	1	1
LW-46	0 + 246.92	2	2
LW-47	0 + 248.69	2	2
LW-48	0 + 253.04	1	5
LC-34	0 + 258.92	2	2
CC-3	0 + 259.16	2	1
RW-29	0 + 275.396	1	1

Figure 4 represents the condition of water leaking on the rock surface before the application of shotcrete lining. The data shows that the most severe area has occurred along the chainage 0+600 to 0+700 which are most likely at rate 4 to 5 based on CIRIA guidelines. It is clearly can be seen that the leaking along the chainage 0+600 to 0+700 have reduced to rate 2 to rate 3 after application of shotcrete lining as shown in Figure 5.

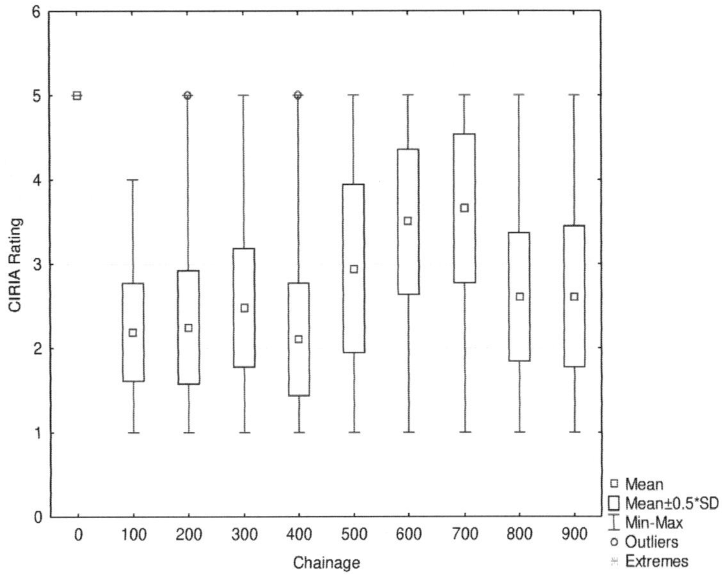

Figure 4. Water leaking condition along the MAT before shotcrete lining was applied

From the box plot analysis of MAT based on the CIRIA and also FHWA leakage scale, the result obtained indicates that chainage 0+300 to chainage 0+400 and chainage 0+900 to chainage 0+1000 were in severe condition as shown in Figure 5. Both chainages represent condition of continuous leak of water drips and they fall at dripping rate of about 300 drops/min which is at scale 5. Meanwhile, less affected area have observed along the chainage 0+500 to chainage 0+600 where most of the defects have been categorized into condition 1 which is representing a damp patch-discoloration of part of the surface of a lining and moist to touch.

Majority the leakages observed along the CVT are in moderate rate which can be said that there are only minor numbers of leakages classified as severe which are needed to be taken into further investigation. Based on box plot analysis as in Figure 6 and histogram in Figure 8 along the CVT, it shows that majority of leakage observed are classified in level 2 (i.e. seeping visible movement of film of water at the surface). It is discovered that leakages are mostly occurred from chainage 0+000 to chainage 0+100 with number of 78 moderate leakages observed. The largest number of severe leakage are discovered occurred at chainage 0+200 to 0+300 and at chainage 0+400 to chainage 0+500 hence the area need to be considered for further action in order to avoid failure.

It is observed that after almost 20 years of application of shotcrete lining, the function of the lining for the purpose of supporting the tunnel and also for diverting ground water leaking through the discontinuity of rock has degraded. Normally the process of degradation of the shotcrete lining is started when the ground water percolate or seep through the weak formation of structure of rock and the tunnel lining. It is also caused by calcination process where it clogged the weeping holes on the lining. Proper maintenance works need to be conducted to avoid possible failures such as falls of rock through or with the shotcrete and inability of shotcrete to support its own weight.

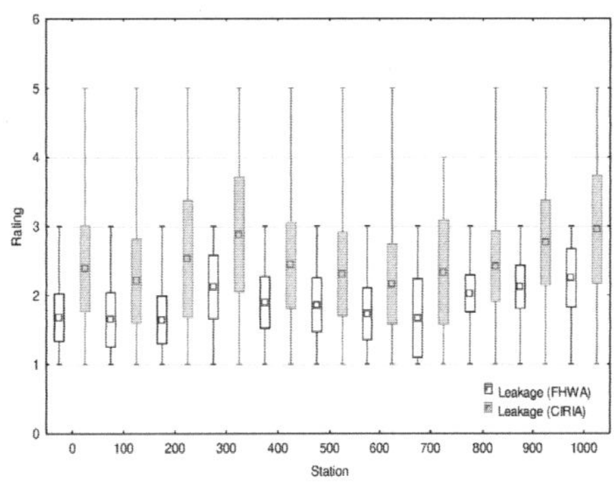

Figure 5. Box-plot of water leakage on the MAT.

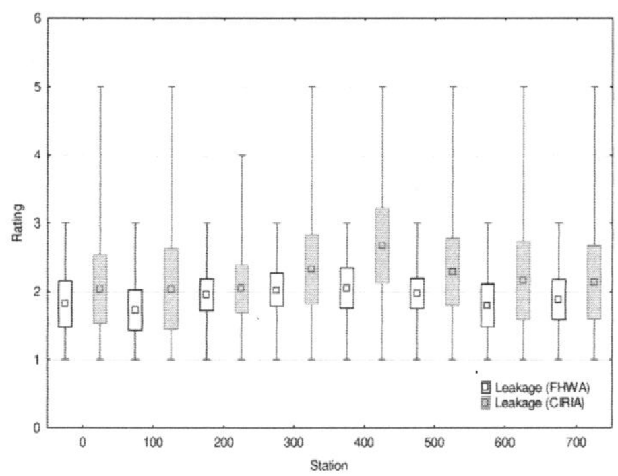

Figure 6. Box-plot of water leakage on the CVT.

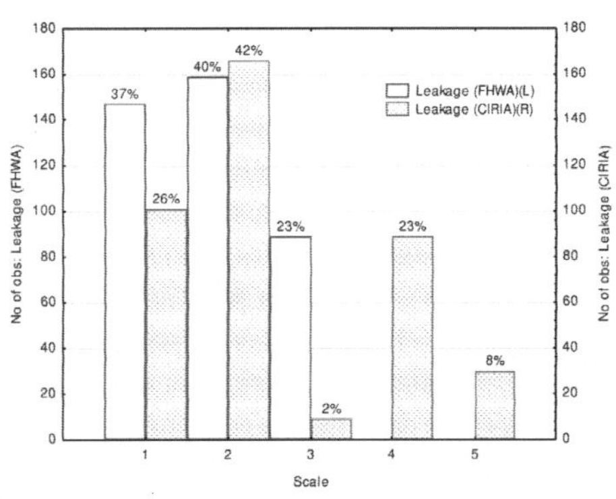

Figure 7. Box-plot of water leakage on the MAT.

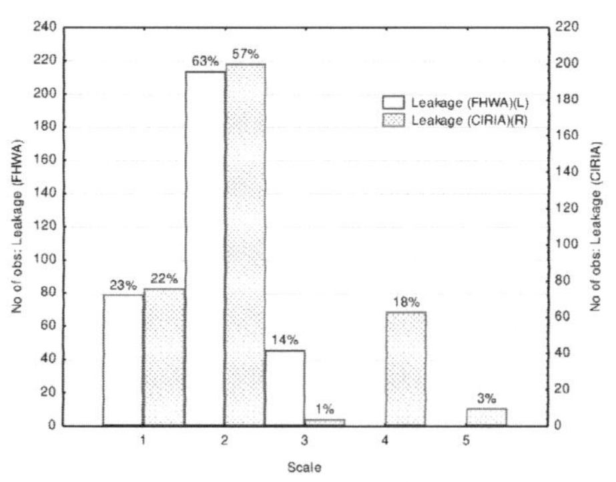

Figure 8. Box-plot of water leakage on the CVT.

Figure 9 shows condition of water leaking at chainage 0+350 to chainage 0+375 of left wall of the MAT. It shows many leaking on the shotcrete lining are found in different rating condition. There are also sign of dried deposited materials indicating there was water leaking from that point but due to calcination the path of the water has clogged. Analysis of the deposit material using SEM EDX found that the mineral content on the deposited material at this location is Carbon (77.06%), Calcium (6.92%) and Ferrum (16.02%). It is also reflected from the color of the deposited material, it varies from white, reddish to black. The red circles are the weeping holes provided on the shotcrete lining to drain the water.

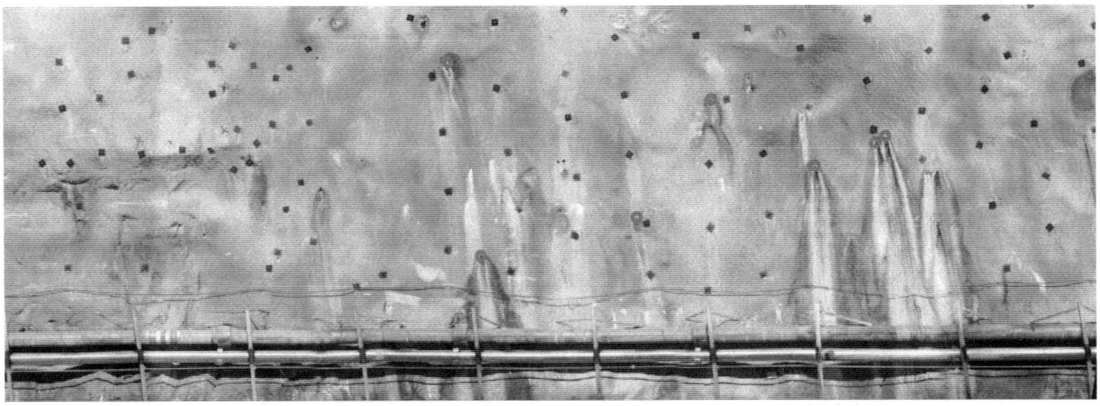

Figure 9. Water leaking at chainage 0+350 to 0+375 left wall of the MAT

4. Conclusion

Most of distress observed is ground water leaking as its condition often appears in wetness and moisture. Result of statistical analysis confirmed that there are severe leakages observed along the right and left wall of the MAT with the presence of deposited mineral thereby could contribute to shotcrete deterioration. Most of the deposited mineral appears in white, green, black, brown and reddish in color and all those colors might indicate high mineral content and chemical reaction occurred behind the underground structure. Meanwhile, based on analysis for the CVT, the area having the highest water leaking rating is from chainage 0+400 to chainage 0+500 as a consequence more prone to shotcrete failure. Thereby require maintenance works needed to avoid any disruption in future.

References

[1] Federal Highway Administration 2005, Highway and Rail Transit Tunnel Inspection Manual, US Department of Transportation, 2005 Edition. US.

[2] L. McKibbins, R. Elmer, K. Roberts 2009, Tunnel: Inspection, Assessment and Maintenance, CIRIA C671, Classic House, London

Journal of Physics: Conference Series **710** (2016) 012036 doi:10.1088/1742-6596/710/1/012036

A simplified nonlinear model of the Marangoni instability in gas absorption

E F Skurygin[1,1] and T A Poroyko[1]
[1]Yaroslavl State Technical University, Yaroslavl, Russia

E-mail: skouryguine@rambler.ru

Abstract. The process of gas absorption into initially motionless liquid layer is investigated. The convective instability caused by the temperature dependence of the surface tension. The critical time of transition of the process to unstable convective regime, as well as the intensity of mass transfer in a surface convection are estimated numerically. The mathematical model includes the equations of convective diffusion, thermal conduction and fluid motion. The problem was solved numerically in the two-dimensional formulation. In the coordinate along the interface the concentration of the absorbed substance is represented by three terms of the trigonometric Fourier series. A difference approximation of equations with an exponentially changing grid in the direction normal to the interface is used. The simulations results agree with the well-known experimental data on the absorption of carbon dioxide in water.

Introduction

In the well-known experiments for the physical gas absorption by liquid layer the process goes in diffusion mode, only to a certain critical time, and then turn into more intense convective [1-3]. A linear stability analysis of the absorption process by frozen-time assumption was performed in [2, 4]. In order to estimate the intensity of mass transfer in the convective mode a semi-empirical formula was used in [3]. Critical time obtained by frozen-time assumption is a stationary point of the process, corresponding to the beginning of the joint growth of perturbations of concentration of absorbable material, temperature and velocity of the fluid. Transition of the process to convective mode occurs at a later time when the disturbances will reach the necessary intensity. For the process of carbon dioxide absorption in water the critical time [1, 3] exceeds a stationary point about 1000 times. In [5] it is shown that the reason for the long delay is the high viscosity of the fluid. Curently no available numerical estimates of the critical time and the intensity of mass transfer in a surface convection obtained without involving semiempirical hypotheses. The purpose of this paper is to construct a simplified non-linear model to assess these characteristics, as well as checking its adequacy by comparison with experimental data.

2. The formulation of the problem. The main equations

We consider non-stationary process of gas absorption by initially fixed bed fluid of infinite thickness. The liquid interacts with the gas phase, wherein component A is contained. The component

[1]Author to whom any correspondence should be addressed

A is absorbed by liquid, its concentration in the gas phase is maintained constant, in the liquid phase, its concentration in the beginning of the process is c_∞.The main resistance to heat and mass transfer during the absorption is concentrated in the liquid phase. The liquid temperature at the beginning of the process is constant and equal T_∞. Convective instability caused by the temperature dependence of surface tension.

The absorption process in the presence of surface convection is described by a system of the equations of convective diffusion, thermal conductivity [9] and the motion of a viscous fluid [10]

$$\frac{\partial c}{\partial t} + \mathbf{v}\nabla c - D\Delta c = 0, \tag{1}$$

$$\frac{\partial T}{\partial t} + \mathbf{v}\nabla T - \chi\Delta T = 0, \tag{2}$$

$$\frac{\partial}{\partial t}\omega + (\mathbf{v}\nabla)\omega \text{ - } \nu\Delta\omega = 0, \quad \omega = -\frac{\partial v_z}{\partial x} + \frac{\partial v_x}{\partial z}, \tag{3}$$

$$\nabla \mathbf{v} = 0, \tag{4}$$

where c *is* the concentration of component A in the liquid, T -temperature, $\mathbf{v} = (v_x, v_z)$ - vector of fluid velocity, D and χ are molecular diffusivity and thermal diffusivity, ν - kinematic viscosity of the fluid, t - time,

The problem is considered in a two-dimensional setting. The x coordinate is directed along the interface, the coordinate z – directed normal to the surface, the value z =0 corresponds to the gas-liquid interface. The boundary conditions for the system of equations (1) - (4) are as follows.

$$c = c_B, \quad D\frac{\partial c}{\partial z} = \Delta H \lambda \frac{\partial T}{\partial z}, \quad v_z = 0, \quad \rho\nu\frac{\partial v_x}{\partial z} = \varepsilon\frac{\partial T}{\partial x} \quad \text{for } z = 0. \tag{5}$$

$$\frac{\partial c}{\partial z} -> 0, \quad T = -> T_\infty, \quad v_x -> 0, \quad v_z -> 0 \quad \text{for } z = \infty. \tag{6}$$

where ΔH - specific heat of solution, $\lambda = \rho c_p \chi$ - thermal conductivity of the fluid, ρ - its density, c_p - specific heat capacity, ε - negative temperature coefficient of surface tension, c_B - equilibrium concentration of component A in the liquid corresponding to its partial pressure in the gas phase [11]: Initial conditions can be represented in the following form:

$$\begin{gathered} c(x,z,t) = c_\infty + f_c(x,z), \quad T(x,z,t) = T_\infty + f_T(x,z), \\ v_x(x,z,t) = f_{vx}(x,z), \quad v_z(x,z,t) = f_{vz}(x,z) \quad for\ t = 0 \end{gathered} \tag{7}$$

Zero values of the functions f_c, f_T, f_{vx}, f_{vz} correspond to the absorption process in a motionless liquid. The flux density of an absorbable material at the interface j is equal to:

$$j(x,t) = -D\frac{\partial c(x,z,t)}{\partial z}\Big|_{z=0}, \tag{8}$$

In the case of motionless liquid we have:

$$j^{(0)}(t) = (c_B - c_\infty)\left(\frac{D}{\pi t}\right)^{1/2}; \tag{9}$$

where the superscript (0) corresponds to a stationary liquid.

The following approximation formulas for the stationary point time t_{st} and the corresponding wavenumber k_{st} are obtained in [4]

$$t_{st} = \pi(1 + 3\pi^{1/3}Le^{-1/3} + 12Le^{-2/3})t_*, \tag{10}$$

$$k_{st} = (2^{-1}\pi^{-2/3}Le^{1/6} - 0.7Le^{-1/6} + 0.9Le^{-1/2})z_*^{-1}. \tag{11}$$

Where $Le = \chi / D$ is Lewis number.

$$z_* = \frac{\lambda\rho\nu}{\varepsilon\Delta Hc_*}\left(\frac{\chi}{D}\right)^{1/2}; \quad t_* = \frac{1}{D}\left(\frac{\lambda\rho\nu}{\varepsilon\Delta Hc_*}\right)^2; \tag{12}$$

are characteristic length and time scales. As already mentioned, the time t_{st} corresponds to the beginning of the joint growth of perturbations in the absorption process.

3. A simplified nonlinear model.

Let us represent the instantaneous values of concentration, temperature and velocity as follows:

$$c(z,x,t) = c_0(z,t) + c_1(z,t)\cos(kx) + c_2(z,t)\cos(2kx) \tag{13}$$

$$T(z,x,t) = T_0(z,t) + T_1(z,t)\cos(kx) + T_2(z,t)\cos(2kx) \tag{14}$$

$$\mathrm{v}_z(z,x,t) = \mathrm{v}_1(z,t)\cos(kx) + \mathrm{v}_2(z,t)\cos(2kx) \tag{15}$$

$$\begin{gathered} \mathrm{v}_x(z,x,t) = \mathrm{v}_{x1}(z,t)\cos(kx) + \mathrm{v}_{x2}(z,t)\cos(2kx), \\ \mathrm{v}_{x1} = \frac{-1}{k}\frac{\partial \mathrm{v}_1}{\partial z}; \quad \mathrm{v}_{x2} = \frac{-1}{2k}\frac{\partial \mathrm{v}_2}{\partial z}; \end{gathered} \tag{16}$$

The wave number k is chosen to be 0.5 k_{st}. This choice of the wavenumber provides the most rapid growth in the second Fourier component of concentration.

The equations for all components of the velocity and temperature linearized. The nonlinearity considered only the equation of concentration. The equations of the model are as follows:

$$\frac{\partial c_0}{\partial t} - D\frac{\partial^2}{\partial z^2}c_0 = -\frac{1}{2}\frac{\partial}{\partial z}(\mathrm{v}_1 c_1 + \mathrm{v}_2 c_2). \tag{17}$$

$$\frac{\partial c_1}{\partial t}-D(\frac{\partial^2}{\partial z^2}-k^2)c_1=-\frac{1}{2}\frac{\partial}{\partial z}(2\mathrm{v}_1c_0+\mathrm{v}_2c_1+\mathrm{v}_1c_2)+\mathrm{c}_0\frac{\partial \mathrm{v}_1}{\partial z}-\frac{1}{2}c_2\frac{\partial \mathrm{v}_1}{\partial z}+\frac{1}{4}C_1\frac{\partial \mathrm{v}_2}{\partial z} \tag{18}$$

$$\frac{\partial c_2}{\partial t}-D(\frac{\partial^2}{\partial z^2}-4k^2)c_2=-\frac{1}{2}\frac{\partial}{\partial z}(2\mathrm{v}_2c_0+\mathrm{v}_1c_1)+\mathrm{c}_0\frac{\partial \mathrm{v}_2}{\partial z}+c_1\frac{\partial \mathrm{v}_1}{\partial z} \tag{19}$$

$$\frac{\partial T_0}{\partial t}-\chi\frac{\partial^2}{\partial z^2}T_0=0 \tag{20}$$

$$\frac{\partial T_n}{\partial t}-\chi\left(\frac{\partial^2}{\partial z^2}T_n+n^2k^2T_n\right)=-\mathrm{v}_n\frac{\partial T_0}{\partial Z};\quad n=1,2 \tag{21}$$

$$\left(\frac{\partial}{\partial t}-\nu(\frac{\partial^2}{\partial z^2}-n^2k^2)\right)(\frac{\partial^2}{\partial z^2}-n^2k^2)\mathrm{v}_n=0,\quad \mathrm{n=1,2.} \tag{22}$$

The initial conditions for zero Fourier components of the concentration and temperature are specified in the following form:

$$c_0(z,t)=c_\infty,\quad T_0(z,t)=T_\infty\quad for\ t=0 \tag{23}$$

Concentration perturbation of different intensities and different spatial scales were set for non-zero components in the initial conditions,

$$c_n(z,t)=\begin{cases}\alpha_n(\mathrm{c}_B\text{-}\mathrm{c}_\infty) & for\ z<Z_n\mathrm{z}_*\\ 0 & for\ z>Z_n z_*\end{cases}\quad for\ t=0,\quad n=1,2 \tag{24}$$

where α_n - intensities of disturbances, Z_n - their spatial scales in the direction normal to the surface. The initial perturbations of velocity and temperature are equal to zero

$$\mathrm{v}_n(z,t)=0,\quad T_n(z,t)=0\quad for\ t=0,\quad n=1,2 \tag{25}$$

4. **The simulations results**

The simulations were performed for the case of *Le = 79, Pr = 7*, which corresponds to the absorption of carbon dioxide by water. Simulations were carried out using Matlab software. Exponentially varying mesh built on the *z* coordinate, Feldberg difference approximation is used [12, 13].

Traditionally, the data of non-stationary mass transfer at the gas-liquid interface are presented in the form according to the amount of the absorbed substance q versus the square root of time:

$$q(t)=\int_0^t j_0(t_1)dt_1,\qquad \mathrm{j}_0(\mathrm{t})=-\mathrm{D}\frac{\partial c_0}{\partial z}(t,z=0), \tag{26}$$

where c_0, j_0 are zero Fourier components of concentration and flux density.

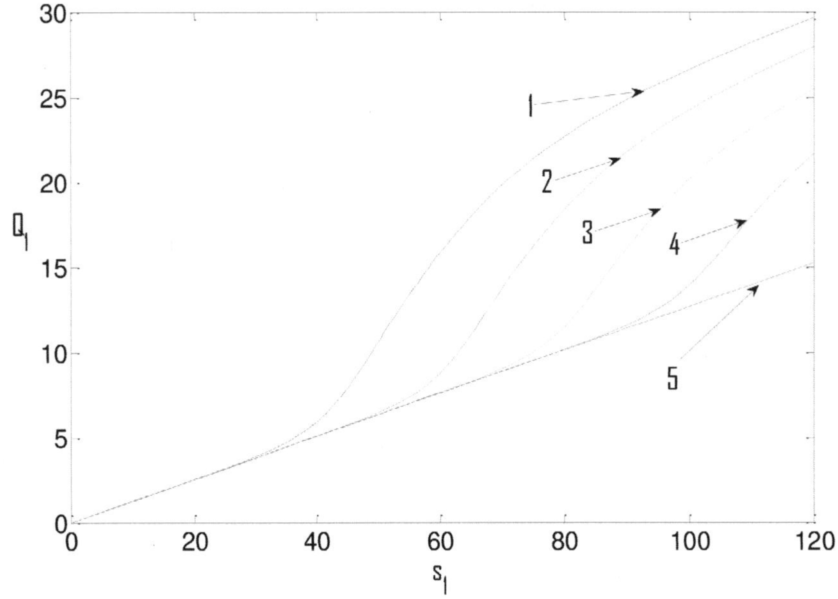

Fig.1. The amount of adsorbed substance versus the square root of time. . Z_1=Z_2=1; 1 - $\alpha_1=\alpha_2=0.1$; 2 - $\alpha_1=\alpha_2=10^{-2}$; 3 - $\alpha_1=\alpha_2=10^{-3}$; 4 – $\alpha_1=\alpha_2=10^{-4}$; 5- mass transfer in the stationary liquid.

The results of simulations for different intensities of the initial perturbations are shown in Fig.1. The data are given in dimensionless variables:

$$Q_1 = \frac{q}{j_* t_*}; \quad s_1 = \sqrt{\frac{t}{t_*}} = \sqrt{\tau} \tag{27}$$

where

$$j_* = (c_B - c_\infty)(D / t_*)^{1/2} \tag{28}$$

is characteristic scale of flux density, the magnitude of t_* defined in (12). The moments of time when the amount of the absorbed substance under the conditions of surface convection begins to exceed the corresponding value in a stationary liquid are quite clearly visible in the chart. We define a critical time as the time at which the amount of absorbed substance under the conditions of surface convection exceeds the corresponding value for the motionless liquid on the threshold ε .

$$(Q_1 - Q_{1D}) / Q_{1D} > \varepsilon \tag{27}$$

where Q_{1D} - amount of absorbed substance in a stationary liquid. The estimations of the critical time for various intensities of initial disturbances are represented in the Table 1.

Table 1. Computational critical time.

The intensity of initial disturbances	Critical time t_{cr}	
	ε=0.01	ε=0.1
$\alpha_1 = \alpha_2 = 0.1$, $Z_1 = Z_2 = 1$	610 t_*	$10^3 t_*$
$\alpha_1 = \alpha_2 = 0.01$, $Z_1 = Z_2 = 1$	$2.2*10^3 t_*$	$3.2*10^3 t_*$
$\alpha_1 = \alpha_2 = 0.001$, $Z_1 = Z_2 = 1$	$4.1*10^3 t_*$	$6*10^3 t_*$
$\alpha_1 = \alpha_2 = 0.0001$, $Z_1 = Z_2 = 1$	$7.5*10^3 t_*$	$9.1*10^3 t_*$

Note that the time of stationary point is $t_{st} = 8.57t_*$. As you can see, for the intensity of $\alpha_1=\alpha_2=0.1$ the critical time t_{cr} exceeds the stationary point t_{st} about 70 times for $\varepsilon = 0.01$ and about 120 times for $\varepsilon = 0.1$. Reducing the initial intensity 1,000 times the critical time estimation increases about 10 times, indicating its weak dependence of intensity on the initial perturbations. For the wide range of initial intensities the critical time is the same order of magnitude with the experimental value for the absorption of carbon dioxide, water, equal to 100 seconds, ie, about 6000t_* [3].

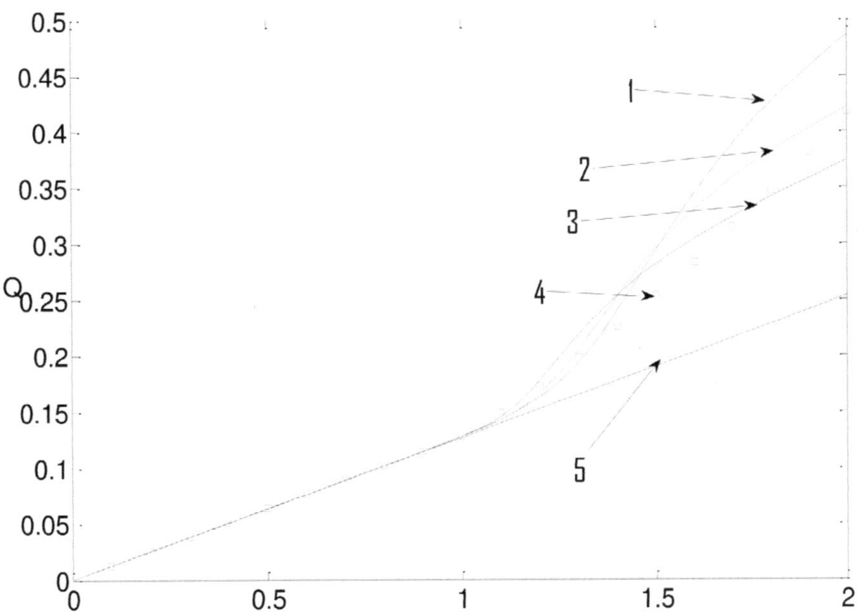

Fig.2. The amount of adsorbed substance depending on the square root of time. The data dimensionless on the critical time. $Z_1=Z_2=1$; 1 - $\alpha_1=\alpha_2=10^{-2}$; 2 - $\alpha_1=\alpha_2=10^{-3}$; 3 - $\alpha_1=\alpha_2=10^{-4}$; 4 - experimental data [3]; 5 - mass transfer in the stationary liquid.

Let us represent data (26) dimensionless by the critical time calculated for ε = 0.01:

$$Q = \frac{q}{j_* \sqrt{t_* t_{cr}}}; \qquad s = (t / t_{cr})^{1/2} \tag{28}$$

Fig.2 shows the results of simulations for the initial disturbances with the intensity of 0.01 and below. The function Q (s) is weakly dependent on the initial conditions, the curves 1 and 3 correspond to the initial intensity of disturbances varies 100 times. Comparing with the well-known experimental data [3] shows that the function Q (s) qualitatively correctly describes the transition of the absorption process from diffusion to convective mode. On the chart you can see quite clearly that at a critical time (s = 1) the dependence of Q (s) is changing its functional. In the beginning of the convection mode, s <1.5, i.e. $t < 2t_{cr}$, the numerical values of Q(s) are consistent with the experimental data [3].

5. Conclusions

The simplified non-linear model agrees with the experimental data for the case of low intensities of initial perturbations of concentration. Accounting only three terms in Fourier series of concentration versus the coordinate along the interface one can correctly describe the intensity of mass transfer in the beginning of the convective regime.

References

[1] Plevan R E , Quin J A 1996, The effect of monomolercular films on the rate of gas absorption into a quiescent liquid *AIChE Journal*, **12**, 894-902.

[2] Dilman V V, Naydenov V I, Olevsky V V 1992 Marangoni diffusive_thermal instability in absorption *Russian Chemical Industry* **69** 8 25

[3] Dil'man V V,Kulov N N,Lotkhov V A,Kaminskii V A, and Naidenov V I 1998 On the differencein rates of absorption and desorption of gases *Theor. Found. Chem. Eng.* **32** 337.

[4] PoroykoT A and Skurygin E F 2013Effect of Thickness of Liquid Layer on Instability of Absorption *Theor. Found. Chem. Eng.* **47** 682.

[5] T.A.Poroyko and E.F.Skurygin. 2014 On the Marangoni Instability in Gas Absorption *Journal of Physics: Conference Series.* **495** 012033.

[6] Truette B, Chenier E, Dumenc F, Declarte C, Guerrier B 2012, Transient, Transient Rayleigh-Bernard –Marangoni solutal convection *Physics of Fluids, American Institute of Physics (AIP)*, **24**, 074108.

[7] Boeck T, Tess A 1998, Turbulent Bénard-Marangoni Convection: Results of Two-Dimensional Simulations. *Journal. Physical Rewiew Letters*, **80**,. 1216-1219.

[8] Boeck T 2005, Bénard–Marangoni convection at large Marangoni numbers: Results of numerical simulations., *Advances in Space Research*, **36**, 4-10.

[9] Bird R,Stewart W and Lightfoot E 2007*Transport Phenomena.* (New York : Wiley).

[10] Landau L.D., Lifshitz E.M. 1987, *Fluid Mechanics*, Pergamon Press.

[11] F.Daniels, R.Alberty 1975, *Physical chemistry*. (New York: Wiley).

[12] Feldberg S.W.1969, Digital simulation: a general method for solving electrochemical diffusion- kinetic problems, in *Electroanalytical Chemistry*, **3**, (A.J.Bard, ed.), Marcel dekker, NY, 199.

[13] M. Rudolph 1993, Digital simulations with the Fast Implicit Finite Difference Algorithm: The Development of a General Simulator for Electrochemical Processes, in *Physical chemistry: Principles, Methods and Applications,* I.Rubinstein, ed, Marcel Dekker, N.Y., 81.

Journal of Physics: Conference Series **710** (2016) 012037 doi:10.1088/1742-6596/710/1/012037

TE and TM modes polaritons in multilayer system comprise of a PML-type magnetoelectric multiferroics and ferroelectrics

Vincensius Gunawan and Hendri Widiyandari

Jurusan Fisika, Fakultas Sains dan Matematika, Universitas Diponegoro, Jl. Prof Soedarto, Tembalang, Semarang, Indonesia

E-mail: goenangie@fisika.undip.ac.id (V. Gunawan)

Abstract. In this paper, we report our study on both bulk and surface polaritons generated in Multilayer system. The multilayer consists of ferroelectric and multiferroic with canted spins structure. The effective medium approximation is employed to derive the dispersion relation for both bulk and surface modes. Surface and bulk polaritons are calculated numerically for the case of Transverse electric (TE) and Transverse magnetic (TM) modes. Example results are presented using parameters appropriate for $BaMnF_4/BaAl_2O_4$. We found in both TE and TM modes, that the region where the surface modes may exist is affected by the volume fraction of the multiferroics. The region of the surface modes decrease when the volume fraction of the multiferroic is reduced. This region decrement suppress the surface polariton curves which result in shortening the surface modes curves.

1. Introduction

Electromagnetic waves which are coupled to the elementary excitations of the materials are called polaritons[1]. Polaritons have received much attention due to its unique properties. Surface polaritons are able to demonstrate localization on the surface. They can also show the non reciprocity, where the two opposite directions of propagation have different frequencies, i.e. $\omega(k) \neq \omega(-k)$[2–4].

Theoretical and experimental treatments for phonon polaritons in dielectrics were published several decades ago[5–7]. The theoretical studies of magnon polaritons in magnetic materials were also reported several years ago [3, 8, 9] and reviewed in Ref.[4]. Experimental studies of magnon polaritons in antiferromagnet had been performed using many methods: transmission for FeF_2[10],reflection for CoF_2[11], MnF_2[12] and attenuated total reflection (ATR)[13].

The study of polaritons in multiferroics is interesting since multiferroics have both phonon polaritons and magnon polaritons simultaneously[14–17]. It is happen since in multiferroics exist at least two types of ferroicity[18], hence there will be more than one resonance frequency in the multiferroics. For example, magnetoelectric multiferroics have both phonons and magnons which lead to the existence of the electric and the magnetic resonance frequencies simultaneously.

Study the multilayer geometry by combining at least two types of different material is important to understand the changes of the elementary excitations of the materials which are represented by the changes of the polaritons. The manipulation of polaritons had been studied theoretically by considering the multilayer configuration which comprise of magnetic/non-magnetic materials[19] or magnetic/magnetic materials[20, 21]. The dispersion relations were calculated by employing transfer matrix which depend on the thickness of the involved materials. However, in long-wavelength limit where the wavelength much longer than multilayer period, the calculation can be approached by applying an effective medium method[20, 22]. In this method, a multilayer configuration is represented by a unit medium having effective permittivity and effective permeability. According to Ref.[20], these effective parameters are calculated by employing Maxwell's continuity of the fields at the interface of the involved materials.

In the present paper we report in detail the study of polaritons in multilayer comprises of multiferroics/electric system. We show how the fraction of multiferroics affects both the bulk and surface polaritons. We provide numerical predictions of dispersion relations for multilayer comprise of $BaMnF_4/BaAl_2O_4$. Those materials are used since the model of multiferroic $BaMnF_4$ is frequently used in previous study[14–17] and there is a complete report about resonance frequencies of ferroelectric $BaAl_2O_4$[23]. We also discuss the properties of polaritons for both the transversal electric (TE) modes and transversal magnetic (TM) modes.

2. Geometry of the Multilayer

The geometry of the multilayer is illustrated in figure 1. The multilayer is semi-infinite which is comprise of multiferroics (MF) with the thickness d_1 and ferroelectric (E) with the thickness d_2. The surface of multilayer is lied in $x-y$ plane. Following the previous studies[16, 17], the multiferroics is ferroelectric-two sub-lattices antiferromagnet with canted spin.

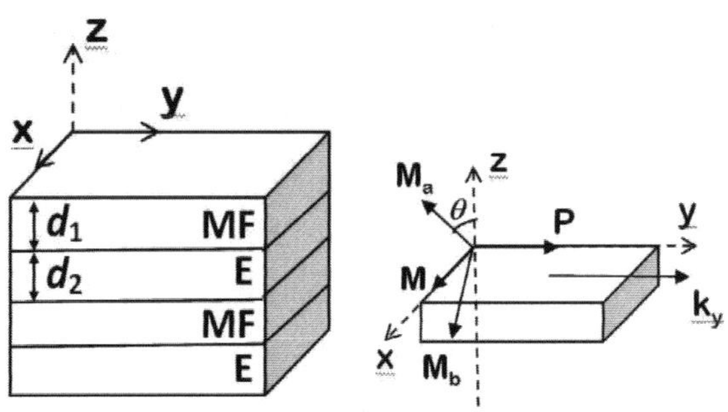

(a) (b)

Figure 1. Geometry of the multilayer. (a) The multilayer is comprised of the ferroelectrics with the thickness d_1 and the multiferroics which is denoted by E with the thickness d_2. (b) The polarisation and the magnetisation of the multiferroics (MF). Polarisasion is directed in the y direction. The two-sublattice magnetisation is canted with canting angle θ yield weak ferrmognetism M.

The susceptibilites are obtained by examining the density energy of material in the equations of motion. In this process, we also obtain magnetic and electric resonance frequencies. The susceptibility components of this type of magnetoelectric multiferroic were formulated into two groups. The first group represents TE modes involving the electric component e_x and the magnetic components h_y and h_z. The susceptibility components for TE modes are[16]:

$$\chi_{yy}^{m} = \frac{2\omega_s\left(\omega_a\cos 2\theta + 2\omega_{me}\sin 2\theta + \omega_\circ\sin\theta\right)}{\left(\omega_M^2 - \omega^2\right)}, \quad (1)$$

$$\chi_{zz}^{m} = \frac{2\omega_s\left(\omega_{me}\sin 2\theta + \omega_o \sin\theta\right)}{\left(\omega_M^2 - \omega^2\right)}, \tag{2}$$

$$\chi_{yz}^{m} = -\chi_{zy}^{m} = \frac{i2\omega_s\omega\sin\theta}{\left(\omega_M^2 - \omega^2\right)}, \tag{3}$$

and

$$\chi_x^e = \epsilon_x^\infty \frac{\omega_{L,x}^2 - \omega^2}{\omega_{T,x}^2 - \omega^2} - 1 \tag{4}$$

where $\omega_M = \left(\omega_{afm}^2\cos^2\theta + \Omega_{me,TE}^2 + \Omega_{o,TE}^2\right)^{1/2}$ represents magnetic resonance frequency in TE modes. The frequencies $\Omega_{me,TE} = \left[\omega_{me}\left(4\omega_{me}\cos\theta + \omega_a\sin 2\theta + 2\omega_{ex}\sin 2\theta\right)\right]^{1/2}$ is contributed from magnetoelectric interaction while $\Omega_{0,TE} = \left[\omega_0\left(\omega_0 - \omega_a\sin\theta + 4\omega_{me}\cos\theta\right)\right]^{1/2}$ is given by external magnetic field.

Here, the frequency $\omega_{ex} = \gamma\lambda\mu_0 M_s$ represent the frequency from exchange interaction while the frequency $\omega_a = \gamma K\mu_0 M_s$ is frequency which is contributed by anisotropy energy. Frequencies ω_{me} and ω_0 are defined as $\omega_{me} = \gamma\mu_0\alpha P M_s$ and $\omega_0 = \gamma\mu_0 H_0$. The frequency from sub-lattice magnetization is defined as $\omega_s = \gamma\mu_0 M_s$. The frequency $\omega_{afm} = \left[\omega_a\left(\omega_a + 2\omega_{ex}\right)\right]^{1/2}$ represents antiferromagnet frequency. In electric susceptibility, $\omega_{L,x}$ and $\omega_{T,x}$ represent the longitudinal and transverse phonon resonance frequency in x direction.

The TM mode which involves a magnetic component h_x and the electric components e_y and e_z has the magnetic and electric susceptibility components[16]

$$\chi_{xx}^{m} = \frac{2\omega_s\left(\omega_{el}^2 - \omega^2\right)\left(\omega_a\cos\theta + 2\omega_{me}\sin\theta\right)}{\left[\left(\omega_{el}^2 - \omega^2\right)\left(\omega_m^2 - \omega^2\right) - \Omega_s^4\right]}, \tag{5}$$

$$\chi_{y,2}^{e} = \frac{\eta/\epsilon_0\left(\omega_m^2 - \omega^2\right)}{\left[\left(\omega_{el}^2 - \omega^2\right)\left(\omega_m^2 - \omega^2\right) - \Omega_s^4\right]}, \tag{6}$$

$$\chi_z^e = \epsilon_z^\infty \frac{\omega_{L,z}^2 - \omega^2}{\omega_{T,z}^2 - \omega^2} - 1 \tag{7}$$

and also the magnetoelectric susceptibility in the form

$$\chi_{xy}^{me} = \chi_{yx}^{em} = \frac{4\alpha\left(\eta/\epsilon_0\right)M_s\omega_s\left(\omega_a\cos\theta + 2\omega_{me}\sin\theta\right)\cos 2\theta}{c\left[\left(\omega_{el}^2 - \omega^2\right)\left(\omega_m^2 - \omega^2\right) - \Omega_s^4\right]} \tag{8}$$

where frequency $\omega_m = \left(\omega_{afm,TM}^2 + \Omega_{me,TM}^2 + \Omega_{0,TM}^2 +\right)^{1/2}$ is defined as magnetic resonance frequency and $\omega_{el} = \left(\frac{\alpha_1}{\epsilon_0} + P_0\frac{\alpha_2}{\epsilon_0^2}\right)\eta$ represents the electric frequency. Here, η represents the inverse of the multiferroics' phonon mass and the frequency Ω_s in magnetoelectric susceptibility is defined as $\Omega_s = \left[8\frac{\eta}{\epsilon_0}\frac{\alpha^2 M_s^2}{c^2}\omega_s\left(\omega_a\cos^2\theta + 2\omega_{ex}\sin^2\theta\right)\right]^{1/4}$.

In this TM modes, $\omega_{afm,TM} = \left(\omega_a\left(\omega_a + 2\omega_{ex}\right)\cos^2\theta + 2\omega_{ex}\left(\omega_a + 2\omega_{ex}\right)\sin^2\theta\right)^{1/2}$ represents the antiferromagnetic resonance frequency in TM modes. The frequency from the magnetoelectric interaction is represented by $\Omega_{me,TM} = \left[2\omega_{me}\left(4\omega_{me} - 2\omega ex\sin 2\theta + \omega_a\sin 2\theta\right)\right]^{1/2}$. The frequency $\Omega_{0,TM} = \left[\omega_0\left(\omega_0 - \omega_a\sin\theta - 4\omega_{ex}\sin\theta + 6\omega_{me}\cos\theta\right)\right]^{1/2}$ is contributed by the field H_0.

In this study, we simply use the electric susceptibilities for ferroelectric in the similar form as Eq.(5) and (7) with the form

$$\chi_{i,e}^{e} = \epsilon_{i,e}^{\infty}\frac{\left(\omega_{L,i}^e\right)^2 - \omega^2}{\left(\omega_{T,i}^e\right)^2 - \omega^2} - 1 \tag{9}$$

where $\omega^e_{T,i}$ represents a transverse optical lattice vibration frequency in the i direction and ω^e_L is cut off frequency of the forbidden band for waves propagation. Once all the magnetic and electric susceptibilites are defined, the permeabilities can easily be obtained through relation $\mu = I + \chi^m$ while the permittivities with the relation $\epsilon = I + \chi^e$.

In this calculation, we exclude intermixing interaction between ferroelectric and the electric part of the multiferroics as proposed in previous study[24]. Since this interaction will only slightly shift the electric resonance frequencies for both material, it will not significantly affect the feature of the dispersion relation curves.

3. Effective Medium

A multilayer system comprised of two or more materials can be approximated as a single material called effective medium[20, 25]. In this approximation, we assume the wavelength much longer than the thickness of each layer or periodicity of the multilayer. According to Ref.[20], the effective permittivity and permeability are calculated by considering the Maxwell's continuity at the interfaces.

A multilayer as illustrated in Fig.1(a) has the average of tangential electric displacement vector as

$$\bar{D}_x = \frac{d_1 D^x_1 + d_2 D^x_2}{(d_1 + d_2)} = f_1 D^x_1 + f_2 D^x_2 \tag{10}$$

where f_1 and f_2 denote the volume fraction of the multiferroics and ferroelectrics. The displacement components D^x_1 and D^x_2 are the x component of the electric displacement vector of the multiferroics and ferroelectrics. In the next step, by considering the continuity of tangential electric fields, we have $E^x_1 = E^x_2 = \bar{E}_x$. Using constitutive equation $\vec{D} = \epsilon \vec{E}$ yield the relation in the form $\bar{D}_x = f_1 \epsilon^x_1 E^x_1 + f_2 \epsilon^x_2 E^x_2 = (f_1 \epsilon^x_1 + f_2 \epsilon^x_2) \bar{E}_x$. Hence, the effective permittivity in the x component is written as

$$\epsilon^{xx}_{eff} = f_1 \epsilon^x_1 + f_2 \epsilon^x_2. \tag{11}$$

Now, analyzing the average of the electric displacement, $\bar{D}_y = f_1 D^y_1 + f_2 D^y_2$ using constitutive relation $D^y_1 = \epsilon^{yy}_1 E^y_1 + \chi^{em} H^x_1$ for multiferroics and $D^y_2 = \epsilon^{yy}_2 E^y_2$ for ferroelectric , also by considering the continuity of the tangential component of magnetic field (H_x) and tangential component of the electric field (E_y), we obtain $\bar{D}_y = (f_1 \epsilon^{yy}_1 + f_2 \epsilon^{yy}_2) \bar{E}_y + f_1 \chi^{em} \bar{H}_x$. Then, the y component of the effective permittivity can be defined as

$$\epsilon^{yy}_{eff} = f_1 \epsilon^{yy}_1 + f_2 \epsilon^{yy}_2 \tag{12}$$

and the effective magnetoelectric susceptibility as

$$\chi^{em}_{eff} = f_1 \chi^{em}. \tag{13}$$

The z component of permittivity is derived by analyzing the average of the electric component $\bar{E}_z = f_1 E^z_1 + f_2 E^z_2$ and the continuity of the normal component of electric displacement (D_z), result in the form

$$\epsilon^{zz}_{eff} = \frac{\epsilon^{zz}_1 \epsilon^{zz}_2}{(f_1 \epsilon^{zz}_2 + f_2 \epsilon^{zz}_1)}. \tag{14}$$

If the same procedure above is performed to the magnetic induction field $\vec{B}$ and magnetic field $\vec{H}$, the effective permeability components are obtained in the form of

$$\mu^{xx}_{eff} = f_1 \mu^{xx}_1 + f_2 \mu^{xx}_2, \tag{15}$$

$$\mu_{eff}^{yy} = \left\{ f_1\mu_1^{yy} + f_2\mu_2^{yy} + f_1^2 \frac{\mu_2^{zz}\mu_1^{yz}\mu_1^{zy}}{\left[f_2\left(\mu_1^{zz}\right)^2 + f_1\mu_1^{zz}\mu_2^{zz}\right]} - f_1\frac{\mu_1^{yz}\mu_1^{zy}}{\mu_1^{zz}} \right\}, \quad (16)$$

$$\mu_{eff}^{zz} = \left(\frac{\mu_1^{zz}\mu_2^{zz}}{f_1\mu_2^{zz} + f_2\mu_1^{zz}} \right) \quad (17)$$

and

$$\mu_{eff}^{yz} = \mu_{eff}^{zy} = f_1 \frac{\mu_2^{zz}\mu_1^{yz}}{(f_1\mu_2^{zz} + f_2\mu_1^{zz})}. \quad (18)$$

4. Theory of Polariton in MF/FE Multilayer

The behavior of electromagnetic waves propagation is illustrated by dispersion relation which is obtained by solving the Maxwell equations using the appropriate electromagnetic waves such as

$$\vec{\psi} = (\psi_x, \psi_y, \psi_z)\exp\left[i\left(k_y y + k_z z - \omega t\right)\right]. \quad (19)$$

Using configuration as illustrated in Fig.1(b), The Maxwell equations result in two groups of coupled equations. The first group which involves the H_y, H_z and E_x is classified as transverse electric (TE) mode. The second group is transverse magnetic (TM) mode which involves the fields components H_x, E_y and E_z.

Using electromagnetic waves as in Eq.(19) into Maxwell equations, the matrix equation for the TE modes which represent the bulk modes is

$$\begin{pmatrix} -k_z & k_y & \epsilon_{eff}^{xx}\frac{\omega}{c} \\ -\mu_{eff}^{yy}\frac{\omega}{c} & \mu_{eff}^{yz}\frac{\omega}{c} & k_z \\ \mu_{eff}^{zy}\frac{\omega}{c} & \mu_{eff}^{zz}\frac{\omega}{c} & k_y \end{pmatrix} \begin{pmatrix} H_y \\ H_z \\ E_x \end{pmatrix} = 0. \quad (20)$$

The solution requires the determinant matrix equal to zero which results in the form of bulk polariton as

$$k_y^2 = \epsilon_{eff}^{xx} \frac{\left[\mu_{eff}^{yy}\mu_{eff}^{zz} + \left(\mu_{eff}^{yz}\right)^2\right]}{\mu_{eff}^{yy}} \left(\frac{\omega}{c}\right)^2. \quad (21)$$

The TE surface modes are derived by assuming that the electromagnetic waves are decaying exponentially as they go into the materials. Hence, the appropriate solutions of the surface modes are

$$\vec{E} \sim e^{\beta z} e^{i(k_y y - \omega t)} \qquad \text{for } z < 0 \quad (22)$$

$$\vec{E} \sim e^{-\beta_o z} e^{i(k_y y - \omega t)} \qquad \text{for } z > 0 \quad (23)$$

where β and β_0 represent attenuation constant for multilayer and vacuum which define as

$$\mu_{eff}^{zz}\beta^2 = \mu_{eff}^{yy}k_y^2 - \epsilon_{eff}^{xx}\left[\mu_{eff}^{yy}\mu_{eff}^{zz} + \left(\mu_{eff}^{yz}\right)^2\right]\frac{\omega^2}{c^2}, \quad (24)$$

$$\beta_0^2 = k_y^2 - \frac{\omega^2}{c^2}. \quad (25)$$

The TE dispersion relation for the surface polaritons is derived by applying the continuity of tangential $\vec{H}$ (H_y) and the continuity of normal $\vec{B}$ (B_z) which result in the form

$$\mu^{zz}_{eff}\beta + \beta_0 \left[\mu^{yy}_{eff}\mu^{zz}_{eff} + \left(\mu^{yz}_{eff}\right)^2\right] + i\mu^{yz}_{eff}k_y = 0. \tag{26}$$

Since the dispersion relation contains odd order of k_y, then surface polariton is non-reciprocal, with $\omega(k_y) \neq \omega(-k_y)$. Hence the non-reciprocality which was found in a pure PML-type multiferroics as reported in Ref.([16]) is still observable in this multilayer geometry.

The TM modes for bulk polaritons which is involving field components H_x, E_y and E_z are represented by a matrix equation resulted from the Maxwell equations as

$$\begin{pmatrix} \mu^{xx}_{eff}\frac{\omega}{c} & \left(k_z + \chi^{me}_{eff}\frac{\omega}{c}\right) & -k_y \\ \left(k_z + \chi^{me}_{eff}\frac{\omega}{c}\right) & \epsilon^{yy}_{eff}\frac{\omega}{c} & 0 \\ k_y & 0 & -\epsilon^{zz}_{eff} \end{pmatrix} \begin{pmatrix} H_x \\ E_y \\ E_z \end{pmatrix} = 0 \tag{27}$$

which result in the dispersion relation for bulk modes as

$$\epsilon^{yy}_{eff}k_y^2 = \left[\epsilon^{yy}_{eff}\mu^{xx}_{eff} - \left(\chi^{me}_{eff}\right)^2\right]\epsilon^{zz}_{eff}\frac{\omega^2}{c^2}. \tag{28}$$

The TM surface polaritons are calculated by assuming the solution is of the form

$$\vec{H} \propto e^{\beta z}e^{i(k_y y - \omega t)}. \tag{29}$$

where the attenuation constant β is defined by an equation such as

$$\epsilon^{zz}_{eff}\left(\beta + i\chi^{me}_{eff}\frac{\omega}{c}\right)^2 = \epsilon^{yy}_{eff}\left(k_y^2 - \mu^{xx}_{eff}\epsilon^{zz}_{eff}\frac{\omega^2}{c^2}\right). \tag{30}$$

The equation above illustrates that the attenuation constant β is complex. It means that the surface polariton modes are leaky modes, where there is an amount of energy leaking into the environment. Performing the similar procedure as in TE modes by analyzing the continuity of the fields (continuity of tangential $\vec{H}$, tangential $\vec{E}$ and normal $\vec{D}$) at the surface, results in

$$\left(\beta + \frac{i\omega}{c}\chi^{me}_{eff}\right) + \epsilon^{yy}_{eff}\beta_0 = 0. \tag{31}$$

The dispersion relation Eq.(31) reflects that the TM surface modes is reciprocal, $\omega(k_y) = \omega(-k_y)$.

5. Application to Multilayer $BaMnF_4/BaAl_2O_4$

In this section, the previous theory is applied to the multilayer comprises of multiferroic $BaMnF_4$ and ferroelectric $BaAl_2O_4$. For $BaMnF_4$, some values of the variables are taken from the references such as: canting angle $\theta = 3$ mrad[26], polarisation $P_\circ = 0,115$ C/m^2[27], inverse phonon mass $f = 3 \times 10^{14}$ A^{2}s^{2}kg^{-1}m^{-3}[14], frequency $\omega_{ey} = 41$ cm^{-1}[28], $\omega_{ez} = 33.7$ cm^{-1}[14] and permittivity $\epsilon_x = 8.2$[29]. The others are estimated as in Ref.[16]. Sub-lattice magnetization $M_s = 2,4333 \times 10^5$ Am^{-1} is estimated using $M_c = 2M_s \sin\theta$ where weak ferromagnetism $M_c = 1460$ Am^{-1}[30]. An exchange constant λ=163,54 is calculated using $H_E = \lambda\mu_0 M_s$ with H_E=50 T[31]. An anisotropy constant K=0.337 is approximated using $\omega_r^2 = \gamma^2\mu_0^2 K(K - 2\lambda)M_s^2$ with $\omega_r = 3$ cm^{-1}[28]. The calculated magnetoelectric coupling α is obtained using relation $\tan(2\theta) = \frac{4\alpha P}{K+2\lambda}$ as in Ref.[16]. The values of variables for $BaAl_2O_4$ are taken from Ref.[23] such as $\epsilon^\infty_\perp = 3.12$, $\epsilon^\infty_{//} = 3.13$, $\epsilon^0_\perp = 9.82$, $\epsilon^0_{//} = 11.09$, frequency $\omega_{t,\perp} = 63$ cm^{-1} and $\omega_{t,//} = 63$ cm^{-1}.

The solutions of the TE dispersion relation for bulk in Eq.(21) and surface modes in Eq.(26) were obtained numerically using root finding technique. For the volume fraction of multiferroics $f_1 = 0.75$, the solutions of the surface dispersion relation in TE modes was found at around the magnetic resonance frequency of $BaMnF_4$ ($\omega_M \approx 3.023$ cm^{-1}) which is illustrated in Fig.2(a). In that figure, the surface modes are represented by thick lines which are denoted 'SP' while the bulk bands are represented by shaded regions.

Since around the magnetic frequency ω_M, permittivity component ϵ_{xx} of multiferroic $BaMnF_4$ is constant at the value 8.2 while the permittivity ϵ_{xx} of ferroelectric $BaAl_2O_4$ is also constant around the value 3.12, hence the value of the effective permittivity ϵ_{eff}^{xx} can be approximated as constant. Then, based on Eq.(21), the bulk bands comprise of only two bands which are separated by an area where the surface modes may exist. This region is limited by magnetic resonant frequency ω_M and the zeros frequency ω_O of the dispersion relation for bulk polaritons (see Fig.2(a)).

Since the magnetic resonance frequency is only contributed by multiferroic $BaMnF_4$, then the bulk equation in Eq.(21) can be brought into

$$k_y^2 \propto (f_1 + f_2\mu_{yy})\,\mu_{zz} - f_2\mu_{yz}^2 \tag{32}$$

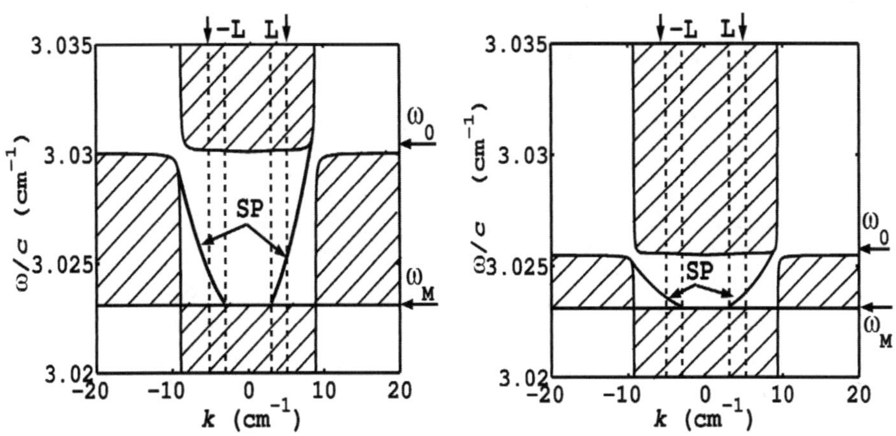

(a) (b)

Figure 2. Dispersion relation of Multilayer with various volume fraction of Multiferroics. In (a) with the fraction $f_1 = 0.75$ and (b) with $f_1 = 0.25$. Shaded region is the area for bulk polaritons. Surface modes are represented by thick lines which are denoted by 'SP"'. The dashed lines which are denoted by L represent Lightlines. The ATR line are illustrated by dashed lines with the downward arrows.

where the denominators of μ_{yy}, μ_{zz} and μ_{yz} have the same resonance frequency. Hence, the volume fraction of the multiferroic layer do not affect the magnetic resonance frequency of the multilayer as illustrated in Fig.2(a)-2(b).

The zeros frequency ω_0 can be calculated by solving the condition

$$\left[\mu_{eff}^{yy}\mu_{eff}^{zz} + \left(\mu_{eff}^{yz}\right)^2\right] = 0 \tag{33}$$

which can be brought into cubic equation $\omega^4 + B\omega^2 + C = 0$ where B and C were defined as $B = -\left(f_1\omega_M^2 + f_2\omega_{oy}^2 + \omega_{oz}^2 - f_2\omega_s^2 \sin^2\theta\right)$ and $C = \left(f_1\omega_M^2 + f_2\omega_{oy}^2\right)\omega_{oz}^2$. The frequencies ω_{oy} and ω_{oz} represent zero frequency of the permeability components μ_{yy} and μ_{zz}. The solution of cubic equation is then $\omega_0 = \left[\frac{-B+\sqrt{B^2-4C}}{2}\right]^{1/2}$.

It was found that by reducing the volume fraction of the multiferroics $BaMnF_4$, the zeros frequency of bulk polariton will decrease as illustrated in Fig.2(a)-2(b). The zeros frequency drop from 3.031 cm^{-1} to 3.026 cm^{-1} when the volume fraction was reduced from 75% to 25%. This zeros frequency reduction result in the decrement of the region where the surface modes may exist, since the magnetic resonance frequency is independent of the volume fraction. Hence, it also shifts the surface modes curves.

Since the attenuation total reflection (ATR) has been successfully used to identify the surface polaritons in antiferromagnet FeF_2[13, 32], in this report we also calculate the ATR reflectance to verify the polaritons generation. ATR reflectance is reflection of the incident waves in ATR system which is took place at the material sample. In this method, the existence of surface modes is represented by sharp dips while the bulk polaritons are illustrated with shallow dips of the ATR reflectance spectrum. The ATR reflectivity is calculated by considering the reflectance at the multilayer surface and also at the bottom of high index prism which results in the formula[33]

$$R = \left| \frac{k_z \left(1 + re^{-2\beta_0 d}\right) - i\beta_0 \left(1 - re^{-2\beta_0 d}\right)}{k_z \left(1 + re^{-2\beta_0 d}\right) + i\beta_0 \left(1 - re^{-2\beta_0 d}\right)} \right| \tag{34}$$

where k_z is propagation normal to the surface of the multilayer. Here, $r = \frac{\beta_0 - \zeta}{\beta_0 + \zeta}$ represents reflectivity at the surface of multilayer with $\zeta = \frac{\left(\mu_{eff}^{zz}\beta - i\mu_{eff}^{yz}k_y\right)}{\left[\mu_{eff}^{yy}\mu_{eff}^{zz} + \left(\mu_{eff}^{yz}\right)^2\right]}$.

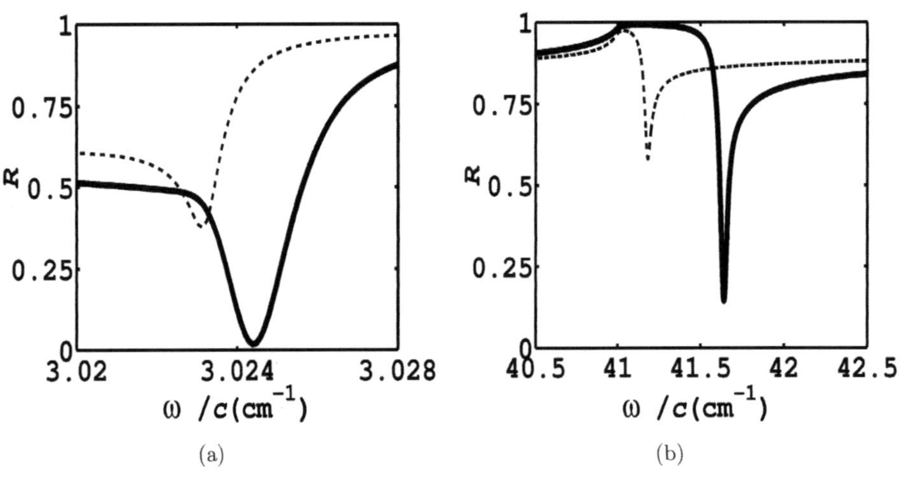

Figure 3. Calculated ATR Reflectivity with incident angle 30^0. In (a), The ATR spectroscopy for TE mode, (b) The ATR for TM mode. The dashed line represents the volume fraction $f_1 = 0.25$, while the solid thick lines is for$f_1 = 0.75$.

In figure 2(a), the dashed lines denoted by arrow '↓' represent ATR line with incident angle at 30^0. This ATR line intersects the surface modes at around 3.025 cm^{-1}, agree with the sharp dip of the solid line in the calculated ATR reflectance at Fig.3(a). The shift of surface modes curves due to the reduction of volume fraction of the multiferroics to $f_1 = 0.25$ can also be observed as illustrated as a dashed line.

The solutions of the dispersion relation in TM modes for bulk in Eq.(28) and surface in Eq.(31) are illustrated in Fig.4(a). Here, the volume fraction of the multiferroics $BaMnF_4$ is 0.75. The shape of polaritons around the phonon frequency is also similar to the previous report for the pure $BaMnF_4$[16]. However, in this report, the 'window' where the surface modes may exist is thinner for the volume fraction $f_1 = 0.75$ than the 'window' in the pure multiferroics.

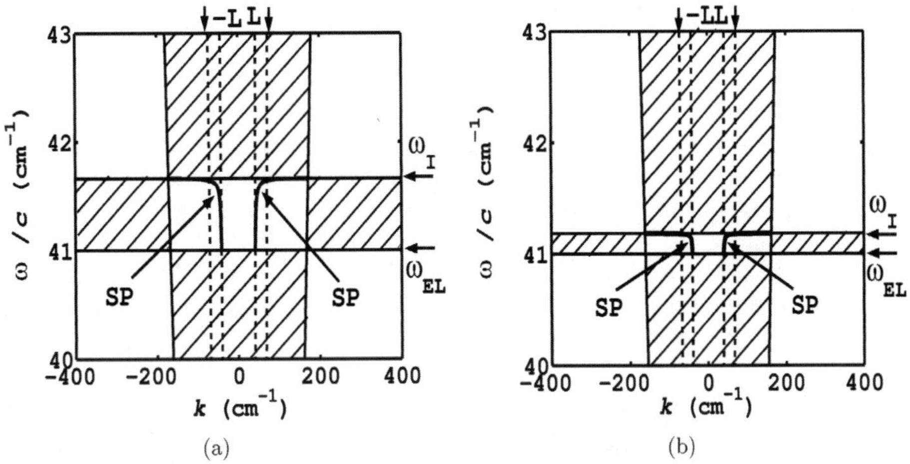

Figure 4. Surface polaritons for TM modes. (a) Surface modes with volume fraction of multiferroics $f_1 = 0.75$. (b) with volume fraction $f_1 = 0.25$. The bulk bands are represented by the shaded regions, while the surface modes are represented by thick lines which denote by 'SP'.

The existence of both surface and bulk modes in this modes is confirmed by the calculated ATR spectrum as illustrated in Fig.3(b). The ATR reflectance for TM mode is also given by formula in Eq.(34) with the reflectivity r is now given as $r = \frac{\epsilon_{eff}^{yy}\beta_0 - \beta}{\epsilon_{eff}^{yy}\beta_0 - \beta}$.

When the volume fraction of the multiferroics is decreased to 0.25, the 'window' where the surface modes are present become very narrow as illustrated at Fig.4(b). This decrement is resulted from the reduction of the zero frequency ω_I while the shifted resonance frequency ω_{EL} is not affected by the reduction of volume fraction of the multiferroics. The zero frequency ω_I obey the condition

$$\epsilon_{eff}^{yy}\mu_{eff}^{xx} - \left(\chi_{eff}^{me}\right)^2 = 0. \tag{35}$$

Near the phonon resonance frequency, the permeability $\mu_{eff}^{xx} \cong 1$ while the value $\epsilon_{eff}^{yy} >> \chi_{eff}^{me}$. Hence the Eq.(35) can be approximated by

$$\epsilon_{eff}^{yy} = f_1\epsilon_1^{yy} + f_2\epsilon_2^{yy} = 0. \tag{36}$$

Then, by considering that permittivity $BaAl_2O_4$ around the phonon frequency of the multiferroics is approximately 17, the Eq.(36) above can be brought into a cubic equation as

$$A\omega^4 + B\omega^2 + C = 0. \tag{37}$$

Here, $A = 1 + \kappa$ with $\kappa = \frac{f_2}{f_1}\frac{\epsilon_2^{yy}}{\epsilon_{yy,1}^{\infty}} \approx \frac{f_2}{f_1}\frac{17}{\epsilon_{yy,1}^{\infty}}$. The parameters B and C are defined as $B = -(1+\kappa)(\omega_{mag}^2 + \omega_{ey}^2) - \omega_f^2$ and $C = (1+\kappa)(\omega_{mag}^2\omega_{ey}^2 - \Omega_c^4) + \omega_f^2\omega_{mag}^2$.
Hence, the zeros frequency ω_I can be solved easily as

$$\omega_I = \left\{\frac{-B + \sqrt{B^2 - 4AC}}{2A}\right\}^{1/2}. \tag{38}$$

The decrease of the region where the surface modes exist will suppress the surface modes curves. As a consequence, the surface modes curves become shorter as the volume fraction of multiferroics is decreased. As it is illustrated in Fig.3(b), the calculated ATR reflectance is also showing a downward shift in surface modes curves when the volume fraction is decreased.

6. Conclusion

We show how the dispersion relations of multiferroic are modified in multilayer configuration for both TE and TM modes, even though the reciprocity or non-reciprocity of the multiferroics does not change by arranging multiferroics into a multilayer configuration with electric material. The dispersion relations of the multilayer in both TE and TM modes are affected by the volume fraction of the multiferroics. The region where the surface mode exist is narrow when the volume fraction of multiferroic in multilayer system is small (around 25%). In multilayer with the volume fraction of multiferroic is relatively big (around 75%), the region for surface modes is wide.

Acknowledgments

We wish to acknowledge the support of Ditlitabmas Ditjen Dikti Kemendikbud through DIPA Diponegoro University.

References

[1] Mills D and Burstein E 1974 *Rep. Prog. Phys.* **37** 817
[2] Camley R E 1987 *Surf. Sci. Rep.* **7** 103
[3] Harstein A, Burstein E, Maradudin A A, Brewer R and Wallis R F 1973 *J. Phys. C: Solid State Physics* **6** 1266
[4] Abraha K and Tilley D R 1996 *Surf. Sci. Rep.* **24** 129
[5] Marschall N and Fischer B 1972 *Phys. Rev. Lett.* **28** 811
[6] Falge H, Otto A and Sohler W 1974 *Phys. Status Solidi B* **63** 259
[7] Borstel G and Falge H 1978 *Appl Phys.* **16** 211
[8] Karsono A D and Tilley D R 1978 *J. Phys. C: Solid State* **11** 3487
[9] Camley R E, Jensen M R F, Feiven S A and Parker T J 1998 *J. Appl. Phys.* **83** 6280
[10] Sanders R W, Belanger R M, Motokawa M, Jaccarino V and Rezende S M 1981 *Phys. Rev. B* **23** 1190
[11] Haussler K M, Brandmuller J, Merten L, Finsterholtzl H and Lehmeyer A 1983 *Phys. Status Solidi B* **117** 225
[12] Remer L, Luthi B, Sauer H, Geick R and Camley R E 1986 *Phys. Rev. Lett* **56** 2752
[13] Jensen M R F, Feiven S, Parker T J and Camley R 1997 *Phys. Rev. B* **55** 2745
[14] Barnas J 1986 *J. Magn. Magn. Mat.* **62** 381
[15] Barnas J 1986 *J. Phys. C: Solid State Physics* **19** 419
[16] Gunawan V and Stamps R 2011 *J. Phys. : Condens Matter* **23** 105901
[17] Gunawan V and Stamps R 2012 *J. Phys. : Condens Matter* **24** 406003
[18] Schmid H 1994 *Ferroelectrics* **162** 317
[19] jia Zhu N and Cao S 1987 *Phys. Lett. A* **124** 515
[20] Raj N and Tilley D R 1987 *Phys. Rev. B* **36** 7003
[21] Barnas J 1986 *J. Phys.: Condens. Matter* **2** 7173
[22] Almeida N S and Tilley D R 1990 *Solid State Commun.* **73** 23
[23] Xie C, Dong D, Gao S, Cai Y and Oganov A R 2014 *Phys. Lett. A* **378** 1867
[24] Ong L H and Chew K 2013 *Ferroelectrics* **450** 7
[25] Elmzughi F G and Camley R E 1997 *J. Phys.: Condens Matter* **9** 1039
[26] Venturini E L and Morgenthaler F R 1975 *AIP Conf. Proc.* **24** 168
[27] Abrahams S C and Keve E T 1971 *Ferroelectrics* **2** 129
[28] Samara G A and Richards P M 1976 *Phys. Rev. B* **14** 5073
[29] Samara G A and Scott J F 1977 *Solid State Commun.* **21** 167
[30] Scott J 1979 *Rep. Prog. Phys.* **12** 1055
[31] Holmes L, Eibschutz M and Guggenheim H J 1969 *Solid State Commun* **7** 973
[32] Jensen M R F, Parker T J, Abraha K and Tilley D R 1995 *Phys. Rev. Lett.* **75** 3756
[33] Abraha K, Smith S R P and Tilley D R 1995 *J. Phys.: Condens. Matter* **7** 6423

Cross-validation method for bivariate measure with certain mixture

Rachid Sabre

LE2I UMR6306, CNRS, ENSAM, Univ. Bourgogne Franche-Comté, Agrosup, Département Sciences Pour l'Ingénieur et Procédés, 26, Bd Docteur Petitjean 21000 Dijon France

E-mail: r.sabre@agrosupdijon.fr

Abstract. We consider a pair of random variables (X, Y) whose probability measure is the sum of an absolutely continuous measure, a discrete measure and a finite number of absolutely continuous measures on several lines. An asymptotically unbiased and consistent estimate of the density of the continuous part is given in [13]. In this work, we focus on the choice of these parameters so that this estimate will be optimal and the rate of convergence will be better. we as well as its rate of convergence. To achieve this we use the cross-validation technics.

1. Introduction

It has been studied in Sabre [13] the estimation of the bivariate density of the continuous part of a certain mixture. More precisely, he considered a pair of the random variables (X, Y) whose probability measure is the sum of an absolutely continuous measure with respect to Lebesgue measure, a discrete measure and a finite number of absolutely continuous measures on several lines:

$$d\mu = f(x,y)dxdy + \sum_{m=1}^{q} a'_m \delta_{(\omega_{1m},\omega_{2m})} + \sum_{k=1}^{q'} \varphi_k(u_1)\delta_{(u_1,a_k u_1+b_k)}, \tag{1}$$

where δ is a Dirac measure, f and φ_i are nonnegative uniformly continuous integrable functions. The motivation for the choice of such model is illustrated thought a concrete example in Sabre [14], It concerns the study of structural fissuration of the agricultural soil. On a homogenous soil, measure of the resistance variable X and the humidity variable Y are taken on serval locations at a depth of 30 cm. The measurement values are distributed according to a continuous law, except in certain locations where the experimentalist finds small galleries where measurement values of resistance and humidity decrease (the presence of jumps). When the measures are made in places where the passage of tractors is frequent, the variable Y becomes linear with respect to the variable X and their measures follow a new distribution noted φ_i (the presence of some measures continuous on the lines determined by frequent passage of tractors).

An asymptotically unbiased and consistent estimate of the density of the continuous part "f" is given in Sabre [13].The estimate was based on the work of Sabre [12]. This estimate uses some kernels with smoothing parameters. Our work is devoted to determining how to choose the smoothing parameters for the best rate of convergence of bias and variance.To do this we will use the method of cross validation allowing us to choose the smoothing parameters of the kernel density estimators from data. This method was introduced by Rudemo [10] and

Bowman [2], for choosing the smoothing parameter to minimize the mean integrated square where the probability measure is absolutely continuous. Hardle and Marron [7] developed the same technic for regression function estimation. Hall and Marron[6], [9]have shown, under reasonable assumptions that the parameter given by the cross validation procedure is optimal. Cross-validation method was used in several works including [1], [3] and [15]...

We briefly indicate the organization of this paper: the second section recalls the estimate and its convergence given in Sabre [13]. The third section provides criteria for selecting parameters by using cross validation method and estimation of integrated square error. In the forth section we give the asymptotic optimality theorems. The last section is reserved to numerical results and simulation.

2. Kernel estimate of the density function and amplitudes of jumps:

As Sabre[13], consider a pair of random variables (X, Y) whose probability measure, μ, is defined by (1). The numbers q and q' are assumed nonnegative integers and known. f is the density of the continuous variable which is assumed to be a nonnegative uniformly continuous function. The real positive number a'_j is the amplitude of the *jump* at $(\omega_{1j}, \omega_{2j})$ assumed unknown. The densities φ_i are nonnegative uniformly continuous functions assumed unknown. The coefficients of the lines a_i, b_i are real numbers assumed unknown. δ is the Dirac measure. Suppose that the *jump points* don't belong to the *jump lines* (ie. $w_{2j} \neq a_i w_{1j} + b_i$ for all i and j).

Sabre [13] has assumed, that any *jump point* $(\omega_{1j}, \omega_{2j})$ can be localized in a small block $[\alpha_{1j}, \beta_{1j}] \times [\alpha_{2j}, \beta_{2j}]$. Thus, the estimate of the density $f(x, y)$ proposed is different according to the position of (x, y):

$$\widehat{h}_H(x,y) = \begin{cases} f_n(x,y) & \text{if } (x,y) \notin \mathcal{A} \\ g_n(x,y) & \text{if } (x,y) \in \mathcal{A} \end{cases}.$$

$$\begin{aligned} \text{with} \quad f_n(x,y) &= \sum_{i=1}^{n} \frac{1}{nh_n^2} K\left(\frac{x-x_i}{h_n}, \frac{y-y_i}{h_n}\right) \\ g_n(x,y) &= \int S_n(x-u_1)R_n(y-u_2)f_n(u_1,u_2)du_1du_2 \end{aligned} \tag{2}$$

$$\begin{aligned} \text{where} \quad \mathcal{A} &= \cup_{j=1}^{q}([\alpha_{1j},\beta_{1j}] \times \mathbb{R}) \cup (\mathbb{R} \times [\alpha_{2j},\beta_{2j}]) \cup \mathcal{B} \\ \text{with } \mathcal{B} &= \left\{(x,y) \in \mathbb{R}^2 \text{ such as } \exists i \in \{1,...,q'\} : y = a_i x + b_i\right\}. \end{aligned}$$

The kernel K is defined by $K(u,v) = K_1(u)K_2(v)$ with K_1 and K_2 two continuous, even, decreasing bounded function such that:

$\int \left\|y^2\right\| K_i(y)dy < \infty \quad i = 1,2$. The smoothing parameter h_n, converges to zero and nh_n^2 converges to the infinite, where

$$S_n(z) = \frac{W_n^{(2)}(z) - \frac{M_n^{(2)}}{M_n^{(1)}}W_n^{(1)}(z)}{1 - \frac{M_n^{(2)}}{M_n^{(1)}}} \text{ and } R_n(t) = \frac{W_n^{(3)}(t) - \frac{L_n^{(1)}}{L_n^{(1)}}W_n^{(4)}(t)}{1 - \frac{L_n^{(2)}}{L_n^{(1)}}} \tag{3}$$

The windows functions are defined as follows:

$$\begin{aligned} W_n^{(1)}(t) &= M_n^{(1)}W^{(1)}(tM_n^{(1)});\; W_n^{(2)}(t) = M_n^{(2)}W^{(2)}(tM_n^{(2)}); \\ W_n^{(3)}(t) &= L_n^{(1)}W^{(3)}(tL_n^{(1)}) \text{ and } W_n^{(4)}(t) = L_n^{(2)}W^{(4)}(tL_n^{(2)}) \end{aligned}$$

where $M_n^{(1)}, M_n^{(2)}, L_n^{(1)}$ and $L_n^{(2)}$ are nonnegative real sequences satisfying:

$$M_n^{(r)} \to +\infty; \quad L_n^{(r)} \to +\infty; \quad M_n^{(r)} h_n \to 0; \quad L_n^{(r)} h_n \to 0; \quad \frac{M_n^{(2)}}{M_n^{(1)}} \to 0 \text{ and } \frac{L_n^{(2)}}{L_n^{(1)}} \to 0$$

The function $W^{(i)}$ is a nonnegative, even, integrable function vanishing outside the interval $[-1,1]$ such that $\int_{-1}^{1} W^{(i)}(x)dx = 1$, $i = 1,2,3,4$. Moreover $W^{(i)}$ satisfying the following equalities:

$$W^{(2)}\left(M_n^{(2)}\theta\right) - W^{(1)}\left(M_n^{(1)}\theta\right) = 0 \qquad \forall \theta \in \left]\frac{-1}{M_n^{(1)}}, \frac{1}{M_n^{(1)}}\right[. \tag{4}$$

$$W^{(4)}\left(L_n^{(2)}\theta\right) - W^{(3)}\left(L_n^{(1)}\theta\right) = 0 \qquad \forall \theta \in \left]\frac{-1}{L_n^{(1)}}, \frac{1}{L_n^{(1)}}\right[. \tag{5}$$

We assume that $\frac{1}{h_n} K_1\left(\frac{1}{h_n M_n^{(1)}}\right)$ and $\frac{1}{h_n} K_2\left(\frac{1}{h_n L_n^{(1)}}\right)$ converge to zero, for example $K_1(x) = \frac{1}{(2\pi)^{\frac{1}{2}}} \exp(\frac{-x^2}{2})$. The estimator $\hat{h}$ depends to the smoothing parameters $M_n^{(1)}, M_n^{(2)}, L_n^{(1)}, L_n^{(2)}, \frac{1}{h_n}$. In the following we denote $\hat{h}_H$ instead $\hat{h}$. Let H the vector of smoothing parameters : $H = \left(M_n^{(1)}, M_n^{(2)}, L_n^{(1)}, L_n^{(2)}, \frac{1}{h_n}\right)$. The goal of the following sections is to gives the best vector H for having the optimal quadratical convergence of the estimator. To simplify without lost the generality we consider in the following of this paper that:

$$d\mu = f(x,y)dxdy + \sum_{m=1}^{q} a'_m \delta_{(\omega_{1m}, \omega_{2m})} \text{ and } \mathcal{A} = \cup_{j=1}^{q}([\alpha_{1j}, \beta_{1j}] \times \mathbb{R}) \cup (\mathbb{R} \times [\alpha_{2j}, \beta_{2j}]).$$

3. Estimation of integrated square error

In this section we use the method of cross validation to select smoothing parameters to have the optimal rate of convergence. First, we study the estimation of integrated square error. In deed, our gaol is to give the vector H, minimizing the integrated square error ($ISE = \int(\hat{h}_H(x,y) - f(x,y))^2 w(x,y)dxdy$) where $w(x,y)$ is a nonnegative bounded weight function.

$$ISE = Q_1(\hat{h}_H) - 2Q_2(\hat{h}_H, f) + \int f^2(x,y)w(x,y)dxdy,$$

where $Q_1(\hat{h}_H) = \int \hat{h}_H^2(x,y)w(x,y)dxdy$ and $Q_2(\hat{h}_H, f) = \int \hat{h}_H(x,y)f(x,y)w(x,y)dxdy$.
Since the last term of the expression of ISE is independent of H, to minimize ISE is equivalent to minimize $Q(\hat{h}_H, f) = Q_1(\hat{h}_H) - 2Q_2(\hat{h}_H, f)$. According to the cross-validation method used in [9], we propose the following estimator of $Q_2\left(\hat{h}, f\right)$:

$$\widehat{Q}_2\left(\widehat{h}, f\right) = \frac{1}{n}\sum_{j=1}^{n} \widehat{h}_{-j}(x_j, y_j)\, w(x_j, y_j) \text{ where } \widehat{h}_{-j}(x,y) = \begin{cases} \widehat{f}_{-j}(x,y) & \text{if } (x,y) \notin \mathcal{A} \\ \widehat{g}_{-j}(x,y) & \text{if } (x,y) \in \mathcal{A} \end{cases}$$

$$\text{with} \widehat{f}_{-j}(x,y) = \frac{1}{(n-1)h_n^2} \sum_{i \neq j}^{n} K\left(\frac{x-x_i}{h_n}, \frac{y-y_i}{h_n}\right) \text{ and } \widehat{g}_{-j}(x,y) = \int S_n(x-u)R_n(y-v)\hat{f}_{-j}(u,v)dudv$$

where S_n and R_n are defined in (3).

Theorem 3.1 *We have*

- *i)* $E\widehat{Q}_2\left(\widehat{h}, f\right) - EQ_2\left(\widehat{h}, f\right) \longrightarrow 0.$
- *ii) The variance of* $\widehat{Q}_2\left(\widehat{h}, f\right)$ *converges to zero:*
$Var\left(\widehat{Q}_2\left(\widehat{h}, f\right)\right) = O\left(\frac{1}{nh_n^4}\left[\sup\left\{K_1\left(\frac{1}{h_n M_n^{(1)}}\right)K_2\left(\frac{1}{h_n L_n^{(1)}}\right); K_1\left(\frac{1}{h_n M_n^{(1)}}\right); K_2\left(\frac{1}{h_n L_n^{(1)}}\right)\right\}\right]^2\right).$

4. Asymptotic optimality theorem

The least squares method used above consists in choosing H which minimizes $LS(H)$ where

$$LS(H) = \int \widehat{h}^2(x,y)w(x,y)dxdy - \frac{2}{n}\sum_{j=1}^{n}\widehat{h}_{j^-}(x_j,y_j)w(x_j,y_j).$$

Before starting the asymptotic optimal theorems we define the following distances:

- Average square error: $d_A(\widehat{h}_H, f) = n^{-1}\sum_{j=1}^{n}\left[\widehat{h}_H(x_j,y_j) - f(x_j,y_j)\right]^2 f(x_j,y_j)^{-1}w(x_j,y_j).$
- Integrated square error: $d_I(\hat{h}_H, f) = \int\left[\hat{h}_H(x,y) - f(x,y)\right]^2 w(x,y)dxdy = ISE.$
- Mean integrated square error: $d_M(\hat{h}_H, f) = E\int\left[\hat{h}_H(x,y) - f(x,y)\right]^2 w(x,y)dxdy = MISE.$

In the following we give some notations that we will use afterwards:

$$\begin{aligned}
R &= \frac{1}{n}\sum_{j=1}^{n} f(x_j,y_j)w(x_j,y_j) + E\left[f(x_j,y_j)w(x_j,y_j)\right] \\
S &= \frac{2}{n}\sum_{j=1}^{n} f(x_j,y_j)w(x_j,y_j)\left[1 - \log f(x_j,y_j)\right] - R \\
T &= -\int f^2(x,y)w(x,y)dxdy - 2R \\
\mathcal{P} &= LS(H) - d_I(\hat{h}_H, f) - T
\end{aligned}$$

$$B_H(x,y) = \begin{cases} \int \frac{1}{h_n^2}K\left(\frac{x-u}{h_n}, \frac{y-v}{h_n}\right)f(u,v)dudv - f(x,y) & \text{if } (x,y)\notin\mathcal{A} \\ \int S_n(x,u_1)R_n(y,u_2)\left(\int\frac{1}{h_n^2}K\left(\frac{u_1-v_1}{h_n}, \frac{u_2-v_2}{h_n}\right)dv_1dv_2\right)du_1du_2 - f(x,y) & \text{if } (x,y)\in\mathcal{A}\end{cases}$$

As in [9], we assume the following assumptions. These assumptions are verified by the examples given in [6]and [7].

- a) $\lim_{n\longrightarrow\infty}\sup_{H\in\Lambda_n}\left|\frac{d_A(\hat{h}_H,f) - d_M(\hat{h}_H,f)}{d_M(\hat{h}_H,f)}\right| = 0 \quad a.s.,$
- b) $\lim_{n\longrightarrow\infty}\sup_{H\in\Lambda_n}\left|\frac{d_I(\hat{h}_H,f) - d_M(\hat{h}_H,f)}{d_M(\hat{h}_H,f)}\right| = 0 \quad a.s.,$ where Λ_n is a finite subset of $\mathbb{R}^5$ whose cardinality grows algebraically fast, with $card(\Lambda_n) \le n^C$, C is some nonnegative constant, and every vector $H = (H_1, H_2, H_3, H_4, H_5) \in \Lambda_n$ verifies the fact that:
$C^{-1}n^\delta \le H_k \le Cn^{1-\delta}$, $k = 1, 2.., 5$; δ is a constant satisfying $0 < \delta < 1$,
$0 < H_2 < n^{-\beta}H_1$ and $0 < H_4 < n^{-\beta'}H_3$ where β and β' are some nonnegative constants.
- c) $w(x,y) \le C\ \forall(x,y)\in\mathbb{R}^2$

- d) $f(x,y) \leq C, \quad \forall (x,y) \in \mathcal{S}$, where $\mathcal{S}$ denotes the support of the weight function w.
- e) $B_H(x,y) \leq Cn^{-\delta}$, $\forall H \in \Lambda_n$, $\forall (x,y) \in \mathcal{S}$.
- f) For all $i \in \{1,2,...n > Q\}$

$$\lim_{n\to\infty} \sup_{\lambda\in\Lambda_n} \left| \frac{\int Var\left[K_{H,i}(x,y)\right] w(x,y)dxdy}{C' H_k} - 1 \right| = 0, \quad k = 1,2,...,5$$

where $$\widehat{K}_{H,i}(x,y) = \begin{cases} \frac{1}{h_n^2} K\left(\frac{x-x_i}{h_n}, \frac{y-y_i}{h_n}\right) & \text{if } (x,y) \notin \mathcal{A} \\ \int S_n(x,u_1) R_n(y,u_2) \frac{1}{h_n^2} K\left(\frac{u_1-x_i}{h_n}, \frac{u_2-x_i}{h_n}\right) du_1 du_2 & \text{if } (x,y) \in \mathcal{A} \end{cases}$$

- g) $\int ... \int K_H(x_{i_1}, x_{j_1})...K_H(x_{i_k}, x_{j_k}) dx_1 dx_2 ... dx_m \leq C_k \left(inf(H_k)\right)^{k-\frac{m}{2}}$, where $i_1, j_1, ... i_k, j_k = 1...m$ subject to $i_1 \neq j_1, ..., i_k \neq j_k$.

Theorem 4.1 *Under the assumptions a)- g)*

$$\lim_{n\longrightarrow\infty} \sup_{H\in\Lambda_n} \left| \frac{LS(H) - d_I(\hat{h}_H, f) - T}{d_M(\hat{h}_H, f)} \right| = 0 \quad a.s.,$$

The immediate consequence of this theorem is the following corollary

Corollary 4.1 *Under the assumptions a)- g), if $\hat{h}_H$ is the minimizer of $LS(H)$ over Λ_n, then*

$$\lim_{n\longrightarrow\infty} \frac{d_I(\hat{h}_H, f)}{\inf_{\lambda\in\Lambda_n} d_I(\hat{h}_H, f)} = 1 \quad a.s.$$

5. Proofs

5.1. Proof of the theorem 3.1

We have $E\left[\widehat{Q}_2\left(\widehat{h}_H, f\right)\right] = F_1 + F_2$ where

$$F_1 = \frac{1}{n} \frac{1}{(n-1)h_n^2} E \sum_{j=1}^{n} \sum_{i\neq j}^{n} \left[K\left(\frac{x_j - x_i}{h_n}, \frac{y_j - y_i}{h_n}\right) w(x_j, y_j) \mathbf{1}_{\mathcal{A}^c}(x_j, y_j) \right]$$

$$F_2 = \frac{1}{n} \frac{1}{(n-1)h} \sum_{j=1}^{n} \sum_{i\neq j}^{n} \rho(i,j)$$

where

$$\rho(i,j) = E \int \mathbf{1}_{\mathcal{A}}(x_j, y_j) S_n(x_j - t_1) R_n(y_j - t_2) K\left(\frac{t_1 - x_i}{h_n^2}, \frac{t_2 - y_i}{h_n}\right) w(x_j, y_j) dt_1 dt_2$$

and $\mathcal{A}^c$ is the complementary of $\mathcal{A}$.

$$F_1 = \frac{1}{(n-1)h_n^2} \sum_{i\neq j}^{n} \left[\int K\left(\frac{u-x}{h_n}, \frac{v-y}{h_n}\right) w(u,v) \mathbf{1}_{\mathcal{A}^c}(u,v) f(u,v) dP_{(x_i,y_i)}(x,y) du dv \right.$$
$$\left. + \int \sum_{m=1}^{q} \mathbf{1}_{\mathcal{A}^c}(\lambda_{1m}, \lambda_{2m}) a_m K\left(\frac{\lambda_{1m} - x}{h_n}, \frac{\lambda_{2m} - y}{h_n}\right) w(\lambda_{1m}, \lambda_{2m}) dP_{(x_i,y_i)}(x,y) \right]$$

As $(\lambda_{1m}, \lambda_{2m}) \notin \mathcal{A}^c$, the second term is null.

$$F_1 = \frac{1}{h_n^2} \left[\int K\left(\frac{u-x}{h_n} \frac{v-y}{h_n}\right) w(u,v) \mathbf{1}_{\mathcal{A}^c}(u,v) f(u,v) f(x,y) dy dx du dv \right.$$
$$\left. + \int \sum_{m=1}^{q} \mathbf{1}_{\mathcal{A}^c}(u,v) a_m K\left(\frac{u - \lambda_{1m}}{h_n}, \frac{v - \lambda_{2m}}{h_n}\right) w(u,v) f(u,v) du dv \right].$$

The second term of the last equality tends to $\sum_{m=1}^{q} a_m \mathbf{1}_{\mathcal{A}^c}(\lambda_{1m}, \lambda_{2m}) f(\lambda_{1m}, \lambda_{2m}) = 0$, because $(\lambda_{1m}, \lambda_{2m}) \notin \mathcal{A}^c$. For F_2, we use the fact that $dP_{(x_j,y_j)}(u,v) = f(u,v)dudv + \sum_{m=1}^{q} a_m \delta_{(\lambda_{1m},\lambda_{2m})}$ for all (x_j, y_j) and interchanging the integrals, we have

$$\begin{aligned} F_2 &= \frac{1}{(n-1)\,h_n^2} \sum_{i \neq j}^{n} \\ &\left\{ \int \left[\int \left(\int K\left(\frac{t_1 - x}{h_n}, \frac{t_2 - y}{h_n}\right) S_n(t_1 - u) R_n(t_2 - v) \mathbf{1}_{\mathcal{A}}(u,v) f(u,v) w(u,v) dudv \right) \right.\right. \\ &\left. dP_{(x_i,y_i)}(x,y) \right] dt_1 dt_2 \\ &+ \int \left[\int \sum_{m=1}^{q} a_m \mathbf{1}_{\mathcal{A}}(\lambda_{1m}, \lambda_{2m}) K\left(\frac{t_1 - x}{h_n}, \frac{t_2 - y}{h_n}\right) S_n(t_1 - \lambda_{1m}) R_n(t_2 - \lambda_{2m}) w(\lambda_{1m}, \lambda_{2m}) \right. \\ &\left.\left. dP_{(x_i,y_i)}(x,y) \right] dt_1 dt_2 \right\} \end{aligned}$$

Thus $F_2 = F_{2,1} + F_{2,2} + F_{2,3} + F_{2,4}$, where

$$\begin{aligned} F_{2,1} &= \int \left[\int S_n(t_1 - u) R_n(t_1 - v) \left(\int \frac{1}{h_n^2} K\left(\frac{t_1 - x}{h_n}, \frac{t_2 - y}{h_n}\right) f(x,y) dxdy \right) dt_1 dt_2 \right] \\ &\quad \mathbf{1}_{\mathcal{A}}(u,v)\, w(u,v) f(u,v) dudv \\ F_{2,2} &= \frac{1}{h_n^2} \int \int \left[\sum_{m=1}^{q} a_m K\left(\frac{t_1 - \lambda_{1m}}{h_n}, \frac{t_2 - \lambda_{2m}}{h_n}\right) \times \right. \\ &\quad S_n(t_1 - u) R_n(t_2 - v) \mathbf{1}_{\mathcal{A}}(u,v)\, w(u,v) f(u,v) dudv]\, dt_1 dt_2 \\ F_{2,3} &= \frac{1}{(h_n^2)} \int \left[\sum_{m=1}^{q} a_m \int S_n(t_1 - \lambda_{1m}) R_n(t_2 - \lambda_{2m}) w(\lambda_{1m}, \lambda_{2m}) \right. \\ &\quad \left. K\left(\frac{t_1 - x}{h_n}, \frac{t_2 - y}{h_n}\right) dt_1 dt_2 \right] f(x,y) dxdy \\ F_{2,4} &= \frac{1}{h_n^2} \sum_{m=1}^{q} a_m \sum_{m'=1}^{q} a_{m'} R(m, m'), \text{ where} \\ R(m,m') &= \int S_n(t_1 - \lambda_{1m}) R_n(t_2 - \lambda_{2m}) w(\lambda_{1m}, \lambda_{2m}) K\left(\frac{t_1 - \lambda_{1m'}}{h_n}, \frac{t_2 - \lambda_{2m'}}{h_n}\right) dt_1 dt_2. \end{aligned}$$

Let us now show for $(x,y) \in \mathcal{A}$ and $(z,t) \in \mathbf{R}^2$, that $\frac{1}{h_n^2} \int S_n(u - x) R_n(v - y) K\left(\frac{u-z}{h_n}, \frac{v-t}{h_n}\right) dudv$ converges to zero Indeed,

$$\begin{aligned} &\frac{1}{h_n^2} \int S_n(u - x) R_n(v - y) K\left(\frac{u - z}{h_n}, \frac{v - t}{h_n}\right) dudv \\ &= \frac{M_n^{(2)}}{1 - \frac{M_n^{(2)}}{M_n^{(1)}}} \frac{L_n^{(2)}}{1 - \frac{L_n^{(2)}}{L_n^{(1)}}} \int \left\{ W^{(2)}\left[M_n^{(2)}(u - x)\right] - W^{(1)}\left[M_n^{(1)}(u - x)\right] \right\} \\ &\times \left\{ W^{(4)}\left[L_n^{(2)}(v - y)\right] - W^{(3)}\left[L_n^{(1)}(v - y)\right] \right\} \frac{1}{h_n^2} K\left(\frac{u - z}{h_n}, \frac{v - t}{h_n}\right) dudv. \end{aligned}$$

Putting $s = M_n^{(2)}(u - x)$ and $r = L_n^{(2)}(v - y)$. Using the fact $W^{(1)}$ is vanishing outside $[-1, 1]$, we get

$$\int S_n(u - x) R_n(v - y) K\left(\frac{u - z}{h_n}, \frac{v - t}{h_n}\right) dudv$$

$$= \frac{1}{1-\frac{M_n^{(2)}}{M_n^{(1)}}}\frac{1}{1-\frac{L_n^{(2)}}{L_n^{(1)}}}\int_{\frac{M_n^{(2)}}{M^{(1)}}}^{1}\int_{\frac{L_n^{(2)}}{L^{(1)}}}^{1} W^{(2)}(s)W^{(4)}(r)K\left(\frac{x+\frac{s}{M_n^{(2)}}-z}{h_n},\frac{y+\frac{r}{L_n^{(2)}}-t}{h_n}\right)dsdr$$

If $x \neq z$ and $y \neq t$, since $\int W^{(2)}(x)dx = 1$ and $\frac{M_n^{(2)}}{M_n^{(1)}}$; $\frac{L_n^{(2)}}{L_n^{(1)}}$ tend to zero, we have

$$\int S_n(u-x)R_n(v-y)K\left(\frac{u-z}{h_n},\frac{v-t}{h_n}\right)dudv = O\left(K\left(\frac{1}{h_n},\frac{1}{h_n}\right)\right).$$

If $x = z$ and $y \neq t$ since $\frac{1}{M^{(1)}} < \frac{s}{M_n^{(2)}} < \frac{1}{M_n^{(2)}}$, K_2 is bounded and and the function K_1 is decreasing, we obtain

$$\int S_n(u-x_2)R_n(v-y_2)K\left(\frac{u-x_1}{h_n},\frac{v-y_1}{h_n}\right)dudv = O\left(K_1\left(\frac{1}{h_nM_n^{(1)}}\right)\right). \quad (6)$$

If $x = z$ and $y = t$, it is easy to show that $\int S_n(u-x_2)R_n(v-y_2)K\left(\frac{u-x_1}{h_n},\frac{v-y_1}{h_n}\right)dudv = O\left(K_1\left(\frac{1}{h_nM_n^{(1)}}\right)K_2\left(\frac{1}{h_nM_n^{(1)}}\right)\right)$. Thus from the sum of the first term of F_1 and the first term of F_2, we have $E\left[\hat{Q}_2(\hat{h}_H,f)\right]$ is asymptotically equal to $E\left[Q_2(\hat{h}_H,f)\right]$.

ii) Let us show that the variance of $\widehat{Q}_2\left(\widehat{h}_H,f\right)$ tends to zero. $Var\left(\widehat{Q}_2\left(\widehat{h}_H,f\right)\right) = V_1 + V2$, where

$$\begin{aligned} V_1 &= \frac{1}{n^2}\sum_{j=1}^{n}\frac{1}{(n-1)^2h_n^4}\left\{E\left(\mathbf{1}_{\mathcal{A}^c}(x_j,y_j)\sum_{j\neq i}^{n}K\left(\frac{x_j-x_i}{h_n},\frac{y_j-y_i}{h_n}\right)w(x_j,y_j)\right)^2\right.\\ &\left.-\left(E\left[\mathbf{1}_{\mathcal{A}^c}(x_j,y_j)\sum_{j\neq i}^{n}K\left(\frac{x_j-x_i}{h_n},\frac{y_j-y_i}{h_n}\right)w(x_j,y_j)\right]\right)^2\right\}\\ V_2 &= \frac{1}{n^2}Var\left(\sum_{j=1}^{n}\mathbf{1}_{\mathcal{A}}(x_j,y_j)\int S_n(u-x_j)R_n(v-y_j)\widehat{f}_{-j}(u,v)\,dudv\right). \end{aligned}$$

We can write $V_1 = \frac{1}{n^2}\sum_{j=1}^{n}\frac{1}{(n-1)^2h_n^4}(T_1 - T_2)$. Since $(\lambda_{1m},\lambda_{2m}) \notin \mathcal{A}^c$, we get

$$\begin{aligned} T_1 &= (n-1)^2\int \mathbf{1}_{\mathcal{A}^c}(x_2,y_2)K^2\left(\frac{x_2-x_1}{h_n},\frac{y_2-y_1}{h_n}\right)w(x_2,y_2)f(x_2,y_2)f(x_1,y_1)dx_1dy_1dx_2dy_2\\ &+(n-1)^2\sum_{m=1}^{q}a_m\int_{\mathcal{A}^c}K^2\left(\frac{x-\lambda_{1m}}{h_n},\frac{y-\lambda_{2m}}{h_n}\right)w(x,y)f(x,y)dxdy. \end{aligned}$$

Let $D(a,b) = \int K^2\left(\frac{x-a}{h_n},\frac{y-b}{h_n}\right)f(x,y)dxdy$. Putting $\frac{x-a}{h_n} = u$ and $\frac{y-b}{h_n} = v$, we have$D(a,b) = h_n^2\int K^2(u,v)f(a+uh_n,b+vh_n)dudv$. Using the Taylor's formula we obtain

$$\begin{aligned} D(a,b) &\simeq h_n^2f(a,b)\int K^2(u,v)\,dudv + h_n^3\frac{df}{da}(a,b)\int uK^2(u,v)\,dudv\\ &+ h_n^3\frac{df}{db}(a,b)\int vK^2(u,v)\,dudv, \end{aligned}$$

with $\int uK^2(u,v)\,dudv < \infty$. Thus, we obtain the following inequality

$$\begin{aligned}
T_1 &\le (n-1)^2 h_n^2 \int f^2(x_2,y_2)dx_2dy_2 \int K^2(u,v)\,dudv \\
&+ (n-1)^2 h_n^3 \int \frac{df}{dx_2}(x_2,y_2) f(x_2,y_2)dx_2dy_2 \int uK^2(u,v)\,dudv \\
&+ (n-1)^2 h_n^3 \int \frac{df}{dy_2}(x_2,y_2) f(x_2,y_2)dx_2dy_2 \int vK^2(u,v)\,dudv \\
&+ (n-1)^2 \sum_{m=1}^{q} a_m h_n^2 f(\lambda_{1m},\lambda_{2m}) \int K^2(u,v)\,dudv \\
&+ (n-1)^2 \sum_{m=1}^{q} a_m h_n^3 \frac{df}{dx}(\lambda_{1m},\lambda_{2m}) \int uK^2(u,v)\,dudv \\
&+ (n-1)^2 \sum_{m=1}^{q} a_m h_n^3 \frac{df}{dy}(\lambda_{1m},\lambda_{2m}) \int vK^2(u,v)\,dudv.
\end{aligned}$$

Since the function w is bounded, we get $\frac{1}{n^2}\sum_{i=1}^{n}\frac{1}{(n-1)^2h_n^4}T_1 = O\left(\frac{1}{nh_n}\right)$. Similarly, we show that $\frac{1}{n^2}\sum_{i=1}^{n}\frac{1}{(n-1)^2h_n^4}T_2 = O\left(\frac{1}{n}\right)$. Thus, $V_1 = O\left(\frac{1}{nh_n}\right)$. On the other hand,

$$V_2 = \frac{1}{n^2} Var\left(\sum_{j=1}^{n} \mathbf{1}_{\mathcal{A}}(x_j,y_j) \int S_n(u-x_j) R_n(v-y_j) \widehat{f}_{-j}(u,v)\,dudv\right)$$

Putting $O_j = \mathbf{1}_{\mathcal{A}}(x_j,y_j) \int S_n(u-x_j) R_n(v-y_j) \sum_{i\neq j}^{n} K\left(\frac{u-x_i}{h_n},\frac{v-y_i}{h_n}\right) dudv$ we have

$$\begin{aligned}
V_2 &= \frac{1}{n^2}\frac{1}{(n-1)^2}\frac{1}{h_n^4}\sum_{j=1}^{n} E\left(O_j - E(O_j)\right)^2 \\
&= \frac{1}{n^2}\sum_{j=1}^{n} E\left(\int \mathbf{1}_{\mathcal{A}}(x_j,y_j) S_n(u-x_j) R_n(v-y_j) \widehat{f}_{-j}(u,v)\,dudv\right)^2 - (E(O_j))^2 \\
&= R_1 - R_2, \text{ where}
\end{aligned}$$

$$\begin{aligned}
R_1 &= \frac{1}{n}\frac{1}{h_n^4}\int\int\left(\int \mathbf{1}_{\mathcal{A}}(x_2,y_2) S_n(u-x_2) R_n(v-y_2) K\left(\frac{u-x_1}{h_n},\frac{v-y_1}{h_n}\right) dudv\right)^2 \\
&\quad f(x_2,y_2)dx_2dy_2dP_{(x_i,y_i)}(x_1,y_1) \\
&\quad + \frac{1}{n}\frac{1}{h_n^2}\int \sum_{m=1}^{q} a_m \left(\int \mathbf{1}_{\mathcal{A}}(\lambda_{1m},\lambda_{2m}) S_n(u-\lambda_{1m}) R_n(v-\lambda_{2m}) K\left(\frac{u-x_1}{h_n},\frac{v-x_2}{h_n}\right) dudv\right)^2 \\
&\quad dP_{(x_i,y_i)}(x_1,y_1) \\
&= \frac{1}{n}\frac{1}{h_n^4}\int\left(\int \mathbf{1}_{\mathcal{A}}(x_2,y_2) S_n(u-x_2) R_n(v-y_2) K\left(\frac{u-x_1}{h_n},\frac{v-y_1}{h_n}\right) dudv\right)^2 \\
&\quad f(x_2,y_2)dx_2dy_2 f(x_1,y_1)dx_1dy_1 \\
&\quad + \frac{1}{n}\frac{1}{h_n^4}\int \sum_{m=1}^{q} a_m \left(\int \mathbf{1}_{\mathcal{A}}(x_2,y_2) S_n(u-x_2) R_n(v-y_2) K\left(\frac{u-\lambda_{1m}}{h_n},\frac{v-\lambda_{2m}}{h_n}\right) dudv\right)^2 \\
&\quad f(x_2,y_2)dx_2dy_2
\end{aligned}$$

$$+\frac{1}{n}\frac{1}{h_n^4}\int\sum_{m=1}^{q}a_m\left(\int S_n\left(u-\lambda_{1m}\right)R_n(v-\lambda_{2m})K\left(\frac{u-x_1}{h_n},\frac{v-y_1}{h_n}\right)dudv\right)^2$$
$$f(x_1,y_1)dx_1dy_1$$
$$+\frac{1}{n}\frac{1}{h_n^4}\sum_{m'=1}^{q}a_{m'}\sum_{m=1}^{q}a_m\left(\int S_n\left(u-\lambda_{1m}\right)R_n(v-\lambda_{2m})K\left(\frac{u-\lambda_{1m'}}{h_n},\frac{u-\lambda_{2m'}}{h_n}\right)dudv\right)^2.$$

Using (6), we get

$$R_1=O\left\{\frac{1}{nh_n^4}\left[\overset{2}{\sup}\left(K_1\left(\frac{1}{h_nM_n^{(1)}}\right)K_2\left(\frac{1}{h_nL_n^{(1)}}\right);K_1\left(\frac{1}{h_nM_n^{(1)}}\right);K_2\left(\frac{1}{h_nL_n^{(1)}}\right)\right)\right]^2\right\}.$$

Since $R_2=O\left(\frac{1}{n}\right)$, we obtain

$$V_2=O\left(\frac{1}{nh_n^4}\left[\sup\left\{K_1\left(\frac{1}{h_nM_n^{(1)}}\right)K_2\left(\frac{1}{h_nL_n^{(1)}}\right);K_1\left(\frac{1}{h_nM_n^{(1)}}\right);K_2\left(\frac{1}{h_nL_n^{(1)}}\right)\right\}\right]^2\right).$$

5.2. Proof of the theorem 4.1
$LS(H)$ can be written as follows

$$LS(H) = \int\left(\widehat{h}(x,y)\right)^2w(x,y)dxdy-\frac{2}{n}\sum_{j=1}^{n}\widehat{h}_{j-}(x_j,y_j)w(x_j,y_j).$$

Let $\mathcal{P}=\mathcal{LS}(\mathcal{H})-\lceil_{\mathcal{I}}(\Lsh_{\mathcal{H}},\{)-\mathcal{T}$, from the definition of $d_I(\hat{h}_H,f,$ P becomes

$$\begin{aligned}\mathcal{P} &= -\frac{2}{n}\sum_{j=1}^{n}\widehat{h}_{j-}(x_j,y_j)w(x_j,y_j)\\ &\quad+2\int\widehat{h}_H(x,y)f(x,y)w(x,y)dxdy\\ &\quad+\frac{2}{n}\sum_{j=1}^{n}f(x_i,y_i)w(x_i,y_i)-2E\left[f(x_i,y_i)w(x_i,y_i)\right]\end{aligned}$$

It is easy to show that $|\mathcal{P}|=2n^{-1}(n-1)^{-1}\left|\sum_{j=1}^{n}\sum_{i\neq j}^{n}U_{ij}\right|$, where

$$\begin{aligned}U_{ij} &= -\frac{1}{h_n^2}K\left(\frac{x_j-x_i}{h_n},\frac{y_j-y_i}{h_n}\right)\mathbf{1}_{\mathcal{A}^c}\left(x_j,y_j\right)w(x_j,y_j)\\
&- \frac{1}{h_n^2}\int S_n\left(x_j-u\right)R_n(y_j-v)K\left(\frac{u-x_i}{h_n},\frac{v-y_i}{h_n}\right)dudv\mathbf{1}_{\mathcal{A}}\left(x_j,y_j\right)w(x_j,y_j)\\
&+ \frac{n-1}{nh_n^2}\int_{\mathcal{A}^c}K\left(\frac{x-x_i}{h_n},\frac{y-y_i}{h_n}\right)w\left(x,y\right)f(x,y)\mathbf{1}_{\mathcal{A}^c}\left(x,y\right)dxdy\\
&+ \frac{1}{nh_n^2}\int K\left(\frac{x-x_j}{h_n},\frac{y-y_j}{h_n}\right)w\left(x,y\right)\mathbf{1}_{\mathcal{A}^c}\left(x,y\right)f(x,y)dxdy\\
&+ \int_{\mathcal{A}}\frac{n-1}{nh_n^2}\left(\int S_n\left(u-x\right)R_n(v-y)K\left(\frac{u-x_i}{h_n},\frac{v-y_i}{h_n}\right)dudv\right)w\left(x,y\right)f(x,y)\mathbf{1}_{\mathcal{A}^c}\left(x,y\right)dxdy\\
&+ \int_{\mathcal{A}}\frac{1}{nh_n^2}\left(\int S_n\left(u-x\right)R_n(v-y)K\left(\frac{u-x_j}{h_n},\frac{v-y_j}{h_n}\right)dudv\right)w\left(x,y\right)f(x,y)\mathbf{1}_{\mathcal{A}^c}\left(x,y\right)dxdy\\
&+ f(x_j,y_j)w(x_j,y_j)-\int f^2(x,y)w(x,y)dxdy+\sum_{m=1}^{q}a_mf(\lambda_{1m},\lambda_{2m})w(\lambda_{1m},\lambda_{2m}).\end{aligned}$$

For $i \neq j$, define $V_{ij} = U_{ij} - E\left(U_{ij}/(x_j, y_j)\right)$ Observe that

$$E\left[V_{ij}/(x_i, y_i)\right] = 0 \quad \text{and} \quad E\left[E\left(U_{ij}/(x_j, y_j)\right)\right] = 0.$$

To finish the proof of the theorem 4.1 it enough to prove that

$$\sup_{H\in\Lambda_n} n^{-2}\left|\sum_{j=1}^{n}\sum_{i\neq j}^{n} V_{ij}\right| d_M^{-1}\left(\widehat{h}_H, f\right) \to 0 \text{ a.s} \tag{19}$$

and that
$$\sup_{H\in\Lambda_n} n^{-1}\left|\sum_{j=1}^{n} E\left(U_{ij}/(x_j, y_j)\right)\right| d_M^{-1}\left(\widehat{h}_H, f\right) \to 0 \text{ a.s.} \tag{20}$$

To verify (20), note that by the Borel-Cantelli lemma, it is enough to show that for $\varepsilon > 0$

$$\sum_{n=1}^{\infty} card\left(\Lambda_n\right) \sup_{H\in\Lambda_n} P\left[\left|n^{-1}\sum_{j=1}^{n}\alpha_j\right| > \varepsilon d_M\left(\widehat{h}_H, f\right)\right] < \infty$$

$$E\left(U_{ij}/(x_j, y_j)\right) = B(x_j, y_j)w(x_j, y_j) - E\left[B(x_j, y_j)w(x_j, y_j)\right] + Q_m.$$

$$\begin{aligned}\text{where } Q_m &= \sum_{m'} a_{m'}\frac{1}{h_n^2}\int S_n(u-\lambda_{1m'})R_n(v-\lambda_{2m'})\left[\int K\left(\frac{u-x}{h_n}, \frac{v-y}{h_n}\right) f(x,y)dxdy\right. \\ &- \left.\sum_m a_m K\left(\frac{u-\lambda_{1m}}{h_n}, \frac{v-\lambda_{2m}}{h_n}\right)\right] dudv\end{aligned}$$

Therefore $Var\left(E\left(U_{ij}/(x_j, y_j)\right)\right) = Var\left(B(x_j, y_j)w(x_j, y_j) - E\left[B(x_j, y_j)w(x_j, y_j)\right]\right)$.

$$\begin{aligned}Var\left(E\left(U_{ij}/(x_j, y_j)\right)\right) &= E\left[B^2(x_j, y_j)(w^2(x_j, y_j))\right] - \left(E\left[B(x_j, y_j)w(x_j, y_j)\right]\right)^2 \\ &\leq \int B^2(x,y)w^2(x,y)f(x,y)dxdy + \sum_{m=1}^{q} a_m B^2(\lambda_{1m}, \lambda_{2m})w^2(\lambda_{1m}, \lambda_{2m})\end{aligned}$$

From the assumptions (c), (d) and (e) we get

$$\begin{aligned}Var\left(E\left(U_{ij}/(x_j, y_j)\right)\right) &\leq C^2 n^{-2\delta}\left(\int w^2(x,y)f(x,y)dxdy + \sum_m a_m w^2(\lambda_{1m}, \lambda_{2m})\right) \\ &\leq cte\ n^{-2\delta}.\end{aligned}$$

From Berstein's inequality(see [8]), we have

$$P\left[\left|n^{-1}\sum_{j=1}^{n}\alpha_j\right| > \varepsilon d_M(\hat{h}_H, f)\right] \leq \exp\left\{\frac{-n\varepsilon^2 d_M^2(\hat{h}_H, f)}{2\left(Var\left(E\left(U_{ij}/(x_j, y_j)\right)\right) + \frac{cn^{-\delta}\varepsilon d_M(\hat{h}_H, f)}{3}\right)}\right\}$$

Since $d_M(\hat{h}_H, f) = \int Bias^2(x,y)dxdy + \int Var(\hat{h}_H(x,y))dxdy$, from b), f),e)and g), we have $n^\delta \leq \int Var\left(\hat{h}_H(x,y)\right)dxdy \leq n^{1-\delta}$. From (e), we have $0 \leq B^2(x,y) \leq n^{-2\delta}$ and $n^\delta \leq \int Var\left(\hat{h}_H(x,y)\right)dxdy \leq n^{1-\delta}$, we obtain $n^\delta \leq d_M(\hat{h}_H, f) \leq cte(n^{1-\delta} + n^{-2\delta})$. Therefore

$$2Var\left(E\left(U_{ij}/(x_j,y_j)\right)\right) + \left(\frac{cn^{-\delta}\varepsilon d_M(\hat{h}_H, f)}{3}\right) \leq cte\left(n^{-2\delta} + n{-}\delta\left(n^{1-\delta} + n^{-2\delta}\right)\right)$$

Then $P\left[\left|n^{-1}\sum_{j=1}^{n}\alpha_j\right| > \varepsilon d_M(\hat{h}_H, f)\right] \leq \exp\left\{-Constant\ \ n^{4\delta}\right\}$.
Thus $\sum_{n=1}^{\infty} card\left(\Lambda_n\right)\sup_{\lambda\in\Lambda_n} P\left[n^{-1}\sum_{j=1}^{n}|\alpha_j| > \varepsilon d_M\left(\hat{h}, f\right)\right] < \infty$.
As in Marron [9], to verify (19), using the proof of (20) and the chebyshev inequality, it is enough to show that there is a constant $\gamma > 0$, such that for $k = 1, 2, ...$ there are constants C_k such that

$$\sup_{\lambda\in\Lambda_n} E\left[n^{-2}\sum_{i\neq j} V_{ij} d_M(\hat{h}_H, f)^{-1}\right]^{2k} \leq C_k n^{-\gamma k}. \tag{21}$$

where cum_k is the kth order cumulant and $\sum$ denotes the summation over $i_1, j_1, ..., i_k, j_k = 1, ..., n$ subject to $i_1 \neq j_1, ..., i_k \neq j_k$. From the inequality (7.5) in Marron [9] and the cumulant expansion of the $2k$th centred moment [see, for example, Kendall an Stuart [5], we obtain the inequality (21). This completes the proof of the theorem.

6. Example and application

A concrete application of this work to the one-dimensional case deals with the process of filling bottles of a 33cl volume each. To control the quality of this process, we check that this process is evenly distributed according to gaussian distribution. The experimenter has taken a sample of 115 bottles and measured the quantity of liquid contained in these bottles. The measures normally stand at around of 33cl. But on account of an abnormal disfunction due to a random slowing or acceleration of the motion of the rolling band. Thus the measure randomly increases to reach a value around of 37cl or decrease to a value around of 29cl. But apart from the problem of the disfunction of the rolling band, we would also like to verify whetter the process of filling bottles respects the norm of 33cl with a standard deviation of 1cl.

We consider the random variable Y representing the quantity of liquid in the bottle. Let $X = \frac{Y-\mu}{\sigma}$ where $\mu = 33cl$ and $\sigma = 0.8$. testing if Y is gaussian variable is equivalent to tests if X is a standard gaussian variable. By using the histogram and density of the variable X we find that we have two points of discontinuity of the density: two peaks one in the interval $[\alpha_1, \beta_2] = [-2.8, -2.5]$ and the second in the interval $[\alpha_2, \beta_2] = [1.8, 2.2]$. We propose the model:

$$d\mu = f(x)dx + a_1\delta_{\lambda_1} + a_2\delta_{\lambda_2},$$

where $\lambda_1 \in [\alpha_1, \beta_1]$ and $\lambda_2 \in [\alpha_2, \beta_2]$.
The kernels K_1 and K_2 are chosen as follows: $K_1(x) = K_2(x) = \frac{1}{\sqrt{2\pi}}\exp\{-\frac{x^2}{2}\}$, and $n = 1000$.
First, choosing $W^{(1)}$ as a nonnegative, even and integrable function. We propose:

$$W^{(1)}(t) = \begin{cases} \frac{64}{63}t + \frac{64}{63} & \text{if} \quad t \in [-1, -1/8[\\ 8/9 & \text{if} \quad t \in [-1/8, 1/8] \\ -\frac{64}{63}t + \frac{64}{63} & \text{if} \quad t \in]1/8, 1] \\ 0 & \text{otherwise} \end{cases}$$

It is easy to show that $\int W^{(1)}(t)dt = 1$. Choosing now a nonnegative, even and integrable function $W^{(2)}$ such that $W^{(1)}$ and $W^{(2)}$ satisfying (4) and (5). We propose:

$$W^{(2)}(t) = \begin{cases} W^{(1)}\left(\frac{M_n^{(1)}}{M_n^{(2)}}t\right) & \text{if} \quad t \in [-1/8, 1/8] \\ 4/7\left(1 - \frac{M_n^{(2)}}{M^{(1)}}\right) & \text{if} \quad t \in [-1, -1/8[\cup]1/8, 1] \\ 0 & \text{if} \quad \text{otherwise} \end{cases}$$

$$\int W^{(2)}(t)dt = \int_{-1}^{-1/8} W^{(2)}(t)dt + \int_{-1/8}^{1/8} W^{(2)}(t)dt + \int_{1/8}^{1} W^{(2)}(t)dt.$$

$$\int W^{(2)}(t)dt = \left(1 - \frac{M_n^{(2)}}{M^{(1)}}\right) + \int_{-1/8}^{1/8} W^{(2)}(t)dt$$

From the definition of $W^{(2)}$, we have $\int_{-1/8}^{1/8} W^{(2)}(t)dt = 2\int_0^{1/8} W^{(1)}\left(\frac{M_n^{(1)}}{M_n^{(2)}}t\right)dt$. Putting $u = \frac{M_n^{(1)}}{M_n^{(2)}}t$, we have $\int_{-1/8}^{1/8} W^{(2)}(t)dt = 2\left(\frac{M_n^{(2)}}{M_n^{(1)}}\right)\int_0^{1/8\left(\frac{M_n^{(1)}}{M_n^{(2)}}\right)} W^{(1)}(u)\,du$. Since $\frac{M_n^{(1)}}{M_n^{(2)}} \to \infty$, for n large enough we have $\frac{M_n^{(1)}}{M_n^{(2)}} > 1$. Thus, we obtain $\int_{-1/8}^{1/8} W^{(2)}(t)dt = 2\left(\frac{M_n^{(2)}}{M_n^{(1)}}\right)\int_0^1 W^{(1)}(u)\,du$. $W^{(1)}$ being even and $\int W^{(1)}(t)dt = 1$, we deduce $\int_{-1/8}^{1/8} W^{(2)}(t)dt = \left(\frac{M_n^{(2)}}{M_n^{(1)}}\right)$. Thus, $\int W^{(2)}(t)dt = 1$.

Let us show that (3) and (5) is satisfied. Indeed, let t a reel number belonging to $\left]-\frac{1}{M_n^{(1)}}, \frac{1}{M_n^{(1)}}\right[$, since $\frac{M_n^{(2)}}{M_n^{(1)}}$ converges to zero, for n large enough, we have $-1/8 < -\frac{M_n^{(2)}}{M_n^{(1)}} \leq M_n^{(2)}t \leq \frac{M_n^{(2)}}{M_n^{(1)}} < 1/8$. Therefore, $W^{(2)}(M_n^{(2)}t) = W^{(1)}\left(\frac{M_n^{(1)}}{M_n^{(2)}}M_n^{(2)}t\right) = W^{(1)}\left(M_n^{(1)}t\right)$. We calculate the parameters which minimize $LS(H)$ (cross-validation) by the function "fmins" implemented in Matlab with a respect the constraints and assumptions a)-f) given in the section 4. We calculate the estimate , $\hat{h}_H(x)$, of the density function f. The graphics of the estimate $\hat{h}_H(x)$ and the kernel density of the variable X:

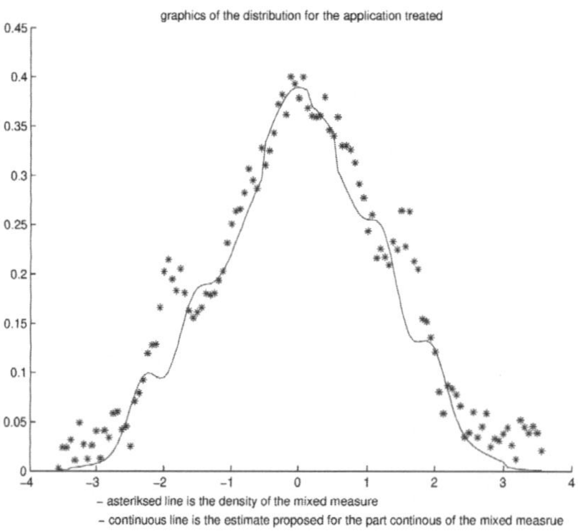

7. Conclusions

We have presented in this paper some results about cross-validation method for choosing the smoothing parameters of density estimate when the measure has certain mixture. A criterion for find the smoothing parameter was given in order to have better convergence of the estimator of the density of the continuous part. The proposed methods can be extended to other applications in several sectors. Indeed, the control of the quality for a product manufactured in the auto industry use the measure of two variables: the consumption of diesel and the pollution. Their joint distribution can follow a continuous law except some observations which are taken when there is fog and reached the constant value (point of the jump). One example in economics, it is the observation of the variables: taxes on income and purchasing power can have a joint distribution contains some point of jumps due to exemption (disabled, former soldier, ...). In oceanography when we observe, by using a camera placed at a certain depth in water, two variables: the length of the fishes and their movement speed. The joint distribution may represent some jumps due to the acceleration of movement during the passage of a predator. In Astronomy the repeated passage of an object preventing the vision of stars (cloud, bird, ...) can create a jump of data.

Acknowledgments

I would like to thank the anonymous referees for their interest in this paper and their valuable comments and suggestions.

8. References

[1] M. I. Ashraf, Z. Zhao, C.A. Bourque, D. A. MacLean, F. Meng 2013 Integrating biophysical controls in forest growth and yield predictions with artificial intelligence technology, *Canadian Journal of Forest Research,* **43**, 12, 1162-1171,

[2] A. W. Bowman, , P. Hall and D. M. Tittrington 1984 Cross-validation in nonparametric estimation of probabilities and probability density , *Biometrika,* **71**, 341-352.

[3] D. M. Hawkins , S.C. Basak and D. Mills 2003 Assessing Model Fit by Cross-Validation, *J. Chem. Inf. Comput. Sci.,* , **43**, 2, 579586.

[4] P. Hall 1983 Large sample optimality of least squares cross-validation in density estimation, *Ann. Statist.***11**, 1156-1174.

[5] M.G. Kendall and A. Stuart 1977 *the Avenced Theory of statistics: Distribution theory1,* (4 th ed. Macmillan, New York).

[6] P. Hall and J.S. Marron 1988 Choice of Kernel Order in Density Estimation, *Ann. of Stat,* **16**, n 1 , 161-173.

[7] W. Hardle and J.S. Marron 1985 Optimal Bandwidth Selection in Nonparametric Regression Function Estimation, *Ann.Statist,* **13**, n 4, 1465-1481.

[8] W. Hoeffding 1963 Probability inequalities for sums of bounded random variables, *J. Amer. Statist. Assoc,* **58**, 13-30.

[9] J.S. Marron 1987 A comparison of cross-validation techniques in density estimation, *Ann.Statist,* **15**, n 1, 152-162.

[10] M. Rudemo 1982 Empirical choice of histogram and kernel density estimation, *Scand. J. Statist.* **9**, 65-78.

[11] G. Wahba and J.Wendelberger 1994 Some New Mathematical Methods for Variational Objective Analysis Using Splines and Cross Validation
, *C. R. Acad. Sci. Paris,* **319**,

[12] R. Sabre 1995 Spectral density estimation for stationary stable random fields, *Journal Applicationes Mathematicae,* **23**, 2, 107-133.

[13] R. Sabre 2003 Nonparametric estimation of the continuous part of mixed measure *Revisita Statistica* **LXIII**, n3, 441-467.

[14] R. Sabre 2010 Tests and Asymptotic normality for mixed Bivariate measure *Revisita Statistica* **LXX**, n 2, 115-136.

[15] A. Vehtari and J. Lampinen 2002 Bayesian Model Assessment and Comparison Using Cross-Validation Predictive Densities, *Neural Computation,* **14**, 10, 2439-2468.

The dynamics of the total output of the fishery sector: The case of Indonesia

Ubaidillah Zuhdi

Faculty of Management and Economics, Gdansk University of Technology, Gdansk 80-233, Poland

zuhdi@zie.pg.gda.pl

Abstract. The purpose of this study is to analyze the dynamics of the total output of the Indonesian fishery industry when the changes of the final demand occur. This study employs a demand-pull Input-Output (IO) quantity model, one of the calculation tools in the IO analysis, as an analysis instrument. Two conditions are noticed in calculations and analysis parts, namely (1) "whole sector change", and (2) "pure change". An initial period in this study is 2008. The results show that, in both conditions, the discussed sector has similar patterns, namely this industry receives the positive impacts from scenarios 1 and 3 while the negative impact is obtained from scenario 2. The results also explain that, in both conditions, the biggest positive impact for the analyzed sector is given by scenario 3, the change of households and non-profit private institutions consumptions.

1. Introduction

[1] explained the role of fisheries in the society through the following statements:

"*Directly and indirectly, fisheries provide employment for hundreds of millions of people. The vast majority of these people are in developing countries where the sector often plays a key role in preventing and reducing poverty; it is likely that millions more people are involved in fishing activities than appear in official statistics.*"

The following exposure, which was given by [2], confirms the role:

"*Around 140 million tonnes of fish and seafood per year are used for human consumption. Set against the global production of cereals of around 2.2 billion tonnes, that figure is comparatively low. Owing to its unique combination of nutrients, fish makes a major contribution to a healthy diet. It supplies proteins, healthy fatty acids, vitamins and other elements essential for health such as iodine and selenium. Furthermore, in developing countries fish is often the only affordable and relatively easily available source of animal protein. In some regions on Earth fish can provide up to 50 per cent of the total animal protein in people's diets. This is the case, for example, in Bangladesh, Cambodia and Ghana.*"

On the other hand, [3] mentioned the importance of the fishery industry in the society through the following explanation:

"*To those involved in fisheries, the importance of the industry to the economy is obvious: (i) exports of fish and fish products earn foreign exchange which helps provide the resources needed to pay for crucial imports; (ii) it provides employment for a substantial number of people; (iii) the catch of commercial and artisanal fishers provide an important source of food for the community; and (iv) the fish, shellfish, seaweed and other aquatic resources collected by families are a crucial part of their livelihood.*"

Meanwhile, [4], through the following description, exposed the values of fish:

"*Historically, economic values of fish have been viewed solely in the context of commercial harvest. However in recent years, others aspects of value are being recognised both in the UK and abroad. Recreational fisheries, principally angling, can have significant economic importance and social benefits while fish can be valued for their own existence as part of the native fauna, regardless of their exploitation.*"

Therefore, based on above facts, one can argue that, from an economic point of view, the fishery is an important aspect in the society.

Many previous studies discussed the fishery topic. For example, [5] evaluated the assumption which is used by stock assessment scientists and fishery managers in their works. They compared industry-generated reports of landed catch to independent observer estimates in their study. [6] explored the crisis involving the Europe Union (EU), Iceland, Norway, and the Faroe Islands regarding the relative allocation and size of Total Allowable Catches (TACs) in the mackerel fishery in the Northeast Atlantic. They employed simple non-cooperative and cooperative game theory in their study. [7] developed a method to objectively measure and characterize the location and size of the patterns of discoloration which are currently appeared in the fillets of commercially harvested yellowtail flounder (*Limanda ferruginea*) on the east coast of Canada. Further, they developed an image processing program to analyze the patterns of the fillet discoloration in order to achieve the goal of their study. [8] assessed the effects of lure color on catch-per-unit-effort (CPUE), size selectivity and hooking injury of largemouth bass, *Micropterus salmoides*, by employing artificial 12.7 cm unscented soft-plastic worms. Their study expressed that while different lure colors might capture the imagination and wallet of the angler, they do not influence CPUE or hooking injury in bass but appear to have a small effect on the size of the captured fish.

Meanwhile, [9] provided the estimations of (1) catches for all marine fishing sectors, (2) the effort of fishing in the major Italian fishing fleets, and (3) the catch per unit of the effort from 1950 to 2010. [10] tested the impact of the three types of the underwater lighting on observable rockfish density and behavior by utilizing an underwater stereo camera. [11] addressed the fish behavior at the mouth of a midwater trawl in two temperate reservoirs. They used a SIMRAD EK60 (38 kHz) split-beam echosounder with the transducer deployed at the water surface, attached to the surface trawl headrope, in their study in monitoring fish distribution and behavior. [12] did the comparison of the estimated life history parameters for *Hyporthodus octofasciatus* from the south-eastern Indian Ocean with the one from the western central Pacific Ocean. [13] conducted the study in order to get a deeper understanding of the relationships between *Doryteuthis gahi* immigrations and size distributions during the seasons of fishing, and related wind, temperature, and geostrophic current conditions. The focused area of their study was Falkland Islands.

The study analyzes the fishery sector of the specific Asian country, from above literatures, however, is still limited. On the other hand, the readers are presented the previous studies discussed the other industrial sectors of Asian countries. For example, [14] investigated the dynamics of Indonesian creative industries. More specifically, the purposes of his study were (1) to obtain the other perspective about the role of creative industry sectors in the Indonesian national economy, and (2) to inquire about the strategies to improve these industries. He employed the Input-Output (IO) analysis in order to achieve these goals. His study focused on the period between 1990 and 2005. Besides, [15] conducted a deeper analysis regarding the impacts of the changes of final demands on the total outputs of Japanese energy sectors. His study focused on two industries, namely (1) petroleum refinery products, and (2) non-ferrous metals.

Meanwhile, [16] forecasted the influences of the Information and Communication Technology (ICT) on the structural changes of the national economics of Japan. They employed IO and statistical approaches as analysis instruments. [17] analyzed the role of ICT sectors on the Japanese national economy using simple household income multiplier, one of the analysis tools in the IO analysis. The analysis period of his study was from 1995-2005. On the other hand, using this multiplier, [18] analyzed this role on the Indonesian national economy. The analysis period of his study was from 1990-2005. [19] investigated the influences of Gross Domestic Product (GDP) and ICT on the structural changes of Indonesian industrial sectors on the period between 1990 and 2005. They employed the statistical tool in analyzing these influences, namely the Constrained Multivariate Regression (CMR) model. [20] exposed the impacts of the modifications of final demands on the total outputs of Indonesian ICT sectors by using the IO analysis as an analysis tool. [21] applied a simple output multipliers method in order to get the other perspective regarding the role of ICT sectors in the Indonesian national economy from 1990-2005. The study investigated the other perspective regarding this role for the Japanese case was conducted by [22]. His analysis focused on the period from 1995-2005. Besides, using the Structural Decomposition Analysis (SDA), [23] compared the role for the cases of Indonesia and Japan.

The study focuses on the analysis of the fishery sector of the specific Asian country is important because it will describe the characteristics of the sector. Further, this analysis can also open the opportunity in observing this sector from an economic point of view. This study is conducted in order to (1) fulfill the gap of the study in the fishery topic, (2) get the characteristics, and (3) open the opportunity.

The purpose of this study is to analyze the dynamics of the fishery industry of the specific Asian country. In other words, the following research question, "how to describe the dynamics of the fishery sector of the specific Asian country?", is tried to be solved by this study. This study focuses on the case of Indonesia. This country is chosen because their ocean area is wide. Using the previous studies as references, this study employs the IO analysis as an analysis tool. The dynamics are represented by the amendments of the total output of the sector. The trigger of these amendments is the changes of the final demand of the sector.

2. Methodology

The methodology of this study refers to the previous study which was conducted by [24]. The methodology of this study is described as follows. The first step is to elaborate the data used. This study uses the IO table of Indonesia for 2008. This table consists of 66 industries. These industries are explained in Appendix.

The second step is to define the Indonesian fishery sector used. This sector is described in Table 1. The third step is to conduct the calculations in order to know the impacts of the changes of the final demand on the total output of the analyzed sector. A demand-pull IO quantity model, one of the calculation tools in the IO analysis, is used in the calculations. [25] described that the following equation is a representation of this model:

$$\mathbf{x}^1 = \mathbf{L}^0\mathbf{f}^1 \tag{1}$$

where **x**, **L**, and **f** are the matrices of the total outputs of sectors, the Leontief inverse, and the final demands of sectors, respectively. 0 and 1 explain initial and future periods, respectively. An initial period in this study is 2008. Table 2 exposes the final demand modification scenarios used. These scenarios, compared with the previous ones, are slightly different. One of the differences can be seen on the name of the scenario 3. The name is changed from "The change of outside households consumption" to "Households and non-profit private institutions consumptions modification".

The conditions of "whole sector change" and "pure change" are included in above calculations. The former situation explains the condition which the changes of the components of the final demand are addressed to all Indonesian industrial sectors while the latter one only focuses on the discussed sector. In this study, the former situation will be called "condition A" while the term of "condition B" is applied to explain the latter one. The analysis regarding above impacts is discussed on the next step. Conclusions of this study and suggestions for further researches are exposed on the final step.

Table 1. Indonesian fishery sector used in this study.

Sector Number	Sector Name
23	Fishery

Table 2. The final demand modification scenarios used in this study.

The Component of the Final Demand	Scenario		
	1	2	3
	Exports Modification	Imports Modification	Households and Non-profit Private Institutions Consumptions Modification
Exports	Increase 30%	Constant	Constant
Imports	Constant	Increase 30%	Constant
Households and Non-profit Private Institutions Consumptions	Constant	Constant	Increase 30%

(Source: [26], with the slight modifications)

3. Results and analysis

Table 3 exposes the total output of the discussed sector for each scenario on condition A. Figure 1 explains in more details the dynamics of the total output of the sector on this condition. Based on the results, one can argue that, on the condition, the biggest positive impact on the total output of the analyzed sector is given by scenario 3, households and non-profit private institutions consumptions modification. Meanwhile, the negative impact is given by scenario 2, the change of imports.

On the other hand, Table 4 describes the total output of the analyzed sector for each scenario on condition B. Figure 2 exposes in more details the dynamics of the sector's total output on this condition. Based on the results, one can say that, on the condition, the biggest positive impact on the total output of the discussed sector is given by scenario 3, households and non-profit private institutions consumptions change. On the contrary, the negative impact is delivered by scenario 2, the imports modification.

Above phenomena show that, in both conditions, the discussed sector has similar patterns, namely this industry receives the positive impacts from scenarios 1 and 3 while the opposite impact is obtained from scenario 2. Above phenomena also explain that, in both conditions, the biggest positive impact for the analyzed sector is given by scenario 3, the change of households and non-profit private institutions consumptions. Based on these results, one can argue that, the effective way to enhance the total output of the Indonesian fishery sector in the future are to increase the rate of households and non-profit private institutions in terms of consuming the products of the industry. Besides, to restrict import activities for these products also will be a good step.

The availability of the sector's products in the especially domestic market is a key point in order to achieve above enhancement. Therefore, the obstacles of the goal should be eliminated. One of the serious obstacles is an illegal fishing activity. This activity can be omitted by executing the strict punishment for the perpetrators. On the other hand, this punishment can be smoothly executed if the law in terms of the activity is clear and enforced. The other essential point in order to achieve the goal is every party who has a relationship with the fishing should notice the importance of the marine ecosystem balance. This balance will ensure that the products are available for long periods of time.

Table 3. The total output of the discussed sector for each scenario on condition A (100 million Rupiah).

Sector Number	Sector Name	X_t, t = 2008	X_{t+1}, Scenario 1	X_{t+1}, Scenario 2	X_{t+1}, Scenario 3
23	Fishery	1,837,672.86	1,878,964.00	1,807,693.94	2,384,981.80

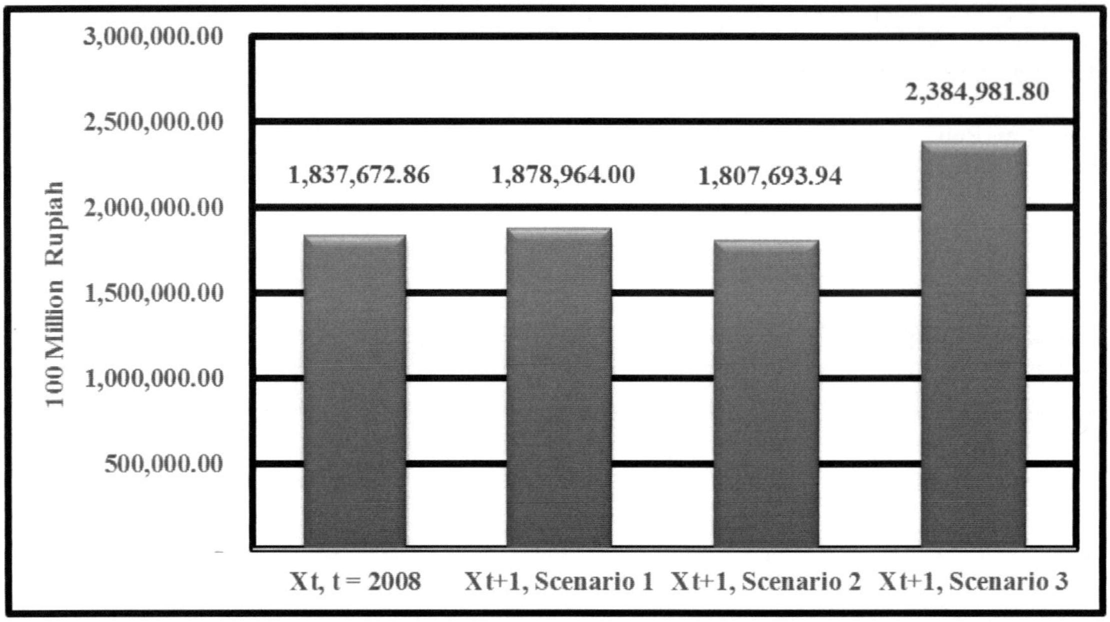

Figure 1. The dynamics of the total output of the Indonesian fishery sector (condition A).

Table 4. The total output of the discussed sector for each scenario on condition B (100 million Rupiah).

Sector Number	Sector Name	X_t, t = 2008	X_{t+1}, Scenario 1	X_{t+1}, Scenario 2	X_{t+1}, Scenario 3
23	Fishery	1,837,672.86	1,847,278.01	1,836,979.96	2,217,310.86

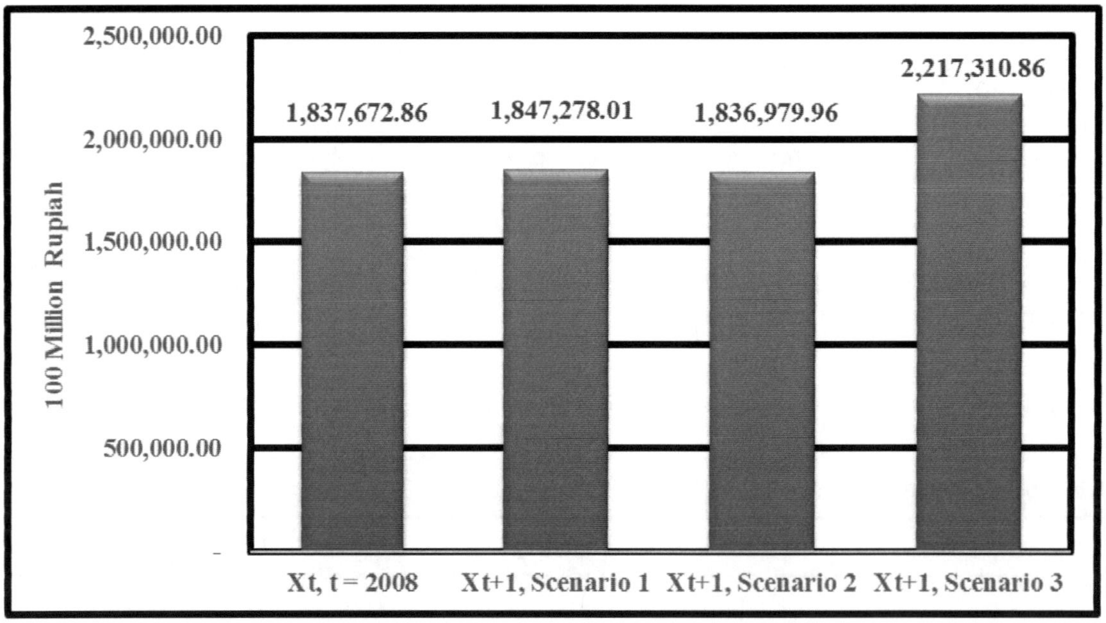

Figure 2. The dynamics of the total output of the Indonesian fishery sector (condition B).

4. Conclusions and further researches

This study investigated the dynamics of the total output of the Indonesian fishery industry when the changes of the final demand occurred. This study employed a demand-pull IO quantity model, one of the calculation tools in the IO analysis, as an analysis instrument. Two conditions were noticed in calculations and analysis parts, namely (1) "whole sector change", and (2) "pure change". An initial period in this study was 2008. One of the differences between current and previous studies could be seen on the name of the scenario 3, households and non-profit private institutions consumptions modification.

The results showed that, in both conditions, the discussed sector had similar patterns, namely this industry received the positive impacts from scenarios 1 and 3 while the negative impact was obtained from scenario 2. The results also explained that, in both conditions, the biggest positive impact for the analyzed sector was given by scenario 3, the change of households and non-profit private institutions consumptions. Based on the results, the suggestions from this study regarding the effective ways to enhance the total output of the fishery sector of Indonesia in the future were to increase the rate of households and non-profit private institutions in terms of consuming the products of the industry, and to restrict import activities for these outputs.

The dynamics of the total output of the Indonesian fishery sector could be investigated from this study. This study, however, did not conduct the deep analysis about how to effectively increase the output in the future. This analysis is needed in order to concretely map out the strategies for achieving the goal. Therefore, as a further research, this study proposes the analysis.

This study also did not conduct the investigations in order to know the dynamics of the total outputs of other Indonesian sectors. The results of these investigations will be good insights especially in order to know the characteristics of these sectors. Hence, this study also suggests the investigations as a further research.

The role of the fishery industry on the Indonesian economy, however, could not be observed from this study. The information about this role is needed in order to sharpen the strategies for increasing the total output of this industry in the future. Based on this reason, this study suggests the exploration in order to know the role as one of the future researches. This exploration will be more interesting if the other data are also included.

The other suggested future research from this study is to conduct the international comparison on the current discussion. This comparison will describe the characteristics of the fishery industries of analyzed countries when the changes of final demands occur. One of the examples is to compare developed and developing countries.

Journal of Physics: Conference Series **710** (2016) 012039 doi:10.1088/1742-6596/710/1/012039

Appendix. Indonesian Industrial Sectors (66 Sectors)

No.	Sector Name
1	Paddy
2	Beans
3	Maize
4	Root crops
5	Vegetables and fruits
6	Other food crops
7	Rubber
8	Sugarcane
9	Coconut
10	Oil palm
11	Tobacco
12	Coffee
13	Tea
14	Clove
15	Fiber crops
16	Other estate crops
17	Other crops
18	Livestock
19	Slaughtering
20	Poultry and its products
21	Wood
22	Other forest products
23	Fishery
24	Coal and metal ore mining
25	Crude oil, natural gas, and geothermal mining
26	Other mining and quarrying
27	Industry of food processing and preservation
28	Industry of oil and fat
29	Industry of rice milling
30	Industry of all kinds of flour
31	Industry of sugar
32	Industry of other food products
33	Industry of beverages
34	Industry of cigarettes
35	Industry of yarn spinning
36	Industry of textile, wearing apparel, and leather
37	Industry of bamboo, wood, and rattan
38	Industry of paper and the products of cardboard
39	Industry of fertilizer and pesticide
40	Industry of chemicals
41	Petroleum refinery
42	Industry of rubber products and plastic
43	Industry of non-metallic mineral products
44	Industry of cement
45	Industry of basic iron and steel
46	Industry of non-ferrous basic metal

Journal of Physics: Conference Series **710** (2016) 012039 doi:10.1088/1742-6596/710/1/012039

47	Industry of fabricated metal products
48	Industry of machine, electrical apparatus, and equipment of electric
49	Industry of transportation equipment and its repairement
50	Industry of other products
51	Electricity, gas, and water supply
52	Construction
53	Trade
54	Restaurant and hotel
55	Railway transport
56	Road transport
57	Water transport
58	Air transport
59	Services for supporting the transports
60	Communication
61	Financial institutions
62	Real estate and services of offices
63	General government and defense institutions
64	Social and community services
65	Other services
66	Activities not elsewhere classified

(Source: [27], with the slight modifications)

References

[1] The Prince's Charities' International Sustainability Unit 2012 *Towards Global Sustainable Fisheries: The Opportunity for Transition* [online] http://www.pcfisu.org/wp-content/uploads/2012/01/ISUMarineprogramme-towards-global-sustainable-fisheries.pdf (accessed 4 January 2016)

[2] The maribus gGmbH 2013 *World Ocean Review 2: The Future of Fish-The Fisheries of the Future* [online] http://worldoceanreview.com/wp-content/downloads/wor2/WOR2_english.pdf (accessed 4 January 2016)

[3] Gillett R and Lightfoot C 2001 *The Contribution of Fisheries to the Economies of Pacific Island Countries* [online] http://www.adb.org/sites/default/files/publication/28819/contribution-fisheries.pdf (accessed 5 January 2016)

[4] The Environment Agency 2009 *Economic Evaluation of Inland Fisheries: Managers Report from Science Project SC050026/SR2* [online] https://www.gov.uk/government/uploads/system/uploads/attachment_data/file/291109/scho0109bpgi-e-e.pdf (accessed 5 January 2016)

[5] Faunce C H, Cahalan J, Bonney J and Swanson R 2015 Can observer sampling validate industry catch reports from trawl fisheries? *Fisheries Research* **172** 34–43

[6] Jensen F, Frost H, Thøgersen T, Andersen P and Andersen J L 2015 Game theory and fish wars: the case of the Northeast Atlantic mackerel fishery *Fisheries Research* **172** 7–16

[7] Kenney J L, Rahman T, Manuel H and Winger P D 2015 Bruising pattern in commercially harvested yellowtail flounder (*Limanda ferruginea*) *Fisheries Research* **172** 79–84

[8] Moraga A D, Wilson A D M and Cooke S J 2015 Does lure colour influence catch per unit effort, fish capture size and hooking injury in angled largemouth bass? *Fisheries Research* **172** 1–6

[9] Piroddi C, Gristina M, Zylich K, Greer K, Ulman A, Zeller D and Pauly D 2015 Reconstruction of Italy's marine fisheries removals and fishing capacity, 1950-2010 *Fisheries Research* **172** 137–47

[10] Rooper C N, Williams K, Robertis A D and Tuttle V 2015 Effect of underwater lighting on observations of density and behavior of rockfish during camera surveys *Fisheries Research* **172** 157–67

[11] Sajdlová Z *et al* 2015 Fish behaviour in response to a midwater trawl footrope in temperate reservoirs *Fisheries Research* **172** 105–13

[12] Wakefield C B, Williams A J, Newman S J, Bunel M, Boddington D K, Vourey E and Fairclough D V 2015 Variations in growth, longevity and natural mortality for the protogynous hermaphroditic eightbar grouper *Hyporthodus octofasciatus* between the Indian and Pacific Oceans *Fisheries Research* **172** 26–33

[13] Winter A and Arkhipkin A 2015 Environmental impacts on recruitment migrations of Patagonian longfin squid (*Doryteuthis gahi*) in the Falkland Islands with reference to stock assessment *Fisheries Research* **172** 85–95

[14] Zuhdi U 2015 The dynamics of Indonesian creative industry sectors: an analysis using input-output approach *Journal of the Knowledge Economy* **6** 1177–90

[15] Zuhdi U 2015 An application of input-output analysis in analyzing the impacts of final demands changes on the total outputs of Japanese energy sectors: a further study *Journal of Physics: Conference Series* **622** 012041

[16] Zuhdi U, Mori S and Kamegai K 2015 Forecasting the influences of information and communication technology on the structural changes of Japanese industrial sectors: a study using statistical analysis *International Journal of Social, Behavioral, Educational, Economic and Management Engineering* **9** 531–7

[17] Zuhdi U 2015 An analysis of the role of information and communication technology sectors on Japanese national economy from 1995 through 2005: an application of multiplier analysis *IOP Conference Series: Earth and Environmental Science* **23** 012014

[18] Zuhdi U 2015 An application of multiplier analysis in analyzing the role of information and communication technology sectors on Indonesian national economy: 1990-2005 *IOP Conference Series: Earth and Environmental Science* **23** 012015

[19] Zuhdi U, Mori S and Kamegai K 2014 Analysis of influences of GDP and ICT on Indonesian industrial structural changes using statistical analysis: 1990-2005 *Journal of Finance and Accountancy* **17** 1–19

[20] Zuhdi U 2014 The impacts of final demand changes on total output of Indonesian ICT sectors: an analysis using input-output approach *IOP Conference Series: Materials Science and Engineering* **58** 012011

[21] Zuhdi U 2014 Using multipliers analysis in order to get another perspective related to the role of ICT sectors in national economy of Indonesia: 1990-2005 *Journal of Physics: Conference Series* **495** 012051

[22] Zuhdi U 2014 The other perspective related to the role of information and communication technologies sectors in national economy: the case of Japan *Advanced Science Letters* **20** 483–6

[23] Zuhdi U, Mori S and Kamegai K 2012 Analyzing the role of ICT sector to the national economic structural changes by decomposition analysis: the case of Indonesia and Japan *Procedia-Social and Behavioral Sciences* **65** 749–54

[24] Zuhdi U 2014 The dynamics of total output of Indonesian information and communication technology sector when final demand changes occur: an analysis using input-output approach *Advanced Science Letters* **20** 2254–7

[25] Miller R E and Blair P D 2009 *Input-Output Analysis: Foundations and Extensions* (Cambridge: University Press)

[26] Zuhdi U, Prasetyo A D and Sianipar C P M 2013 Analyzing the dynamics of total output of Japanese creative industry sectors: an input-output approach *Procedia Economics and Finance* **5** 827–35

[27] Zuhdi U 2014 The role of information and communication technology sectors in Indonesian national economy from 1990 through 2008: an analysis using input-output approach *Advanced Science Letters* **20** 1932–5

The dynamics of the total output of the Japanese fisheries sector: An analysis using input-output approach

Ubaidillah Zuhdi

Faculty of Management and Economics, Gdansk University of Technology, Gdansk 80-233, Poland

zuhdi@zie.pg.gda.pl

Abstract. The purpose of this study is to investigate the dynamics of the total output of the fisheries industry when the changes of the final demand occur. This study focuses on the case of Japan. This study employs a demand-pull Input-Output (IO) quantity model, one of the calculation instruments in the IO analysis, as an analysis tool. Two conditions are included in calculations and analysis parts, namely (1) "whole sector change", and (2) "pure change". An initial period in this study is 2005. The results show that, in both conditions, the analyzed sector has similar patterns, namely this industry obtains the positive impacts from scenarios 1, 3, and 4 while the negative impact is received from scenario 2. The results also expose that, in both conditions, the biggest positive impact for the discussed sector is given by scenario 4, the modification of the consumption expenditures of the private.

1. Introduction

[1] exposed the importance of the fishery industry in the society through the following explanation:

"To those involved in fisheries, the importance of the industry to the economy is obvious: (i) exports of fish and fish products earn foreign exchange which helps provide the resources needed to pay for crucial imports; (ii) it provides employment for a substantial number of people; (iii) the catch of commercial and artisanal fishers provide an important source of food for the community; and (iv) the fish, shellfish, seaweed and other aquatic resources collected by families are a crucial part of their livelihood."

On the other hand, [2], through the following description, mentioned the values of fish:

"Historically, economic values of fish have been viewed solely in the context of commercial harvest. However in recent years, others aspects of value are being recognised both in the UK and abroad. Recreational fisheries, principally angling, can have significant economic importance and social benefits while fish can be valued for their own existence as part of the native fauna, regardless of their exploitation."

Meanwhile, [3] described the role of fisheries in the society through the following statements:

"Around 140 million tonnes of fish and seafood per year are used for human consumption. Set against the global production of cereals of around 2.2 billion tonnes, that figure is comparatively low. Owing to its unique combination of nutrients, fish makes a major contribution to a healthy diet. It supplies proteins, healthy fatty acids, vitamins and other elements essential for health such as iodine and selenium. Furthermore, in developing countries fish is often the only affordable and relatively easily available source of animal protein. In some regions on Earth fish can provide up to 50 per cent of the total animal protein in people's diets. This is the case, for example, in Bangladesh, Cambodia and Ghana."

The following exposure, which was given by [4], reaffirms the role:

"*Directly and indirectly, fisheries provide employment for hundreds of millions of people. The vast majority of these people are in developing countries where the sector often plays a key role in preventing and reducing poverty; it is likely that millions more people are involved in fishing activities than appear in official statistics.*"

Therefore, based on above facts, one can say that, from an economic point of view, the fishery is an essential part in the society.

Many previous studies discussed the fishery topic. For example, [5] addressed the fish behavior at the mouth of a midwater trawl in two temperate reservoirs. They employed a SIMRAD EK60 (38 kHz) split-beam echosounder with the transducer deployed at the water surface, attached to the surface trawl headrope, in their study in monitoring fish distribution and behavior. [6] assessed the impacts of lure color on catch-per-unit-effort (CPUE), size selectivity and hooking injury of largemouth bass, *Micropterus salmoides*, by employing artificial 12.7 cm unscented soft-plastic worms. Their study exposed that while different lure colors might capture the imagination and wallet of the angler, they do not influence CPUE or hooking injury in bass but appear to have a little impact on the size of the captured fish. [7] conducted the estimations of (1) catches for all marine fishing sectors, (2) the effort of fishing in the major Italian fishing fleets, and (3) the catch per unit of the effort from 1950 to 2010. [8] did the comparison of the estimated life history parameters for *Hyporthodus octofasciatus* from the south-eastern Indian Ocean with the one from the western central Pacific Ocean.

Meanwhile, [9] conducted the study in order to obtain a deeper understanding of the relationships between *Doryteuthis gahi* immigrations and size distributions during the seasons of fishing, and related wind, temperature, and geostrophic current conditions. The focused area of their study was Falkland Islands. [10] measured the impact of the three types of the underwater lighting on observable rockfish density and behavior by utilizing an underwater stereo camera. [11] explored the crisis involving the Europe Union (EU), Iceland, Norway, and the Faroe Islands regarding the relative allocation and size of Total Allowable Catches (TACs) in the mackerel fishery in the Northeast Atlantic. They used simple non-cooperative and cooperative game theory in their study. [12] evaluated the assumption which is used by stock assessment scientists and fishery managers in their works. They compared industry-generated reports of landed catch to independent observer estimates in their study. [13] developed a method to objectively measure and characterize the location and size of the patterns of discoloration which are currently appeared in the fillets of commercially harvested yellowtail flounder (*Limanda ferruginea*) on the east coast of Canada. Further, they developed the program of image processing to analyze the patterns of the fillet discoloration in order to achieve the goal of their study.

The study analyzes the fishery sector of the specific Asian country, from above literatures, however, is still limited. On the other hand, the readers are provided the previous studies discussed the other industrial sectors of Asian countries. For example, [14] forecasted the influences of the Information and Communication Technology (ICT) on the structural changes of the national economics of Japan. They used Input-Output (IO) and statistical approaches as analysis instruments. [15] analyzed the role of ICT sectors on the Japanese national economy using simple household income multiplier, one of the analysis devices in the IO analysis. The analysis period of his study was from 1995-2005. On the other hand, using this multiplier, [16] analyzed this role on the Indonesian national economy. The period of the analysis of his study was from 1990-2005. [17] inquired the influences of Gross Domestic Product (GDP) and ICT on the changes of the structures of Indonesian industrial sectors from 1990-2005. They used the statistical instrument in analyzing these influences, namely the Constrained Multivariate Regression (CMR) model. [18] exposed the impacts of the modifications of final demands on the total outputs of Indonesian ICT sectors by employing the IO analysis as an analysis device. [19] employed a simple output multipliers method in order to obtain the other perspective regarding the role of ICT sectors in the Indonesian national economy from 1990-2005. The study analyzed the other perspective regarding this role for the case of Japan was conducted by [20]. His analysis focused on the period between 1995 and 2005. Besides, [21] did the comparison between Indonesia and Japan regarding the role by using the Structural Decomposition Analysis (SDA).

Meanwhile, [22] analyzed the dynamics of Indonesian creative industries. More specifically, the purposes of his study were (1) to obtain the other perspective about the role of creative industry sectors in the Indonesian national economy, and (2) to inquire the strategies to improve these industries. He employed the

IO analysis in order to achieve these goals. His study focused on the analysis period between 1990 and 2005. Besides, [23] conducted a deeper analysis regarding the effects of the changes of final demands on the total outputs of Japanese energy sectors. His study focused on two industries, namely (1) petroleum refinery products, and (2) non-ferrous metals.

The study focuses on the analysis of the fishery sector of the specific Asian country is needed because it will expose the characteristics of the sector. Further, this analysis can also open the chance in observing this sector from an economic point of view. This study is conducted in order to fulfill the gap of the research in the fishery topic.

The purpose of this study is to analyze the dynamics of the fisheries industry of the specific Asian country. This study focuses on the Japanese case. This country is chosen because their rate of fish consuming is high. Using the previous studies as references, this study uses the IO analysis as an analysis instrument. The dynamics are represented by the amendments of the total output of the industry. The trigger of these amendments is the changes of the final demand of the sector.

2. Methodology

The methodology of this study refers to the previous study which was conducted by [23]. The methodology of this study is explained as follows. The first step is to describe the data used. This study uses the aggregated IO table of Japan for 2005. This table consists of 89 sectors. These sectors are exposed in Appendix.

The second step is to define the Japanese fishery sector used. This sector is explained in Table 1. The third step is to conduct the calculations in order to observe the impacts of the changes of the final demand on the total output of the analyzed sector. A demand-pull IO quantity model, one of the calculation instruments in the IO analysis, is employed in the calculations. [24] explained that the following equation is a representation of this model:

$$\mathbf{x}^1 = \mathbf{L}^0\mathbf{f}^1 \tag{1}$$

where **x**, **L**, and **f** are the matrices of the total outputs of sectors, the Leontief inverse, and the final demands of sectors, respectively. 0 and 1 describe initial and future periods, respectively. An initial period in this study is 2005. Table 2 explains the final demand modification scenarios used.

The conditions of "whole sector change" and "pure change" are noticed in above calculations. The former situation explains the condition which the changes of the components of the final demand are addressed to all Japanese industrial sectors while the latter one only focuses on the discussed sector. In this study, the former situation will be called "condition A" while the term of "condition B" is used to explain the latter one. The analysis regarding above impacts is conducted on the next step. Conclusions of this study and suggestions for further researches are described on the final step.

3. Results and analysis

Table 3 describes the total output of the discussed sector for each scenario on condition A. Figure 1 exposes in more details the dynamics of the total output of the sector on this condition. Based on the results, one can argue that, on the condition, the biggest positive impact on the total output of the analyzed sector is given by scenario 4, the change of the consumption expenditures of the private. Meanwhile, the negative impact is delivered by scenario 2, the modification of imports.

Table 1. Japanese fisheries sector used in this study.

Sector Number	Sector Name
5	Fisheries

Table 2. The final demand modification scenarios used in this study.

The Component of the Final Demand	Scenario			
	1	2	3	4
	Exports Modification	Imports Modification	The Modification of the Consumption Expenditures of Outside Households	The Modification of the Consumption Expenditures of the Private
Exports	Increase 30%	Constant	Constant	Constant
Imports	Constant	Increase 30%	Constant	Constant
The Consumption Expenditures of Outside Households	Constant	Constant	Increase 30%	Constant
The Consumption Expenditures of the Private	Constant	Constant	Constant	Increase 30%

(Source: [23], with the slight modifications)

On the other hand, Table 4 explains the total output of the analyzed sector for each scenario on condition B. Figure 2 describes in more details the dynamics of the total output of the sector on this condition. Based on the results, one can argue say, on the condition, the biggest positive impact on the total output of the analyzed sector is given by scenario 4, consumption expenditures of the private modification. Meanwhile, the negative impact is delivered by scenario 2, the change of imports.

Above phenomena show that, in both conditions, the discussed sector has similar patterns, namely this industry obtains the positive impacts from scenarios 1, 3, and 4 while the opposite impact is received from scenario 2. Above phenomena also describe that, in both conditions, the biggest positive impact for the analyzed sector is given by scenario 4, the change of the consumption expenditures of the private. Based on these results, one can argue that, the effective way to increase the total output of the Japanese fisheries sector in the future are to enhance the rate of private institutions in terms of consuming the products of the industry. Besides, to restrict import activities for these products also will be a good strategy.

The marine ecosystem balance, however, should be noticed by all parties who have relationships with the fishing. Besides, they must realize that the overfishing activity is a cruel thing. The following explanation, which was mentioned by [25], confirms the negative side of this activity:

"*The market for seafood is huge, and expanding rapidly, but wild stocks of fish are declining dramatically due to overfishing, pollution and other human impacts.*"

The importance of the balance should be understood because we, humans, will experience the huge pains when the ecosystem is damaged. In other words, they have to know and conduct the marine ecosystem maintenance activities too. One of the pains is a food shortage phenomenon.

Table 3. The total output of the discussed sector for each scenario on condition A (100 million Yen).

Sector Number	Sector Name	X_t, t = 2005	X_{t+1}, Scenario 1	X_{t+1}, Scenario 2	X_{t+1}, Scenario 3	X_{t+1}, Scenario 4
5	Fisheries	16,101.68	16,349.53	14,168.71	16,775.16	21,561.09

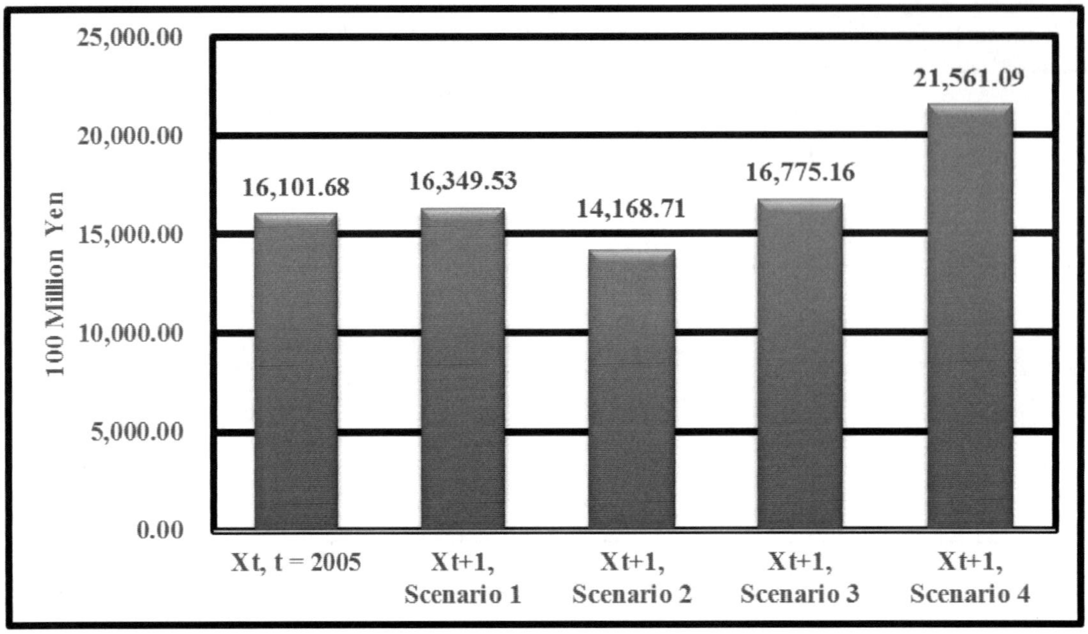

Figure 1. The dynamics of the total output of the Japanese fisheries sector (condition A).

Table 4. The total output of the discussed sector for each scenario on condition B (100 million Yen).

Sector Number	Sector Name	X_t, t = 2005	X_{t+1}, Scenario 1	X_{t+1}, Scenario 2	X_{t+1}, Scenario 3	X_{t+1}, Scenario 4
5	Fisheries	16,101.68	16,230.05	15,091.61	16,162.85	17,260.46

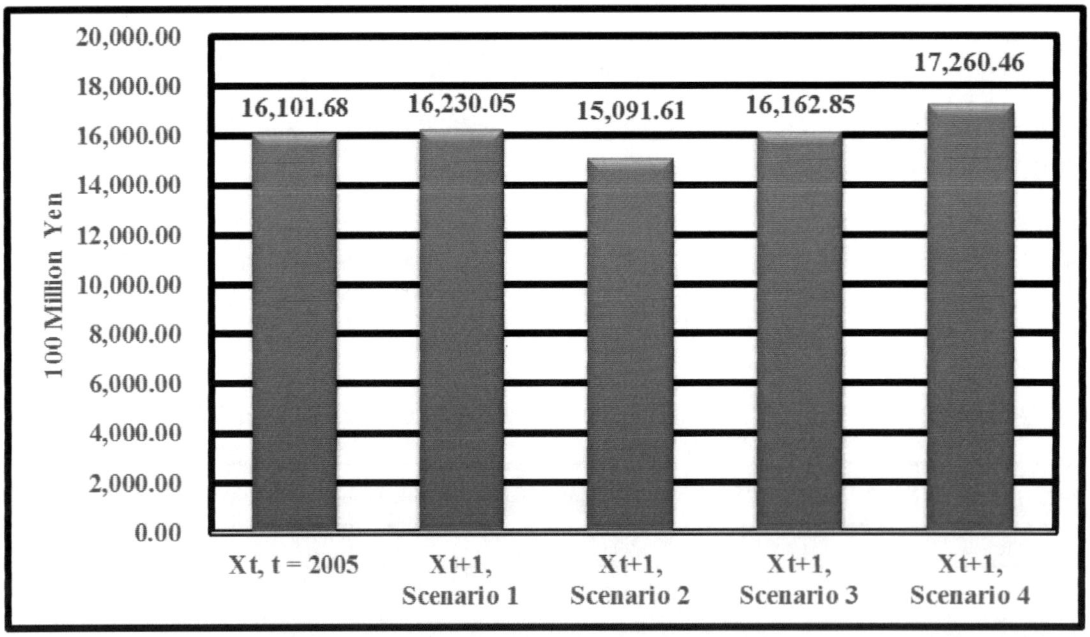

Figure 2. The dynamics of the total output of the Japanese fisheries sector (condition B).

4. Conclusions and further researches

This study examined the dynamics of the total output of the Japanese fisheries industry when the changes of the final demand occurred. This study employed a demand-pull IO quantity model, one of the calculation instruments in the IO analysis, as an analysis tool. Two conditions were included in calculations and analysis parts, namely (1) "whole sector change", and (2) "pure change". An initial period in this study was 2005.

The results showed that, in both conditions, the analyzed sector had similar patterns, namely this industry obtained the positive impacts from scenarios 1, 3, and 4 while the negative impact was received from scenario 2. The results also exposed that, in both conditions, the biggest positive impact for the discussed sector was given by scenario 4, the modification of the consumption expenditures of the private. Based on the results, the suggestions from this study regarding the effective ways to escalate the total output of the fisheries sector of Japan in the future were to increase the rate of private institutions in terms of consuming the products of the industry, and to limit import activities for these outputs.

The dynamics of the total output of the Japanese fisheries sector could be examined from this study. This study, however, did not conduct the deep analysis about how to effectively enhance the output in the future. This analysis is needed in order to concretely map out the steps for achieving the objective. Therefore, as a further research, this study suggests the analysis.

This study also did not execute the investigations in order to know the dynamics of the total outputs of other Japanese sectors. The outcomes of these investigations will be good insights especially in order to know the characteristics of these sectors. Hence, this study also proposes the investigations as a further research.

The role of the fisheries industry on the Japanese economy, however, could not be seen from this study. The information about this role is needed in order to sharpen the strategies for escalating the total output of this industry in the future. Based on this reason, this study recommends the exploration in order to know the role as one of the future researches. This exploration will be more comprehensive if the other data are also included.

The other suggested future research from this study is to execute the international comparison on the current discussion. This comparison will expose the characteristics of the fishery sectors of analyzed countries when the changes of final demands occur. One of the examples is to compare developed and developing countries. More specifically, a good example is to compare Japan and one of the developing countries, such as Indonesia.

Journal of Physics: Conference Series **710** (2016) 012040 doi:10.1088/1742-6596/710/1/012040

Appendix. Japanese Industrial Sectors (89 Sectors)

No.	Sector Name
1	Crop cultivation
2	Livestock
3	Agricultural services
4	Forestry
5	Fisheries
6	Metallic ores
7	Non-metallic ores
8	Coal mining, crude petroleum, and natural gas
9	Foods
10	Beverage
11	Feeds and organic fertilizer, n.e.c.
12	Tobacco
13	Textile products
14	Wearing apparel and other textile products
15	Timber and wooden products
16	Furniture and fixtures
17	Pulp, paper, paperboard, and building paper
18	Paper products
19	Printing, plate making, and book binding
20	Chemical fertilizer
21	Industrial inorganic chemicals
22	Basic petrochemical and intermediate chemical products
23	Synthetic resins
24	Synthetic fibers
25	Medicaments
26	Final chemical products, n.e.c.
27	Petroleum refinery products
28	Coal products
29	Plastic products
30	Rubber products
31	Leather, fur skins, and miscellaneous leather products
32	Glass and glass products
33	Cement and cement products
34	Pottery, china, and earthenware
35	Other ceramic, stone, and clay products
36	Pig iron and crude steel
37	Steel products
38	Steel castings and forgings, and other steel products
39	Non-ferrous metals
40	Non-ferrous metal products
41	Metal products for construction and architecture
42	Other metal products
43	General industrial machinery
44	Special industrial machinery
45	Other general machines
46	Machineries for office and service industry

47	Electrical appliance
48	Motor vehicles
49	Ships, and the repairement of ships
50	Other transportation equipment, and the repairement of transportation equipment
51	Precision instruments
52	Miscellaneous manufacturing products
53	Building construction
54	Construction repairement
55	Civil engineering
56	Electricity
57	Gas and heat supply
58	Water supply
59	Waste management services
60	Commerce
61	Finance and insurance
62	Real estate agencies and rental services
63	House rent
64	Railway transport
65	Road transport (except transport by private cars)
66	Self-transport by private cars
67	Water transport
68	Air transport
69	Freight forwarding
70	Storage facility services
71	Services related to transport
72	Communication
73	Broadcasting and information services
74	Public administration
75	Education
76	Research
77	Medicals services and health
78	Social security
79	Other public services
80	Advertising, survey, and information services
81	Goods rental and leasing services
82	Repairement of motor vehicles and machine
83	Other business services
84	Amusement and recreational services
85	Eating and drinking places
86	Accommodations
87	Other personal services
88	Office supplies
89	Activities not elsewhere classified

(Source: [26], with the slight modifications)

References

[1] Gillett R and Lightfoot C 2001 *The Contribution of Fisheries to the Economies of Pacific Island Countries* [online] http://www.adb.org/sites/default/files/publication/28819/contribution-fisheries.pdf (accessed 5 January 2016)

[2] The Environment Agency 2009 *Economic Evaluation of Inland Fisheries: Managers Report from Science Project SC050026/SR2* [online] https://www.gov.uk/government/uploads/system/uploads/attachment_data/file/291109/scho0109bpgi-e-e.pdf (accessed 5 January 2016)

[3] The maribus gGmbH 2013 *World Ocean Review 2: The Future of Fish-The Fisheries of the Future* [online] http://worldoceanreview.com/wp-content/downloads/wor2/WOR2_english.pdf (accessed 4 January 2016)

[4] The Prince's Charities' International Sustainability Unit 2012 *Towards Global Sustainable Fisheries: The Opportunity for Transition* [online] http://www.pcfisu.org/wp-content/uploads/2012/01/ISUMarineprogramme-towards-global-sustainable-fisheries.pdf (accessed 4 January 2016)

[5] Sajdlová Z *et al* 2015 Fish behaviour in response to a midwater trawl footrope in temperate reservoirs *Fisheries Research* **172** 105–13

[6] Moraga A D, Wilson A D M and Cooke S J 2015 Does lure colour influence catch per unit effort, fish capture size and hooking injury in angled largemouth bass? *Fisheries Research* **172** 1–6

[7] Piroddi C, Gristina M, Zylich K, Greer K, Ulman A, Zeller D and Pauly D 2015 Reconstruction of Italy's marine fisheries removals and fishing capacity, 1950-2010 *Fisheries Research* **172** 137–47

[8] Wakefield C B, Williams A J, Newman S J, Bunel M, Boddington D K, Vourey E and Fairclough D V 2015 Variations in growth, longevity and natural mortality for the protogynous hermaphroditic eightbar grouper *Hyporthodus octofasciatus* between the Indian and Pacific Oceans *Fisheries Research* **172** 26–33

[9] Winter A and Arkhipkin A 2015 Environmental impacts on recruitment migrations of Patagonian longfin squid (*Doryteuthis gahi*) in the Falkland Islands with reference to stock assessment *Fisheries Research* **172** 85–95

[10] Rooper C N, Williams K, Robertis A D and Tuttle V 2015 Effect of underwater lighting on observations of density and behavior of rockfish during camera surveys *Fisheries Research* **172** 157–67

[11] Jensen F, Frost H, Thøgersen T, Andersen P and Andersen J L 2015 Game theory and fish wars: the case of the Northeast Atlantic mackerel fishery *Fisheries Research* **172** 7–16

[12] Faunce C H, Cahalan J, Bonney J and Swanson R 2015 Can observer sampling validate industry catch reports from trawl fisheries? *Fisheries Research* **172** 34–43

[13] Kenney J L, Rahman T, Manuel H and Winger P D 2015 Bruising pattern in commercially harvested yellowtail flounder (*Limanda ferruginea*) *Fisheries Research* **172** 79–84

[14] Zuhdi U, Mori S and Kamegai K 2015 Forecasting the influences of information and communication technology on the structural changes of Japanese industrial sectors: a study using statistical analysis *International Journal of Social, Behavioral, Educational, Economic and Management Engineering* **9** 531–7

[15] Zuhdi U 2015 An analysis of the role of information and communication technology sectors on Japanese national economy from 1995 through 2005: an application of multiplier analysis *IOP Conference Series: Earth and Environmental Science* **23** 012014

[16] Zuhdi U 2015 An application of multiplier analysis in analyzing the role of information and communication technology sectors on Indonesian national economy: 1990-2005 *IOP Conference Series: Earth and Environmental Science* **23** 012015

[17] Zuhdi U, Mori S and Kamegai K 2014 Analysis of influences of GDP and ICT on Indonesian industrial structural changes using statistical analysis: 1990-2005 *Journal of Finance and Accountancy* **17** 1–19

[18] Zuhdi U 2014 The impacts of final demand changes on total output of Indonesian ICT sectors: an analysis using input-output approach *IOP Conference Series: Materials Science and Engineering* **58** 012011

[19] Zuhdi U 2014 Using multipliers analysis in order to get another perspective related to the role of ICT sectors in national economy of Indonesia: 1990-2005 *Journal of Physics: Conference Series* **495** 012051

[20] Zuhdi U 2014 The other perspective related to the role of information and communication technologies sectors in national economy: the case of Japan *Advanced Science Letters* **20** 483–6

[21] Zuhdi U, Mori S and Kamegai K 2012 Analyzing the role of ICT sector to the national economic structural changes by decomposition analysis: the case of Indonesia and Japan *Procedia-Social and Behavioral Sciences* **65** 749–54

[22] Zuhdi U 2015 The dynamics of Indonesian creative industry sectors: an analysis using input-output approach *Journal of the Knowledge Economy* **6** 1177–90

[23] Zuhdi U 2015 An application of input-output analysis in analyzing the impacts of final demands changes on the total outputs of Japanese energy sectors: a further study *Journal of Physics: Conference Series* **622** 012041

[24] Miller R E and Blair P D 2009 *Input-Output Analysis: Foundations and Extensions* (Cambridge: University Press)

[25] University of Otago 2011 *Aquaculture and Fisheries: Meeting Market Needs while Protecting Our Sea-life* [online] http://www.otago.ac.nz/study/otago032657.pdf (accessed 5 January 2016)

[26] Zuhdi U 2014 Analyzing the role of creative industries in national economy of Japan: 1995-2005 *Open Journal of Applied Sciences* **4** 197–211

The dynamics of the total outputs of Japanese information and communication technology sectors: A further study

Ubaidillah Zuhdi

Faculty of Management and Economics, Gdansk University of Technology, Gdansk 80-233, Poland

zuhdi@zie.pg.gda.pl

Abstract. The purpose of this study is to continue the previous studies which discussed the impacts of the changes of final demands on the total outputs of the Information and Communication Technology (ICT) sectors of the specific country. More specifically, this study aims to conduct a deeper analysis regarding these impacts. This study focuses on the case of Japan. This study employs a demand-pull Input-Output (IO) quantity model, one of the calculation tools in the IO analysis, as an analysis instrument. Two conditions are included in calculations and analysis parts, namely (1) "whole sector change", and (2) "pure change". An initial period in this study is 2005. The results show that, in both conditions, the discussed sectors have similar patterns, namely these industries receive the positive impacts from scenarios 1, 3, and 4 while the opposite impacts are obtained from scenario 2. This negative impact also appeared in the previous studies. The results also expose that, in both conditions, the biggest positive impacts for analyzed sectors are given by scenario 4, the modification of the consumption expenditures of the private. Compared with the previous studies, these are new findings.

1. Introduction

[1] explained the role of the Information and Communication Technology (ICT) in the society through the following statements:

"*ICT technologies are already having a role in driving productivity improvements and innovation in sectors such as financial services, as well as in some aspects of mining, agriculture. They are also transforming the delivery of key government services (including health and education).*"

Meanwhile, [2] described the role using the following exposure:

"*As we have seen, information and communications technologies help expand economic opportunity by enabling people to enhance their knowledge and skills; identify, apply, and qualify for better-paying jobs; use their disposable income more wisely; manage their own businesses efficiently; and tap into broader markets for their goods and services. ICTs also enhance capacity in industries and institutions of all kinds.*"

The following explanation, which was given by [3], confirms the role:

"*The role of ICT in the dynamic process of allocation and selection is that it may increase the flexibility of firms and as such their ability to cope with economic shocks, increasing relative productivity and chances of survival. ICT can be a major factor in determining the winners in the competitive process of reallocation and exit.*"

Therefore, based on above facts, one can argue that, from an economic point of view, ICT is an important part in the society.

Many previous studies discussed the ICT topic. For example, [4] forecasted the influences of the ICT on the structural changes of Japanese national economies. They applied Input-Output (IO) and statistical approaches as analysis tools. [5] analyzed the role of ICT sectors on the Japanese national economy using simple household income multiplier, one of the analysis devices in the IO analysis. The analysis period of his study was from 1995 to 2005. On the other hand, using this multiplier, [6] analyzed this role on the Indonesian national economy. The period of the analysis of his study was from 1990 to 2005. [7] investigated the influences of Gross Domestic Product (GDP) and ICT on the changes of the structures of Indonesian industrial sectors on the period between 1990 and 2005. They applied the statistical instrument in analyzing these influences, namely the Constrained Multivariate Regression (CMR) model. [8] described the impacts of the modifications of final demands on the total outputs of Indonesian ICT sectors by employing the IO analysis as an analysis instrument. On the other hand, these impacts on the case of Japan were analyzed by [9] and [10].

The study analyzes the impacts of the changes of final demands on the total outputs of the ICT sectors of the specific country, from above literatures, however, is still limited. This analysis is important because it will expose the characteristics of these sectors when the changes of national economic conditions occur. Further, the analysis can give the insights about the improvements needed for enhancing the industries. This study is conducted in order to fulfill the gap of the research in ICT topic.

The purpose of this study is to continue the previous studies which discussed the impacts of the changes of final demands on the total outputs of the ICT sectors of the specific country. More specifically, this study aims to conduct a deeper analysis regarding these impacts. This study focuses on the case of Japan. This country is chosen because their ICT aspects have developed rapidly. Using the previous studies as references, this study employs the IO approach as an analysis instrument.

2. Methodology

The methodology of this study refers to the previous studies which were conducted by [9] and [10]. The methodology of this study is described as follows. The first step is to expose the data used. This study uses the aggregated IO table of Japan for 2005. This table consists of 89 sectors. These sectors are explained in Appendix.

The second step is to define the Japanese ICT sectors used. These sectors are described in Table 1. The third step is to conduct the calculations in order to observe the impacts of the changes of final demands on the total outputs of discussed sectors. A demand-pull IO quantity model, one of the calculation tools in the IO analysis, is employed in the calculations. [11] explained that the following equation is a representation of this model:

$$\mathbf{x}^1 = \mathbf{L}^0\mathbf{f}^1 \tag{1}$$

where **x**, **L**, and **f** are the matrices of the total outputs of sectors, the Leontief inverse, and the final demands of sectors, respectively. 0 and 1 explain initial and future periods, respectively. An initial period in this study is 2005. Table 2 exposes the final demand modification scenarios used. The difference between current and previous studies can be seen on these scenarios. More specifically, the current study has the scenario 4, the modification of the consumption expenditures of the private, while the previous ones did not include this plot.

Table 1. Japanese ICT sectors used in this study.

No.	Sector Number	Sector Name
1	72	Communication
2	73	Broadcasting and information services
3	80	Advertising, survey, and information services

(Source: [12])

Table 2. The final demand modification scenarios used in this study.

The Component of the Final Demand	Scenario			
	1	2	3	4
	Exports Modification	Imports Modification	The Modification of the Consumption Expenditures of Outside Households	The Modification of the Consumption Expenditures of the Private
Exports	Increase 30%	Constant	Constant	Constant
Imports	Constant	Increase 30%	Constant	Constant
The Consumption Expenditures of Outside Households	Constant	Constant	Increase 30%	Constant
The Consumption Expenditures of the Private	Constant	Constant	Constant	Increase 30%

(Source: [13], with the slight modifications)

The conditions of "whole sector change" and "pure change" are infused in above calculations. The former situation explains the condition which the changes of the components of the final demand are addressed to all Japanese industrial sectors while the latter one only focuses on the analyzed sectors. In this study, the former situation will be called "condition A" while the term of "condition B" is utilized to explain the latter one. The analysis regarding above impacts is addressed on the next step. Conclusions of this study and suggestions for further researches are exposed on the final step.

3. Results and analysis

Table 3 describes the total outputs of discussed sectors for each scenario on condition A. Figures 1, 2, and 3 explain in more details the dynamics of the total outputs of analyzed sectors on this condition. Based on the results, one can argue that, on this condition, the biggest positive impacts on the total outputs of analyzed sectors are given by scenario 4, the modification of the consumption expenditures of the private. On the contrary, the negative impacts are given by scenario 2, the change of imports.

On the other hand, Table 4 explains the total outputs of discussed sectors for each scenario on condition B. Figures 4, 5, and 6 expose in more details the dynamics of the total outputs of analyzed sectors on this condition. Based on the results, one can say that, on this condition, the biggest positive impacts on the total outputs of analyzed sectors are given by scenario 4, consumption expenditures of the private change. Meanwhile, the negative impacts are given by scenario 2, the modification of imports.

Table 3. The total outputs of discussed sectors for each scenario on condition A (100 million Yen).

Sector Number	Sector Name	X_t, t = 2005	X_{t+1}, Scenario 1	X_{t+1}, Scenario 2	X_{t+1}, Scenario 3	X_{t+1}, Scenario 4
72	Communication	163,581.94	167,969.65	158,989.08	165,040.06	200,546.89
73	Broadcasting and information services	295,777.63	305,626.75	285,514.21	297,980.65	335,887.37
80	Advertising, survey, and information services	90,833.06	96,221.67	85,652.04	91,992.82	107,103.75

Table 4. The total outputs of discussed sectors for each scenario on condition B (100 million Yen).

Sector Number	Sector Name	X_t, t = 2005	X_{t+1}, Scenario 1	X_{t+1}, Scenario 2	X_{t+1}, Scenario 3	X_{t+1}, Scenario 4
72	Communication	163,581.94	163,875.13	163,204.22	164,013.34	187,385.37
73	Broadcasting and information services	295,777.63	296,856.35	293,299.36	296,092.46	310,074.07
80	Advertising, survey, and information services	90,833.06	91,239.76	90,088.38	90,855.02	91,932.59

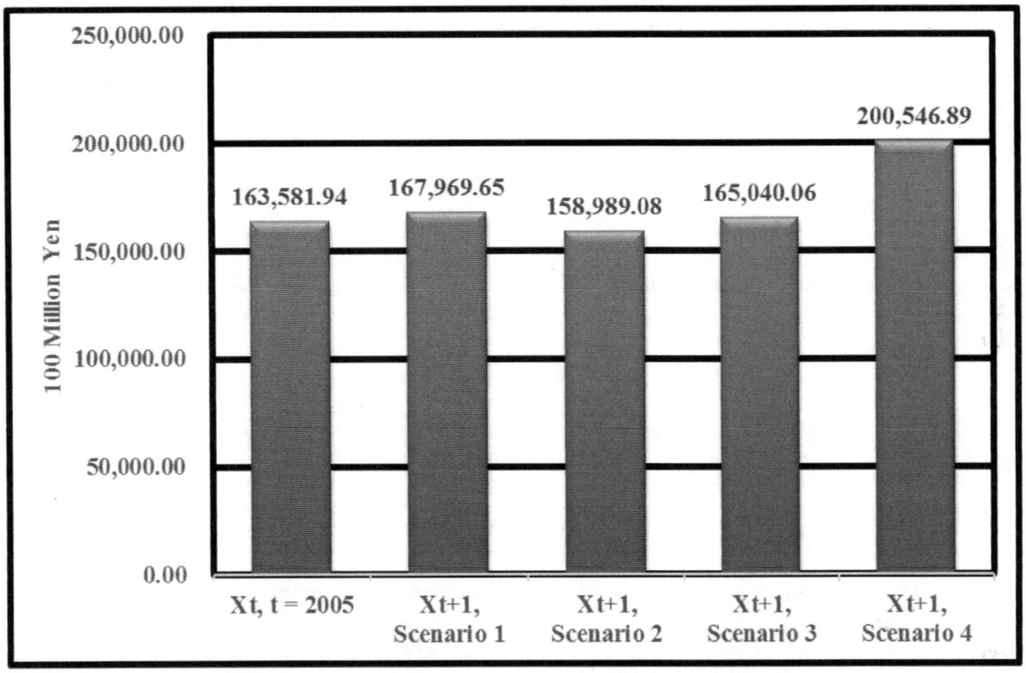

Figure 1. The dynamics of the total output of the communication sector (condition A).

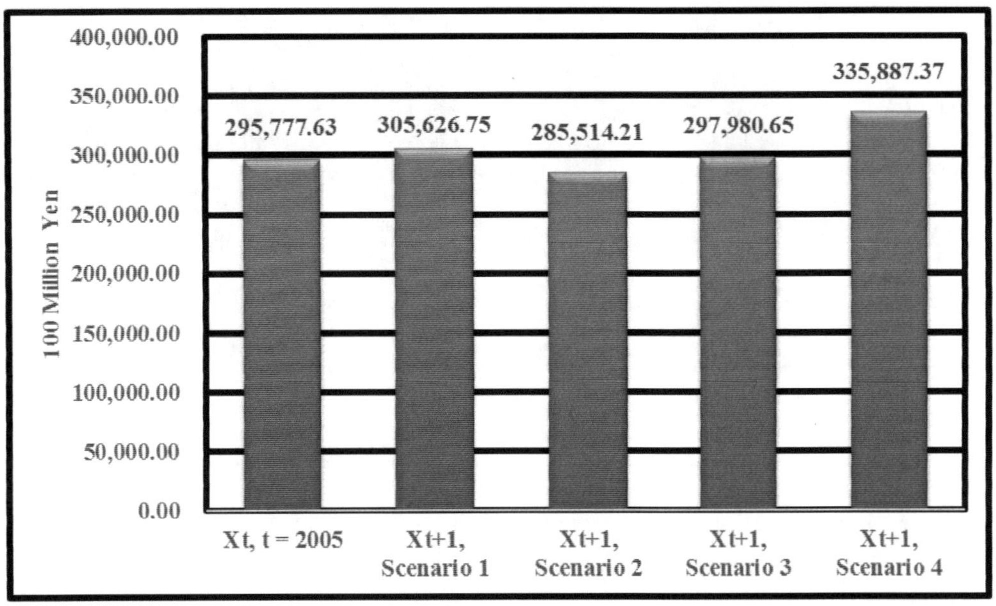

Figure 2. The dynamics of the total output of the broadcasting and information services sector (condition A).

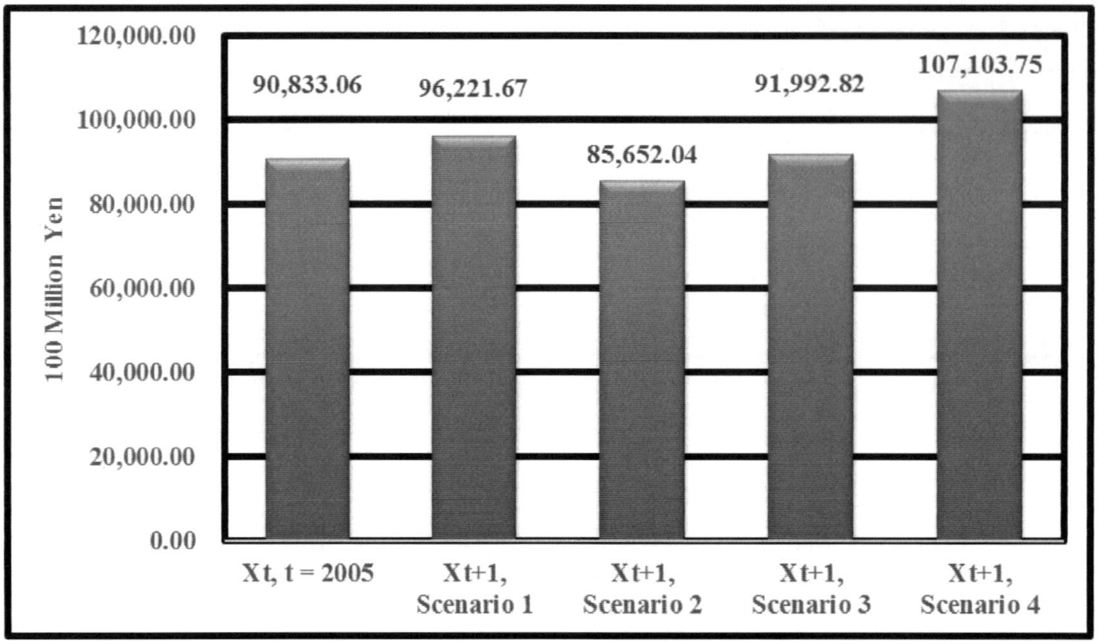

Figure 3. The dynamics of the total output of the advertising, survey, and information services sector (condition A).

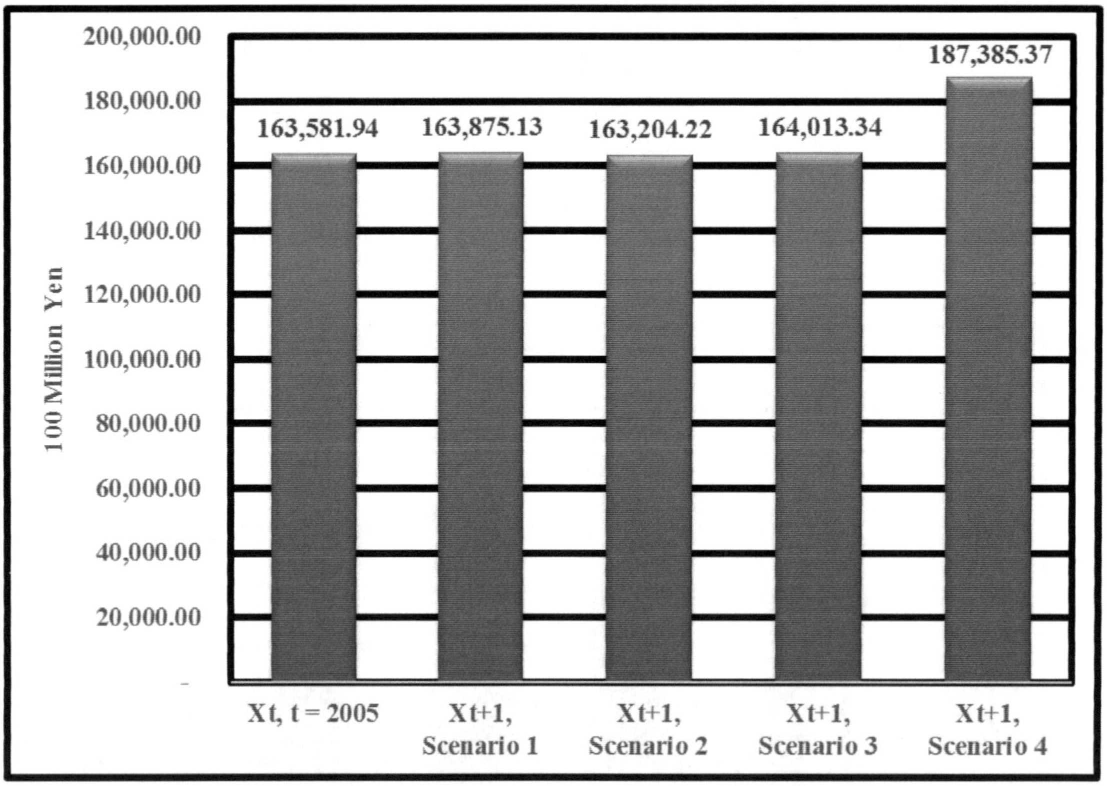

Figure 4. The dynamics of the total output of the communication sector (condition B).

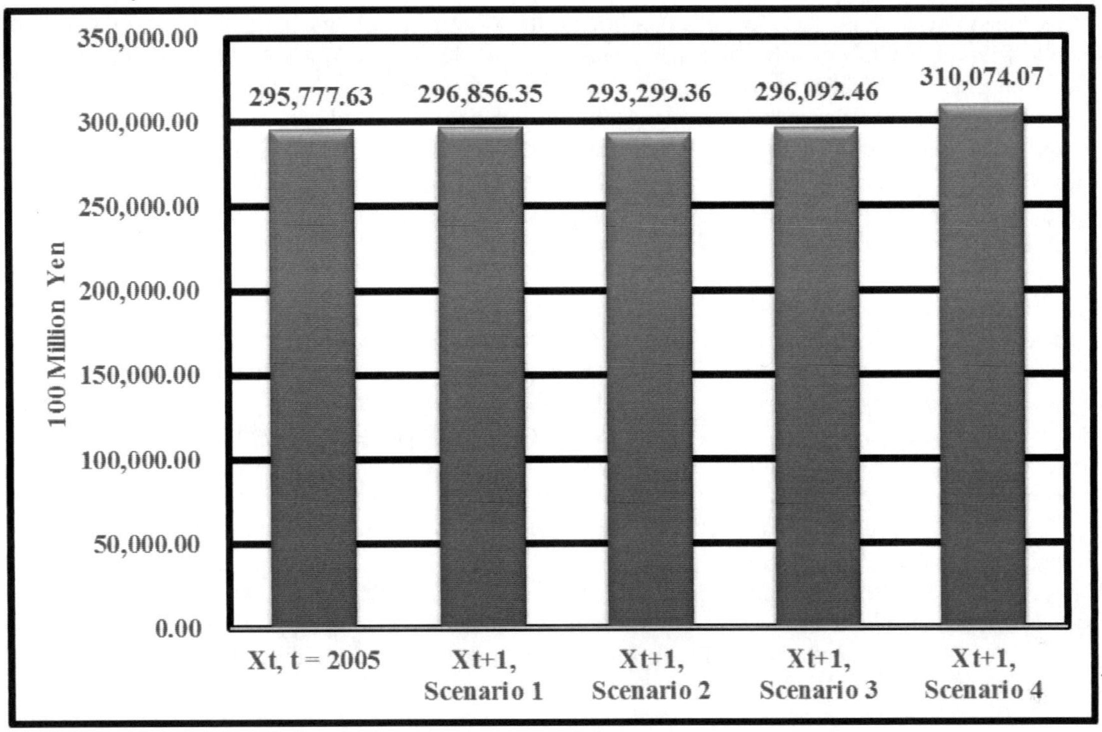

Figure 5. The dynamics of the total output of the broadcasting and information services sector (condition B).

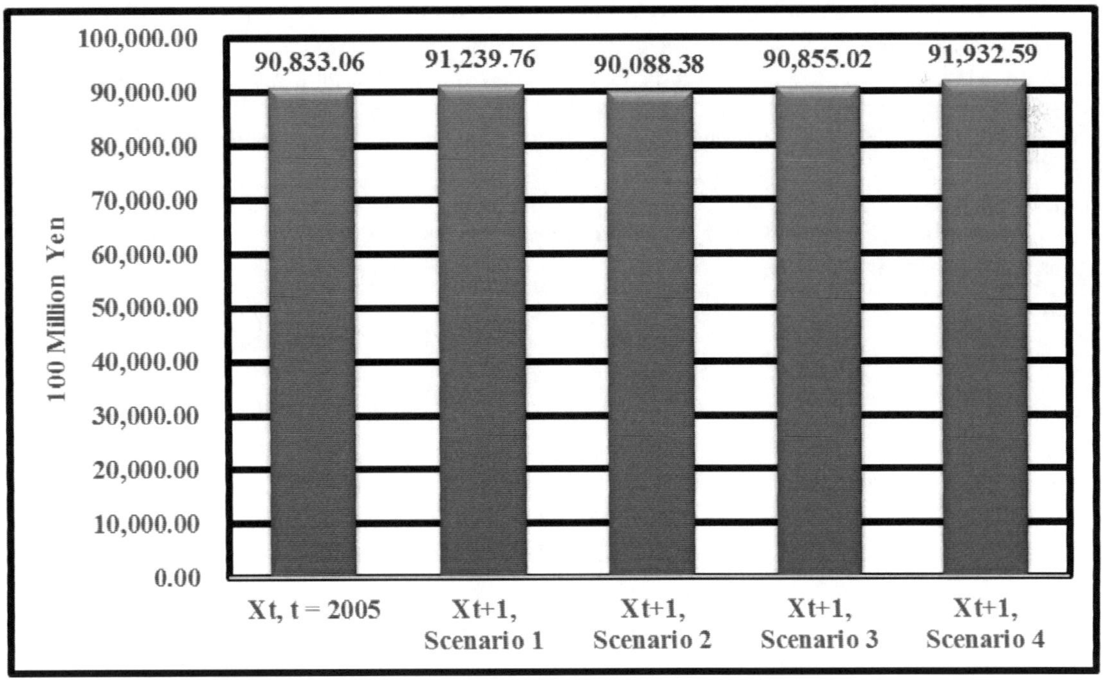

Figure 6. The dynamics of the total output of the advertising, survey, and information services sector (condition B).

Journal of Physics: Conference Series **710** (2016) 012041 doi:10.1088/1742-6596/710/1/012041

Above phenomena show that, in both conditions, the discussed sectors have similar patterns, namely these industries obtain the positive impacts from scenarios 1, 3, and 4 while the opposite impacts are received from scenario 2. This negative impact also appeared in the previous studies. Above phenomena also explain that, in both conditions, the biggest positive impacts for analyzed sectors are given by scenario 4, consumption expenditures of the private modification. Compared with the previous studies, these are new findings. Based on these results, one can argue that the effective way to increase the total outputs of focused sectors in the future are to provide more opportunities and facilities for private sectors in getting the ICT products. Besides, to restrict the import activities regarding these products will also be a good strategy.

4. Conclusions and further researches

This study, as a continuation of previous studies, conducted the deeper analysis regarding the impacts of the modifications of final demands on the total outputs of Japanese ICT sectors. This study employed a demand-pull IO quantity model, one of the calculation tools in the IO analysis, as an analysis instrument. This study focused on three sectors, namely (1) communication, (2) broadcasting and information services, and (3) advertising, survey, and information services. Two conditions were included in calculations and analysis parts, namely (1) "whole sector change", and (2) "pure change". An initial period in this study was 2005. The difference between current and previous studies could be observed on the existence of scenario 4, the modification of the consumption expenditures of the private.

The results showed that, in both conditions, the discussed sectors had similar patterns, namely these industries received the positive impacts from scenarios 1, 3, and 4 while the opposite impacts were obtained from scenario 2. This negative impact also appeared in the previous studies. The results also explained that, in both conditions, the biggest positive impacts for analyzed sectors were given by scenario 4, the change of the consumption expenditures of the private. Compared with the previous studies, these were new findings. Based on these outcomes, the suggestions from this study regarding the effective ways to enhance the total outputs of focused sectors in the future were to provide more opportunities and facilities for private sectors in getting the ICT products, and to limit the import activities regarding these commodities.

The deeper understanding regarding the impacts of the changes of final demands on the total outputs of Japanese ICT sectors was obtained from this study. However, this study only analyzed the specific industries of Japan. In other words, the comprehensive view about the impacts on the Japanese national economy could not be achieved from this study. This view is needed in order to make comprehensive policies for increasing the economic condition of Japan in the future. Therefore, as a further research, this study proposes the same analysis for the other Japanese industries.

The other suggested further research from this study is to conduct the international comparison on the current discussion. This comparison will describe the characteristics of the ICT industries of analyzed countries when the changes of final demands occur. An interesting example is to compare developed and developing countries. More specifically, a good example is to compare Japan and one of the developing countries, such as Thailand.

Journal of Physics: Conference Series **710** (2016) 012041 doi:10.1088/1742-6596/710/1/012041

Appendix. Japanese Industrial Sectors (89 Sectors)

No.	Sector Name
1	Crop cultivation
2	Livestock
3	Agricultural services
4	Forestry
5	Fisheries
6	Metallic ores
7	Non-metallic ores
8	Coal mining, crude petroleum, and natural gas
9	Foods
10	Beverage
11	Feeds and organic fertilizer, n.e.c.
12	Tobacco
13	Textile products
14	Wearing apparel and other textile products
15	Timber and wooden products
16	Furniture and fixtures
17	Pulp, paper, paperboard, and building paper
18	Paper products
19	Printing, plate making, and book binding
20	Chemical fertilizer
21	Industrial inorganic chemicals
22	Basic petrochemical and intermediate chemical products
23	Synthetic resins
24	Synthetic fibers
25	Medicaments
26	Final chemical products, n.e.c.
27	Petroleum refinery products
28	Coal products
29	Plastic products
30	Rubber products
31	Leather, fur skins, and miscellaneous leather products
32	Glass and glass products
33	Cement and cement products
34	Pottery, china, and earthenware
35	Other ceramic, stone, and clay products
36	Pig iron and crude steel
37	Steel products
38	Steel castings and forgings, and other steel products
39	Non-ferrous metals
40	Non-ferrous metal products
41	Metal products for construction and architecture
42	Other metal products
43	General industrial machinery
44	Special industrial machinery
45	Other general machines
46	Machineries for office and service industry
47	Electrical appliance
48	Motor vehicles
49	Ships, and the repairement of ships
50	Other transportation equipment, and the repairement of transportation equipment
51	Precision instruments
52	Miscellaneous manufacturing products
53	Building construction
54	Construction repairement
55	Civil engineering
56	Electricity
57	Gas and heat supply
58	Water supply
59	Waste management services
60	Commerce
61	Finance and insurance
62	Real estate agencies and rental services
63	House rent
64	Railway transport
65	Road transport (except transport by private cars)
66	Self-transport by private cars
67	Water transport
68	Air transport
69	Freight forwarding
70	Storage facility services
71	Services related to transport
72	Communication
73	Broadcasting and information services
74	Public administration
75	Education
76	Research
77	Medicals services and health
78	Social security

79	Other public services
80	Advertising, survey, and information services
81	Goods rental and leasing services
82	Repairement of motor vehicles and machine
83	Other business services
84	Amusement and recreational services
85	Eating and drinking places
86	Accommodations
87	Other personal services
88	Office supplies
89	Activities not elsewhere classified

(Source: [14], with the slight modifications)

References

[1] The National Office for the Information Economy 2002 *Contribution of ICT to Economic Growth* [online] http://ict-industry-reports.com/wp-content/uploads/sites/4/2013/10/2002-Contribution-of-ICT-to-Economic-Growth-DCITA.pdf (accessed 5 January 2016)

[2] Kramer W J, Jenkins B and Katz R S 2007 *The Role of the Information and Communications Technology Sector in Expanding Economic Opportunity* [online] http://www.hks.harvard.edu/m-rcbg/CSRI/publications/report_22_EO%20ICT%20Final.pdf (accessed 5 January 2016)

[3] Statistics Netherlands 2015 *ICT and Economic Growth* [online] http://www.cbs.nl/NR/rdonlyres/E2024F47-51B5-44B1-9A3B-A0C5DFA05069/0/2015ICTandeconomicgrowthpub.pdf (accessed 5 January 2016)

[4] Zuhdi U, Mori S and Kamegai K 2015 Forecasting the influences of information and communication technology on the structural changes of Japanese industrial sectors: a study using statistical analysis *International Journal of Social, Behavioral, Educational, Economic and Management Engineering* **9** 531–7

[5] Zuhdi U 2015 An analysis of the role of information and communication technology sectors on Japanese national economy from 1995 through 2005: an application of multiplier analysis *IOP Conference Series: Earth and Environmental Science* **23** 012014

[6] Zuhdi U 2015 An application of multiplier analysis in analyzing the role of information and communication technology sectors on Indonesian national economy: 1990-2005 *IOP Conference Series: Earth and Environmental Science* **23** 012015

[7] Zuhdi U, Mori S and Kamegai K 2014 Analysis of influences of GDP and ICT on Indonesian industrial structural changes using statistical analysis: 1990-2005 *Journal of Finance and Accountancy* **17** 1–19

[8] Zuhdi U 2014 The impacts of final demand changes on total output of Indonesian ICT sectors: an analysis using input-output approach *IOP Conference Series: Materials Science and Engineering* **58** 012011

[9] Zuhdi U and Prasetyo A D 2014 Examining the total output changes of ICT sectors of Japan: an approach of input-output *Procedia-Social and Behavioral Sciences* **109** 659–63

[10] Zuhdi U 2014 Analyzing the impacts of final demand changes on total output using input-output approach: the case of Japanese ICT sectors *IOP Conference Series: Earth and Environmental Science* **19** 012016

[11] Miller R E and Blair P D 2009 *Input-Output Analysis: Foundations and Extensions* (Cambridge: University Press)

[12] Zuhdi U, Mori S and Kamegai K 2012 Analyzing the role of ICT sector to the national economic structural changes by decomposition analysis: the case of Indonesia and Japan *Procedia-Social and Behavioral Sciences* **65** 749–54

[13] Zuhdi U 2015 An application of input-output analysis in analyzing the impacts of final demands changes on the total outputs of Japanese energy sectors: a further study *Journal of Physics: Conference Series* **622** 012041
[14] Zuhdi U 2014 Analyzing the role of creative industries in national economy of Japan: 1995-2005 *Open Journal of Applied Sciences* **4** 197–211

Corrosion Inhibition by *Tithonia diversifolia* (Hemsl) A. Gray leaves extract for 304 SS in hydrochloric acid solution

S Firdausi[1] and F Kurniawan[2*]

Laboratory of Instrumentation and Analytical Sciences, Chemistry Department, Faculty of Mathematics and Natural Sciences, Institut Teknologi Sepuluh Nopember, Arief Rahman Hakim, Surabaya, 60111, Indonesia

*fredy@chem.its.ac.id

Abstract. The inhibition effect of *Tithonia diversifolia* (Hemsl) A. Gray leaves extract on the corrosion phenomena of 304 SS in 1 M HCl has been studied by polarization potentiodynamic. The powder of *T. diversifolia* was extracted by demineralized water which was cultivated in East Java, Indonesia. The extract was characterized by FTIR spectrophotometer. The presence of *T. diversifolia* can inhibit the corrosion rate of 304 SS. The efficiency inhibition value of 2 g/L *T. diversifolia* leaves extract reached up to 77.27% at room temperature.

1. Introduction

Many metals and alloy which used in human activities are susceptible to different mechanisms of corrosion due to their exposure in different corrosive media or environment to form a stable compound [1,2]. Each metal is subjected to its own unique corrosion process and it occurs very fast and continuously. Sometimes it is difficult to eliminate completely [3,4]. Corrosion problems cause economic losses in industry [5]. Stainless steel (SS) is a material which frequently used duet to its resistant properties against corrosion. It was widely applied in industries and machinery [6-10]. In the other hand, acid solution can accelerate corrosion process on the surface of stainless steel and it had been a concern in industry [11]. Hydrochloric acid solutions are a chemical which commonly used for the removal of undesirable scale and rust in metal finishing industries, i.e. cleaning of boilers and heat exchanger [8,12-17]. So in order to prevent corrosion process in cleaning step, inhibitor corrosion is inevitably needed. It can be one of the best method in practice to protect metal surface from aggressive solution [12,18-21]. *T. diversifolia* is one of shrub plant which abundantly can be found in Indonesia. There is no report which uses the leaves as green corrosion inhibitor for 304 Stainless Steel in HCl. The aim of this research is to study the inhibition effect *T. diversifolia* leaves extract on the surface of 304 SS in 1M HCl.

2. Experimental

2.1. Specimen Preparation

The composition (wt%) of 304 SS specimens was as follows : 0.08% C, 2.00% Mn, 0.045% P, 1.00% Si, 18.00% Cr, 8.00% Ni and the balance Fe. The specimen was cut into dimension 4.0 cm × 1.0 cm × 1.0 mm with only 1 cm^2 exposed surface areas were used as working electrode in polarization measurement.

2.2. Electrolytic Solution
An aggressive solution was made from AR grade hydrochloric acid. 1 M HCl was prepared by dilution of 37% HCl (purchased from Merck, Germany) with demineralized water. A 50 mL aggresive solution was used in each polarization measurement.

2.3. Inhibitor
Fresh *T. diversifolia* leavess were collected from Institut Teknologi Sepuluh Nopember garden, Surabaya, Indonesia. The collected leavess were cleaned with water to remove small particle on the surface of it prior to use. Then they were dried for a weeks in air at room temperature. The dry leaves were grounded to powder. All the leaves powder obtained was dissolved in 90 °C of demineralized water (800 mL). These aqueous extract was stirred at 80 rpm for 30 minutes, then filtered using Buchner. The brown filtrate obtained was evaporated by hot plate for 4 h to remove the water as small as possible (not completely dry). Then the result obtained was dried in freeze dryer (FD-1A-50 Series) for 24 h. The extract obtained was stored in desiccator at room temperature before used.

2.4. Characterization of *T. diversifolia* leavess extract
The *T. diversifolia* extract, which was obtained from 2.3, was characterized using FTIR (Shimadzu-FTIR-8400S). The small amount of the extract was mixed with KBr powder until homogeneous. The mixture otained was flattened and placed onto the sample holder for analysis.

2.5. Electrochemical measurement
Three electrode cell system were carried out for electrochemical measurement. It consists of specimens (304 SS) as working electrode (WE) with exposed area of 1cm^2, Ag/AgCl (KCl 3 M) as reference electrode (RE) and Platinum as counter electrode (CE). All electrochemical measurement was performed by Autolab PGSTAT128N which is provided by Nova 1.11 software. The potential was scanned from -700 mV to 700 mV with scan rate of 1 mV/s. The inhibitor effect was monitored by performing polarization potentiodynamic measurement on 304 SS in the absence and presence of 2 g/L of the inhibitor in 1 M HCl at room temperature.

3. Result and discussion

3.1. Characterization of T. diversifolia leaves extract
FTIR spectrum of *T. diversifolia* leaves extract is shown in **Figure 1.** The spectrum shows that *T. diversifolia* leaves extract has absorption of several functional groups. The 3398.8 cm^{-1} band is assigned to O-H functional group. The 1124.54 cm^{-1} is assigned to C-O functional groups, both of them indicate of alcohol groups and assumed as phenolic compounds. The peak at 2974.33, 1384.94 cm^{-1} can be assigned to stretching and bending of C-H groups respectively [22]. It can be observed that there is an absorption at 1599.04 cm^{-1} which assumed as C=C stretching which indicates for aromatic groups. The presence of electronegative atom (i.e O) and the double bond of aromatic group give potency of *T. diversifolia* as corrosion inhibitor. The free electron pairs from hydroxyl groups are able to interact with 304 SS surface to form thin protective layer which can inhibit the corrosion process.

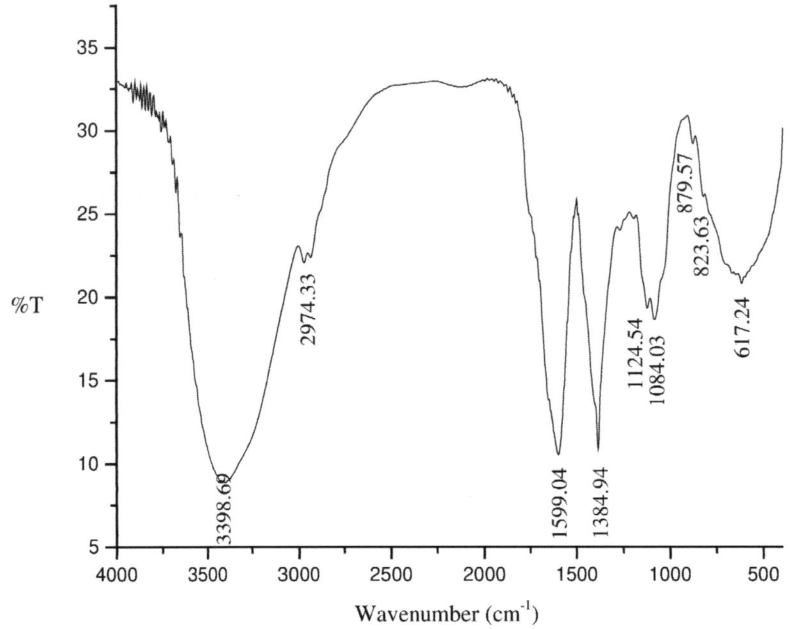

Figure 1. FTIR spectra of *T. diversifolia* leaves extract

3.2. Polarization potentiodynamic study

The study of *T. diversifolia* leaves extract inhibition on the surface of 304 SS was conducted using polarization potentiodynamic measurement. The Tafels obtained in the presence and in the absence were compared to see the effect of the inhibitor. The result can be seen at **Figure 2**. The corresponding electrochemical parameters including Tafel slopes (b_a, b_c), corrosion potential (E_{corr}), corrosion current density (i_{corr}) and inhibition efficiency (η%) are given in **Table 1**. The inhibition efficiency (η%) was calculated using expression given in Eq. **(1)** from the values of corrosion current density in the absence (i^0_{corr}) and presence (i_{corr}) of *T. diversifolia* [23]. The inhibition efficiency (η%) is defined as :

$$\eta\% = \frac{i^0_{corr} - i_{corr}}{i^0_{corr}} \times \mathbf{100} \qquad (1)$$

The inhibition efficiency value indicates the capability of inhibitor to reduce the corrosion rate on the surface of 304 SS in 1 M HCl. There is the significant difference between corrosion current density of blank and inhibitor. The i^0_{corr} is corrosion current density for the blank, which has a higher value (109.830 μA) and the i_{corr} is the corrosion current density for 304 SS in the presence of inhibitor (24.962 μA).

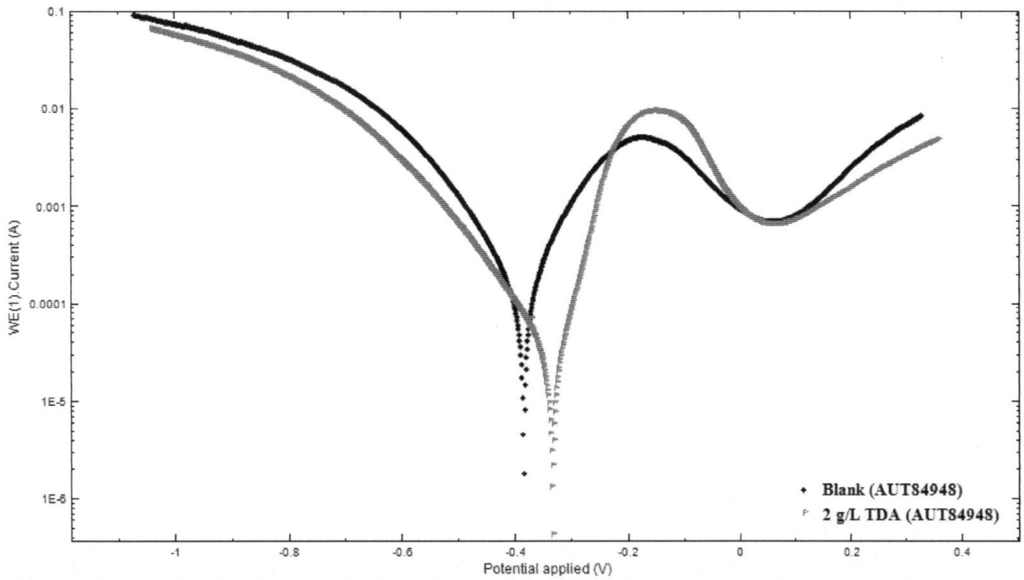

Figure 2. Polarization Curves for 304 SS in 1 M HCl in the absence and presence of inhibitor.

The polarization curves show a clear difference between the 304 SS in absence and presence of inhibitor. The black line shows the polarization curve of 304 SS in the absence of inhibitor. In the other hand, the red line shows the polarization curve of 304 SS in the presence of *T. diversifolia* leaves extract. The corrosion potential (E_{corr}) shifts to cathodic as 48.41 mV. Furthermore, it has been observed that the corrosion rate decrease in the presence of 2 g/L *T. diversifolia* leaves extract. The *T. diversifolia* leaves extract provides protection up to 77.27% at room temperature.

Table 1. Potentiodynamic polarization parameters for 304 SS in 1M HCl in the absence and presence of inhibitor

Concentration (g/L)	b_a (mV/dec)	b_c (mV/dec)	E_{corr} (mV)	J_{corr} (mV)	I_{corr} (μA)	C_R (mmpy)	η%
Blank	982.240	74.666	-384.060	109.830	109.830	1.27620	-
2	956.060	600.662	-335.650	24.962	24.962	0.29006	77.27

4. Conclusion

The *T. diversifolia* leaves extract can be used as green corrosion inhibitor for 304 SS in 1 M HCl at room temperature. The presence of electronegative atom (i.e O) and the double bond of aromatic group give potency of *T. diversifolia* as corrosion inhibitor. The inhibition efficiency of *T. diversifolia* leaves extract reach up to 77,27% which achieved by addition 2 g/L *T. diversifolia* leaves extract.

References

[1] A.S. Yaro, A.A. Khadom, R.K. Wael, Apricot juice as green corrosion inhibitor of mild steel in phosphoric acid, Alex. Eng. J. 52 (2013) 129–135. doi:10.1016/j.aej.2012.11.001.

[2] P.B. Raja, M.G. Sethuraman, Natural products as corrosion inhibitor for metals in corrosive media — A review, Mater. Lett. 62 (2008) 113–116. doi:10.1016/j.matlet.2007.04.079.

[3] k. Krishnaveni, j. Ravichandran, Effect of aqueous extract of leaves of Morinda tinctoria on corrosion inhibition of aluminium surface in HCl medium, Trans. Nonferrous Met. Soc. China. 24 (2014) 2704–2712. doi:10.1016/S1003-6326(14)63401-4.

[4] B.. A. Rani, B.B.J. Basu, Green Inhibitors for Corrosion Protection of Metals and Alloys : An Overview, Int. J. Corros. (2011) 1–15. doi:10.1155/2012/380217.

[5] M. Bello, N. Ochoa, V. Balsamo, F. López-Carrasquero, S. Coll, A. Monsalve, et al., Modified cassava starches as corrosion inhibitors of carbon steel: An electrochemical and morphological approach, Carbohydr. Polym. 82 (2010) 561–568. doi:10.1016/j.carbpol.2010.05.019.

[6] M. Abdallah, Rhodanine azosulpha drugs as corrosion inhibitors for corrosion of 304 stainless steel in hydrochloric acid solution, Corros. Sci. 44 (2002) 717–728.

[7] M.A. Amin, K.F. Khaled, S.A. Fadl-Allah, Testing validity of the Tafel extrapolation method for monitoring corrosion of cold rolled steel in HCl solutions – Experimental and theoretical studies, Corros. Sci. 52 (2010) 140–151. doi:10.1016/j.corsci.2009.08.055.

[8] L.R. Chauhan, G. Gunasekaran, Corrosion inhibition of mild steel by plant extract in dilute HCl medium, Corros. Sci. 49 (2007) 1143–1161. doi:10.1016/j.corsci.2006.08.012.

[9] M.S. Al-Otaibi, A.M. Al-Mayouf, M. Khan, A.A. Mousa, S.A. Al-Mazroa, H.Z. Alkhathlan, Corrosion inhibitory action of some plant extracts on the corrosion of mild steel in acidic media, Arab. J. Chem. 7 (2014) 340–346. doi:10.1016/j.arabjc.2012.01.015.

[10] Y.A. Albrimi, A.A. Addi, J. Douch, R.M. Souto, M. Hamdani, Inhibition of the pitting corrosion of 304 stainless steel in 0.5 M hydrochloric acid solution by heptamolybdate ions, Corros. Sci. 90 (n.d.) 522–528.

[11] M. Abdallah, Corrosion behaviour of 304 stainless steel in sulphuric acid solutions and its inhibition by some substituted pyrazolones, Mater. Chem. Phys. 82 (2003) 786–792. doi:10.1016/S0254-0584(03)00367-5.

[12] M. Znini, L. Majidi, A. Bouyanzer, J. Paolini, J.-M. Desjobert, J. Costa, et al., Essential oil of Salvia aucheri mesatlantica as a green inhibitor for the corrosion of steel in 0.5 M H2SO4, Arab. J. Chem. 5 (2012) 467–474. doi:10.1016/j.arabjc.2010.09.017.

[13] C. Kamal, M.G. Sethuraman, Spirulina platensis – A novel green inhibitor for acid corrosion of mild steel, Arab. J. Chem. 5 (2012) 155–161. doi:10.1016/j.arabjc.2010.08.006.

[14] N.O. Obi-Egbedi, I.B. Obot, Xanthione: A new and effective corrosion inhibitor for mild steel in sulphuric acid solution, Arab. J. Chem. 6 (2013) 211–223. doi:10.1016/j.arabjc.2010.10.004.

[15] N.A. Negm, N.G. Kandile, I.A. Aiad, M.A. Mohammad, New eco-friendly cationic surfactants: Synthesis, characterization and applicability as corrosion inhibitors for carbon steel in 1N HCl, Colloids Surf. Physicochem. Eng. Asp. 391 (2011) 224–233. doi:10.1016/j.colsurfa.2011.09.032.

[16] N.S. Ayati, S. Khandandel, M. Momeni, M.H. Moayed, A. Davoodi, M. Rahimizadeh, Inhibitive effect of synthesized 2-(3-pyridyl)-3,4-dihydro-4-quinazolinone as a corrosion inhibitor for mild steel in hydrochloric acid, Mater. Chem. Phys. 126 (2011) 873–879. doi:10.1016/j.matchemphys.2010.12.023.

[17] S. Zhang, Z. Tao, W. Li, B. Hou, The effect of some triazole derivatives as inhibitors for the corrosion of mild steel in 1M hydrochloric acid, Appl. Surf. Sci. 255 (2009) 6757–6763. doi:10.1016/j.apsusc.2008.09.089.

[18] N. Soltani, N. Tavakkoli, M. Khayatkashani, M.R. Jalali, A. Mosavizade, Green approach to corrosion inhibition of 304 stainless steel in hydrochloric acid solution by the extract of Salvia officinalis leaves, Corros. Sci. 62 (2012) 122–135. doi:10.1016/j.corsci.2012.05.003.

[19] N. Soltani, N. Tavakkoli, M. Khayat Kashani, A. Mosavizadeh, E.E. Oguzie, M.R. Jalali, Silybum marianum extract as a natural source inhibitor for 304 stainless steel corrosion in 1.0 M HCl, J. Ind. Eng. Chem. 20 (2014) 3217–3227. doi:10.1016/j.jiec.2013.12.002.

[20] X. Li, S. Deng, H. Fu, Inhibition of the corrosion of steel in HCl, H2SO4 solutions by bamboo leaves extract, Corros. Sci. 62 (2012) 163–175. doi:10.1016/j.corsci.2012.05.008.

[21] M. Srimathi, R. Rajalakshmi, S. Subhashini, Polyvinyl alcohol–sulphanilic acid water soluble composite as corrosion inhibitor for mild steel in hydrochloric acid medium, Arab. J. Chem. 7 (2014) 647–656. doi:10.1016/j.arabjc.2010.11.013.

[22] A.K. Satapathy, G. Gunasekaran, S.C. Sahoo, K. Amit, P.V. Rodrigues, Corrosion inhibition by Justicia gendarussa plant extract in hydrochloric acid solution, Corros. Sci. 51 (2009) 2848–2856. doi:10.1016/j.corsci.2009.08.016.
[23] F. Kurniawan, K.A. Madurani, Electrochemical and optical microscopy study of red pepper seed oil corrosion inhibition by self-assembled monolayers (SAM) on 304 SS, Prog. Org. Coat. 88 (2015) 256–262. doi:10.1016/j.porgcoat.2015.07.010.

Determination of free radical scavenging activity from aqueous extract of *Curcuma mangga* by DPPH method

N A Indis[1] and F Kurniawan[2]

Laboratory of Instrumentation and Analytical Science, Chemistry Department, Faculty of Mathematics and Natural Sciences, Institut Teknologi Sepuluh Nopember, Arif Rahman Hakim, Surabaya, 60111, Indonesia

fredy@chem.its.ac.id

Abstract. *Curcuma mangga* (mango ginger) belongs to the family of Zingiberaceae. The rhizome of *C. mangga* are morphologically similar to ginger (Zingiber officinale) with a little mango flavour. *C. mangga* can growth in tropical areas and easy found in Indonesia. The rhizomes of *C. mangga* were washed and cut into the small piece, then drying at room temperature for 6 days, and then grinded until get the powder of *C. mangga*. The powder of *C. mangga* was extracted with demineralized water by maceration for 6 hours. *C. mangga* extract was analysed with FTIR spectrophotometer to determine its functional groups. *C. mangga* extract was diluted at various of concentration (5, 10, 25, 50, 100, 250, 500 mg/L) using demineralized water. *C. mangga* extracts were tested the antioxidant activity using 0.002% DPPH at 517nm with UV-Vis spectrophotometer, and the IC_{50} value of *C. mangga* extract is 212.70 mg/L.

1. Introduction

C. Mangga is the one kind of herbal medicinal plants. In long time ago, *C. mangga* was used as a traditional herbal medicine. The *C. mangga* extract was very useful for health, as antimicrobial, antidiarrheal, anti-inflammatory, anticancer, analgesic, and antioxidants [1-5]. That is because the *C. mangga* extract contain some active compounds such as curcuminoid and phenolic. The curcuminoid compounds are curcumin, bisdimethoxycurcumin, bismethoxycurcumin, and phenolic compounds are ferulic acid, p-coumaric acid, cinnamic acid, cafeic acid. In this research is focused into antioxidant activity of *C. mangga* extract. The general solvent for antioxidant extraction is using the organic solvents such as methanol, ethanol, and n-hexane, but in this research we use demineralized water, because the *C. mangga* extract will be applied for herbal medicine. The solvent must be kept safe and halal for consumption. In the previous research, Policegoudra (2007) reported about antioxidant activity from *C. mangga* extract [6]. The *C. mangga* rhizomes were washed, sliced and dried in a hot air oven at 50°C for 72 hours and powdered to 60 meshes using an apex grinder. About 100 g of dry *C. mangga* powder was extracted with chloroform at room temperature and at atmospheric pressure, for 48 hours by shaking at 100 rpm/min speed. The *C. mangga* extract was filtered and concentrated using rotary evaporator. The concentrated extract of *C. mangga* was freeze-dried and stored in refrigerator. The *C. mangga* extracts (10–100 µL) were mixed with 0.8 mL of Tris–HCl buffer (pH 7.4) to which 1 mL of DPPH (0.05 mM) in ethanol was added. The mixture was shaken vigorously and left to stand for 30 min to complete the reaction. And then absorbance of the resulting solution was measured at 517 nm using UV–vis spectrophotometer. The antioxidant activity of *C. mangga* extract was measured as a decrease in the absorbance of DPPH. Lower absorbance of the mixture, indicated the higher antioxidant activity. The antioxidant activity was expressed as IC_{50} value, which showed as concentration of sample at which 50% of the DPPH radicals scavenged. The IC_{50} value from the previous research is 178 mg/L [6]. The aim of this research is to study the free radical scavenging activity from aqueous extract from *C. mangga* by DPPH method.

2. Experimental

2.1. Materials and chemicals

Fresh and healthy mango ginger (*C. mangga*) rhizomes were purchased from the local market, Pucang, Surabaya, Indonesia, during July 2015. The rhizomes were washed, sliced and dried in a room temperature for 6 days and grinded until get the powder. The chemicals are demineralized water, ethanol absolute (Merck, Germany), 1,1-Diphenyl-2-picryl-hydrazyl (DPPH) (Merck, Germany).

2.2. Extraction of *C. mangga* with demineralized water

The rhizomes of *C. mangga* were washed and cut into the small piece, then drying at room temperature for 6 days, and then grinded using mixer grinder (Philips, Indonesia) until getting the powder. 100 g of *C. mangga* powder was extracted with 1000 mL demineralized water by maceration for 6 hours. The *C. mangga* extract was filtered and concentrated using hot plate at temperature 50°C for 12 hours. The concentrated extract were drying by vacuum freeze dryer (FD-1A-50 Series) and stored in desiccator. The extraction process of *C. mangga* according to the method was described earlier [7].

2.3. Characterization of *C. mangga* extract

The concentrate of *C. mangga* extract was characterizated by FTIR (Shimadzu-FTIR-8400S) using the KBr-disk technique. The little part of *C. mangga* extract was mix with KBr powder until homogenous. The mixture obtained, was flattened and placed onto the sample holder for analysis.

2.4. DPPH free radical scavenging activity of *C. mangga* extracts

1,1-Diphenyl-2-picryl-hydrazyl radical scavenging activity was determined according to the method described earlier [7]. The concentrate of *C. mangga* extract (100 mg) was mixed with 100 mL of demineralized water (1000 mg/L of stock solution). From the stock solution, diluted to various concentration (5, 10, 25, 50, 100, 250, 500 mg/L). 0.002% DPPH (w/w) made from 2 mg DPPH was mixed with 126.75 mL ethanol absolute (ρ = 0,789g/mL). 2 mL of *C. mangga* extracts (5–500 mg/L) were mixed with 2 mL of 0.002% DPPH. The mixture was shaken and left to stand for 10 min to complete the reaction. Absorbance of the resulting solution was measured at 517 nm using UV–Vis spectrophotometer (Genesys 10S UV-Vis). The radical scavenging activity of *C. mangga* extract was measured as a decrease in the absorbance of 0.002% DPPH. Lower absorbance of the mixture, indicated the higher of radical scavenging activity. The radical scavenging activity was expressed as IC_{50} value, which represents the sample concentration at 50% of the DPPH radicals scavenged.

3. Result and Discussion

3.1. Characterization of C. mangga extract

The functional groups of *C. mangga* extract was characterization using FTIR spectrophotometer. The spectr FTIR of *C. mangga* extract was shown at **Figure 1**. It shows N-H stretching (amine group) at 3712 cm^{-1}, O-H stretching and C-O stretching (alcohol group) at 3493 cm^{-1} and 1210 cm^{-1}, C-H sp^3 stretching (alkyl group) at 2947 cm^{-1}, C=O stretching (carbonyl group) at in 1702 cm^{-1}, and C=C stretching (aromatic group) at 1597 cm^{-1}. FTIR spectrum of *C. mangga* extract from the previous research was shown at Figure 2. It shows O-H stretching and C-O stretching (alcohol group) at 3403cm^{-1} and 1202 cm^{-1}, C-H sp^3 stretching and C-H sp^3 bending (alkyl group) at 2912 cm^{-1} and 2996

cm^{-1}, C=O stretching (carbonyl group) at 1657 cm^{-1}. There are 2 differences on FTIR spectra at Figure 1 and Figure 2. Figure 2 does not show N-H and C=C functional groups, because the extraction method between the previous research and in this work is different.

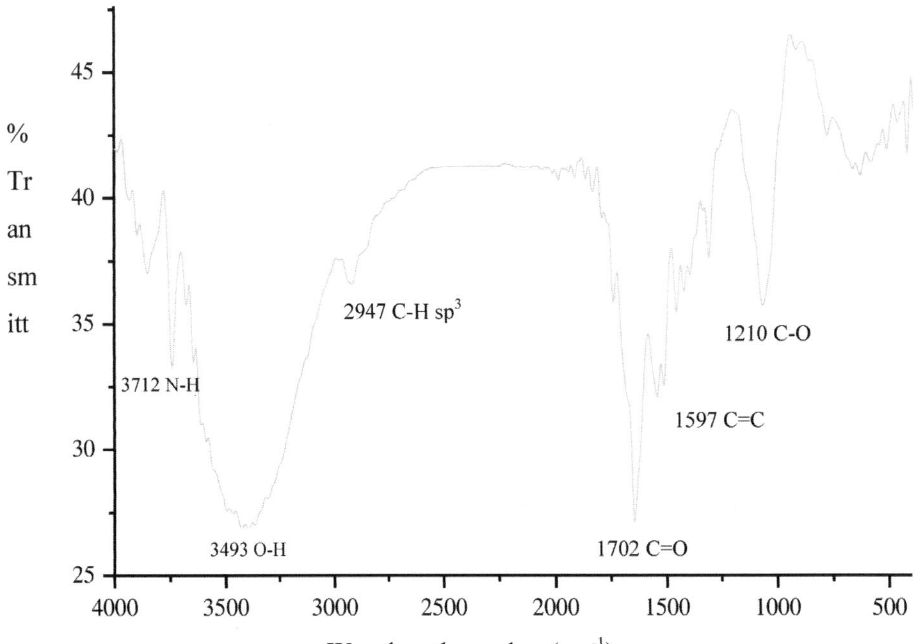

Fig. 1. FTIR spectra of *C. mangga* extract

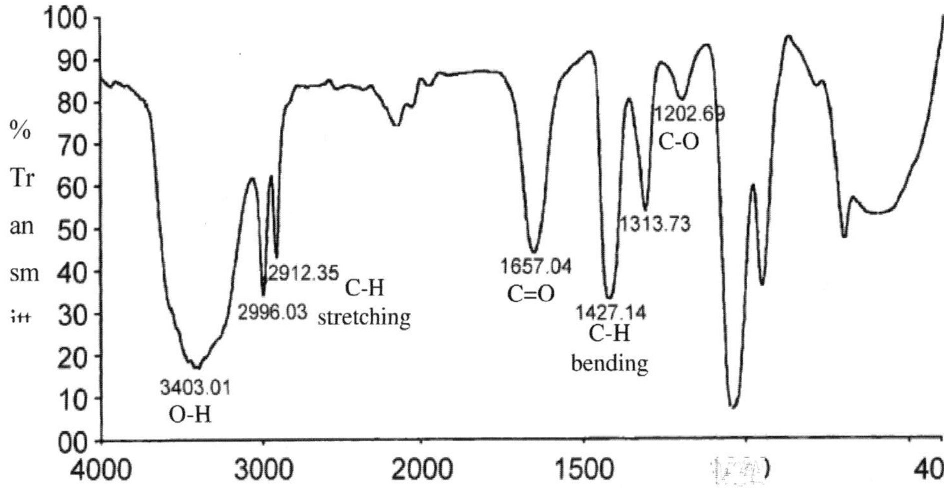

Fig. 2. FTIR spectra of *C. mangga* extracts from the previous research [6]

3.2. DPPH free radical scavenging activity of C. mangga extracts

The maximum wavelength of 0.002% DPPH measured using UV-Vis spectrophotometer. The result is shown at **Figure 3**. It shows that the maximum wavelength of 0.002% DPPH is 517 nm. The maximum wavelength of 0.004% and 0.1 mM DPPH from the previous research are 517 nm [8,9]. The result is DPPH solutions at different concentrations have the same wavelength. The results of free radical scavenging activity from *C. mangga* extracts (5–500 mg/L) using 0.002% DPPH is shown at **Figure 4**. It shows the IC_{50} value of *C. mangga* extract is 212.70 mg/L, with the linear regression equation is $y = 0.0919x + 30.453$ with linear correlation of 0.9906.

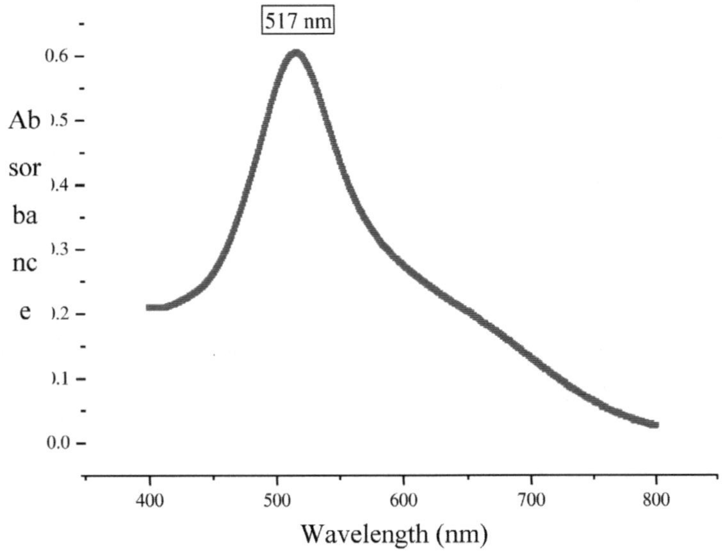

Fig. 3. UV-Vis spectra of 0.002% DPPH solution

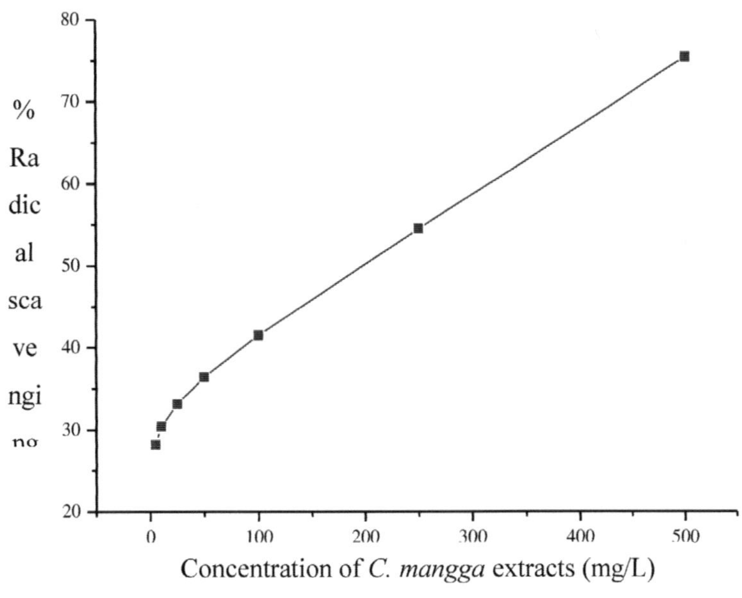

Fig. 4. 0.002% DPPH free radical scavenging activity of *C. mangga* extracts

Smaller IC_{50} value indicates higher free radical scavenging activity [10]. The result obtained in this research provide information about the IC_{50} value from the aqueous extract of *C. mangga*, it can be used as a standard to determine the dose of traditional herbal medicine from *C. mangga* extract.

4. Conclusion

The aqueous extracts of *C.mangga* have a good free radical scavenging activity with the IC_{50} value is 212.70 mg/L, the linear regression equation is $y = 0.0919\times + 30.453$ with linear correlation of 0.9906. This result can be used as a standard to determine the dose of traditional herbal medicine from *C. mangga* extract

References

[1] Tripathi, Mridula, Priyanka Chawla, Ruby Upadhyay, and Shivang Trivedi. 2013. "Essential Oils from Family Zingiberaceae for Antimicroba Activity." *International Journal of Pharmacy and Bio Science* 4 (4): 149–162.

[2] Nuratmi, Budi, Yun Astuti Nugroho, and Sundari Dian. 2006. "Efek Antidiare Jus Temu Putih (Curcuma Zedoaria) Dan Temu Mangga (Curcuma Mangga) Pada Tikus Putih." *Media Litbang Kesehatan* 16 (1): 29–34.

[3] Kaushik, Madan L., and Sunil S. Jalalpure. 2011. "Anti-Inflammatory Efficacy of Curcuma Zedoaria Rosc Root Extract." *Asian Journals of Pharmaceutical and Clinical Research* 4 (3): 90–92.

[4] Liu, Yunbao, and Muraleedharan G. Nair. 2011. "Labdane Diterpenes in Curcuma Mangga Rhizomes Inhibit Lipid Peroxidation, Cyclooxygenase Enzymes and Human Tumour Cell Proliferation." *Food Chemistry* 124 (2): 527–32. doi:10.1016/j.foodchem.2010.06.064.

[5] Das, Kamanashis, and Ashikur Mohammad Rahman. 2012. "Analgesic and Antimicroba Activities of Curcuma Zedoaria." *International Journal of Pharmacy and Pharmaceutical Science* 4 (5): 321–26.

[6] Policegoudra, R. S., K. Abiraj, D. Channe Gowdab, and S. M. Aradhya. 2007. "Isolation and Characterization of Antioxidant and Antibacterial Compound from Mango Ginger (Curcuma Amada Roxb.) Rhizome." *Journal of Chromatography B* 825: 40–48. doi:10.1016/j.jchromb.2006.12.036.

[7] Jalip, Ikna Suyatna, Suprihatin, Ida Wiryanti, and Ernawati Sinaga. 2013. "Antioxidant Activity and Total Flavonoids Content of Curcuma Rhizome Extract." *Proceeding International Conference*, 93–99.

[8] Singh, Pratibha, Sunita Singh, I.P.S. Kapoor, G. Singh, Valery Isidorov, and Lech Szczepaniak. 2013. "Chemical Composition and Antioxidant Activities of Essential Oil and Oleoresins from Curcuma Zedoaria Rhizomes." *Food Bioscience* 3 (0): 42–48. doi:10.1016/j.fbio.2013.06.002.

[9] Nahak, Gayatri, and Rajani Kanta Sahu. 2011. "Evaluation of Antioxidant Activity in Ethanolic Extracts of Five Curcuma Species." *International Research Journal of Pharmacy* 2 (12): 243–48.

[10] Molyneux, Philip. 2004. "The Use of The Stable Free Radical Diphenylpicrylhydrazyl (DPPH) for Estimating Antioxidant Activity." *Journal of Sience and Technology* 26 (2): 212–18.

Temperature-dependence of Threshold Current Density-Length Product in Metallization Lines: A Revisit

Rahmat Saptono Duryat[1], Choong-Un Kim[2]

[1]Universitas Indonesia, Depok 16424 Indonesia
[2]University of Texas at Arlington, Texas 76019 USA

E-mail: rahmat.saptono@ui.ac.id

Abstract. One of the important phenomena in Electromigration (EM) is Blech Effect. The existence of Threshold Current Density-Length Product or EM Threshold has such fundamental and technological consequences in the design, manufacture, and testing of electronics. Temperature-dependence of Blech Product had been thermodynamically established and the real behavior of such interconnect materials have been extensively studied. The present paper reviewed the temperature-dependence of EM threshold in metallization lines of different materials and structure as found in relevant published articles. It is expected that the reader can see a big picture from the compiled data, which might be overlooked when it was examined in pieces.

1. Introduction

Electromigration (EM) is a complex current-induced mass transport phenomenon in metallic interconnecting lines. The phenomenon had been recognized as one of the most significant factors in the failure of electronic circuits and hence become the central issue in reliability. Many efforts have been done to understand the underlying physics and describe the mechanisms of EM. The earliest theoretical study of EM was known to be established by Fiks[1] and Huntington and Grone[2] whereas the simple semi-empirical description of EM phenomena was attributed to Black[3] based on the established theory. A comprehensive review on the theoretical frameworks was presented by Sorbello[4] starting from very general phenomenological equations of irreversible thermodynamics ended up with the dynamical non-adiabatic considerations. A recent introductory review on the physically-based models of EM was written by Orio et.al.[5]. It is generally known that the established models was developed based on a general assumption that EM is a Diffusion-Convection Problem in which the transport phenomena is dictated by several driving forces. Two boundary problems are introduced to provide a simple solution of EM-induced mass transport equation that has a practical use[6]. The first boundary condition refers to a metallic line connecting two infinite vacancies reservoir subjected to an electric current. The second one was derived from a line packed in two diffusion barriers (perfectly blocking boundary condition). Such a model considering vacancy accumulation due to the flux divergence of EM-induced flux and normal Fickian diffusion flux was proposed by Shatzkes and Lloyd[7]. The examination of the model solution from the second boundary value has lead to two main deficiencies involving a very short time scale to reach the maximum vacancy supersaturation and a very low value of it. In line with Rosenberg and Ohring[8] who revealed that the lifetime is mainly dictated by the void growth mechanism. Orio et.al.[5] revealed that it will prevent vacancy condensation-aided void formation so that a critical vacancy concentration cannot be employed to predict the interconnect failure. It was assumed that the situation can only be resolved by introducing the mechanical stresses into the equation[5].

The existence of mechanical stresses in EM dated back to the Blech Experiment on EM in thin Aluminum films on Titanium Nitride. Blech[9] proposed that the deceleration of EM in the experiment was associated with the generation of compressive stress around the anode alligned with the possible explanation of EM retardation in Al suggested by Ainslie et.al.[10]. The presence of the stress was experimentally proven by the crack of the overlying SiN as well as the formation of long aluminum extrusion through etched holes in the area of cathode. Similar to the pressure effect investigated in SiN-covered film, it was also assumed to operate in uncovered Al having only their nature oxide.

Referring to Blech[9], the retardation effects took place when the electrical driving force was counter balanced by mechanical driving forces associated with the stress gradient. Since the event cannot be exist at any current, it was assumed that there must be a critical stress the film can withstand before permanently deforming as manifested by the formation and growth of hillocks. Moreover, it was also assumed that the super-saturation of vacancies or deficiency of atoms might decrease the chemical potential resulting in the chemical potential difference. The average chemical potential gradient opposing the EM then was calculated by dividing it with the stripe line.

Applying Ainslie et.al.[10] assumption that the free energy difference between the both ends of the stripe is proportional to the product of atomic volume and the normal stress difference between stripe ends and neglecting any other chemical potential gradients, Blech[9] proposed a threshold equation involving the stress. The equation, however, cannot be used to differentiate the relative contribution of anode and cathode area to the free energy difference between the both ends of the stripe. The effect of strain rate was not studied as the equation was developed based on the assumption that the materials behave plastically ideal. The temperature-dependence of the threshold-length product revealed from the experiment was qualitatively assumed to be associated with the temperature-dependence of Aluminum flow stress.

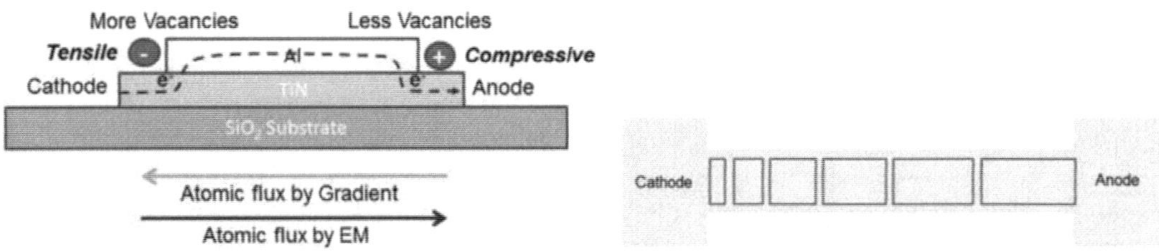

Figure 1. Schematic Illustration of Blech Experiment as Explained in Ref.[9]

2. Stress and Electromigration

More quantitative discussion of the stress and EM in metallization lines were presented by Tu[11], Ross[12], Kirchheim[13], and Korhonen et.al.[14]. Tu[11] has employed the thermodynamic of irreversible process to analyze the interaction of stress and EM, assuming that it is basically a creep process. Ross[12] developed a numerical model for the formation of EM damage in thin metallization which was able to calculate stress developed around the EM flux divergence sites and also to predict time to failure. Physical-based model of the transient stress build up during EM was developed by Kirchheim[13] who calculated for the first time time-dependence tensile and compressive stresses. A pair of partial differential equation for vacancy concentration and stress was developed from the physical basis, employed to solve the problem numerically and, for 3 limiting cases, analytically, and compared with the experimental results. One of the significant drawbacks of the model came from the approach that neglects the direct contribution of the vacancy flux to the stress development as discussed later by Sarychev et.al.[14]. The second physically-based analytical model was developed by Korhonen et.al.[15]. Despite the model was proposed to improve previous works and provide a closed analytical solution, it contains such important flaws as revealed later by Llyod[16] and Sarychev et.al.[14]. The major limitations came from the abandonment of vacancy relaxation and the assumption of vacancy equilibrium. Clement and Thompson[17] re-examined the formulation of Korhonen model comprising the assumptions and approximations in the Korhonen treatment and was able to show that the analytical solution for a semi-finite line with a blocking boundary condition only provide a good approximation when the stress build up is small. More accurate analytic formulation was derived to express the maximum stress evolution in the confined interconnect. Sarychev et.al.[14] proposed a new model using an approach similar to that used for thermal stress model. The key different of the approach compared to that of the established models[13,15,17] is the comprehensive consideration of the total deformation at given position (x_1,x_2,x_3). Employing this approach, the entire stress tensor evolution can be described and related to the boundary value of the line. By doing so, a general 3D equation linking the transport phenomena and stress evolution had been derived to describe the distribution of mechanical stress within the metallization line attached on the surface and confined by passivation layer. Prior to Sarychev et.al.[17], Gleixner and Nix[18] had proposed a physically-based model of EM and stress-induced evolution in realistic microstructures. Unlike Sarychev et.al.[14], the model to solve diffusion-stress equations was developed in 2D while taking into account the 3D effects of passivation constraints. Sukharev, Zschech, and Nix[19] made an improvement by involving mass transport along the grain boundary in addition to that along the interface. It was assumed that the mass transport path was depending on the competing activation energy along it.

3. Blech Effect

One of the important phenomena in EM is the existence of the threshold current density-length product. The existence EM Threshold or Blech Product had been proven and well accepted. The threshold was able to be quantified by measuring drift velocity and explained by the opposing gradient caused by the accumulation and depletion of atoms at the strips end.

Drift velocity was formerly defined by Huntington and Grone[2] as:

$$v_d = \frac{D_a |Z^*|}{kT} . e\rho j \tag{1}$$

$$D_a = D_0 \exp\left(\frac{-\Delta H}{kT}\right) \tag{2}$$

where:
v_d = drift velocity
eZ^* = effective charge of ion
ρ = electrical resistivity
j = electrical current density
D_0 = diffusion coefficient
ΔH = activation energy
k = Boltzman's constant
T = temperature

Blech[9] designed an edge displacement experiment from which drift velocity was measured by observing the movement of strips end. The threshold was able to be determined by assuming that the net drift velocity of EM becoming zero whenever the value was reached.

The varying drift velocity is caused by diffusional back flows during EM associated with the free energy different of moving species between the strip ends. The backflow is hence proportional free energy gradient and reciprocally proportional to the length. The net drift velocity can also be expressed in term of EM velocity and backflow velocity[20].

$$v = v_{EM} - v_{BF} = \frac{D}{kT}|Z^*|e\rho j - \frac{D}{kT}\frac{\Delta G}{l} \tag{3}$$

where:
v = net drift velocity
v_{EM} = EM drift velocity
v_{BF} = backflow drift velocity
ΔG = free energy different
l = length of strip

When the net drift velocity is zero, then:

$$\frac{D}{kT}|Z^*|e\rho j = \frac{D}{kT}\frac{\Delta G}{l} \tag{4}$$

Re-arranging the equation will result in:

$$(jl)_c = j_c l = jl_c = \frac{\Delta G}{|Z^*|e\rho} \tag{5}$$

where:
$(jl)_c$ is a constant at any given temperature known as "threshold current density-length product" or simply "Blech product".

The threshold current density-length product has a significant technological importance as the sub-threshold line will not suffer EM because of the zero net flux. As the Blech product is constant and stress current densities are typical for any given applications, the critical length l_c is known as the Blech length. Both Blech product and length has a significant role in the design and manufacture of electronic circuits and testing structure.

One of the particular interests in electronic packaging is the dependence of Blech product on the temperature. Temperature-dependence of Blech product was firstly established from Blech experiment[9]. The phenomenological equations for EM threshold as a function of temperature can be derived from the effective charge number of an ion in EM, Z*, that consist of two parts [1,2,4,11]:

$$Z^* = Z^*_{el} + Z^*_{wd} \tag{6}$$

Referring to Tu[11], Z^*_{el} can be recognized as the nominal valence Z of the metal while ignoring the dynamical screening effect around the ion. Z^*_{wd}, on the other hand, corresponds to the momentum change effect between the carriers and the ions.

Huntington and Grone[2] expressed Z^*_{wd} in terms of specific resistivity ratio.

$$Z^*_{wd} = -Z\frac{\frac{\rho_d}{N_d}m_0}{\frac{\rho}{N}m^*} \tag{7}$$

where:
$\frac{\rho_d}{N_d}$ = specific resistivity of a diffusing atom

$\frac{\rho}{N}$ = specific resistivity of a normal atom
m_0 = free electron mass
m^* = effective electron mass

Assuming the specific resistivity of an atom in a metal is proportional to the scattering elastic cross section, which is then proportional to the average square displacement from equilibrium or $\langle x^2 \rangle$, Einstein's model of atomic vibration can be employed to estimate the cross section of normal atom[11].

$$\frac{1}{2} m\omega^2 \langle x^2 \rangle = \frac{1}{2} kT \tag{8}$$

where:
m = atomic mass
ω = angular vibrational frequency

Assuming that the atom and its surrounding has acquired the temperature-independence kinetic energy of diffusion ΔH_m, Tu[11] was able to specified the cross section of scattering of the diffusing atom $\langle x_{xd}^2 \rangle$ as:

$$\frac{1}{2} m\omega^2 \langle x_d^2 \rangle = \Delta H_m \tag{9}$$

The ratio of equation (8) and (9) was obtained as:

$$\frac{\langle x_d^2 \rangle}{\langle x^2 \rangle} = \frac{2\Delta H_m}{kT} \tag{10}$$

It can be seen from this ratio that the resistivity of a normal metal changes reciprocally with temperature. This change is attributed to the equation (8) reflecting the well-established findings that the resistivity of normal metal varies linearly with temperature above Debye temperature.

Substitution of (10) to (7) and (6) will result in:

$$Z^* = -Z\left[\frac{\Delta H_m}{kT}\frac{m_0}{m^*} - 1\right] \tag{11}$$

Assuming 12 $\langle 110 \rangle$ paths in FCC lattice, the factor of $\frac{1}{2}$ was cancelled out (see Fig.2).

Using available data of ΔH_m Tu[11] was able to show the agreement between calculated and measured data. Furthermore, the calculated Z* for Au was also found to agree well with the measured Z* from Huntington and Grone[2] as depicted in Fig. 3.

The temperature dependence of Blech Product, therefore, can be generally examined by combining equation (11) with (5).

$$(jl)_c = \frac{\Delta G}{-Z\left[\frac{\Delta H_m}{kT}\frac{m_0}{m^*} - 1\right] e\rho} \tag{12}$$

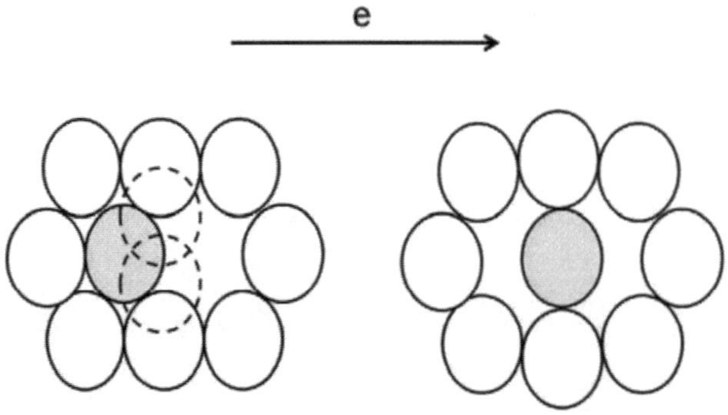

Figure 2. Sketch of the diffusing atom (shaded) to a neighboring vacancy (After Tu[11])

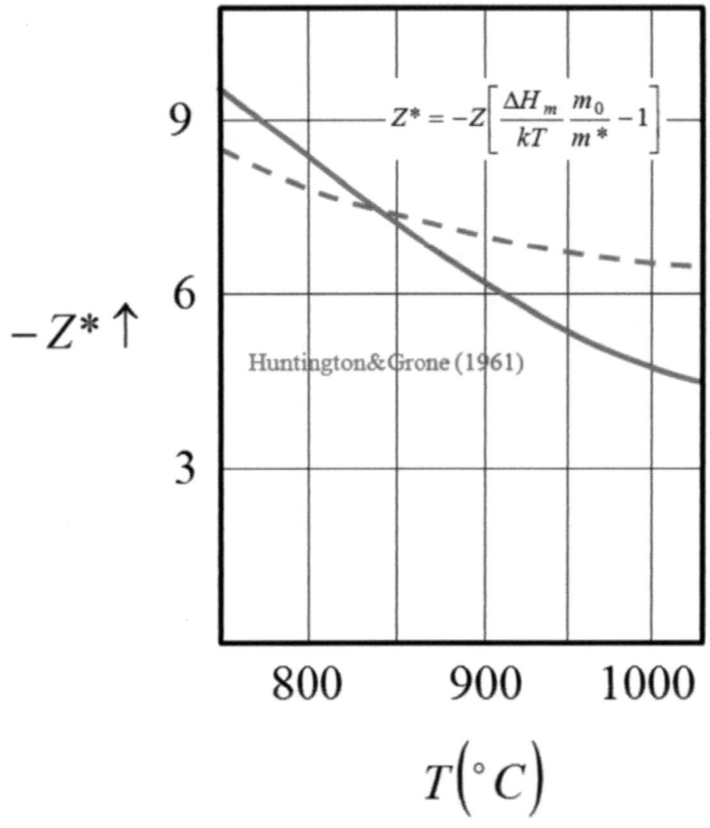

Figure 3. Schematic Plot of Z* vs. T for Au (After[2,11])

Having examined the temperature-dependence of Blech Product thermodynamically, we can always assume that the EM Threshold is a function of temperature. The relationship, however, is not straight forward, as there exist in the equation other factors that are also temperature-sensitive. Therefore, it is important to study the real temperature-dependence characteristics of EM threshold in common interconnect materials. The present paper reviews the temperature-dependence of EM threshold of different interconnect materials and structures as found in relevant publications. It is expected that the reader can see the entire perspective on the compilation of scattered data, which might be overlooked when it was examined separately in more details.

4. Temperature-Dependence of EM Threshold

Temperature-dependence of EM Threshold in various metallization lines had been studied extensively since Blech's Edge Displacement Experiment in 1976[9] and the results had been published accordingly. It was revealed in this remarkable paper that, along with *v/j*, $(jl)_c$ of thin Al films on Titanium Nitride (Ti-N) were temperature-dependence. Furthermore, it was also learned from Blech's Experiment that the drift velocity increased with temperature rise whereas the threshold-length product decreased with temperature. Blech[9] speculated that the increase of the threshold with the decrease of temperature might be related to the increase of Al flow stress. In addition to this common sense, it was assumed that there could be further increase in stress gradient due to the tensile stress near the cathode. Of the important phenomena in EM is the existence of the threshold current density-length product. The existence of the EM Threshold or Blech Product had been soundly proven and well accepted. The threshold was able to be quantified by measuring drift velocity and explained by the opposing gradient caused by the accumulation and depletion of atoms at the strips end.

Temperature-dependence characteristics of Al, Au, and Cu based metallization lines had been studied extensively. The edge displacement method and its variants had been long established in the studies of EM Threshold in thin film metallization lines[9,21-27]. Important modifications of this method were introduced by Ogawa et.al.[28]. Technologically more realistic methods were developed in the integration package level starting from single inlaid to three levels of interconnect structure. Various structure and configurations of line, via, and contact had been developed to study EM phenomena of metals and alloys at various temperature [28-32]. Studies were typically performed in the range between room temperature and 550°C.

Dual Damascene Cu structures had become a field of interest since early 2000s[25-37]. In 2003s, the effects of porous low-k on the temperature-dependent EM Threshold had also become an attractive area[32,34,36,37]. In addition to the key interconnect materials, solder materials including the lead free ones were also studied[38,39]. Edge displacement test was nevertheless used in the latest study[39]. Optical and electron microscopy methods had been long established in characterizing the product. The use of such a non-conventional characterization method as AFM in the area was also introduced. Figure 4 summarizes the experimental data on the temperature-dependence of EM threshold since Blech's experiment in 1976[9].

The linear temperature-dependence of the EM Threshold had been clearly shown by Blech and his co-workers [9,21]. Linear dependency of EM Threshold with temperature can be derived from Equation (12). Similar with the argument found in Tu[11], Equation (12) can be simplified if ΔH_m is much larger than kT. Then the equation is simplified and the relationship is made clear.

$$(jl)_c = \frac{\Delta G}{-Z\dfrac{\Delta H_m}{kT}\dfrac{m_0}{m^*}e\rho} \quad (13)$$

$$(jl)_c \propto -T \quad (14)$$

Based on such assumptions, the only factor that is temperature-sensitive is electrical resistivity, ρ. Provided the temperature is below Debye Temperature, in which the electrical resistivity does not increase with temperature, the linear relationship is maintained. Blech experiment on Al and Au in the selected range of temperature seems to follow this assumption very well. This simple linear relationship, however, is not always maintained.

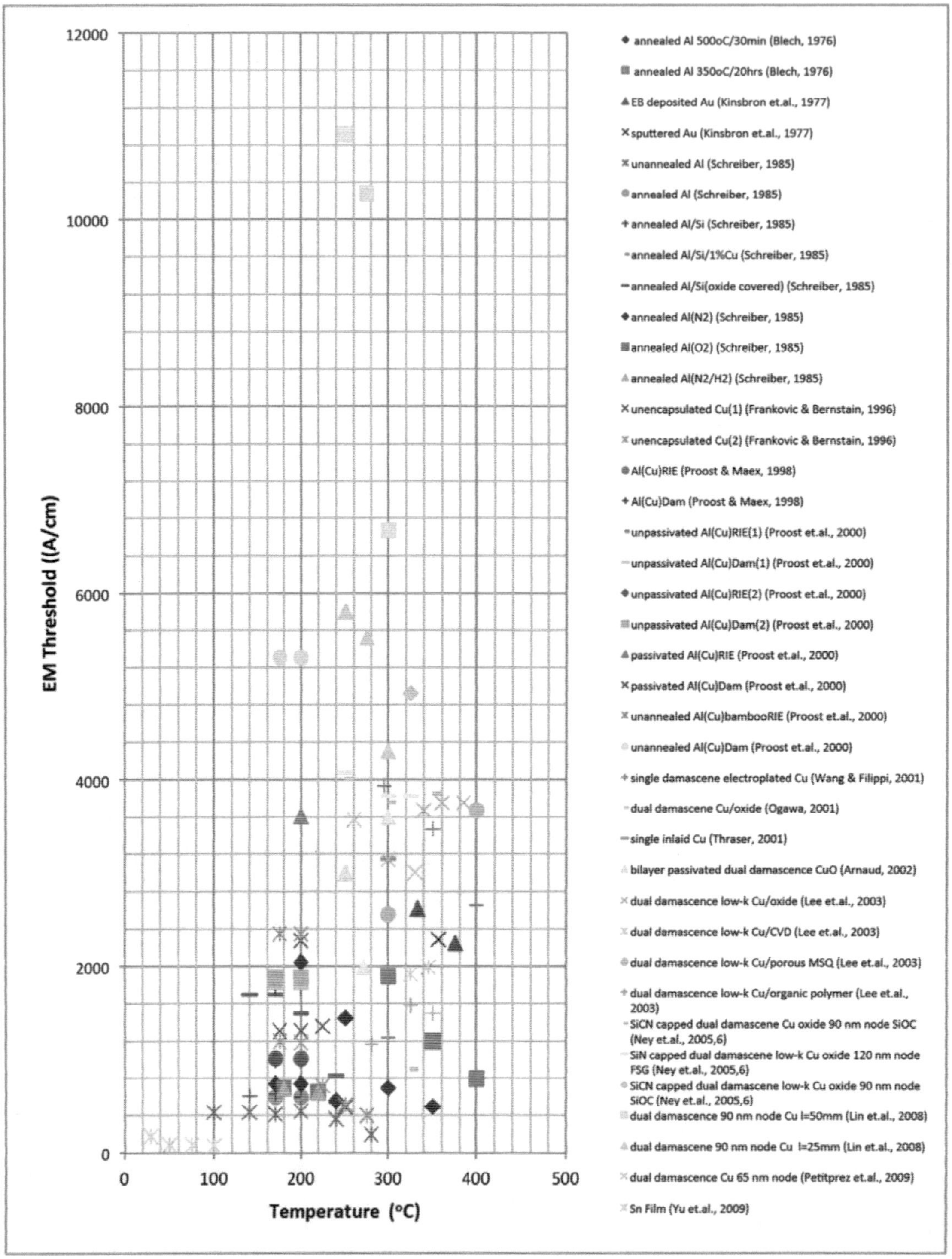

Figure 4. Plot of EM Threshold (A/cm) vs.Temperature (°C) Experimental Data[9,21-36,38]

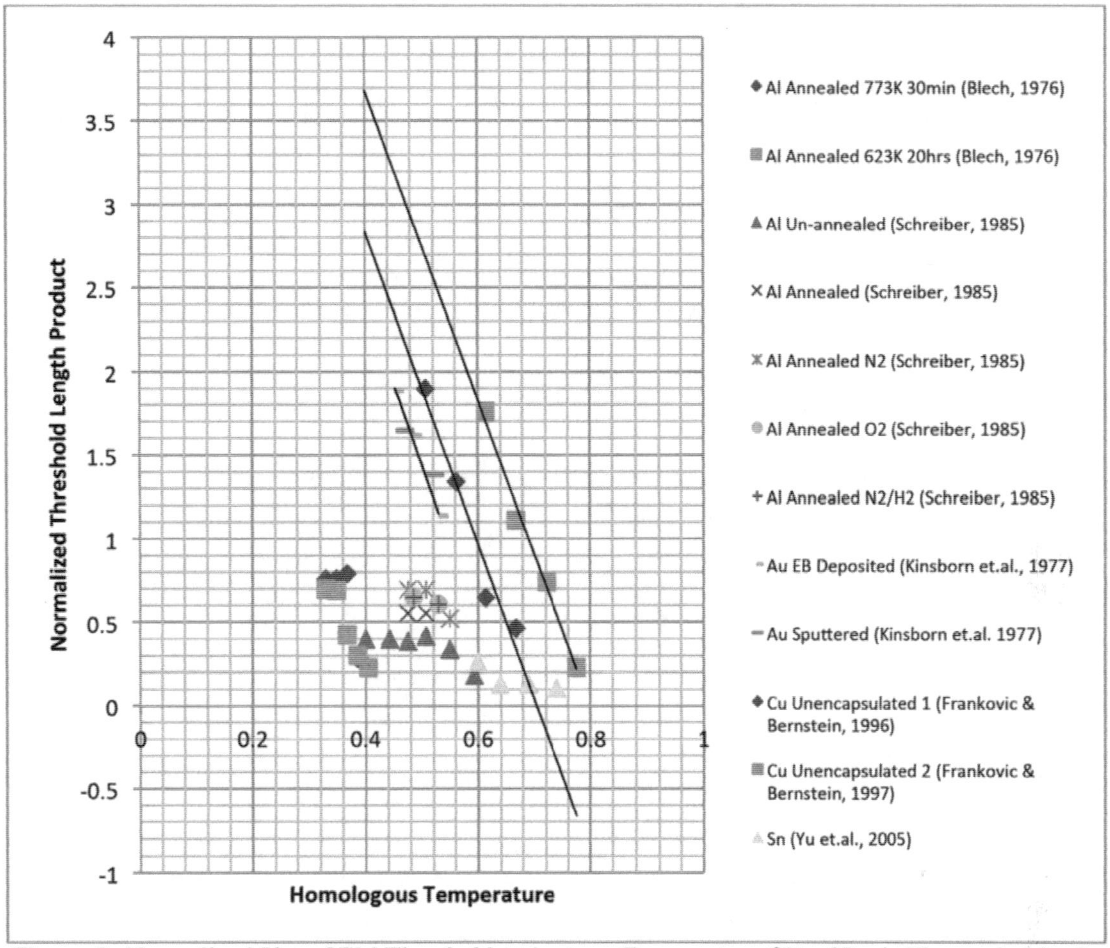

Figure 5. Normalized Plot of EM Threshold (A/cm) vs. Temperature (°C) of Basic Electronic Metals obtained from Edge Displacement Test

Brief analysis on most of data obtained from Schreiber[22] has shown that there is a transition temperature at which the EM Threshold is no longer sensitive to the temperature change. Furthermore, the characteristic curve of EM Threshold vs. Temperature can be generally divided into two region i.e. (1) the region where the product is constant with the temperature increase (time independent region), and (2) the region where the product is decreasing with the increase of temperature (time dependent region).

Similar phenomena was found in many experiment with Cu-based interconnect as shown in Fig. 6. The two regions and transition temperature for example had been clearly seen in Frankovic and Bernstein[23,24]. Different assumptions, however, were proposed to explain such similar phenomena. Frankovic and Bernstein[22] assumed that Z* to be inversely proportional to ρ and therefore the product of Z*ρ is temperature-independence. It was proposed that the gradient should be responsible for any variations of Blech Product. Blech[9], like many others[11-18], assumed that there exists the gradient of stress corresponding to chemical potential. Moreover, in line with Blech[9] who assumed the temperature-dependence behavior is related to temperature-dependence flow stress as well as Tu[11] who stated that EM is essentially a creep problem, Frankovic and Bernstein[23] suggested to use Ashby Deformation Map to explain transition temperature.

Other lessons learned from the big picture of EM Threshold vs. Temperature are the influence of initial material structure, geometry and surrounding as well as testing structure and condition on the temperature-dependence behavior. It can be learned from the map that different initial structure of material, which is also related to the fabrication or assembling process. The difference between annealed/un-annealed, bamboo/polycrystalline, and Reactive Ion Etched/Damascene, dielectric constant, for instance, were well recognized. Effect of surroundings such as encapsulation, cover, capping layer, passivation layer were also be recognized. The variation of testing structures and conditions may also affect the product. There existed edge displacement and via testing of many variants. Most of them are self-made. As the testing system is not well standardized, it is not a simple task (if not impossible) to compare data from different experiments apple to apple. Idealized curve and effect of initial materials and surroundings is depicted in Fig. 6 and Fig. 7.

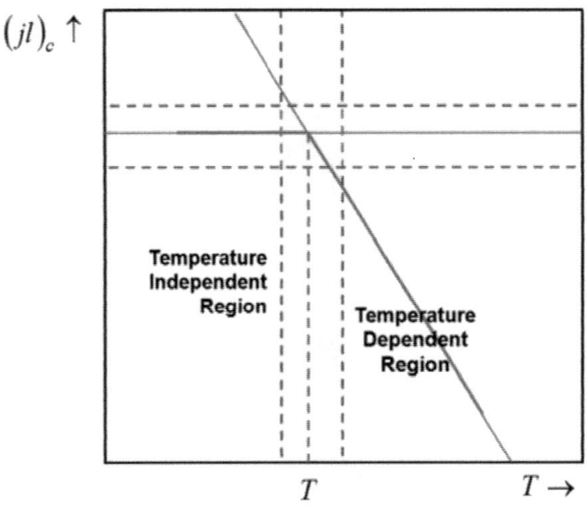

Figure 6. Idealized Curve of $(jl)_c$ *vs*. T

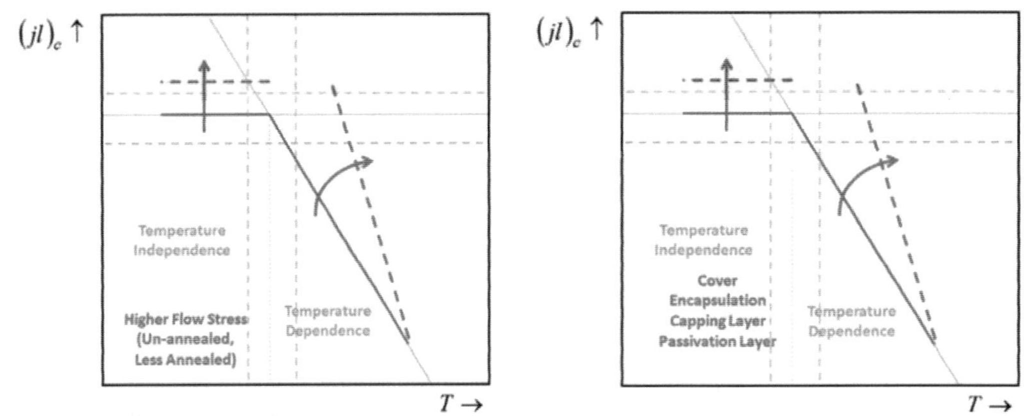

Figure 7. Idealized Effect of The Initial Material (Process History, Structure and Properties) (Left) and Its Surrounding (Right)

Journal of Physics: Conference Series **710** (2016) 012044 doi:10.1088/1742-6596/710/1/012044

5. Concluding Remarks

Temperature-dependence of EM-threshold has been revisited. Phenomenological threshold equation had been derived from the velocity drift and examined thermodynamically to find a constitutive law for temperature-dependence of EM threshold. Experimental data from relevant publication had been reviewed and studied. The plot in general can be idealized into (a) temperature-independent region and (b) temperature-dependent region within it a temperature transition may exist. This temperature may be related to either competing temperature-sensitive thermodynamic factor or deformation mechanisms map based on established assumptions.

General effects of initial materials (process history, geometry, and structure) and surroundings (cover, encapsulation, capping layer, passivation layer) were generalized and idealized despite of the fact that not all of the available data can be compared apple to apple due to variations in testing structure and condition. Since EM threshold and its temperature-dependence has significant implications on design, manufacture, and testing of electronics, it is suggested to develop a more general and well standardized testing system that is able to produce comparable and more reliable EM Threshold Data at various temperatures.

References

[1] V.B. Fiks 1959 *Sov. Phys. Solid State* **1** 14
[2] H.B. Huntington and A.R. Grone 1961 *J. Phys. Chem. Solids* **20** 76-87
[3] James R. Black, Mass Transport of Aluminum by Momentum Exchange with Conducting Electron, 6th Annual Reliability Physics, IEEE Symposium Proceedings C58, IEEE, LA, 1968, pp.7-15 (reprinted in 2005).
[4] R.S. Sorbello 1998 *Solid State Physics* **51** 159-231
[5] R.L. de Orio, H. Ceric, S. Selberherr 2010 *Microelectron. Reliab.* **50** 775–789
[6] D. G. Pierce and P. G. Brusius 1997 *Microelectron. Reliab.* **37** 1053 -1072
[7] M. Shatzkes and J. R. Lloyd 1986 *J. Appl. Phys.* **59** 3890
[8] R. Rosenberg and M. Ohring 1971 *J. Appl. Phys.* **42** 5671
[9] I. A. Blech 1976 *J. Appl. Phys.* **47** 1203
[10] N. G. Ainslie, F. M. d'Heurle, and O. C. Wells 1972 *Appl. Phys. Lett.* **20** 173
[11] K.N. Tu 1992 *Phys. Rev. B* **45** 1409-1413
[12] C.A. Ross 1991 *Mat. Res. Soc. Symp. Proc.* **225** 35-46
[13] R. Kirchheim 1992 *Acta Metall. Mater.* **40** 309-323
[14] M.E. Sarychev, Yu. V. Zhitnikov, L. Borucki, C.-L. Liu, T.M. Makhviladze 2000 *Thin Solid Films* **365** 211-218
[15] M.A. Korhonen, P. Børgesen, K.N. Tu, and CheYu Li 1993 *J. Appl. Phys.* **73** 3790
[16] J.R. Lloyd 1999 *Microelectronic Engineering* **49** 51-64
[17] J.J. Clement and C. V. Thompson 1995 *J. Appl. Phys.* **78** 900
[18] R.J. Gleixner and W. D. Nix 1999 *J. Appl. Phys.* **86** 1932
[19] V. Sukharev, E. Zschech, and W.D. Nix 2007 *J. Appl. Phys.* **102** 053505
[20] I.A. Blech 1997 *Acta. Mater.* **48** 3717-3723.
[21] E. Kinsborn, I. A. Blech and Y. Komem 1977 *Thin Solid Films* **46** 139-150
[22] Schreiber 1985 *Sol. State. Electron.* **28** 617-616
[23] R. Frankovic and G. H. Bernstein 1996 *IEEE Trans. El. Dev.* **43** 2233-2239
[24] R. Frankovic and G. H. Bernstein 1997 *J. Appl. Phys.* **81** 1604
[25] J. Proost and K. Maex 1998 *App. Phys. Lett.* **73** 2748-2749
[26] J. Proost, A. Witvrouw, K. Maex, J. D'Haen, and P. Cosemans 2000 *J. Appl. Phys.* **87** 86
[27] J. Proost, K. Maex, and L. Delacy 2000 *J. Appl. Phys.* **87** 99
[28] E.T. Ogawa, A.J. Bierwag, K.D. Lee, H. Matsuhashi, P.R. Justison et al. 2001 *Appl. Phys. Lett.* **78** 2652
[29] S. Thrasher, C. Capasso, L. Zhao, R. Hernandez, P. Mulski, S. Rose, T. Nguyen, and H. Kawasaki 2001 *Proc. of IEEE Int. Intercon. Tech. Conf* 2001 177-179

[30] P.-C. Wang and R. G. Filippi 2001 *Appl. Phys. Lett.* **78** 3598

[31] L. Amaud, *40th Annual Int. Rel. Phys. Symp., IEEE Symp. Proc.*, Dallas, Texas, 433-434 (2002)

[32] K.D. Lee, E.T. Ogawa, S. Yoon, X. Lu and P.S. Ho 2003 *Proc. of IEEE Int. Intercon. Tech. Conf.* 2003 259-261

[33] D. Ney, X. Federspiel, V. Girault, 0. Thomas, P. Gergaud, IIRW Final Report, IEEE International, 2005

[34] D. Ney, X. Federspiel, V. Girault, O. Thomas, P. Gergaud 2006 *IEEE Trans. Dev. Mat. Rel.* **6** 175-180

[35] M.H. Lin, M.T. Lin, T. Wang 2008 *Microelectron. Rel.* **48** 569–577

[36] E. Petitprez, L. Doyen and D. Ney *47th Ann. Int. Rel. Phys. Symp., IEEE Symp. Proc.*, Montreal, 865-868 (2009)

[37] P.-C. Wang, R.G. Filippi, C.J Christiansen, B. Li, H. Ding 2009 *ECS Trans.* **16**(19) 41-48

[38] H.C. Yu, S.H. Liu, and Chi Chen, 2006 *J. Appl. Phys.* **98** 013540.

[39] Y-C Hsu, C-K Chou, P.C. Liu, and C. Chen 2005 *J. Appl. Phys.* **98** 033523